Handbook of

Nonmedical Applications of Liposomes

From Gene Delivery and Diagnostics to Ecology

VOLUME IV

Handbook of Nonmedical Applications of Liposomes

From Gene Delivery and Diagnostics to Ecology

VOLUME IV

Edited by
Danilo D. Lasic
MegaBios Corporation
Burlingame, California
Yechezkel Barenholz
Department of Biochemistry
Hebrew University–Hadassah Medical School
Jerusalem, Israel

CRC Press
Taylor & Francis Group
Boca Raton London New York

CRC Press is an imprint of the
Taylor & Francis Group, an **informa** business

Cover:
Artist's impression of liposome and DNA complex. While the exact topology of these systems is still not well understood, this picture represents the case in which DNA molecules did not condense and liposomes did not disintegrate and restructure. Created in mixed media by artist Alenka Dvorzak Lasic, Autumn 1995.

CRC Press
Taylor & Francis Group
6000 Broken Sound Parkway NW, Suite 300
Boca Raton, FL 33487-2742

Reissued 2019 by CRC Press

A Library of Congress record exists under LC control number:

Publisher's Note
The publisher has gone to great lengths to ensure the quality of this reprint but points out that some imperfections in the original copies may be apparent.

Disclaimer
The publisher has made every effort to trace copyright holders and welcomes correspondence from those they have been unable to contact.

ISBN 13: 978-0-367-26102-3 (hbk)
ISBN 13: 978-0-429-29147-0 (ebk)

Visit the Taylor & Francis Web site at http://www.taylorandfrancis.com and the CRC Press Web site at http://www.crcpress.com

GENERAL PREFACE

To introduce liposomes, we shall look backwards to the origin of the name liposome. As shown in the last chapter of this series, in the fourth volume, liposomes were produced in scientific laboratories at least from 1847 on. The colloidal properties of lipid dispersions (emulsions, sols, micelles, etc.) were well known. However, it was only Bangham in the early 1960s who realized that they encapsulate part of the solvent and that their membrane presents a permeability barrier. So, they were at that time called "banghasomes", while the name "liposome" can occasionally be found in the literature between the World Wars as a description of colloidal fat (mostly triglyceride) particles. In the late 1960s, Weismann, proposed the name liposomes for these structures. Due to the very general meaning of this word (i.e., fat body), Papahadjopoulus and others introduced the classification according to size and number of lamellae. In addition, and in contrast to this topological and structural classification, and following the development of liposomes as drug delivery vehicles, a new — functional — liposome classification could be proposed. With respect to functional characterization and the interaction characteristics, one can divide liposomes into those with nonspecific reactivity (conventional liposomes which are exposed to interactions of components from the suspending medium), inert liposomes (sterically stabilized or Stealth® liposomes which have minimal reactivity with the medium), liposomes with specific reactivity (Stealth® liposomes with attached specific ligands in which nonspecific repulsion is upgraded with specific antibodies or lectins), and liposomes that are very reactive under particular conditions. The latter are polymorphic liposomes, or "activosomes", which are topologically stable, but change their structure when triggered by a particular stimulus, such as polyion (with examples of cationic liposomes for DNA and antisense molecule delivery) or pH-sensitive liposomes. Fusogenic and target-specific liposomes also fall into this category.

Liposomes are lipid bilayer-based colloidal particles which can encapsulate part of the aqueous medium into their interior, forming a "milieu interne" (internal microenvironment) which can be different from the extraliposomal medium. Nowadays, these assemblies are becoming a very important system in basic research, from theoretical concepts to serving as a tool, model, or reagent in basic studies of many processes. Parallel to numerous medical uses, a subject to which many books and reviews are dedicated, many other applications are being investigated. The aim of these four books is to present these lesser-known applications in depth. This series, *Handbook of Nonmedical Applications of Liposomes,* is split into four rather different collections of related groups of papers, so that many researchers will be able to find the most pertinent articles concentrated in a single volume or two.

The first volume presents theoretical studies on thermodynamics and molecular dynamics simulations as well as some engineering applications of liposomes. The second volume presents liposomes as model systems for many biophysical, biochemical, and cellular processes, as well as protein studies using reconstitution into bilayer membranes. It concludes with a group of chapters related to the origin of life. The third volume commences with the "nuts and bolts" of "liposomology", from rational design, raw materials, large-scale preparation and active agent loading, stability, and stabilization, to sterility. It concludes with the use of liposomes as bioreactors and in catalysis. The fourth volume presents additional liposome applications in life sciences, with a group of chapters on gene therapy, cosmetics, diagnostics, and ecology.

Although these four volumes deal with nonmedical applications of liposomes, their content is also very relevant to their medical applications, through many specific subjects such as all the design, preparation, loading, preservation, and characterization of liposomal formulations, gene delivery, and diagnostics, as well as comparison with other recently developed colloidal carriers. Therefore, we believe that this series should appeal to everyone who has an interest in liposomes, whether from academia or from industry.

It was a pleasure for both of us to undertake the daunting project of putting this series together because obtaining such a broad spectrum of expertise required working with top academic and industrial researchers. We hope that this series will be useful to scientists from many disciplines and to students who are looking for new frontiers. From our personal experience as teachers, we know that such a series is missing, and we hope that this will fill the gap.

We would like to thank the many authors of this series who did such an excellent job in writing and submitting their papers to us. It was enjoyable to interact with them, and there is no question that the pressure we put on many of them was worthwhile.

Finally, we would like to thank Mrs. Beryl Levene for her excellent secretarial work and for sending so many letters to so many scientists and Mr. Sigmund Geller for his editorial assistance. CRC personnel should be acknowledged for their encouragement and patience, as well as their high level of proficiency.

Yechezkel Barenholz
Jerusalem, Israel
Danilo D. Lasic
Burlingame, California

INTRODUCTION TO VOLUME IV
LIPOSOMES: FROM GENE DELIVERY AND DIAGNOSTICS TO ECOLOGY

This volume begins with a theme from the life sciences — gene delivery by using liposomes—which may have important consequences not only in novel medical treatments and preventive medicine, but also in diagnostics, in basic studies of cell function, and in veterinary medicine. Additionally, plant protoplasts can be transfected using liposomes, and this opens new opportunities in agriculture. Currently the hottest area in liposome research is complexation of recombinant DNA plasmids or short, single-stranded antisense oligonucleotides and their insertion into living cells. After very encouraging *in vitro* results, the first successful *in vivo* transfections were reported, with the ultimate goal of gene therapy. These topics are discussed in papers by Lasic (general introduction), Minsky et al., who concentrate more on physicochemical aspects of liposome-DNA complexes, and by *in vitro* studies described by Maccarone et al. and *in vitro/in vivo* studies by Huang and Felgner et al. who, in addition to groups at GeneMedicine, MegaBios, and Genzyme, are the major players in this rapidly evolving field of non-viral gene therapy.

The greatest impact liposomes have at present in our everyday life is undoubtedly in cosmetics. The first liposomal cosmetic products were introduced by L'Oreal and Christian Dior in 1986 and 1987, respectively. Since then several hundred products have been introduced in Europe, North America, and Japan with estimated annual sales of over one billion dollars. Vanlerberghe, who developed nonionic liposomes and the first products, presents his views of the field. The major enigma of topical applications of liposomes is their penetration into skin. Claims differ from no enhancement at all to significantly increased penetration, and Šentjurc presents a sophisticated and reliable method for these measurements. Along similar lines, liposomes were also used in replacement of some lipids in some cell membranes in order to improve particular cell membranes.

Many people are skeptical about the real value of these cosmetic applications. Although we are aware of a lack of well-controlled and well-defined comparative studies, we do not agree with all the skepticism. At least, in the worst case, liposomes can deliver the needed hydrophobic agents solely in the lecithin–water-based carrier system, in the absence of any detergents, oils, alcohols, and similar solvents. We believe that it is preferable to apply on the skin lecithin and water as opposed to these organic solvents or detergents. Furthermore, "quick skin creams" prepared in the lab and consisting of simple dense MLV dispersions of lecithin, vitamin C, and α-tocopherol were quite popular among the women in the liposome labs.

One often hears of other uses, such as self-healing paints, photographic emulsions, and one-component glues (which are based on liposome-encapsulated second components with a triggered release), but it is difficult to find reliable scientific references. Nevertheless, Wallach and Mahur present some of these applications as well.

Most of the nonconventional applications of liposomes are grouped in the next section. Quinn and Perrett describe liposomal formulations of agricultural pesticides which are superior to conventional formulations.

In food production, liposomes were tried as dispersants, controlled-release systems, and protective capsules for entrapped enzymes, and positive results were reported in the early 1980s. The continuation of this work is presented by Simard and co-workers. De la Maza et al. describe the use of liposomes in wool dyeing, while Kalmanzon et al. model the origin of shark repellency and its relationship to pore formation in membranes.

A very important and promising area is the use of liposomes in diagnostics. Recently, sterically stabilized liposomes were introduced to reduce nonspecific interactions (Emmanuel et al.). The use of liposomes in diagnostics is covered in papers by Katoh and Carbonel. Reeves et al. use liposomes for immunodetermination of environmental contaminants, while Miller et al. describe liposome applications in biodegradability testing. Gregoriadis et al. describe microbe encapsulation in giant liposomes, which may have applications in various fields such as vaccination and ecology. Various lipids exhibit various affinities to various metallic cations. If enough specificity can be found for the metallic cations, then one can devise marker systems as well as purification systems.

We conclude this volume by presenting applications of liposomes in ecology. We are using a few simple examples to show the utility of the liposomes. For cleaning of nuclear reactors, for instance, it is rather inconvenient to remove dissolved radioactive ions in large volumes of water. Adding liposomes with surface-attached chelators allows easy separation of liposomes containing radionucleotides by subsequent precipitation of liposomes with bound ions. Another application would be purification of proteins by liposomes. In addition, we have reviewed the past and present state in the field and speculate on what the future has to offer.

We hope that this series will yield valuable information for all the scientists working in the field because in contrast to medical uses, which have been extensively reviewed, this has not been the case for nonmedical applications of liposomes. We believe that the lack of communication between the two groups was also responsible for somewhat slow progress in the development of pharmaceutical products.

Danilo D. Lasic
Burlingame, California
Yechezkel Barenholz
Jerusalem, Israel

THE EDITORS

Danilo D. Lasic, Ph.D., is Principal Scientist at MegaBios, Burlingame, California, where he is currently working with cationic liposomes in gene delivery. Previously, he was a senior scientist at Liposome Technology, Inc. (now SEQUUS), where he led studies for theoretical understanding of long-circulating liposomes as well as developed the first formulations for preclinical studies. In addition, he also actively participated in the scale-up of the preparation of stealth liposomes laden with the anticancer agent, doxorubicin.

Dr. Lasic graduated from the University of Ljubljana, Slovenia, in 1975 with a degree in physical chemistry. He received his M.Sc. in 1977 from the University of Ljubljana. He obtained his Ph.D. at the Institute J. Stefan, Solid State Physics Department in Ljubljana in 1979. After postdoctoral work with Dr. Charles Tanford at Duke and Dr. Helmut Hausser at ETH Zurich, he was a research fellow at the Institute J. Stefan in Ljubljana, and a visiting lecturer in the Department of Chemistry and a visiting scientist in the Department of Physics at the University of Waterloo in Canada. Then he joined Liposome Technology, Inc., in Menlo Park, California. Currently, he is studying cationic liposomes, DNA-liposome interactions, and DNA–lipid complexes for gene delivery and gene therapy.

He has published more than 120 research papers as well as a monograph on liposomes (Liposomes: from Physics to Applications, Elsevier, 1993) and co-edited a book, *Stealth Liposomes* (CRC Press, 1995). His best known papers are the ones dealing with thermodynamics and the mechanism of vesicle formation, stability of liposomes, the origin of liposome stability in biological environments as well as the applications of drug-laden liposomes. He is a member of the editorial board of the *Journal of Liposome Research* and of *Current Opinion in Colloid and Interface Science.*

Yechezkel Barenholz, Ph.D., is Professor of Biochemistry and head of the Department of Biochemistry at the Hebrew University-Hadassah Medical School, Jerusalem, Israel. Professor Barenholz received his undergraduate, M.Sc., and Ph.D. degrees in Biochemistry from the Hebrew University, Jerusalem in 1965, 1968, and 1971, respectively. While working on his Ph.D. thesis, he studied at the Animal Research Council Institute at Babraham, Cambridge, England with Dr. R.M.C. Dawson and Dr. A.D. Bangham. He was on the permanent academic staff of the Hebrew University since 1974 and was a professor there since 1983. He has been on the staff of the University of Virginia in Charlottesville from 1973–1976 and remained a Visiting Professor at the University of Virginia since 1983. Professor Barenholz was a Donders Chair professor at the University of Utrecht in The Netherlands in 1992.

Professor Barenholz's interests are in basic and applied science. In basic research, he is involved in many fields related to the biochemistry and biophysics of lipids, especially sphingolipids, phospholipids, and sterols, and including synthesis, chemical and physical characterization, and the relationship between membrane lipid composition, and function with special focus on aging. In applied research, Professor Barenholz's main interests are in amphiphile-based drug carriers, especially liposomes — from the design of the drug carrier basic aspects through animal studies to clinical trials. Professor Barenholz was involved in clinical trials of Amphocil®, an amphothoricin B micelle-like assembly, and the development of DOX-SL®, a doxorubicin remote loaded sterically stabilized liposome. At present he is also studying the application of liposomes for vaccination and in gene therapy. Professor Barenholz is an author of 200 publications and an editor of a special issue of *Chemistry and Physics of Lipids* on "Quality Control of Liposomes". He is on the editorial board of *Chemistry and Physics of Lipids,* and the *Journal of Liposome Research.* Professor Barenholz is a recipient of the 1995 Kay Award for innovation.

CONTRIBUTORS

Alessandro Finazzi Agrò
Experimental Medicine and Biochemical Science
University of Rome
Rome, Italy

Sophia G. Antimisiaris
Laboratory of Pharmaceutical Technology
School of Health Sciences
University of Patras
Patras, Greece

Yechezkel Barenholz
Department of Biochemistry
Hebrew University–Hadassah Medical School
Jerusalem, Israel

Richard Bartha
Department of Biochemistry and Microbiology
Lipman Hall, Cook College
Rutgers University
New Brunswick, New Jersey

Elijah Bolotin
Department of Biochemistry
Hebrew University–Hadassah Medical School
Jerusalem, Israel

Pilar Bosch
Departmento de Tensioactivos
CID/CSIC
Barcelona, Spain

Ruben G. Carbonell
Department of Chemical Engineering
North Carolina State University
Raleigh, North Carolina

Luisa Coderch
Departmento de Tensioactivos
CID/CSIC
Barcelona, Spain

Patricia Dufour
Centre de Recherche STELA
Universite Laval
Quebec, Canada

Richard A. Durst
Analytical Labs
Cornell University
Geneva, New York

Noam Emanuel
Lautenberg Center of Immunology
Hebrew University–Hadassah Medical School
Jerusalem, Israel

Hassan Farhood
Department of Pharmacology
University of Pittsburgh School of Medicine
Pittsburgh, Pennsylvania

Philip L. Felgner
Vical, Inc.
San Diego, California

Jiin H. Felgner
Vical, Inc.
San Diego, California

Hideki Fukada
Department of Applied Chemistry
Kobe University
Kobe, Japan

Violeta Gabrijelčič
Research and Development
LEK/Pharmaceutical-Chemical Company
Ljubljana, Slovenia

Hezi Gershon
Department of Organic Chemistry
Weizmann Institute of Science
Rehovot, Israel

Rodolfo Ghirlando
Laboratory of Molecular Biology
NIDDK-NIH
Bethesda, Maryland

Gregory Gregoriadis
Centre for Drug Delivery Research
University of London School of Pharmacology
London, England

Luke S. S. Guo
SEQUUS Pharmaceuticals, Inc.
Menlo Park, California

Ishan Gursel
Department of Biological Sciences
Middle East Technical University
Ankara, Turkey

Leaf Huang
Department of Pharmacology
University of Pittsburgh
School of Medicine
Pittsburgh, Pennsylvania

Ilia Y. Jimenez
Department of Biochemistry
and Microbiology
Lipman Hall, Cook College
Rutgers University
New Brunswick, New Jersey

Eliahu Kalmanzon
Steinitz Marine Biological Lab
Inter-University-Institute of Eliat
Eilat, Israel

Shigeo Katoh
Synthetic Chemistry/Biological
Chemistry
Kyoto University
Kyoto, Japan

Eli Kedar
Lautenberg Center of Immunology
Hebrew University–Hadassah Medical
School
Jerusalem, Israel

Masaaki Kishimura
Engineering Research Laboratories
Kaneka Corporation
Hyogo, Japan

Edith Laloy
Centre de Recherche STELA
Universite Laval
Quebec, Canada

Danilo D. Lasic
MegaBios Corp.
Burlingame, California

Mauro Maccarrone
Experimental Medicine and
Biochemical Science
University of Rome
Rome, Italy

Ming Man
Focal, Inc.
Lexington, Massachusetts

Albert M. Manich
Departmento de Tensioactivos
CID/CSIC
Barcelona, Spain

Rajiv Mathur
Micro Vesicular Systems, Inc.
Nashua, New Hampshire

Alfonso de la Maza
Departmento de Tensioactivos
CID/CSIC
Barcelona, Spain

Raina M. Miller
Soil, Water, and Environmental Science
University of Arizona
Tuscon, Arizona

Abraham Minsky
Department of Organic Chemistry
Weizmann Institute of Science
Rehovot, Israel

Jose Luis Parra
Departmento de Tensioactivos
CID/CSIC
Barcelona, Spain

Steven F. Perrett
Division of Life Sciences
King's College London
London, England

Peter J. Quinn
Division of Life Sciences
King's College London
London, England

Ramachandran Radhakrisnan
Chiron Corporation
Emeryville, California

Stuart G. Reeves
Food Science and Technology
Cornell University
Geneva, New York

CONTENTS

Chapter 1

Liposomes in Gene Therapy

Danilo D. Lasic

CONTENTS

I. INTRODUCTION

Gene therapy and diagnostics may become the major development in medicine in the next century/millennium due to extremely rapid developments in recombinant DNA technology, a greater understanding of genetic and some acquired diseases on the molecular level and mapping of the human genome, as well as improved ways to deliver recombinant genetic codes into appropriate cells.

II. GENE THERAPY

Several thousand diseases can be linked to defective or missing genes and the concept is that by bringing the appropriate gene into appropriate cells, the mutated or missing proteins can be synthesized and alleviate the signs of the disease. Furthermore, in some infectious diseases, inflammation states, or in cancer, one can stimulate the immune system to produce and secrete more cytokines (such as various colony stimulating factors and tumor necrosis factor, interleukins, interferons) and via enhanced cytotoxicity and/or enhanced number of killer cells, exterminate the malignancy. In addition to a "turn on" concept as described above (and in which gene therapy aims at replacing nonfunctional gene and/or adding its functional copy), there is also a therapy possible by switching genes off. This includes the so-called antisense (oligonucleotides) technology whose aim is to deliver a short, normally single strand oligonucleotide with a complementary sequence, to the part of the unwanted gene or messenger RNA, or to form a triple helix with DNA, and to stop the process, such as cancer, scar tissue, or other undesired cell growth. Gene therapy in a broader sense encompasses also enhancement of the immune system, tagging neoplastic cells for autoimmune destruction and vaccination in which the body itself produces antigens to induce an immune response. The aim of the treatment is to transfect cells with genes reconstituted into plasmids without incorporating genetic material into chromosomes — in order to avoid possible side effects such as cancer or viral infection. This approach therefore requires either rather frequent dosing or development of self-replicating plasmids which remain in the cell nucleus during cell division. Alternatively, integration of DNA can be accomplished by site-specific integration of cDNA flanked by specific DNA sequences, homologous recombination of genomic DNA, or random integration.

It seems that functional plasmids containing powerful gene expression enhancers and with cell specific promoters can be now routinely constructed, while the delivery, (not only *in vitro* and *ex vivo*, but especially *in vivo*), presents the largest challenge. Currently, in most therapies in humans, appropriate cells are taken out and after transfection in a Petri dish, injected back into the patient. Obviously, the goal of these treatments is *in vivo* administration of appropriate plasmids or their constructs with various carriers. The targeted diseases are cancer, cardiovascular diseases, cystic fibrosis, and many others.

Large size and charge of plasmids (3–15 kilobase pairs = 2–10 million Da = 6–30 thousand negative charges with linear length of molecule 1–5 μm) make the intracellular delivery of plasmids extremely difficult. In addition, free DNA is quickly degraded when applied systemically. *Ex vivo* approaches can use physical and chemical transfection methods such as electroporation, direct injection, and calcium

0-8493-4013-6/96/$0.00+$.50

precipitation. Delivery systems include also gene gun and precipitation with some polymers and differ mostly with respect to transfection efficacy, safety, plasmid loading capacity, immunogenicity, and nature of their interactions with cells. The most frequently used delivery systems *in vivo* are viral constructs, i.e., genes inserted into viral genes and reconstituted into noninfective viruses. Another approach involves complexation of the plasmid with liposomes, (cationic) polymers and oligomers, polyvalent cations, polycations, or any combination of these approaches. Naked DNA can be also injected locally. In addition to intravenous administration, direct intramuscular, subcutaneous, localized, and intraperitoneal injection, and instillation into airways as well as inhalation of an aerosol of such particles, can be used.

DNA is inserted either in a viral vector or in a bacterial plasmid which can thus be introduced into cells either via viral or internalizing (endocytosis, fusion) pathway. The most widely used are retroviruses. They are single-strand RNA viruses which incorporate the gene directly into chromosome of the target cell. This presents a safety concern due to a potential carcinogenicity and infectivity despite the fact that several viral protein codes are deleted. Also, contamination with intact retroviruses, which can be highly oncogenic, cannot be ruled out. Adenoviruses are less dangerous (an example is a common cold virus). They are double-stranded viruses which can incorporate larger genes than retroviruses. They do not incorporate into the genome and this reduces the risk of malignant transformations but also reduces the persistence of expression. Also, an immune response against viral proteins may develop, besides the fact that many people possess an immunity against these viruses. Other viruses, such as adeno-associated viruses or herpes simplex virus, are also employed.[1,2]

Because of some inherent problems associated with viral vectors, such as potential carcinogenicity, infectivity, and development of an immune response, many researchers believe that liposome- or lipid-based transfection, possibly in conjunction with some other DNA condensing agent, will become the mainstream of gene therapy.

III. HISTORICAL PERSPECTIVE

The first attempts to encapsulate and transfect DNA using liposomes were undertaken in the late 1970s.[3] However, despite some successes, the protocols were cumbersome and practically inapplicable *in vivo*, and with the introduction of electroporation these approaches were largely phased out.[4,5]

Following successful DNA complexation with cationic polymers, cationic lipids were used to coat DNA. This lipid coating was followed by complexation with cationic liposomes and several cationic lipids were introduced.[6-14] A variety of different cationic lipids was synthesized which showed great differences in transfection activity. With the exception of positively charged cholesterols and some variation on DOTMA,[9,12] at present, however, no structure–activity relationships are known. Another goal of lipid synthesis is also to produce the safest cationic compounds. For that reason it is desirable that they contain biodegradable chemical bonds. It was realized that in order to improve transfection efficacy, liposomes have to contain also a neutral lipid. Most of the studies report that cationic lipid plus dioleoyl phosphatidyl ethanolamine (DOPE) at approximately 50 mol% yields the highest transfection efficiency. This effect was explained because of its ability to form a hexagonal II phase and to facilitate either complexation or internalization of the DNA–liposome complex (genosomes) into cells and perhaps, facilitate the release of the complex/DNA from the endosome after endocytosis.[9,10]

IV. CATIONIC LIPOSOMES

Even more than the influence of neutral lipid, the nature of cationic lipid and its structure–activity relation were studied.[9,11,12] Various formulations, containing lipids such as DODAB, DOTAP, DOTMA, DOGS, DOSPA, and DC-Chol (formulae are shown in the following papers), resulted in several commercial transfection formulations which showed rather good transfection efficacies in various cell models but were practically ineffective for *in vivo* applications.

In the first 20 years of liposome research, cationic liposomes were not studied extensively due to their toxicity. Colloidal studies have shown that DLVO theory can be applied to explain their stability, although in the case of sonicated DODAC liposomes their stability was found to be lower than expected and observed with anionic liposomes.[15] In general, pK values of cationic groups are closer to the working/physiological pH values and anions may be associated with bilayers to a larger extent than cations in the case of anionic liposomes, also due to nonelectrostatic interactions. Also, entropy effects of anions are rather different than of normally well-hydrated cations. As an example we can state that in many cases the same cationic amphiphile may form different structures depending on the nature of anion: for instance, chloride, bromide, or hydroxide salt forms bilayers and liposomes and tosilate/acetate salts form micelles. The lower-than-theoretically-predicted stability of sonicated DODAB liposomes was explained by the presence of hydrophobic defects on the membranes of these liposomes.[15]

The molecular geometry of these surfactants is often less ideal to form (i.e., pack into) bilayers as compared to phospholipids and therefore liposomes can be less stable. Especially in the case of high surface charges in low ionic strength media where Debye length can exceed the dimensions of the liposome, the optimal shape of a self-closed liposome may not be well defined.[16] In general, higher charges and lower ionic strengths make bilayers more rigid. Furthermore, it was shown that in such cases Gaussian curvature may become negative and in the presence of sufficient surface charge the total bending energy becomes negative and may induce fission and spontaneous vesiculation.[17] Furthermore, high charges may stabilize some bilayer defects. Indeed, in pure DODAB solutions many lens-like structures (a = 30–50 nm, b = 10–15 nm) and micellar structures (open fragments) were observed by cryo electron microscopy (Frederik, Stuart and Lasic, in preparation).

V. TRANSFECTION AND GENE EXPRESSION

The first successful *in vivo* results were recently also reported.[18,19] Plasmids containing appropriate and specific promoter and enhancer sequences and containing largely marker/reporter genes (such as β-galactosidase, luciferase, CAT, etc.) as well as codes for therapeutic proteins (CFTR — cystic fibrosis regulatory transport protein, genes for cytokines, histocompatibility complex proteins, etc.) are normally injected (locally or systemically) or applied via a nebulizer as an aerosol, and the expression of gene is measured in various organs/cells by immunohistochemical or immunological assays (ELISA) of the synthesized protein. Due to the complex interactions, not much is known about the transfection process, the mechanisms involved, or the influence of various parameters on it.[20]

The structure, stability and transfection activity of genosomes are still not understood. Electron microscopic observations reveal globular aggregates with a corona of fibrilar structures, and names such as "spaghetti with meatballs," "medusas," "sea urchins," etc., were used to describe them.[21-24] The enhancing effect of DOPE is believed to be in stimulating an early endosomal release of DNA. DOPE complexes are also characterized by the presence of long, thin fibrilar structures, attached to the aggregated/fused liposome and which are believed to increase transfection in some cases.[25] Some complexes show condensed, periodic lamellar stacks, either flat or concentric. The effect of cationic lipid is probably in DNA encapsulation and protection, condensation, enhanced adhesivity of the complex, possibly enhanced internalizability, and helping in endosome disintegration and perhaps adsorption ("targeting") onto the cell nucleus.

Avoiding nonaqueous media, DNA can be condensed also by using polymers, polycations, positively charged polypeptides, and any combination of the above. Complexation with polyvalent cations does not seem to produce complexes of adequate stability for *in vivo* applications. Following these arguments as well as the mechanism of transfection shown below, the author believes that effective condensation is a necessary condition for efficacious transfection and that elongated structures as the above-described "spaghetti" are not too effective; in addition, they contribute to the colloidal instability of genosomes due to attractive bridging interactions.

VI. MECHANISM OF TRANSFECTION

Assuming an optimal DNA vector construct (containing appropriate enhancer, promoter, introns, cDNA, poly A sequences, untranslated regions, and part responsible for bacterial amplification) one can think, in a Gedanken experiment, of a mechanism of transfection as a complex, multistep process. Because losses at each step can be enormous we refer to the process as an inverse cascade which determines the transfection yield and consists of the following steps:

1. Physicochemical properties of the DNA-lipid complex (genosome), including size, surface charge, DNA protection
2. Stability of genosomes *in vitro* (colloidal) as well as *in vivo* (stability in plasma or other biological milieu)
3. Adsorptivity of genosomes to (appropriate) cells and/or recognition by specific receptors; this depends on surface charge, presence of specific ligands, and opsonins

4. Internalizability of genosomes: for receptor-mediated endocytosis and coated pit entry, size should not be too large, while fusion is less likely; it can be aided by adding fusion proteins but the use of viral proteins may induce immune response
5. Release from the endosome: it seems that the major uptake mechanism is endocytosis and in order that DNA is released before (lysozomal) degradation, several approaches are possible for the membrane disruption, from dissolution (cationic lipids), fusion using DOPE, special lytic peptides (pH triggered random coil–helix transition), or endosomal pH buffering
6. Targeting of nucleus, plasmid persistance in the nucleus, and timely DNA decondensation.

Such a general mechanism can be applied to transfection *in vivo* as well as *in vitro*. In addition to gene expression (DNA transcription and mRNA translation), some of the above steps may be improved also by specific DNA sequences. Other potential expression boosters are fusogenic peptides and lytic polypeptides to induce fusion and/or endosomal release[26] and which can be incorporated into genosomes.

It is known that correlation between various experiments is very weak, if apparently existing at all, but the author believes that this can be traced to changed properties of the cells in the culture (increased phagocytic activity, congruence, etc.), while *in vitro* and *in vivo* differences will eventually be understood and correlated with *in vitro* via plasma stability, pharmacokinetics, and biodistribution, which, of course, may depend on specific, direct or indirect (via opsonins), recognition.

In order that effective gene expression occurs, all steps, as indicated above, must be fulfilled. Different aspects, such as nature of cationic lipid, presence of neutral lipid, condensing agents, size and charge of the genosome, presence of ligands, nuclear targeting mechanisms, and the like, act at specific stage(s) of the process and can improve effectiveness in a particular phase of the process. Assuming that efficacies of 1 in 10,000 are not too bad, we see that the transfection process is an inverse cascade in which slight improvements in some steps may improve the overall expression rate. We hope that with studies of each step in this process additional improvements will be achieved and that gene therapy, after some ethical issues have been solved (genetic manipulation of plants, animals, their breeding, release into environment, cloning of human embryos, germ line manipulation vs. somatic therapy), will not become only a current hype but a viable hope for humanity.

REFERENCES

1. **Anderson, W.F.,** Prospects for human gene therapy, *Science* 226, 401 (1984); and Human gene therapy, *Science* 256, 808 (1992).
2. **Felgner, P.L.,** Particulate systems and polymers for *in vitro* and *in vivo* delivery of polynucleotides, *Adv. Drug Del. Rev.* 5, 163 (1990).
3. **Fraley, R. and Papahadjopoulos, D.,** Liposomes: the development of a new carrier system for introducing nucleic acid into plant and animal cells, *Curr. Top. Microbiol. Immunol.* 96, 171 (1982).
4. **Lasic, D.D.,** Liposomes: from Physics to Applications, Elsevier, Amsterdam (1993).
5. **Lasic, D.D. and Papahadjopoulos, D.,** Liposomes revisited, *Science* 267, 1275 (1995).
6. **Felgner, P.L., Gadek, T., Holm, M., Roman, R., Chan, W., Wenz, M., Northrop, J.P., Ringold, G.M., and Danielsen, M.,** Lipofection: a highly efficient, lipid-mediated DNA-transfection procedure, *Proc. Natl. Acad. Sci. U.S.A.* 84, 7413 (1987).
7. **Behr, J.P., Demenieux, B., Loeffler, J., and Perz, M.J.,** Efficient gene transfer into mammalian primary endocrine cells with lipopolyamine-coated DNA, *Proc. Natl. Acad. Sci. U.S.A.* 86, 6982 (1989).
8. **Gao, X. and Huang, L.,** A novel cationic liposome reagent for efficient transfection of mammalian cells, *Biophys. Biochem. Res. Commun.* 179, 280 (1991).
9. **Felgner, J.H., Kumar, R., Sridhar, C.N., Wheeler, C., Tsai, J., Border, R., Ramsay, P., Martin, M., and Felgner, P.L.,** Enhanced gene delivery and mechanism studies with a novel series of cationic lipid formulations, *J. Biol. Chem.* 269, 2550 (1994).
10. **Litzinger, D. and Huang, L.,** Phosphatidylethanolamine liposomes: drug delivery, gene transfer and immunodiagnostic applications, *Biochim. Biophys. Acta* 1113, 201 (1992).
11. **Leventis, R. and Silvius, J.R.,** Interactions of mammalian cells with lipid dispersions containing new metabolizable cationic lipids, *Biochim. Biophys. Acta* 1023, 124 (1990).
12. **Farhood, H., Bottega, R., Epand, R.M., and Huang, L.,** Effect of cholesterol derivatives on gene transfer and protein kinase C activity, *Biochim. Biophys. Acta* 1111, 239 (1992).
13. **Ballas, N., Zakai, N., Sela, I., and Loyter, A.,** Liposomes bearing a quarternary ammonium detergent as an efficient vehicle for functional transfer of TMV-RNA into plant protoplasts, *Biochim. Biophys. Acta* 939, 8 (1988).

14. **Legendre, J.Y. and Szoka, F.C.,** Delivery of plasmid DNA into mammalian cell lines using pH sensitive liposomes: comparison with cationic liposomes, *Pharm. Res.* 9, 1235 (1992).
15. **Carmona-Ribeiro, A.M. and Chaimovich, H.,** Salt-induced aggregation and fusion of dioctadecyldimethylammonium chloride and sodium dihexadecylphosphate vesicles, *Biophys. J.* 50, 621 (1986).
16. **Helfrich, W.,** private communication, 1994.
17. **Winterhalter, M. and Lasic, D.D.,** Liposome stability and formation: experimental parameters and theories on size distribution, *Chem. Phys. Lipids* 64, 35 (1993).
18. **Brigham, K.L., Meyrick, B., Christman, B., Magnusson, M., King, C., and Berry, L.C.,** Rapid communication: *in vivo* transfection of murine lungs with a functional prokaryotic gene using a liposome vehicle, *Am. J. Med. Sci.* 298, 278, 1989.
19. **Zhu, N., Liggit, D., Liu, Y., and Debs, R.,** Systemic gene expression after intravenous DNA delivery into adult mice, *Science* 261, 209 (1993).
20. **Lasic, D.D.,** *Chim. Oggi/Chem. Today,* 1995.
21. **Gerhson, H., Ghirlando, R., Guttman, S.B., and Minsky, A.,** Mode of formation and structural features of DNA-cationic liposome complexes used for transfection, *Biochemistry* 32, 7143 (1993).
22. **Gustaffson, J., Almgrem, M., and Arvidson, G.,** A cryoTEM study of cationic liposome used for transfection, in *Liposomes, Nineties and Beyond, Book of Abstracts* (G. Gregoriadis, Ed.), London (1993); and *Biochim. Biophys. Acta,* in press (1995).
23. **Sternberg, B., Sorgi, F., and Huang, L.,** New structures in complex formation between DNA and cationic liposomes visualized by FFEM, *FEBS Lett.* 356, 361 (1994).
24. **Frederik, P., Podgornik, R., and Lasic, D.D.,** Unpublished.
25. **Xu, Y., Hui, S.-W., and Szoka, F.C., Jr.,** Effect of lipid composition and lipid-DNA charge ratios on physical properties and transfection activity of cationic lipid-DNA complexes, *Biophys. J.* 61, A432 (1994).
26. **Parente, R.A., Nir, S., and Szoka, F.C.,** Mechanism of leakage of phospholipid vesicle contents induced by the peptide GALA, *Biochemistry* 29, 8720, 1990.

Chapter 2

Structural Features of DNA–Cationic Liposome Complexes and Their Implication for Transfection

Abraham Minsky, Rodolfo Ghirlando, and Hezi Gershon

CONTENTS

I. INTRODUCTION

A. NUCLEIC ACIDS TRANSFECTION: BACKGROUND AND GENERAL CONSIDERATIONS

Recombinant DNA technology depends upon three basic processes — isolation of specific genes, *in vitro* manipulation of the resulting nucleic acids sequences and the reintroduction, or transformation, of the DNA constructs into living cells. The possibility to deliver DNA molecules into living systems and thus to effect a stable, heritable modification, represents an indispensable genetic tool that provides a link between the *in vitro* analysis of nucleic acid segments and their *in vivo* functions. It allows detailed studies of the mechanisms and cellular factors involved in gene regulation within particular cell lines and enables the expression — and overproduction — of foreign genes.

Transformation into prokaryotic systems is straightforwardly achieved by exploiting recombinant bacteriophages as vectors,[1-3] or by using methods that induce, through simple chemical treatments[4,5] or exposure to high-voltage electric fields (electroporation),[6,7] a transient state of "competence" in the recipient bacteria, during which they are capable of uptaking DNA molecules. The analysis of eukaryotic genes has been significantly boosted by the ease with which DNA from eukaryotes can be cloned in bacterial systems. Yet, many fundamental functions common to eukaryotic cells, such as specific gene regulation mechanisms, RNA processing, posttranslational modifications and intracellular transport of proteins, as well as cell differentiation, are either absent in prokaryotic systems or occur through different pathways. The recognition that analysis of gene regulation and of their translation products in eukaryotes must be conducted in a eukaryotic environment elicited an intensive search for techniques that would enable transfection of yeast, plant and mammalian cells. For a stable, heritable genetic manipulation of such systems to occur, exogenous DNA segments have to be uptaken, delivered into the nucleus through

0-8493-4013-6/96/$0.00+$.50

the cytoplasmic compartment without being degraded by DNA-digesting enzymes (nucleases), and subsequently integrated into the host chromosomes.

In yeast cells the entire process is relatively simple.[8] Transfer through the cell membrane is effected either through the formation of spheroplasts (i.e., wall-less yeast cells) which are capable of uptaking exogenous DNA,[9] or by means of electroporation, believed to induce transient pores in the cell membrane through which DNA molecules are able to enter.[10] Depending on the vector, the transfected DNA integrates into the yeast chromosome or resides in the nucleus as an additional, "artificial" chromosome (YAC),[11] that can be isolated and transferred into higher eukaryotic systems.[12,13] Transfection into plant cells is performed by exploiting genetically manipulated plant viruses[14] or specific soil bacteria that harbor a tumor-inducing plasmid (Ti) which, upon infection, is transformed into the plant cell.[15,16] A segment of this plasmid, coined T-DNA, integrates into the host chromosomes and may, consequently, serve as a DNA-cloning vehicle. The two methods are, however, restricted by the limited size of foreign DNA that can be inserted into the viral genome, as well as by the low host range exhibited by both the plant viruses and the soil bacteria. A rather exotic transfection method found to partially solve these difficulties is the biolistic (or "gene-gun") process which employs high-velocity DNA-coated micro-projectiles that are being shot into the tissue.[17] Those few cells which survive the bombardment damage and in which the DNA has been delivered directly to the nucleus reveal a stable exogenous DNA integration.

Calcium phosphate[18,19] and DEAE-dextran[20] are being widely used for nucleic acids transfection into cultured mammalian cells. The co-precipitates or complexes which are formed when these agents are mixed with nucleic acids are phagocytosed by the cells, leading to a stable DNA integration into the nuclear genome. Although theses techniques can be applied on a relatively large diversity of cell types, they are severely limited by the extensive lysosomal DNA degradation which is intrinsic to phagocytosis-mediated uptake pathways and results in a low transfection efficiency. The extent of DNA enzymatic degradation is reduced when nucleic acids are transfected by electroporation,[21-23] or by a direct micro-injection of DNA into the nucleus,[24-27] since in both methods the exogenous DNA is not initially delivered into the lysosomes. The generality and efficiency of the electroporation and microinjection methods are, however, restricted by the harsh conditions associated with these transfection procedures, which result in low cell survival rates. An additional, quite effective gene transfer system utilizes recombinant retroviruses as vectors.[28,29] The use of such vectors raises, however, unassessable safety risks and they have a rather limited carrier capacity due to constraints on the size of the retroviral genome.

Recently, positively charged liposomes have been intensively used as nucleic acids delivery vehicles both *in vitro* and *in vivo*. In this article we present our results concerned with the factors which affect the complexation processes between DNA molecules and positively charged amphiphiles and provide a detailed structural characterization of the resulting species. The study was conducted in two stages; it has been initiated by a comprehensive investigation of the interactions occurring between DNA molecules and amphiphiles which aggregate into micellar clusters but do not form liposomes. The results allowed for the progression to the interaction modes between DNA and positively charged liposomes used for transfection.

B. LIPOSOME-MEDIATED TRANSFECTION

The above-presented brief survey of the various transfection methods underlines the problems and limitations associated with these methods. Specifically, a given DNA delivery pathway might be relatively efficient for a particular cell type, but completely inadequate for other types. Some transfection procedures are limited either by their intrinsic harsh conditions or by a high carrier toxicity, both of which result in low cell survival rates; other methods are severely restricted by a low capacity of the carriers or by an extensive nucleic acids degradation caused by the targeting of the carrier into the lysosomes. The very broad flexibility available in the design of liposome composition, coupled to the diversity of methods for their preparation and to the findings that vesicle-mediated gene transfection can indeed be effected both *in vitro* and *in vivo*, had promoted the notion that liposomes might be efficiently used as custom-designed, cell-type-specific gene transfer vehicles.

The earliest procedures for preparing liposomes yielded multilamellar vesicles (MLV) which could be converted into small unilamellar vesicles (SUV) by extended sonication. Neither MLV nor SUV could be effectively used as vehicles for gene delivery since the multilamellar structure of MLV, as well as the sonication-related damages intrinsic to the preparation of SUV and their small volumes greatly reduced transfection efficiencies. The development of methods for preparing large unilamellar vesicles (LUV), including Ca^+-EDTA chelation, ether injection and, in particular, reverse-phase evaporation,[30,31] provided a means to overcome these drawbacks.[32] Indeed, large unilamellar vesicles were used to effect transfection of RNA molecules (e.g., poliovirus RNA[33] or globin mRNA[34]), DNA fragments,[35,36] bacterial plasmids,[37-39] and even metaphase chromosomes.[40] Liposomes composed of negatively charged phospholipids, mainly phosphatidylserine, were found to be substantially more effective as nucleic acids carriers than neutral vesicles,[32,41] due to their relatively low rate of content leakage upon incubation with cells; the presence of cholesterol in the vesicle membranes resulted in a further decrease of the cell-induced leakage.

The most significant advantage of liposome-mediated gene transfer over all other transfection methods, is associated to its *in vivo* applications, particularly in adult animals. The first successful *in vivo* transfection has been achieved upon intravenous injection of rat preproinsulin I gene encapsulated in LUV, which resulted in a transient expression of the gene in recipient animals.[42] This significant success prompted an intensive search for strategies that would allow the targeting of liposome-encapsulated nucleic acids to specific cells. Towards this aim, cell-surface-specific antibodies, as well as polypeptides and carbohydrate moieties have been coupled to gene-containing liposomes, thus specifically enhancing interactions of the vesicles with those cells in which receptors for these factors are expressed.[38,41,43-45]

C. POSITIVELY CHARGED LIPOSOMES AS GENE CARRIERS

Several observations have indicated that neutral or negatively charged liposomes interact at specific sites on the cell surface and are, subsequently, uptaken into the cells by endocytosis.[32,36,46] These findings reflect a major and general drawback of the liposome-mediated gene transfer technique, since empty, "ghost" liposomes will compete with nucleic acids-containing vesicles for binding sites on the cell surface.[41,46] Moreover, because liposome binding to the cell is saturable, transfection efficiency depends upon the number of DNA molecules within the liposome.[36,46] Yet, encapsulation by negatively charged vesicles is intrinsically limited, being the result of the entrapment of the aqueous phase as the liposomes are formed. Finally, endocytosed liposomes are transported into lysosomes and undergo an enzymatic degradation; only those — relatively few — vesicles that escape lysosomal degradation would be capable of mediating a successful delivery of their content.[46] These considerations lead to the development of transfection methods in which positively charged liposomes are used as carriers.[47-52] Cationic vesicles bind nucleic acids through electrostatic interactions, thus capable of capturing virtually all polynucleotide segments present in the solution at a relatively low liposome-to-nucleic acids ratios. Moreover, due to the electrostatic nature of the interactions, the DNA or RNA molecules are not present during the preparation of the cationic liposomes. Thus, complexation occurs between nucleic acids and *preformed* vesicles, and hence any damages associated with the particular method used for liposome preparation (such as sonication-induced degradation) are prevented. Also, positively charged liposomes are believed to be uptaken into the cells through membrane fusion processes — as opposed to endocytosis. Consequently, they do not require specific cell surface receptors, do not reveal saturation effects and, most significantly, are not delivered into lysosomes, thus capable of escaping lysosomal degradation processes.[53] An additional advantage exhibited by cationic vesicles is related to the observation, discussed in detail in this study, that positively charged amphiphiles effect a cooperative DNA packaging into highly condensed structures, thus allowing the encapsulation of very long DNA segments at a relatively high copy number per vesicle.

Indeed, liposomes composed of the cationic amphiphile *N*-[1-(2,3-dioleyloxy)propyl]-*N,N,N*-trimethylammonium chloride (DOTMA), as well as other, related, positively charged detergents, were found to mediate an efficient delivery of both DNA and RNA molecules into a wide variety of eukaryotic cell types and to result in relatively high levels of expression of the exogenous nucleic acids.[54-65] It is generally assumed that, in contrast to the negatively charged liposomes, the cationic vesicles do not encapsulate or entrap DNA or RNA, but bind it at their surface while maintaining their original size and shape.[50,60,66,67] This hypothetical model is inconsistent with the observations presented in the current study that point towards the occurrence of a DNA-induced fusion of the cationic vesicles into elongated lipid bilayers which encapsulate the nucleic acids. The membrane fusion is accompanied by a cooperative, liposome-induced DNA collapse which plays a crucial role in facilitating and enhancing the encapsulation processes. Significantly, agents which promote DNA packaging are shown to promote encapsulation and may, as such, increase the overall efficiency of the cationic liposome-mediated transfection.

II. METHODOLOGY

A. MATERIALS

- Phosphatidylethanolamine (PE), ethidium bromide, L-glutamate, sodium poly-L-glutamate (poly-Glu; degree of polymerization: 120), and poly-L-lysine (poly-Lys; degree of polymerization: 90) were purchased from Sigma and used as is. Cetyltrimethylammonium bromide (CTAB, 98%) was purchased from BDH. DNAase I (grade II, lyophilized, 2000 units/mg, Boehringer-Mannheim) from bovine pancreas was dissolved in 50 m*M* Tris (pH 7.7), 10 m*M* DTT and 30% glycerol. The solution (100,000 units/ml) was stored at –20°C.
- *N*-[1-(2,3-dioleyloxy)propyl]-*N,N,N*-trimethylammonium chloride (DOTMA) was synthesized following the procedure of Felgner et al.[47] and stored under nitrogen at –70°C.
- Nucleic acids. Highly polymerized calf thymus DNA (type I, Sigma) was dissolved in 20 m*M* Tris buffer, pH 7.5, and sonicated for 4 × 30 sec using an Ultratip Labsonic System (model 9100) sonicator from Lab-Line Instruments Inc. DNA fragments were loaded on a Sephacryl S-400 (Pharmacia LKB Biotechnology, Inc.), and eluted with 20 m*M* Tris (pH 7.5), 0.25 *M* NaCl solution. Fractions of 5 ml were collected, and the size distribution of the fragments was determined by 0.75% agarose gel electrophoresis. Samples were dialyzed against 20 m*M* NaCl, 5 m*M* Tris buffer, 1 m*M* EDTA, and concentrated by ultrafiltration. DNA concentrations were determined by measuring the absorption at 260 nm, applying the relationship: 1.0 OD = 40 μg/ml. The *Bluescript* plasmid used in this study was prepared according to Sambrook et al.[68]

B. PROCEDURES

1. *Liposome preparation.* The liposomes used in this study were prepared either from DOTMA or from an equimolar mixture of DOTMA and PE. Dry PE and DOTMA were dissolved in chloroform. The solvent was evaporated under a stream of nitrogen followed by 30 min exposure to high vacuum. The resulting lipid films were re-suspended in 20 m*M* NaCl (in DDW) by vortex mixing and sonicated in a bath-type sonicator (Ultratip-Labsonic-System, model 9100 from Lab-Line Inc.), until turbidity had cleared. Aliquots of the liposome solution were kept for up to 1 week at –4°C.
2. *Circular dichroism of condensed DNA-micelle aggregates.* Circular dichroism spectra were recorded on a Jasco J-500C spectropolarimeter equipped with a DP-500N data processor, using either 0.1- or 1.0-cm path length quartz cells.
3. *Fluorescence studies.* Fluorescence measurements were carried out on a Shimadzu RF-540 spectrofluorophotometer, using 1-cm light path cell with slits of excitation and emission of 5 nm. DNA-liposome mixtures were prepared by mixing the two components (both dissolved in 20 m*M* NaCl) followed by incubation for 60 min at room temperature prior to the addition of the fluorescence probe. Fluorescence was monitored immediately after the addition of ethidium-bromide to the various DNA-liposome mixtures (λ_{ex} = 260 nm, λ_{em} = 600 nm) using a 395-nm filter.
4. *Lipid mixing.* Lipid mixing was determined by using the resonance energy transfer (RET) methodology.[69] Fluorescently labeled PE/DOTMA (1/1, in molar ratio) vesicles, containing 1 mol% of each *N*-(7-nitro-2,1,3-benzoxadiazol-4-yl) phosphatidylethanolamine (energy donor: λ_{ex} = 467 nm, λ_{em} = 534 nm) and *N-(lissamine rhodamine B sulfonyl) phosphatidylethanolamine (energy acceptor:* λex = 560 nm, λem = 585 nm), (N-NBD-PE and N-Rh-PE, respectively) were prepared in the same procedure as the nonlabeled liposomes. Lipid mixing determinations were conducted on mixtures of 1:9 labeled and nonlabeled liposomes, by following the changes in fluorescence intensity at 530 nm (λ_{ex} = 467 nm) and using 2-nm filter slits to reduce light scattering interference. Total fluorescence (equivalent to minimal energy transfer or dequenching) was obtained by solubilizing the liposomes with 0.2% Triton X-100 and correcting for the quenching effect of Triton X-100, i.e., ×1.5.[70] Zero percent lipid mixing (background fluorescence) was taken as the fluorescence intensity of the liposome mixtures in the absence of DNA.
5. *Agarose-gel-electrophoresis.*
   a. DNA-liposome samples. Samples containing 2.0 μg calf thymus DNA and liposomes in various liposome-to-DNA ratios were incubated at room temperature for 1 h and loaded on 1.5% agarose gel. Samples were run at 3.5 V/cm with TBE buffer and stained by exposure to 10 μg/ml ethidium bromide.

b. Samples digested with the nuclease DNAase I. Samples containing 6.0 μg DNA (500–2000 base pairs) and liposomes in various liposome-to-DNA ratios were incubated at room temperature for 1 h in a buffer containing 100 m*M* Tris (pH 7.5) and 1.0 m*M* $MnCl_2$. Each mixture (corresponding to a given liposome-to-DNA ratio) was divided into two samples, and 20 units of DNAase I were added to one of the two samples. The mixtures were incubated for 1 min at room temperature and then quenched with phenol (added to all samples). Phenol was extracted with 2 × 400 μl of chloroform: isoamylalcohol, 24:1 solution, which also solubilizes the liposomes and extracts the lipids. Following ethanol precipitation, all samples were loaded on 1.5% agarose gel.

6. *Electron microscopy studies.* Samples for electron microscopy were prepared by the Kleinschmidt method of DNA spreading followed by metal rotary shadowing.[71] Aliquots of 10 μl containing DNA (at 3.5 μg/ml) and liposomes mixture were made up to 100 μl with 0.5 *M* ammonium acetate, 0.1 *M* Tris (pH 8.0) and 2.5 m*M* EDTA. To this solution, 10 μl cytochrome C (2.5 mg/ml) was added and samples were spread on the surface of a solution of the above buffer diluted in a ratio of 1:20 . The resulting monolayers were lifted onto parlodion coated grids which were stained for 1 min in ethanolic uranyl acetate, washed and blotted dry. Grids were subsequently rotary shadowed at an angle of 8° with platinum-palladium (80:20) and visualized in a Phillips 410 electron microscope operated at 80 kV.

III. RESULTS

A. MICELLE-MEDIATED DNA PACKAGING

When treated with poly-Glu (sodium poly-L-glutamate) and CTAB (cetyltrimethylammonium bromide) in the presence of a dehydrating agent such as EtOH, DNA molecules undergo a cooperative condensation to yield structures which reveal very large negative nonconservative CD (circular dichroism) signals (Figure 1).[72] Such spectra are known to characterize closely packed DNA species in which the DNA helices assume a left-handed tertiary organization.[73] When CTAB was present at concentrations lower than its cmc (critical micellar concentration), or upon replacing CTAB with OTAB (octadecyltrimethylammonium bromide) which is incapable of forming spherical micellar aggregates, only conservative CD spectra characteristic of B-DNA were obtained. This observation suggests that the ability of the surfactant to aggregate into micelles is required for DNA packaging. The micelle-mediated DNA packaging process was further indicated by X-ray scattering measurements conducted on pellets which were obtained from DNA molecules treated with sodium poly-L-glutamate, CTAB and EtOH. Two peaks were observed: a peak at 44 Å, reflecting the contribution of a DNA-CTAB complex, and a peak at 25 Å which characterizes DNA-DNA interactions and is usually exhibited by closely packed DNA species.[74] Neither this X-ray scattering peak nor a nonconservative CD signal was observed when sodium poly-L-glutamate was omitted from the mixture or when replaced by the monomeric L-glutamate, thus pointing towards the essential role of the negatively-charged polyelectrolyte in effecting an ordered DNA packaging.[72]

B. DNA–LIPOSOME COMPLEXES: FLUORESCENCE STUDIES

Fluorescence experiments were performed by exposing DNA molecules to positively charged liposomes, followed by addition of ethidium bromide which — upon intercalation in between the DNA base pairs — acts as a fluorescence probe. Notably, fluorescence is effected through an energy transfer from the nitrogen base pairs, induced into excited states by irradiation at 260 nm, to a *bound, intercalated* ethidium bromide molecule. The fluorescence properties which are revealed by a given concentration of DNA segments exposed to increasing amounts of cationic liposomes (DOTMA:PE, 1:1[47]) are presented in Figure 2 as a function of both the DNA length and the positive-to-negative charge ratios (in terms of DOTMA to nucleotide molarities). The fluorescence intensity is not affected by increasing concentrations of the cationic liposomes until a specific liposome-to-DNA ratio is reached, upon which a large and very sharp decrease of the intensity is observed. Such a decrease is attributed to an attenuated binding of the probe to the DNA, presumably associated with a liposome-mediated process which results in DNA protection against the fluorescence probe. The specific liposome-to-DNA ratio at which the decrease of the fluorescence intensity occurs, corresponding to a positive-to-negative charge ratio of 1.1, is found to be independent of the DNA size, in the range of 100 to 23,000 base pairs (Figure 2A). Neither is this value affected by the *absolute* concentrations of the DNA or the liposomes: a fivefold decrease of the

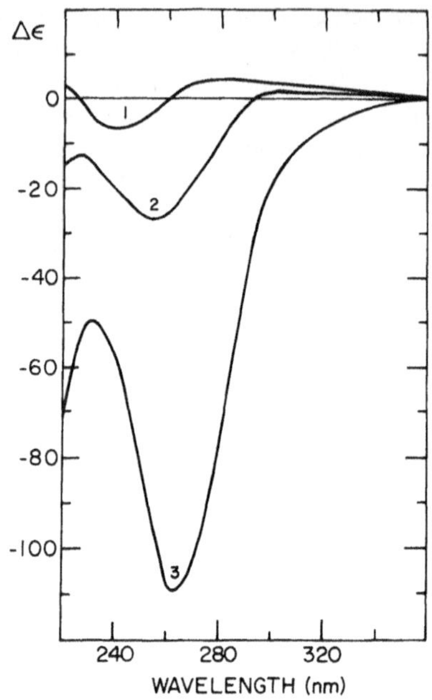

Figure 1 Effects of different cationic amphiphiles on DNA condensation; CD spectra of DNA aggregates obtained under the following sample conditions: DNA ($5{\cdot}10^{-5}$ in base pairs, 4000–9500 base pairs), poly-Glu ($5{\cdot}10^{-5}$ in amino acid residues), 25% aqueous EtOH, and the following amphiphiles: (1) DTAB, (2) TTAB, and (3) CTAB. (From Ghirlando, R., Wachtel, E. J., Arad, T., and Minsky, A., *Biochemistry,* 31, 7110, 1992. With permission.)

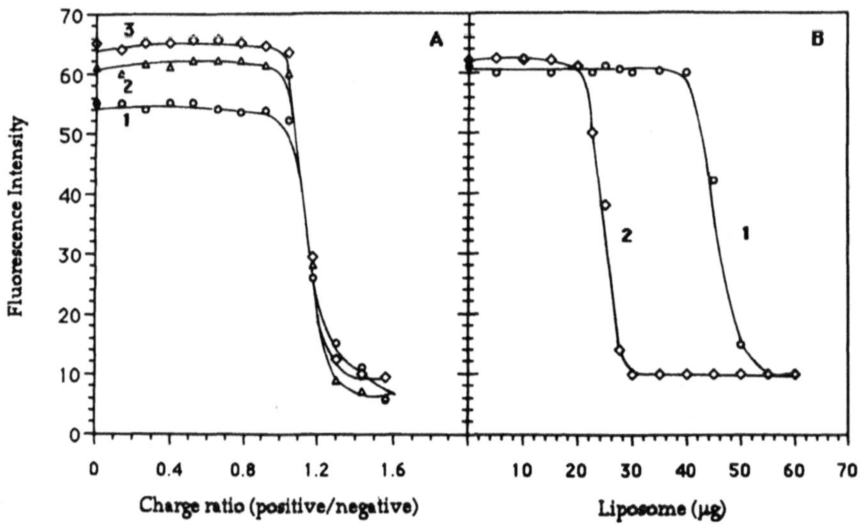

Figure 2 Effects of DNA length and liposome-to-DNA ratio (in terms of positive-to-negative charge ratio) on the fluorescence intensity. (A) Length of DNA molecules (in base pairs): (1) 100–300; (2) 500–8000; (3) 23,000. (B) Different cationic liposomes (1) DOTMA/PE (1:1 molar ratio); (2) DOTMA. DNA concentration, $2.5{\cdot}10^{-5}$ (in nucleotides), ethidium bromide (1:50 molar ratio of probe to nucleotides) was added immediately prior to the measurements. (From Gershon, H., Ghirlando, R., and Minsky, A., *Biochemistry,* 32, 7143, 1993. With permission.)

concentrations of both substances or a twofold increase relative to the concentrations shown in Figure 2A did not modify the step-function shape of the fluorescence curves (data not shown).

The notion that the properties of the liposome-DNA system crucially depend upon the positive-to-negative charge ratios is supported by the fluorescence characteristics exhibited by DNA segments which are exposed to liposomes that are composed *only* of the positively charged lipid DOTMA. In this case, precisely half the amount of lipid molecules are required to affect a decrease of the fluorescence intensity which indicates DNA protection (Figure 2B). Notably, this amount corresponds to the *same positive to negative charge ratio* (lipids-to-nucleotides) that is exhibited by the PE-DOTMA mixed-lipids liposomes at the fluorescence attenuation point. Fluorescence measurements of a closed-circular supercoiled plasmid (*Bluescript*; 2960 base pairs), incubated with the positively charged PE-DOTMA liposomes provided similar results (Figure 3A), namely, a sharp decrease of the fluorescence intensity which occurs at a somewhat higher positive-to-negative charge ratio than that characterizing the systems composed of linear DNA segments (1.3 vs. 1.1, respectively). A clear difference in the fluorescence parameters is, however, observed when very short double-stranded DNA segments are used. As shown in Figure 3B, incubation of oligonucleotides composed of 10 base pairs with increasing amounts of liposomes resulted in a sharp initial increase of the fluorescence intensity, followed by a decrease which occurs at substantially lower positive-to-negative charge ratio values.

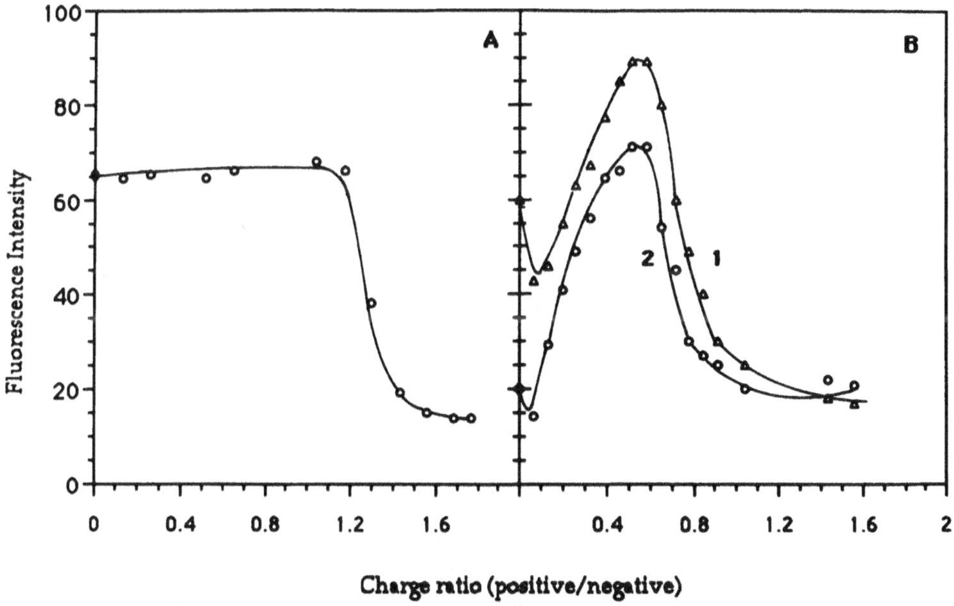

Figure 3 Effects of supercoiled plasmid DNA and of short, self-complementary oligonucleotides on fluorescence intensities. (A) Closed circular Bluescript plasmid. (B) Double-stranded oligonucleotides: (1) 5′-CAAAATTTTG-3′; (2) 5′-CTTTTAAAAG-3′. DNA and ethidium bromide concentrations were as in Figure 2.

Kinetic fluorescence measurements of the DNA–liposomes interactions were performed by addition of ethidium bromide to DNA–liposome mixtures of various charge ratios, following different incubation times. At low positive–to–negative charge ratios, namely, those exhibiting large fluorescence values, no time dependence is observed and the fluorescence remains high and constant. At large liposomes-to-DNA ratios, characterized by low fluorescence values, the decrease of the fluorescence intensity to a background level is found to occur very fast (Figure 4A). At intermediate ratios of liposomes to DNA, corresponding to the value at which the sharp fluorescence decrease is affected, a fluctuation of the fluorescence intensity as function of time is observed, culminating in background fluorescence values (Figure 4B).

The effects of positively charged species upon the fluorescence properties of the liposome–DNA complexes were studied by including lysine, sodium glutamate, poly-L-lysine (poly-Lys) and sodium poly-L-glutamate (poly-Glu) in the vesicles–DNA systems. The presence of poly-Lys — but not of the monomeric lysines — results in a shift of the liposome-to-DNA ratio at which DNA protection occurs

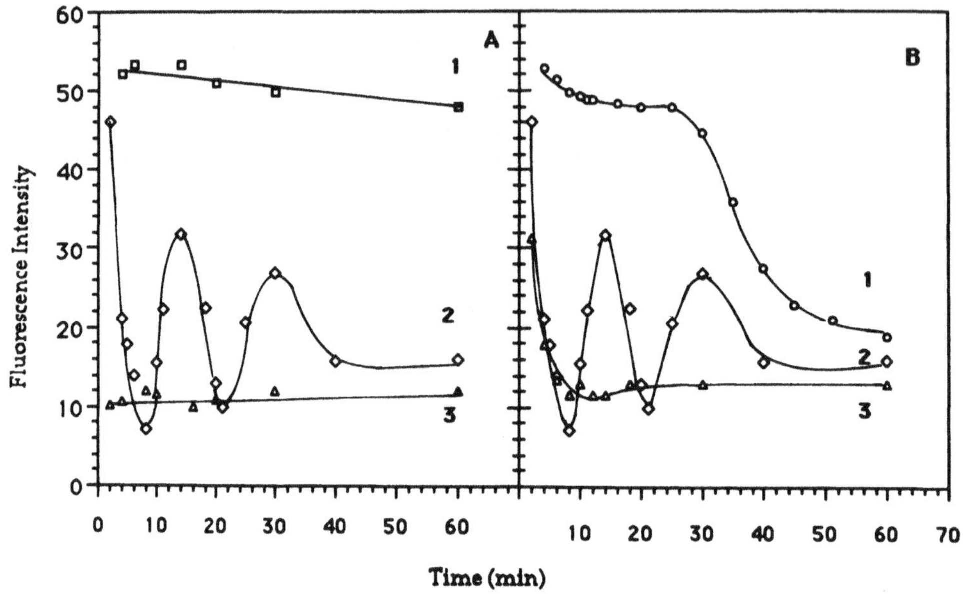

Figure 4 (A) Time dependence of the fluorescence intensity at "extreme" positive (DOTMA residues) to negative (nucleotides) charge ratios: (1) 0.5, (2) 1.1, (3) 1.5. (B) Time dependence of the fluorescence intensity at charge ratios near the fluorescence quenching point: (1) 1.0; (2) 1.1; (3) 1.2. DNA concentrations and ethidium bromide to DNA ratio were as in Figure 2. (From Gershon, H., Ghirlando, R., and Minsky, A., *Biochemistry,* 32, 7143, 1993. With permission.)

towards lower values: as the concentration of poly-Lys is gradually increased, lower amounts of liposomes are required to effect a complete DNA protection against the fluorescence probe (Figure 5A). Significantly, the value of overall positive (DOTMA + poly-Lys) to negative (DNA) charge ratio at which the fluorescence decrease occurs is found to be constant and equal to that observed in the absence of poly-Lys. Thus, in order to induce the sharp decrease of the fluorescence intensity, an increase of the concentration of positive charges supplied by poly-Lys *equals and compensates* for a decrease in the positive charges derived from the liposomes, as is clearly indicated in Figure 5B. In the absence of liposomes, the fluorescence of DNA–ethidium bromide complexes is found to be unaffected by poly-Lys, thus indicating that the observed effects of the polypeptide are not associated with its mere interference upon the fluorescence properties.

The effects of the negatively charged poly-Glu on the DNA–liposome system were studied through the addition of the polypeptide, in increasing amounts, to a given liposome-to-DNA ratio. At those liposome-to-DNA ratios which are too low to induce DNA protection against binding of the fluorescence probe, no effect of the poly-Glu could be observed: the fluorescence intensity remained large and constant (Figure 6, curve 1). However, addition of the polypeptide to liposome–DNA mixtures that exhibit, initially, low fluorescence intensities resulted in a large increase of the intensities; high poly-Glu concentrations restored the fluorescence intensity that characterizes systems completely devoid of liposomes (Figure 6, curves 2 and 3). Notably, the fluorescence values shown in Figure 6 were obtained whether the poly-Glu was added before the addition of the vesicles or to a system which already contains the liposomes. As was the case in the experiments involving poly-Lys, the fluorescence of the DNA–ethidium bromide complexes was found to be unaffected by the poly-Glu in the absence of the liposomes, or upon replacing the polypeptide by the monomeric sodium glutamate.

C. LIPID MIXING EXPERIMENTS

Lipid mixing experiments, which enable the assessment of the occurrence of liposome-fusion processes and differentiation of such processes from mere aggregation, were conducted in order to evaluate the effects exerted upon the cationic vesicles by the DNA molecules. The methodology is based on a resonance energy transfer which occurs between two fluorescence probes that are in close physical proximity if the emission band of the energy donor overlaps the excitation band of the energy acceptor.[69,70]

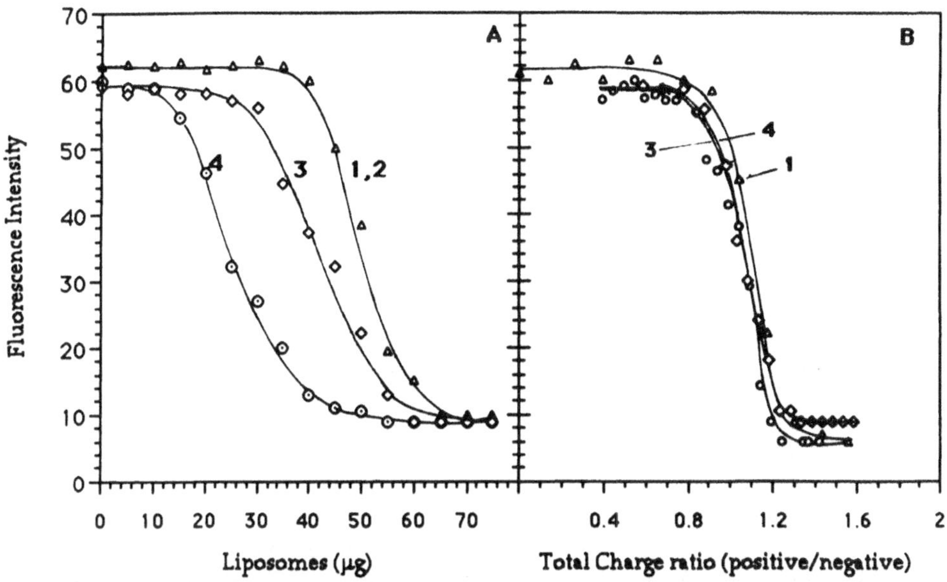

Figure 5 Effects of poly-Lys upon the fluorescence of DNA–liposome complexes, depicted as function of: A. Increasing amounts of liposomes; B. total positive (poly-Lys + DOTMA) to negative (nucleotides) ratio. (1) No poly-Lys included; (2) monomeric lysine residues ($2.5 \cdot 10^{-5}$ *M*); (3, 4) poly-Lys ($1.25 \cdot 10^{-5}$ *M* and $2.5 \cdot 10^{-5}$ *M* in amino acid residues, respectively). DNA concentration was kept at $2.5 \cdot 10^{-5}$ *M*, in nucleotides. (From Gershon, H., Ghirlando, R., and Minsky, A., *Biochemistry,* 32, 7143, 1993. With permission.)

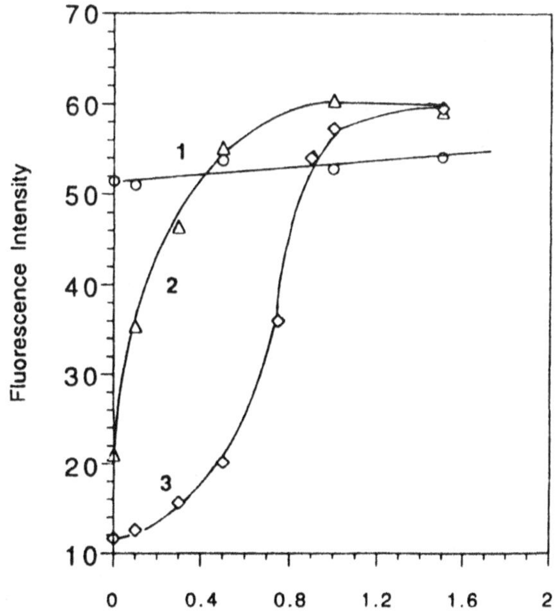

Figure 6 Effects of increasing amounts of poly-Glu upon the fluorescence of the DNA–liposome complexes obtained from the following liposome-to-DNA ratios (in terms of positive to negative charge ratios); (1) 0.8; (2) 1.0; (3) 1.2. Poly-Glu was added to the preformed complexes to yield the indicated poly-Glu (in amino acid residues) to DNA (in nucleotides) ratios. DNA concentrations and ethidium bromide to DNA ratio were as in Figure 2.

Dilution of the probes due to membrane fusion or to a solubilization of the membranes results in a dequenching of the donor emission band, which is the monitored parameter. The fluorescence exhibited by *N*-(7-nitro-2,1,3-benzoxadiazol-4-yl) Phosphatidylethanolamine (energy-donor) and *N*-(lissamine rhodamine B sulfonyl) Phosphatidylethanolamine (energy-acceptor) labeled liposomes, mixed at a 1:9 ratio with nonlabeled liposomes to final vesicle concentrations identical to those used in the above described experiments, was initially studied in the absence of DNA (Figure 7A, curve 1). The values thus obtained, which did not change after an incubation of 6 h, serve as background fluorescence and correspond to maximal quenching of the donor emission. A given amount of DNA molecules was added to a set containing increasing concentrations of labeled liposome mixtures. A substantial increase in the fluorescence intensities at 530 nm (i.e., fluorescence dequenching) relative to the background fluorescence values obtained in the absence of DNA is observed (Figure 7A, curve 2), indicating the occurrence of membrane fusion processes. The DNA-induced membrane fusion is found to be fast, occurring within 1 min after the addition of the DNA molecules. Notably, the increase of the fluorescence intensity as a function of increasing ratios of liposome to DNA is not linear. As is shown in Figure 3B, in which the difference between the fluorescence of the labeled liposome mixture in the presence and absence of DNA is plotted, a clear change in the slope occurs at the specific ratio of liposome to DNA which has been found to affect a liposome-mediated DNA protection in the liposome–DNA–ethidium bromide complexes (i.e., positive to negative charge ratio of 1.1). Total fluorescence dequenching values (Figure 7A, curve 3), corresponding to a complete lipid mixing, were obtained by exposing the labeled–unlabeled liposome mixtures to 0.2% (w/v) of the liposome-solubilizing agent Triton X-100.[70]

Lipid mixing processes induced by poly-Glu were studied by exposing the labeled liposome mixtures to the negatively charged polypeptide at the same positive-to-negative charge ratios (DOTMA to amino acid molarities) used in the liposome-DNA experiments. As indicated in Figure 7A, curve 4, the presence of poly-Glu results in a fluorescence dequenching that is virtually identical to that affected by Triton X-100. Specifically, no change in the slope of the curve describing the fluorescence dequenching is observed, in clear contrast to the effect exerted by the DNA molecules.

D. GEL ELECTROPHORESIS STUDIES

The nature of the processes occurring upon exposure of DNA molecules to the positively charged liposomes was further investigated by means of agarose-gel electrophoresis. DNA molecules (500–10,000 base pairs) and liposomes were incubated at the same liposome-to-DNA molar ratios used in the fluorescence experiments depicted in Figure 2. As shown in Figure 8, at low liposome-to-DNA ratios, in which no fluorescence quenching occurs, all the amount of DNA loaded on the gel could be detected following the exposure of the gel to the fluorescent probe ethidium bromide (compare lanes 2–5 to lane 1 which contains only DNA). In lanes 6 and 7, loaded with mixtures corresponding to those liposome-to-DNA ratios in which the sharp decrease of the fluorescence has been observed, substantially lower amounts of DNA are detected. No ethidium bromide staining of DNA occurs in lanes 8–11, loaded with mixtures composed of high liposome-to-DNA ratios. Thus, both fluorescence and gel studies of the DNA-liposome complexes point toward the existence of a specific, critical ratio of vesicles to DNA below which the nucleic acids are fully accessible to the fluorescence probe, whereas above it they are completely sequestered. Notably, both techniques indicate that the efficiency of DNA encapsulation is independent of its length, in the range of 500 to 10,000 base pairs.

The susceptibility of the complexed DNA molecules within the DNA–liposome species towards enzymatic degradation was studied by exposing the complexes, obtained at various liposome-to-DNA ratios, to the nuclease DNAse I. Samples containing four different ratios which correspond to high, medium and low fluorescence intensity according to Figure 2 (0, 0.75, 1.0 and 1.25, in terms of DOTMA to nucleotide molarities) were exposed to the cleavage activity. At the end of the incubation, the enzymatic activity was terminated with phenol, the liposomes were solubilized with chloroform-isobutanol mixture and the DNA was extracted and loaded on the gel. As a control for the cleavage processes, the four liposome-DNA mixtures at the above indicated ratios — but without the nuclease — were treated with phenol, solubilized, extracted and loaded. As shown in Figure 9, free DNA molecules are totally digested (compare lanes 2 and 1); an almost complete DNA cleavage is observed in mixtures of low liposome-to-DNA ratio (lanes 4 and 3). Only a partial digestion occurs at medium liposome-to-DNA ratio (lanes 6 and 5), whereas a virtually complete protection of the DNA molecules is obtained at the highest molar ratio (compare lanes 8 and 7).

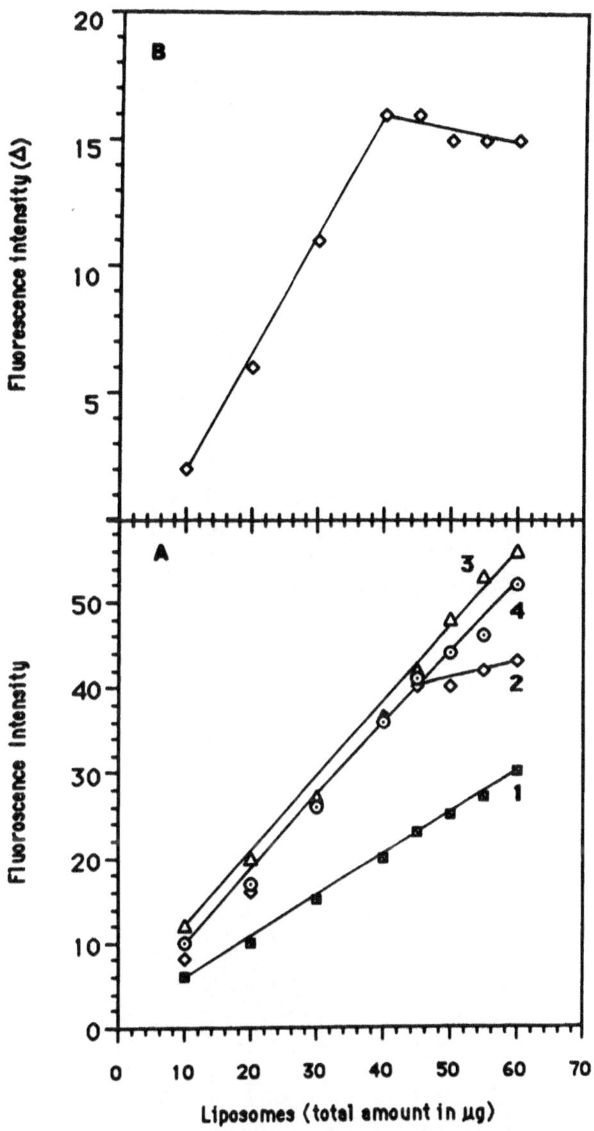

Figure 7 Lipid mixing of labeled vesicles induced by anionic polymers. (1) Background fluorescence of labeled liposomes; (2) exposure to DNA ($2.5 \cdot 10^{-5}$ *M*, in nucleotides); (3) exposure to Triton X-100 (0.2%, w/v); (4), exposure to poly-Glu ($2.5 \cdot 10^{-5}$ *M* in amino acid residues). B. Specific effects of DNA obtained by subtracting the background fluorescence from the fluorescence exhibited in the presence of DNA. (From Gershon, H., Ghirlando, R., and Minsky, A., *Biochemistry*, 32, 7143, 1993. With permission.)

E. ELECTRON MICROSCOPY

The shape of the DNA–liposome complexes, formed at various liposome-to-DNA ratios, was studied by using the Kleinschmidt method for DNA spreading.[71] Spread and metal-shadowed DNA molecules (500–2000 base pairs), as well as uncomplexed liposomes, are shown in Figure 10. When the liposomes are added in increasing amounts to a given concentration of DNA, gradually increasing regions along the DNA molecules become covered by liposome aggregates (Figure 11). Significantly, the DNA-bound liposomes do not appear as separated from each other and randomly spread along the DNA molecules, but seem to aggregate into closely packed clusters. At a 1/1 liposome-to-DNA ratio, approximately half of the DNA molecules are detected as liposome-bound species, and upon a very slight further increase of this ratio almost all the DNA appears to be covered. It should be noted that whereas at low liposome-to-DNA ratios the DNA-bound liposomes exhibit, in most cases, distinct, roughly spherical shapes

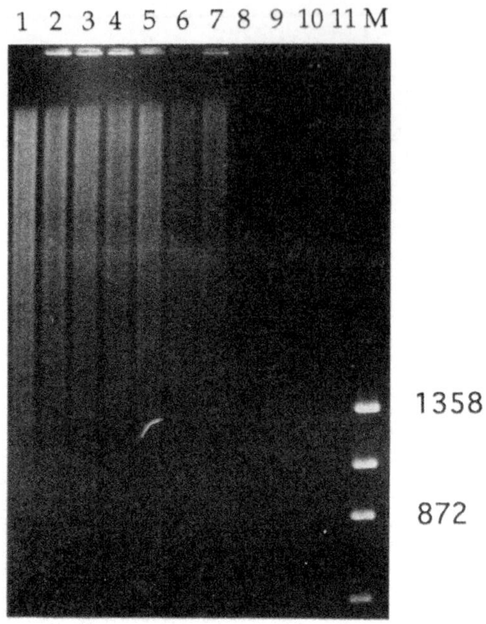

Figure 8 Effects of increasing liposome-to-DNA ratios exhibited in agarose-gel electrophoresis. Complexes were formed at the following positive to negative charge ratios: lane (1) DNA without liposomes; (2) 0.2; (3) 0.4; (4) 0.8; (5) 0.9; (6) 1.0; (7) 1.1; (8) 1.2; (9) 1.3; (10) 1.4; (11) 1.5; lane (M) contains molecular weight markers. Note the independence of the sharp fluorescence quenching between lanes 5 and 6 upon DNA length. Samples contained 2.0 μg DNA.

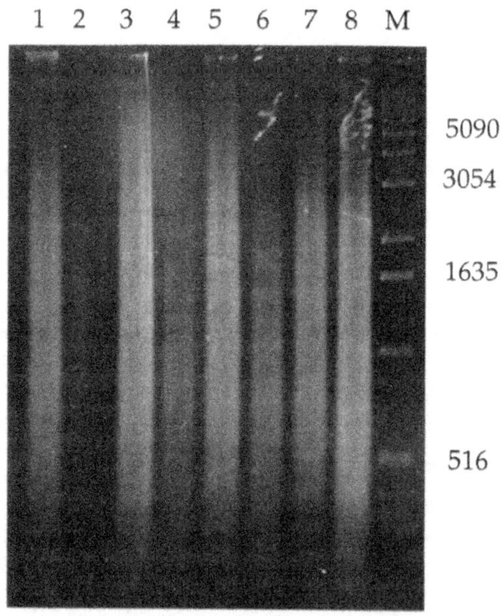

Figure 9 Effects of nuclease (DNAse I) upon DNA–liposome complexes prepared at increasing liposome-to-DNA ratios. Positive-to-negative charge ratios were: lanes (1, 2) DNA without liposomes; (3, 4) 0.75; (5, 6) 1.0; (7, 8) 1.25. Samples to which nuclease was added were loaded in lanes 2, 4, 6, 8; lane (M) contains molecular weight markers.

(Figure 11, panels B and C), at high ratios the DNA–liposome complexes are detected as smooth rod-like structures (Figure 11, panels D and E). When plasmid DNA species are visualized by the Kleinschmidt method in the presence of liposomes, a difference in the affinity of the closed-circular supercoiled molecules and the nicked-circular conformations (present in the plasmid samples) towards the liposomes is observed; apparently, the supercoiled molecules interact preferentially with the positively charged vesicles (Figure 12B,C). Electron microscopy studies were also conducted on samples containing short oligonucleotides (10 base pairs) and increasing amounts of the vesicles. While "naked" short oligomers cannot be visualized, a DNA-induced liposome aggregation into elongated clusters within which the vesicles maintain their spherical shape is observed at low liposome-to-DNA ratio (Figure 13A). As the ratio is increased, both aggregated and fused forms are detected, and finally only smooth, rod-shape structures, similar to those obtained with long DNA molecules, are detected.

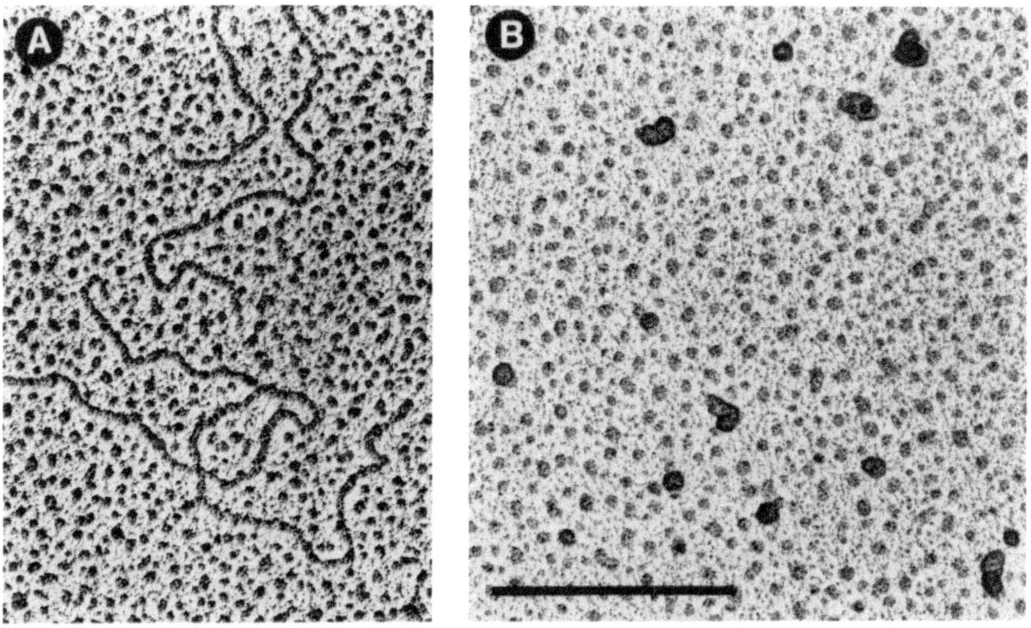

Figure 10 Electron microscopy of metal-shadowed samples containing: (A) DNA molecules (3.5 µg/ml); (B) PE-DOTMA liposomes. It should be noted that visualization of the DNA and liposomes by the Kleinschmidt methodology does not allow a size comparison of these species, since DNA molecules are detected as cytochrome C-DNA complexes. Scale bar in all electron micrographs represents 0.5 µm. (From Gershon, H., Ghirlando, R., and Minsky, A., *Biochemistry,* 32, 7143, 1993. With permission.)

DNA–poly-Lys complexes (1/1 molar ratio in terms of nucleotides and amino acid residues) exposed to increasing concentrations of liposomes are shown in Figure 13. DNA–poly-Lys samples devoid of liposomes reveal similar structures to those characterizing the "naked" DNA species. Addition of small amounts of liposomes leads, initially, to the formation of "spider-like" forms (Figure 14A), in which the thicker regions correspond to DNA–poly-Lys species covered by liposomes, and the thin coils represent unbound, liposome-free DNA. As the liposome-to-DNA ratio is increased, the thick, liposome-bound, segments are elongated while the thin regions progressively disappear (Figure 14B,C).

IV. DISCUSSION

A. DNA–MICELLES SYSTEMS

Interactions between anionic polyelectrolytes such as nucleic acids or poly-Glu with cationic amphiphiles which do not form liposomes are driven mainly by attractive electrostatic forces and are cooperative due to the association of the hydrophobic chains. Literature reports have focused on the measurements

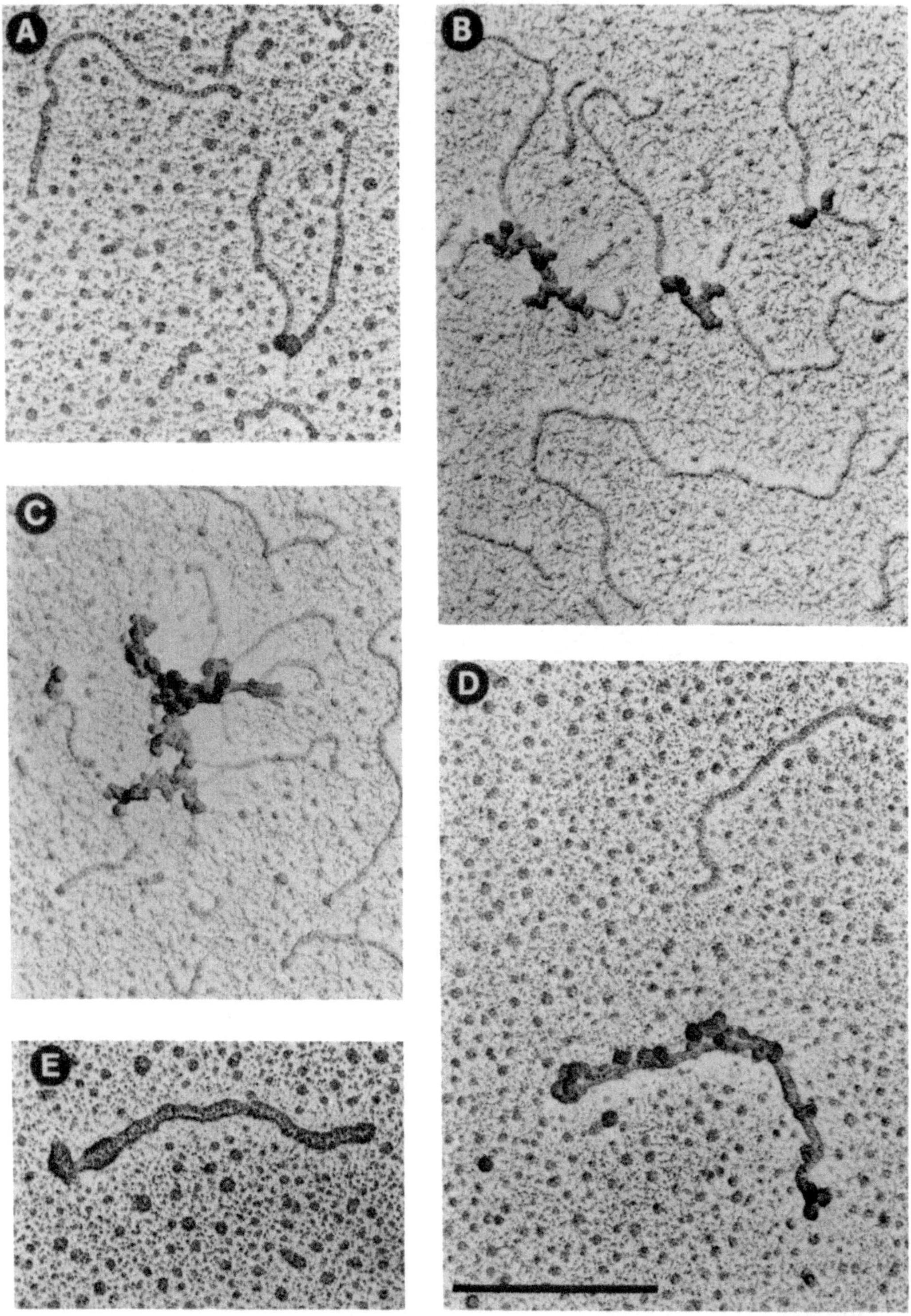

Figure 11 Electron microscopy of DNA–positively charged liposome complexes. A–E: complexes prepared from a constant amount of DNA (3.5 µg/ml), and a gradually increasing amount of cationic liposomes. Liposome-to-DNA ratios (in terms of positive-to-negative charges) are: (A) 0.2; (B) 0.4; (C) 0.6; (D) 1.0; (E) 1.5. Note the aggregated (B–D) vs. fused (E) complexes. (From Gershon, H., Ghirlando, R., and Minsky, A., *Biochemistry*, 32, 7143, 1993. With permission.)

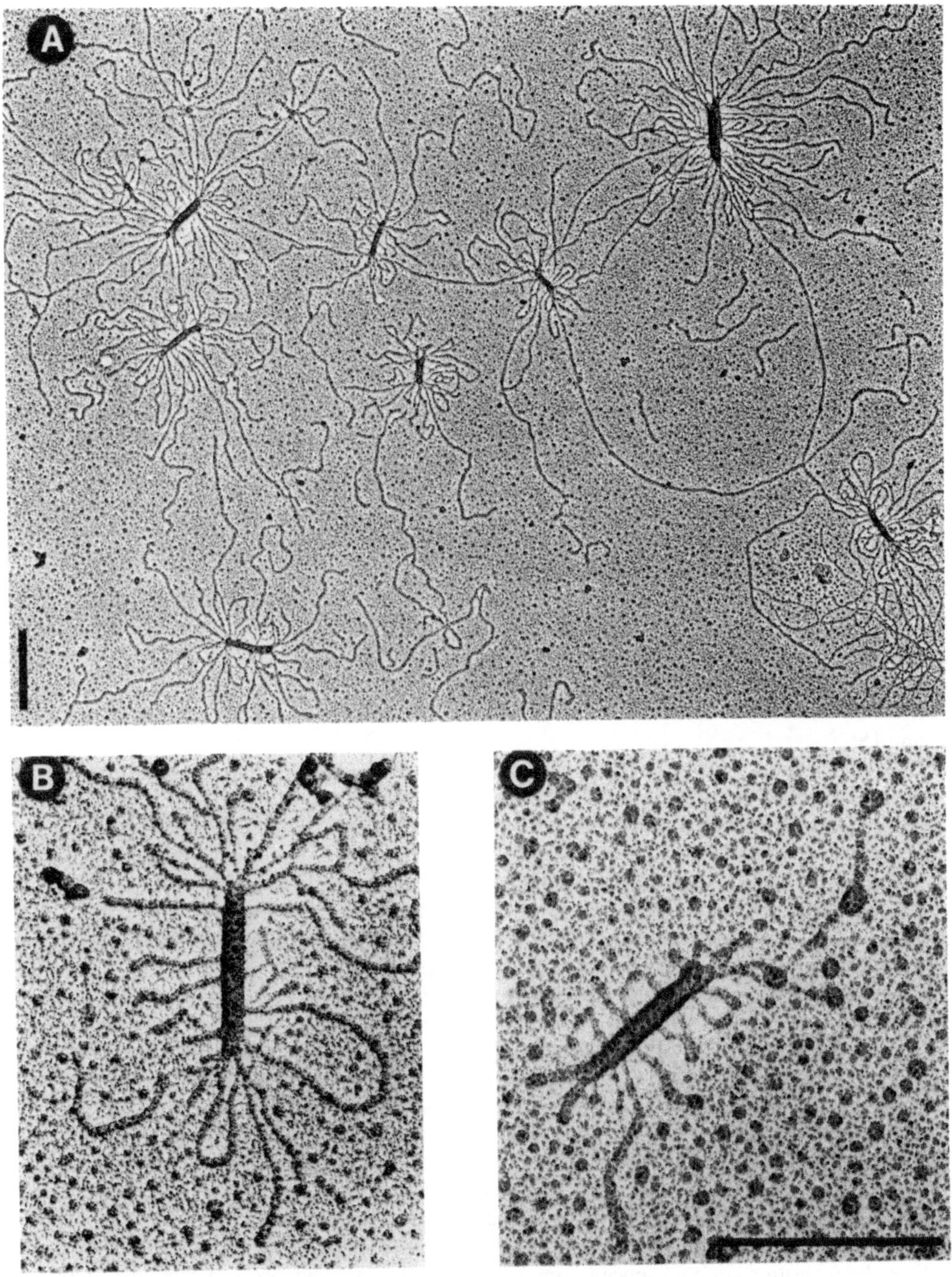

Figure 12 Electron microscopy of DNA–poly-Lys–liposome complexes. A–C: complexes prepared from constant amounts of DNA and poly-Lys (3.5 μg/ml) and a gradually increasing amount of liposomes. Liposome-to-DNA ratios (in terms of positive-to-negative charges) are: (A) 0.2; (B) 0.4; (C) 0.6. (From Gershon, H., Ghirlando, R., and Minsky, A., *Biochemistry*, 32, 7143, 1993. With permission.)

of the binding constant K and the cooperativity parameter μ, at amphiphiles concentrations below the cmc.[75] Above the cmc, micellar aggregates represent the thermodynamically stable species, and it is expected that the interactions of the polyelectrolyte with the amphiphilic species will be influenced by the stability of the micellar aggregates. Electron microscopy, X-ray scattering and fluorescence-transfer experiments indicate that conformationally flexible anionic polyelectrolytes such as poly-Glu induce the formation of loosely packed domains of *spherical* micelles.[72] These clusters are specifically stabilized

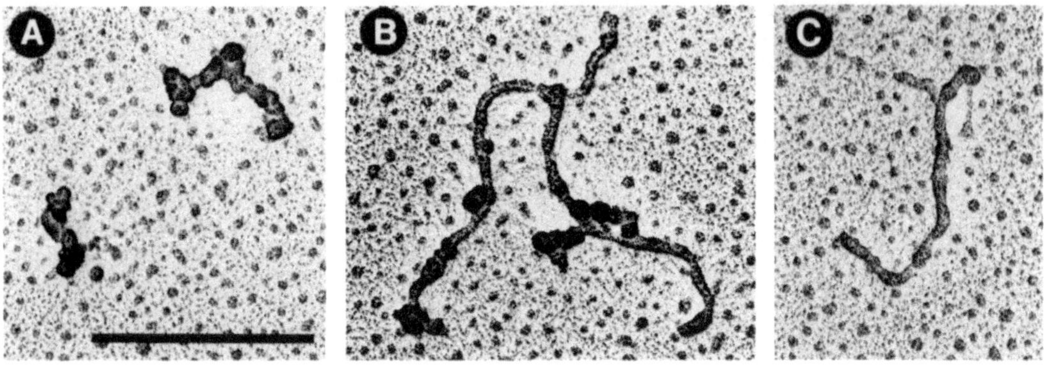

Figure 13 Electron microscopy of complexes formed between short oligonucleotides and liposomes. Liposome-to-DNA ratios are (A) 0.4; (B) 0.6; (C) 0.8. Note the relatively low liposome-to-DNA ratios required for encapsulation.

by the interweaving flexible polyelectrolytes which act as a nucleus for aggregation of spherical micelles in a manner similar to the "beads on a string" model proposed for the interactions between flexible uncharged polymers and anionic amphiphiles.[76]

The interaction between cationic amphiphiles, at concentrations higher than their cmc and the relatively rigid double-stranded DNA molecules have been proposed to result in the formation and stabilization of elongated, rod-like micelles.[72] Since collapsed DNA molecules[74] and DNA–polypeptide complexes,[77] as well as rod-like CTAB micelles[78] pack into a two-dimensional hexagonal symmetry, it is assumed that the DNA–micelle complex forms a two-dimensional hexagonal lattice, as depicted in Figure 15.

The organization of DNA in tightly packed phases which exhibit a long-range left-handed chirality (Figure 1) requires, in addition to specific cationic amphiphiles, a conformationally flexible anionic polymer. The condensation process is proposed to represent the outcome of an interplay between three main interactions within the DNA–amphiphile–flexible anionic polymer system. Whereas the flexible polyelectrolytes stabilize close-packed clusters of spherical micelles, the more rigid DNA molecules effect the elongation of the micelles into rod-like species. In addition, under conditions of charge neutralization, DNA molecules reveal a large intrinsic tendency to collapse into highly condensed phases, characterized by a 25 Å X-ray scattering peak (arising from DNA–DNA interactions[74]), as well as by the nonconservative circular dichroism spectra shown in Figure 1.[73,79] On the basis of the strict requirement for flexible anionic polymers in the DNA condensation system, and following the "bead on a string" model, according to which clusters of spherical micelles represent a particularly stable form of association in the presence of conformationally flexible anionic polyelectrolytes,[76] it has been suggested that the DNA-induced elongation of the micelles is arrested by these polymers, leading to the formation of *capped* rod-like micelles. As a consequence of the energetically favorable polyelectrolyte-induced capping, regions of the DNA molecules are no longer surrounded by the micelles and, at these regions, are free to converge into the ordered, tightly packed structures responsible for the nonconservative ellipticities and the 25 Å scattering peak. Thus, the well-demonstrated *intrinsic* tendency of DNA molecules to form ordered, tightly packed aggregates under appropriate conditions,[79-81] plays a crucial role in the formation of DNA–micelle–anionic polyelectrolyte complexes. A simplified representation of the resulting complex is depicted in Figure 16.

We argue that the three main forces that characterize the DNA–polyanion positively charged micelle system, namely, the DNA-induced fusion of the amphiphilic species into rod-like micelles, the interactions between micelles and conformationally flexible polyanions, as well as the intrinsic tendency of DNA molecules to undergo condensation processes, dominate the complexation processes between DNA and positively charged liposomes as well.

B. DNA–LIPOSOME COMPLEXATION

Two general models for DNA–liposome complexation processes have been proposed:

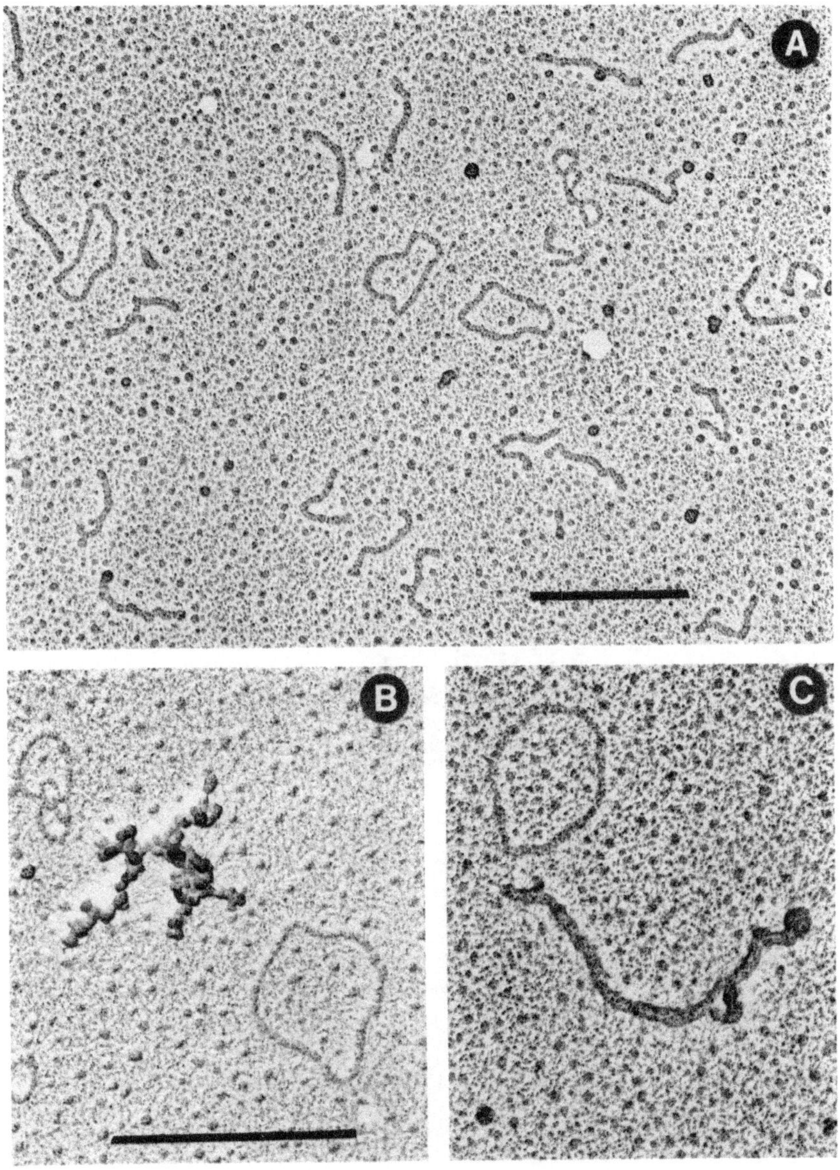

Figure 14 Electron microscopy of supercoiled DNA-liposome complexes. Liposome-to-DNA ratios are (A) 0; (B) 1.0; (C) 1.2. Note the preferential binding of the liposomes to supercoiled plasmids relative to nicked-circular conformations.

1. Felgner and Ringold have suggested that the complexation processes are dominated solely by electrostatic interactions.[50] A complex is formed in which the nucleic acids are progressively surrounded by liposomes that *maintain their original size and their — roughly spherical — shape,* until a charge neutralization of the DNA molecules is achieved. The resulting complex is likely to serve as an efficient transfecting agent due to its positively charged surface which promotes interactions and subsequently fusion processes with the negatively charged cell membrane. A variation of this model consists of the hypothesis that the DNA molecules are externally adsorbed to the liposome surface;[67] presumedly, the partial positive charge remaining on the vesicles still enables the fusion of the complex to the cell membrane.

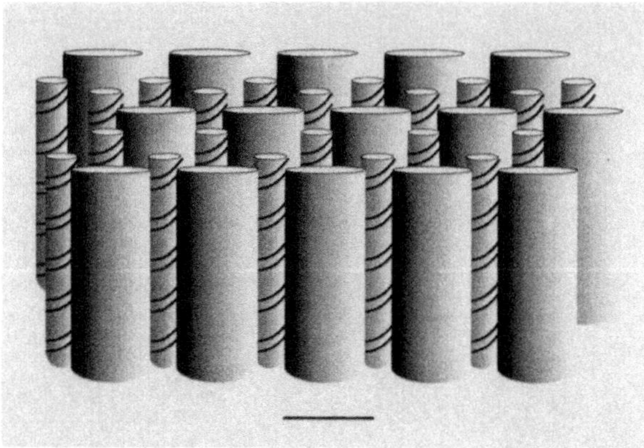

Figure 15 Schematic representation of DNA–induced rod-like CTAB micellar organization. The elongated CTAB micelles are represented as a broad rods, while the DNA is depicted as striped rods. Note the hexagonal arrangement. The scheme is approximately to scale with scale bar representing 50 Å. (From Ghirlando, R., Wachtel, E. J., Arad, T., and Minsky, A., *Biochemistry*, 31, 7110, 1992. With permission.)

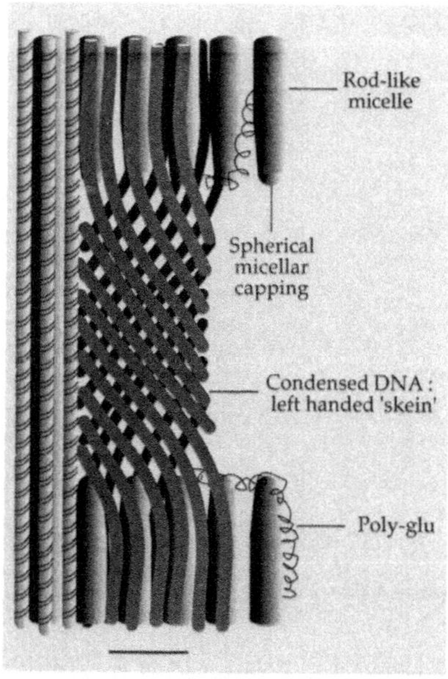

Figure 16 A simplified representation of DNA-condensation mechanism within the DNA-CTAB complex, induced in the presence of poly-Glu. The left-handed section of the scheme illustrates the regular DNA–CTAB hexagonal complex. The remainder highlights the rod-to-spherical micellar capping induced by the helical flexible poly-Glu. The formation of left-handed chiral DNA condensate is shown in the center. The scheme is approximately to scale with scale bar representing 100 Å. (From Ghirlando, R., Wachtel, E. J., Arad, T., and Minsky, A., *Biochemistry*, 31, 7110, 1992. With permission.)

2. According to the second model, the initial interaction between the negatively charged DNA or RNA molecules and the positively charged liposomes leads to a fusion of the liposomes. Ballas et al., who proposed this model,[48] suggested that the nucleic acids can be entrapped within the fused vesicles.

The fluorescence as well as the gel electrophoresis experiments described in the Results section provide a measure to the accessibility of a probe, i.e., ethidium bromide or a DNA-digesting enzyme, to the DNA molecules in the presence of increasing amounts of liposomes. Thus, the decrease of fluorescence intensity and the attenuated enzymatic degradation indicate the formation of a complex in which the nucleic acids are effectively protected and sequestered. The following conclusions can be derived from these experiments:

a. DNA encapsulation, implied by the decrease in binding of the fluorescence probe, occurs over a very narrow range of liposome concentrations and only at a specific, well-defined liposome-to-DNA ratio (Figure 2A), indicating that DNA encapsulation within the liposomes is a highly cooperative process. This ratio is independent of the DNA size, in the range of 100 to 23,000 base pairs (Figures 2A and 8), as well as of the *absolute* concentrations of the DNA or the liposomes.
b. The liposome-to-DNA ratio at which the fluorescence quenching is observed is determined by the positive-to-negative charge ratio between the DOTMA and the nucleotide residues. This is clearly indicated by the finding that when liposomes composed solely of DOTMA are used, as opposed to mixed vesicles composed of both DOTMA and PE in a 1/1 molar ratio, precisely half the amount of liposomes is required to effect complete encapsulation (Figure 2A). The fact that DNA encapsulation is initiated at a specific charge ratio seems to indicate that the process is causally associated with charge neutralization. This assumption is supported by the observation that supercoiled DNA conformations, whose negative charge density is higher than that characterizing linear or nicked-circular DNA molecules, interact preferentially with the positively charged liposomes (Figure 12), but require a higher liposome concentration to undergo encapsulation (Figure 3A).
c. Lipid mixing (Figure 7) and electron microscopy (Figure 11) experiments indicate that DNA–liposome complexation is associated with lipid fusion processes. Indeed, multivalent anions such as citrate or phosphate, as well as DNA molecules have been found to induce fusion of positively charged liposomes.[70,82,83] The conspicuous independence of the DNA encapsulation process upon the DNA length, inferred from both the fluorescence and gel electrophoresis results, supports the notion that lipid fusion and DNA encapsulation are causally related: mere DNA entrapment within the limited space of a *single* SUV would become progressively inefficient as the length of the DNA molecules increases.

These conclusions and considerations, combined with the results obtained from the DNA–micelles system, point towards the occurrence of two interrelated processes: a DNA-induced liposome fusion and a liposome-induced DNA packaging.[84] The lipid mixing results taken on their own do not provide an interpretation for the highly cooperative mode of the DNA-liposome interactions, indicated by the sharp decrease of the fluorescence intensity which is observed in both the fluorescence and electrophoresis experiments. Notably, such a cooperative process is inconsistent with the currently prevailing model for DNA–liposome interactions according to which the DNA is being progressively surrounded by cationic liposomes — which maintain their original structure — as their concentration is increased. The lipid mixing results point, however, towards a different effect of DNA and the conformationally flexible poly-Glu upon membrane fusion. While both negatively charged polymers induce fusion, the process effected by the polypeptide is continuous, in contrast to that induced by the DNA molecules (compare curves 2 and 4 in Figure 7A). The difference may be attributed to a unique structural phase modulation which is specifically exhibited by DNA and occurs precisely at the liposome-to-DNA ratio leading to the sharp decrease of the fluorescence intensity.

It is well established that at a given degree of charge neutralization, DNA molecules collapse into packed forms in a highly cooperative process.[80,81] Since charge neutralization can be effected by binding of DNA molecules to the positively charged vesicles, we claim that a liposome-induced cooperative DNA collapse is a key event in the course of the DNA–cationic vesicles complexation. Notably, it has been suggested that plasmid DNA species assume a condensed form in the presence of specific lipids (lipopolyamines).[66,85] The notion that a liposome-induced, charge-neutralization-dependent DNA collapse is indeed intrinsically associated with its encapsulation provides a straightforward interpretation for the observation that the lipid concentration required for DNA encapsulation within liposomes composed only of the positively charged DOTMA is precisely half that of DOTMA-PE liposomes. This notion is further buttressed by results obtained from the supercoiled DNA conformations, found to interact preferentially with liposomes (Figure 12), but to undergo a cooperative encapsulation at an apparently higher positive-to-negative charge ratio (Figure 3A). Both observations can be assigned to the fact that supercoiled conformations are characterized by a higher negative charge density than linear or nicked-circular species.

The collapsed DNA structures, whose exposed surface is substantially smaller than that of the fully extended forms, undergo particularly efficient encapsulation by the lipid bilayer. Thus, a very slight increase of the liposome concentration above the critical value required for DNA collapse is sufficient to complete the encapsulation in a fast process which is indicated by the kinetic fluorescence experiments conducted at high liposome-to-DNA ratios (Figure 4). The time-dependent fluctuations of the fluorescence intensities observed at intermediate charge-ratios can be attributed to the large sensitivity of DNA collapse processes to minor changes in the environmental conditions which characterizes the initiation of these processes,[81] and may therefore lead to an equilibrium between encapsulated and free DNA regions. The packaging of short double-stranded DNA oligonucleotides, occurs exclusively through intermolecular pathways, as opposed to long sequences which can collapse by either inter- or intramolecular mechanisms. Due to their short length, very few positively charged liposomes are required to neutralize the negative charges per one segment and thus to initiate such collapse. The lower liposome-to-DNA ratio needed to effect encapsulation (Figure 3B) is interpreted in terms of enhanced DNA packaging processes of the short oligonucleotides which consequently facilitate the fusion.

A significant observation concerned with the relation between charge neutralization, DNA collapse and DNA encapsulation is related to the effects exerted by the charged polypeptides poly-Lys and poly-Glu. When poly-Lys is included in the DNA–liposome system, the overall positive charge density (poly-Lys + liposomes) required for DNA encapsulation is found to remain identical to the critical positive charge density which effects this process and provided *only* by the liposomes. Poly-Lys has been shown to induce and promote charge-neutralization-dependent DNA packaging processes,[86,87] and its presence in DNA–liposome mixtures results, consequently, in the formation of packed DNA regions which facilitate DNA encapsulation. Addition of the negatively charged poly-Glu to the DNA–liposome system results in the release of DNA molecules, as evidenced by their accessibility to the probe (Figure 6). Since poly-Glu has been shown to interact avidly with positively charged amphiphiles and to effectively induce liposome fusion, its addition to the DNA–liposome complexes may lead to the destabilization of the fused lipid bilayer that surrounds and sequesters the DNA molecules by mere competition.

The occurrence of DNA-dependent liposome aggregation and fusion and the liposome-induced DNA collapse is corroborated by electron microscopy observations. At low ratios of liposomes-to-DNA the liposomes appear as clusters bound to the DNA molecules in which the distinct, spherical shape of the vesicles can still be discerned (Figure 11). The short, rod-like structures observed above the critical ratio are proposed to reflect complexes in which DNA molecules are packed (and hence the short, rod-like appearance which usually characterizes condensed DNA phases) and completely encapsulated within a smooth lipid bilayer. Moreover, at high liposome-to-DNA ratios the overall number of the rod-like shapes detected on the grid is substantially lower than the number of the unbound DNA molecules, indicating that these structures represent a complex in which *several* DNA molecules are packed together within the lipid bilayer. Intermolecular packaging pathways that result in lipid-encoated species containing many DNA molecules are clearly pointed out in samples composed of liposomes and short oligonucleotides (Figure 13).

C. PROPOSED MODEL FOR DNA–LIPOSOME COMPLEXES

Based on the observations obtained from the DNA-positively-charged micelle and DNA-positively-charged liposome systems, the following model for cationic liposome-DNA complexation is proposed (Figure 17).[84] At low ratios of liposomes-to-DNA, positive vesicles are adsorbed through electrostatic interactions to the nucleic acids to form aggregates that gradually surround larger segments of the DNA. As the amount of liposomes is increased, the aggregated liposomes along the DNA reach critical concentration and charge densities at which lipid fusion and cooperative DNA collapse processes are initiated. Following an additional increase of the liposome concentration, the collapsed DNA structures are efficiently and completely coated by the lipid bilayers, thus becoming fully sequestered from nuclease activities or fluorescent probes. The resulting complexes can be disrupted if negatively charged species characterized by a particularly avid interaction with the lipids, such as poly-Glu, are allowed to interact with the rod-like lipid species.

A distinctive characteristic of the liposome–DNA complexes is that such complexes keep the DNA molecules in solution and hence available for transfection, in contrast to the micelle–DNA structures which effect nucleic acids precipitation.[72] A significant tenet of the proposed model concerns the mutual effects exerted by the DNA molecules and the cationic liposomes. DNA molecules induce aggregation and fusion of vesicles, and the resulting positively charged fused lipid bilayers enable cooperative DNA

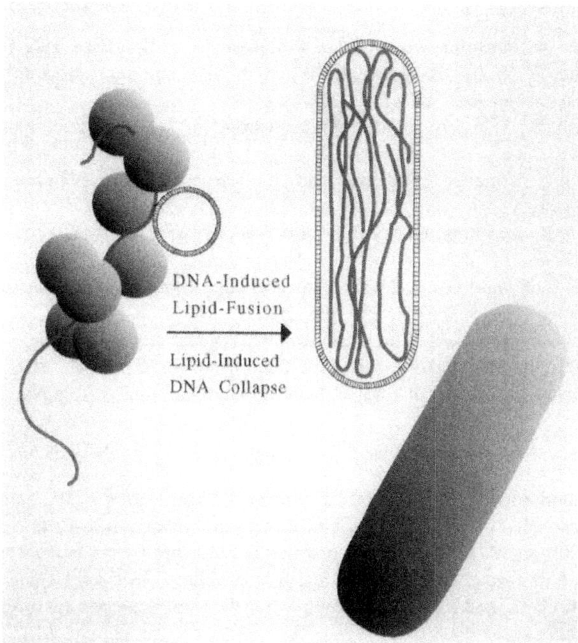

Figure 17 A simplified schematic representation of DNA–positively charged liposome complexation process and of the resulting complexes.

packaging in a manner similar to that revealed by cationic polymers such as poly-Lys. The packed phases, in turn, facilitate and enhance the encapsulation processes by the lipid bilayers. Indeed, it has recently been shown that positively charged amphiphiles that were specifically designed to effect DNA condensation (lipopolyamines) are particularly effective as gene carriers.[85,88] Thus, DNA packaging processes, membrane fusion and nucleic acids encapsulation represent causally related key events in the formation of DNA–liposome complexes used for transfection.

The interactions between DNA and positively charged amphiphiles — either in a free form, such as the lipopolyamines, or in the form of micelles and liposomes — lead to the formation of complexes within which the (tightly packed) DNA molecules are encapsulated, and hence protected, by a sheath composed of the amphiphiles. The unique efficiency of the resulting complexes as gene carriers is a direct outcome of their positively charged surface which induces fusion to the cellular membranes, the packed DNA conformation which allows transfection of very long DNA segments, and the amphiphile-mediated DNA protection. Conceivably, these features, combined with the plastic structure of liposome–DNA complexes indicated by the above described observations, as well as with the possibility to include specific targeting signals within the liposome membranes, will render the positively charged liposome-mediated transfection into the method of choice for *in vivo* gene transfer processes.

ACKNOWLEDGMENT

This work was supported by the Kimmelman Center for Biomolecular Structure & Assembly, and by a grant from the Minerva Foundation, Germany.

REFERENCES

1. **Murray, N. E. and Murray, K.,** Manipulation of restriction targets in phage λ to form receptor chromosomes for DNA fragments, *Nature,* 251, 476, 1974.
2. **Murray, K. and Murray, N. E.,** Phage lambda receptor chromosomes for DNA fragments made with restriction endonuclease III of *H. Influenzae* and restriction endonuclease I of *E. coli, J. Mol. Biol.,* 98, 551, 1975.
3. **Old, R. W. and Primrose, S. B.,** *Principles of Gene Manipulation,* 3rd ed., Blackwell Scientific Publications, Oxford, 1985, chap. 4.
4. **Cohen, S. N., Chang, A. C., and Hsu, L.,** Nonchromosomal antibiotic resistance in bacteria: Genetic transformation of *E. coli* by R-factor DNA, *Proc. Natl. Acad. Sci. U.S.A.,* 69, 2110, 1972.

5. **Hanahan, D.**, Studies on transformation of *E. coli* with plasmids, *J. Mol. Biol.*, 166, 557, 1983.
6. **Dower, W. J., Miller, J. F., and Ragsdale, C. W.**, High efficiency transformation of *E. coli* by high voltage electroporation, *Nucleic Acids Res.*, 16, 6127, 1988.
7. **Fiedler, S. and Wirth, R.**, Transformation of bacteria with plasmid DNA by electroporation, *Anal. Biochem.*, 170, 38, 1988.
8. **Rothstein, R.**, Targeting, disruption, replacement, and allele rescue: integrative DNA transformation in yeast, *Methods Enzymol.*, 194, 281, 1991.
9. **Hinnen, A., Hicks, J. B., and Fink, G. R.**, Transformation of yeast, *Proc. Natl. Acad. Sci. U.S.A.*, 75, 1929, 1978.
10. **Simon, J. H.**, Transfection of intact yeast cells by electroporation, *Methods Enzymol.*, 217, 478, 1993.
11. **Burke, D. T., Carle, G. F., and Olson, M.V.**, Cloning of large segments of exogenous DNA into yeast by means of artificial chromosome vectors, *Science*, 236, 806, 1987.
12. **Pachnis, V., Pevny, L., Rothstein, R., and Costantini, F.**, Transfer of yeast artificial chromosome carrying human DNA from *S. cerevisiae* into mammalian cells, *Proc. Natl. Acad. Sci. U.S.A.*, 87, 5109, 1990.
13. **Huxley, C. and Gnirke, A.**, Transfer of yeast artificial chromosome from yeast to mammalian cells, *BioEssays*, 13, 545, 1991.
14. **Hull, R. and Davies, J. W.**, Genetic engineering with plant viruses and their potential as vectors, *Adv. Virus Res.*, 28, 1, 1983.
15. **Chilton, M. D., Drummond, M. H., Merlo, D. J., Sciaky, D., and Nester, E. W.**, Stable incorporation of plasmid DNA into higher plant cells: the molecular basis of crown gall tumorigenesis, *Cell*, 11, 263, 1977.
16. **Matzke, A. J. and Chilton, M. D.**, Site-specific insertion of genes into T-DNA of the *Agrobacterium* tumor-inducing plasmids: an approach to genetic engineering of higher plant cells, *J. Mol. Appl. Genet.*, 1, 39, 1981.
17. **Sanford, J. C., Smith, F. D., and Russel, J. A.**, Optimizing the biolistic process for different biological applications, *Methods Enzymol.*, 217, 483, 1993.
18. **Graham, F. L. and van der Eb, A. J.**, A new technique for the assay of infectivity of human adenovirus 5 DNA, *Virology*, 52, 456, 1973.
19. **Chen, C. and Okayama, H.**, High-efficiency transformation of mammalian cells by plasmid DNA, *Mol. Cell. Biol.*, 7, 2745, 1987.
20. **Warden, D. and Thorne, H. V.**, Infectivity of polyoma virus DNA for mouse embryo cells in presence of diethylaminoethyl-dextran, *J. Gen. Virol.*, 3, 371, 1968.
21. **Neumann, E., Ridder, M. S., Wang, Y., and Hofschneider, P. H.**, Gene transfer into mouse lyoma cells by electroporation in high electric fields, *EMBO J.*, 1, 841, 1982.
22. **Ohtani, K., Nakamura, M., Saito, S., Nagata, K., and Hinuma, Y.**, Electroporation: application to human lymphoid cell lines for stable introduction of a transactivator gene of human T-cell leukemia virus type I, *Nucleic Acids Res.*, 17, 1589, 1989.
23. **Potter, H.**, Application of electroporation in recombinant DNA technology, *Methods Enzymol.*, 217, 461, 1993.
24. **Mertz, J. E. and Gurdon, J. B.**, Purified DNAs are transcribed after microinjection into *Xenopus* oocytes, *Proc. Natl. Acad. Sci. U.S.A.*, 74, 1502, 1977.
25. **Harbers, K., Jahner, D., and Jaenisch, R.**, Microinjection of cloned retroviral genomes into mouse zygotes: integration and expression in the animal, *Nature (London)*, 293, 540, 1981.
26. **Palmiter, R. D. et al.**, Dramatic growth of mice that develop from eggs microinjected with metallothionein-growth hormone fusion genes, *Nature (London)*, 300, 611, 1982.
27. **Old, R. W. and Primrose, S. B.**, *Principles of Gene Manipulation*, 3rd ed., Blackwell Scientific Publications, Oxford, 1985, chap. 12.
28. **Tabin, C. J., Hoffman, J. W., Groff, S. P., and Weiberg, R. A.**, Adaption of a retrovirus as a eukaryotic vector transmitting the herpes simplex virus thymidine kinase gene, *Mol. Cell. Biol.*, 2, 426, 1982.
29. **Miller, A. D., Miller, D. G., Garcia, J. V., and Lynch, C. M.**, Use of retroviral vectors for gene transfer and expression, *Methods Enzymol.*, 217, 581, 1993.
30. **Szoka, F. and Papahdjopoulos, D.**, Procedure for preparation of liposomes with large internal aqueous space and high capture by reverse-phase evaporation, *Proc. Natl. Acad. Sci. U.S.A.*, 74, 4194, 1978.
31. **Straubinger, R. M. and Papahdjopoulos, D.**, Liposomes as carriers for intracellular delivery of nucleic acids, *Methods Enzymol.*, 101, 512, 1983.
32. **Fraley, R. and Papahdjopoulos, D.**, Liposomes: the development of a new carrier system for introducing nucleic acids into plant and animal cells, *Curr. Top. Microbiol. Immunol.*, 96, 171, 1982.
33. **Wilson, T., Papahdjopoulos, D., and Taber, R.**, The introduction of poliovirus RNA into cells via lipid vesicles (liposomes), *Cell*, 17, 77, 1979.
34. **Dimitriadis, G. J.**, Translation of rabbit globin mRNA introduced by liposome into mouse lymphocytes, *Nature (London)*, 274, 923, 1978.
35. **Wong, T. K., Nicolau, C., and Hofschneider, P. H.**, Appearance of β-lactamase activity in animal cells upon liposome-mediated gene transfer, *Gene*, 10, 87, 1980.
36. **Fraley, R., Subramani, S., Berg. P., and Papahdjopoulos, D.**, Introduction of liposome-encapsulated SV40 DNA into cells, *J. Biol. Chem.*, 255, 10431, 1980.

37. **Fraley, R., Fornari, C. S., and Kaplan, S.,** Entrapment of a bacterial plasmid in phospholipid vesicles: potential for gene transfer, *Proc. Natl. Acad. Sci. U.S.A.,* 76, 3348, 1979.
38. **Wang, C.-Y. and Huang, L.,** pH-sensitive immunoliposomes mediate target-cell-specific delivery and controlled expression of a foreign gene in mouse, *Proc. Natl. Acad. Sci. U.S.A.,* 84, 7851, 1987.
39. **Wang, C.-Y. and Huang, L.,** Plasmid DNA adsorbed to pH-sensitive liposomes efficiently transforms the target cells, *Biochem. Biophys. Res. Comm.,* 147, 980, 1987.
40. **Mukherjee, A. B., Orloff, S., Butler, J. D., Triche, T., Lalley, P., and Schulman, J. D.,** Entrapment of metaphase chromosomes into phospholipid vesicles. Carrier potential in gene transfer, *Proc. Natl. Acad. Sci. U.S.A.,* 75, 1361, 1978.
41. **Mannino, R. J. and Gould-Fogerite, S.,** Liposome-mediated gene transfer, *BioTechnique,* 6, 682, 1988.
42. **Nicolau, C., Legrand, A., and Grosse, G. E.,** Liposomes as carriers for *in vivo* gene transfer and expression, *Methods Enzymol.,* 149, 157, 1987.
43. **Gregoriadis, G., Senior, J., Wolff, B., and Kirby, C.,** Targeting of liposomes to accessible cells *in vivo, Ann. N. Y. Acad. Sci.,* 446, 319, 1985.
44. **Lapidot, A. and Loyter, A.,** Fusion-mediated microinjection of liposome-enclosed DNA into cultured cells with the aid of influenza virus glycoproteins, *Exp. Cell Res.,* 189, 241, 1990.
45. **Leonetti, J.-P., Machy, P., Degols, G., Lebleu, B., and Leserman, L.,** Antibody-targeted liposomes containing oligodeoxyribonucleotides complementary to viral RNA selectively inhibit viral replication, *Proc. Natl. Acad. Sci. U.S.A.,* 87, 2448, 1990.
46. **Fraley, R. and Papahdjopoulos, D.,** New generation liposomes: the engineering of an efficient vehicle for intracellular delivery of nucleic acids, *Trends Biochem. Sci.,* 6, 77, 1981.
47. **Felgner, P. L., Gadek, T. R., Holm, M., Roman, R., Chan, H. W., Wenz, M., Northrop, J. P., Ringold, G. M., and Danielsen, M.,** Lipofection: a highly efficient, lipid-mediated DNA-transfection procedure, *Proc. Natl. Acad. Sci. U.S.A.,* 84, 7413, 1987.
48. **Ballas, N., Zakai, N., Sela, I., and Loyter, A.,** Liposomes bearing a quaternary ammonium detergent as an efficient vehicle for functional transfer of TMV-RNA into plant protoplasts, *Biochim. Biophys. Acta,* 939, 8, 1988.
49. **Malone, R. W., Felgner, P. L., and Verma, I. M.,** Cationic liposome-mediated RNA transfection, *Proc. Natl. Acad. Sci. U.S.A.,* 86, 6077, 1989.
50. **Felgner, P. L. and Ringold, G. M.,** Cationic liposome-mediated transfection, *Nature (London),* 337, 387, 1989.
51. **Pinnaduwage, P., Schmitt, L., and Huang, L.,** Use of ammonium detergent in liposome mediated DNA transfection of mouse L-cells, *Biochim. Biophys. Acta,* 985, 33, 1989.
52. **Leventis, R. and Silvius, J. R.,** Interactions of mammalian cells with lipid dispersions containing novel metabolizable cationic amphiphiles, *Biochim. Biophys. Acta,* 1023, 124, 1990.
53. **Felgner, P. L. and Holm, M.,** Cationic liposome-mediated transfection, *Focus,* 11, 21, 1989.
54. **Welsh, N., Oberg, C., Hellerstrom, C., and Welsh, M.,** Liposome-mediated *in vitro* transfection of pancreatic islet cells, *Biomed. Biochim. Acta,* 12, 1157, 1990.
55. **Innes, C. L., Smith, P. B., Langenbach, R., Tindall, K. R., and Boone, L. R.,** Cationic liposomes (lipofectin) mediate retroviral infection in the absence of specific receptors, *J. Virol.,* 64, 957, 1990.
56. **Holt, C. E., Garlick, N., and Cornel, E.,** Lipofection of cDNA in the embryonic vertebrate central nervous system, *Neuron,* 4, 203, 1990.
57. **Sporlein, B. and Koop, H.-U.,** Lipofection: direct gene transfer to higher plants using cationic liposomes, *Theor. Appl. Genet.,* 83, 1, 1991.
58. **Gnirke, A., Barnes, T. S., Patterson, D., Schild, D., Featherstone, T., and Olson, M. V.,** Cloning and *in vivo* expression of the human GART gene using yeast artificial chromosomes, *EMBO J.,* 10, 1629, 1991.
59. **Ponder, K. P., Dunbar, R. P., Wilson, D. R., Darlington, G. J., and Woo, S.,** Evaluation of relative promoter strength in primary hepatocytes using optimized lipofection, *Hum. Gene Ther.,* 2, 41, 1991.
60. **Bertling, W. M., Gareis, M., Zimmer, A., Kreuter, J., Nurenberg, E., and Harrer, P.,** Use of liposomes, viral capsids, and nanoparticles as DNA carriers, *Biotechnol. Appl. Biochem.,* 13, 390, 1991.
61. **Jarnagin, W. R., Debs, R. J., Wang, S.-S., and Bissell, D. M.,** Cationic lipid-mediated transfection of liver cells in primary culture, *Nucleic Acids Res.,* 20, 4205, 1992.
62. **Ray, J. and Gage, F. H.,** Gene transfer into established and primary fibroblast cell lines: comparison of transfection methods and promoters, *BioTechniques,* 13, 598, 1992.
63. **Bennet, C. F., Chiang, M.-Y., Chan, H., Shoemaker, J. E., and Mirabelli, C. K.,** Cationic lipids enhance cellular uptake and activity of phosphorothioate antisense oligonucleotides, *Mol. Pharmacol.,* 41, 1023, 1992.
64. **Gao, X. and Huang, L.,** Cytoplasmic expression of reporter gene by co-delivery of T7-RNA polymerase and T7 promoter sequence with cationic liposomes, *Nucleic Acids Res.,* 21, 2867, 1993.
65. **Dwarki, V. J., Malone, R. W., and Verma, I. M.,** Cationic liposome-mediated RNA transfection, *Methods Enzymol.,* 217, 644, 1993.
66. **Behr, J.-P.,** DNA strongly binds to micelles and vesicles containing lipopolyamines or lipointercalants, *Tetrahedron Lett.,* 27, 5861, 1986.
67. **Maccarrone, M., Dini, L., Marzio, L., Giulio, A., Rossi, A., Mossa, G., and Agro, A. F.,** Interaction of DNA with cationic liposomes: ability of transfecting lentil protoplasts, *Biochem. Biophys. Res. Commun.,* 186, 1417, 1992.

68. **Sambrook, J., Fritsch, E. F., and Maniatis, T.,** in *Molecular Cloning,* Cold Spring Harbor Laboratory Press, Cold Spring Harbor, NY, 1989, 1.33–1.39.
69. **Struck, D. K., Hoekstra, D., and Pagano, R. E.,** Use of resonance energy transfer to monitor membrane fusion, *Biochemistry,* 20, 4093, 1981.
70. **Beigel, M., Keren-Zur, M., Laster, Y., and Loyter, A.,** Poly(aspartic acid)-dependent fusion of liposomes bearing the quaternary ammonium detergent [[[(1,1,3,3-tetramethylbutyl) cresoxy] ethoxy] ethyl] dimethylbenzylammonium hydroxide, *Biochemistry,* 27, 660, 1988.
71. **Coggins, L. W.,** Preparation of nucleic acids for electron microscopy, in *Electron Microscopy in Molecular Biology,* Sommerville, J. and Scheer, U., Eds., IRL Press, Washington, D.C., 1987.
72. **Ghirlando, R., Wachtel, E. J., Arad, T., and Minsky, A.,** DNA packaging induced by micellar aggregates: a novel *in vitro* DNA condensation system, *Biochemistry,* 31, 7110, 1992.
73. **Maestre, M. F. and Reich, C.,** Contribution of light scattering to the circular dichroism of deoxyribonucleic acid films, deoxyribonucleic acid-polylysine complexes, and deoxyribonucleic acid particles in ethanolic buffers, *Biochemistry,* 19, 5214, 1980.
74. **Maniatis, T., Venable, J. H., and Lerman, L. S.,** The structure of Ψ-DNA, *J. Mol. Biol.,* 84, 37, 1974.
75. **Goddard, E. D.,** Polymer-surfactant interactions; polymers and surfactants of opposite charge, *Colloids Surfaces,* 19, 301, 1986.
76. **Goddard, E. D.,** Polymer-surfactant interactions; uncharged water-soluble polymers and charged surfactants, *Colloids Surfaces,* 19, 255, 1986.
77. **Azorin, F., Vives, J., Campos, J. L., Jordan, A., Subirana, J. A., Mayer, R., and Brack, A.,** Interaction of DNA with lysine-rich polypeptides and proteins. The influence of polypeptide composition and secondary structure, *J. Mol. Biol.,* 185, 371, 1985.
78. **Ekwall, P.,** Composition, properties and structures of liquid crystalline phases in systems of amphiphilic compounds, *Adv. Liq. Cryst.,* 1, 1, 1975.
79. **Keller, D. and Bustamante, C.,** Theory of the interaction of light with large inhomogeneous molecular aggregates. Psi-type circular dichroism, *J. Chem. Phys.,* 84, 2972, 1986.
80. **Manning, G. S.,** Thermodynamic stability theory for DNA doughnut shapes induced by charge neutralization, *Biopolymers,* 19, 37, 1980.
81. **Reich, Z., Ghirlando, R., and Minsky, A.,** Secondary conformational polymorphism of nucleic acids as a possible link between environmental parameters and DNA packaging processes, *Biochemistry,* 30, 7828, 1991.
82. **Keren-Zur, M., Beigel, M., and Loyter, A.,** Induction of fusion in aggregated and nonaggregated liposomes bearing cationic detergents, *Biochim. Biophys. Acta,* 983, 253, 1989.
83. **Düzgünes, N., Goldstein, J. A., Friend, D. S., and Felgner, P. L.,** Fusion of liposomes containing a novel cationic lipid, N-[2,3-(dioleyloxy)propyl]-N,N,N-trimethylammonium: induction by multivalent anions and asymmetric fusion with acidic phospholipid vesicles, *Biochemistry,* 28, 9179, 1989.
84. **Gershon, H., Ghirlando, R., and Minsky, A.,** Mode of formation and structural features of DNA-cationic liposome complexes used for transfection, *Biochemistry,* 32, 7143, 1993.
85. **Behr, J.-P., Demeneix, B., Loeffler, J.-P., and Perez-Mutul, J.,** Efficient gene transfer into mammalian primary endocrine cells with lipopolyamine-coated DNA, *Proc. Natl. Acad. Sci. U.S.A.,* 86, 6982, 1989.
86. **Carroll, D.,** Optical properties of deoxyribonucleic acid-polylysine complexes, *Biochemistry,* 11, 421, 1972.
87. **Reich, Z., Ittah, Y., Weinberger, S., and Minsky, A.,** Chiral and structural discrimination in binding of polypeptides with condensed nucleic acid structures, *J. Biol. Chem.,* 265, 5590, 1990.
88. **Loeffler, J.-P. and Behr, J.-P.,** Gene transfer into primary and established mammalian cell lines with lipopolyamine-coated DNA, *Methods Enzymol.,* 217, 599, 1993.

Chapter 3

Delivery of DNA, RNA, and Proteins by Cationic Liposomes

Hassan Farhood and Leaf Huang

CONTENTS

I. HISTORY AND BACKGROUND

Introduction of genetic material (DNA) into cells has been the goal of numerous and diverse methods which were physical, chemical, and biological in nature. However, the ultimate application of gene transfer, i.e., gene therapy of human diseases, limits the applicable methodologies to only a few.[1] Physical methods for DNA transfer include: direct injection of DNA into organs of live animals,[2] with successful applications in DNA vaccination against disease,[3] and micro projectile biolistics using a gene-propelling gun mostly for gene therapy purposes.[4] These physical delivery methods have not yet been tested on humans. Biological methods of gene delivery include viral[5] and retroviral vectors,[6] both of which have been effectively used for gene therapy in animal models and humans.[7-9] However, certain concerns such as safety, toxicity, and immunogenicity are still unresolved for some of these gene vectors.[10] Chemical methods for DNA delivery are also among the promising available technology for gene tranfer to combat disease. Targeted polymers[11] and liposomes,[12-14] cationic[15,16] and noncationic,[17] are the only chemical delivery techniques potentially available for gene therapy. In practice, however, only cationic liposomes have reached the realm of human clinical trial.[18-20]

Liposomes are lipid vesicles composed of a lipid bilayer membrane entrapping a volume of aqueous solution inside the vesicle. The lipid bilayer is hydrophobic in the interior and hydrophilic on the surface. The vesicle can be multilamellar, composed of multiple concentric lipid bilayers, or unilamellar, composed of a single lipid bilayer, depending on the method of liposome preparation.[21] Unilamellar vesicles can entrap larger volume of aqueous solution per unit lipid mass than the multilamellar vesicles of the same vesicle diameter. Also depending on the preparation method, the size of liposomes can vary widely between 30–50 nm up to several microns in diameter.[21] Liposomes can carry and protect molecules such as conventional drugs or macromolecules (DNA, RNA, or protein) which can be entrapped in the aqueous volume of the liposome. Once the carrier liposome fuses with or enters the target cell, a portion of the entrapped molecules are released into the cytoplasm.[22] Water-insoluble drugs can be made soluble by entrapment of the drug in the hydrophobic interior of the lipid bilayer with relatively high entrapment efficiency.[21] An alternative method for delivering molecules by liposomes is through adsorption of charged molecules to the surface of liposomes with an overall opposite charge from that of the carried

0-8493-4013-6/96/$0.00+$.50

molecules. Negatively charged macromolecules such as DNA can adsorb to, and complex with, cationic liposomes without entrapment of DNA inside the liposome.[14]

In 1987 the first formulation of a cationic liposome, Lipofectin™, was introduced by Felgner et al.[23] Lipofectin™* is composed of a sonicated preparation of equal amounts (by weight) of a synthetic cationic lipid, N [1-(2,3,dioleoyloxy)propyl]-N,N,N trimethylammonium chloride (DOTMA) and a membrane destabilizing phospholipid dioleoyl phosphatidylethanolamine (DOPE) (Figure 1). Sonication of the dried and hydrated lipids, with buffer or water, produces small unilamellar vesicles with an approximate size range of 50–300 nm in diameter. The net positive charge of the liposome allows it to complex with the negatively charged DNA by electrostatic interaction, without the cumbersome and inefficient encapsulation of DNA inside the liposomes. The result of the electrostatic binding of DNA to cationic liposomes is a complex with a net positive charge, at an optimal ratio of liposome to DNA, which enables the complex to adsorb to the negatively charged cell surface. Following the adsorption, cellular uptake of the complexed DNA facilitate intracellular DNA delivery and subsequent gene expression.[23]

The composition of Lipofectin™ was the prototype for most of the subsequent cationic liposome formulations prepared in different laboratories.[14-16] Basically, all cationic liposomes used for successful gene transfer are made of an amphipathic cationic lipid with or without DOPE. In contrast to the nonvariable phospholipid component (DOPE) of cationic liposomes, the cationic amphiphile of the different available liposome formulations differs markedly and can be a cationic cholesterol,[24-26] cationic detergent,[27,28] single or multiple tertiary and quaternary amine-diacyl conjugates,[23,24,29-32] lipospermine,[33] and lipopolylysine.[34,35] These cationic amphiphiles have only two components in common: the net cationic charge on the hydrophilic headgroup of the amphiphile, and the hydrophobic tail that anchors the molecule to the liposome lipid bilayer.[14] The chemical structure of the cationic lipids is quite diverse, and each molecule can have a single or multiple cationic charges. In addition, and in contrast to the natural phospholipid component of cationic liposomes, the cationic lipid component is synthetic and is not found naturally among the lipids in biological membranes. This and the nondegradable bond that connects the cationic headgroup to the hydrophobic tail, found in some liposome formulations, may be the main reason behind the toxic effects seen with high concentrations of virtually all cationic liposome formulations. Fortunately, the cationic liposome concentration used for successful gene transfer is below the toxic level in the majority of applications.

II. LIPID COMPOSITION OF CATIONIC LIPOSOMES

DNA delivery by cationic liposomes to target cells is dependent on the quantitative content of both the cationic lipid and DOPE in the liposome. Three exceptions to this general rule have been reported.[29,33,34] These three different liposome formulations lack DOPE and are composed of 100% cationic amphiphile. The mechanism of DNA delivery using these formulations is unknown. One cationic lipid, i.e., lipopoly-L-lysine, was found to deliver DNA only after scraping the transfected cells in culture.[34] However, scraping was unnecessary if DOPE was added to the cationic lipid to form liposomes.[35] However, the other two cationic lipid formulations, i.e., lipospermine[33] and the cationic detergent *O,O'*-didodecyl-*N*-[*p*-(2-trimethylammonioethyloxy)-benzoyl]-(L)-glutamate bromide,[29] do not require any mechanical assistance for transfection, confirming a role for DNA delivery by the cationic lipid independent of DOPE.

A. THE ROLE OF THE CATIONIC LIPID IN DNA DELIVERY

As a result of the large number of phosphodiester anionic charges on DNA, the majority of the cationic lipid in cationic liposomes is required for DNA complexing, and hypothetically for DNA condensation. A small amount of excess cationic lipid is required for binding the liposome to the cell surface. These two requirements probably create a critical quantitative ratio of DNA to cationic liposomes. The critical ratio must be reached for successful gene transfer. This ratio is optimal when the net ionic charge of the DNA/liposomes complex is positive. The negative charges exposed on the DNA molecule must be completely neutralized by the addition of sufficient amount of cationic liposomes. There is a highly reproducible distinctive sharp increase in transfection activity when this critical ratio is reached.[25,26,33,34] Any subsequent addition of cationic lipids to the complex does not enhance DNA delivery and only increases the toxicity to the treated cells.[25]

---

* Registered Trademark of Life Technologies Inc., Gaithersburg, MD.

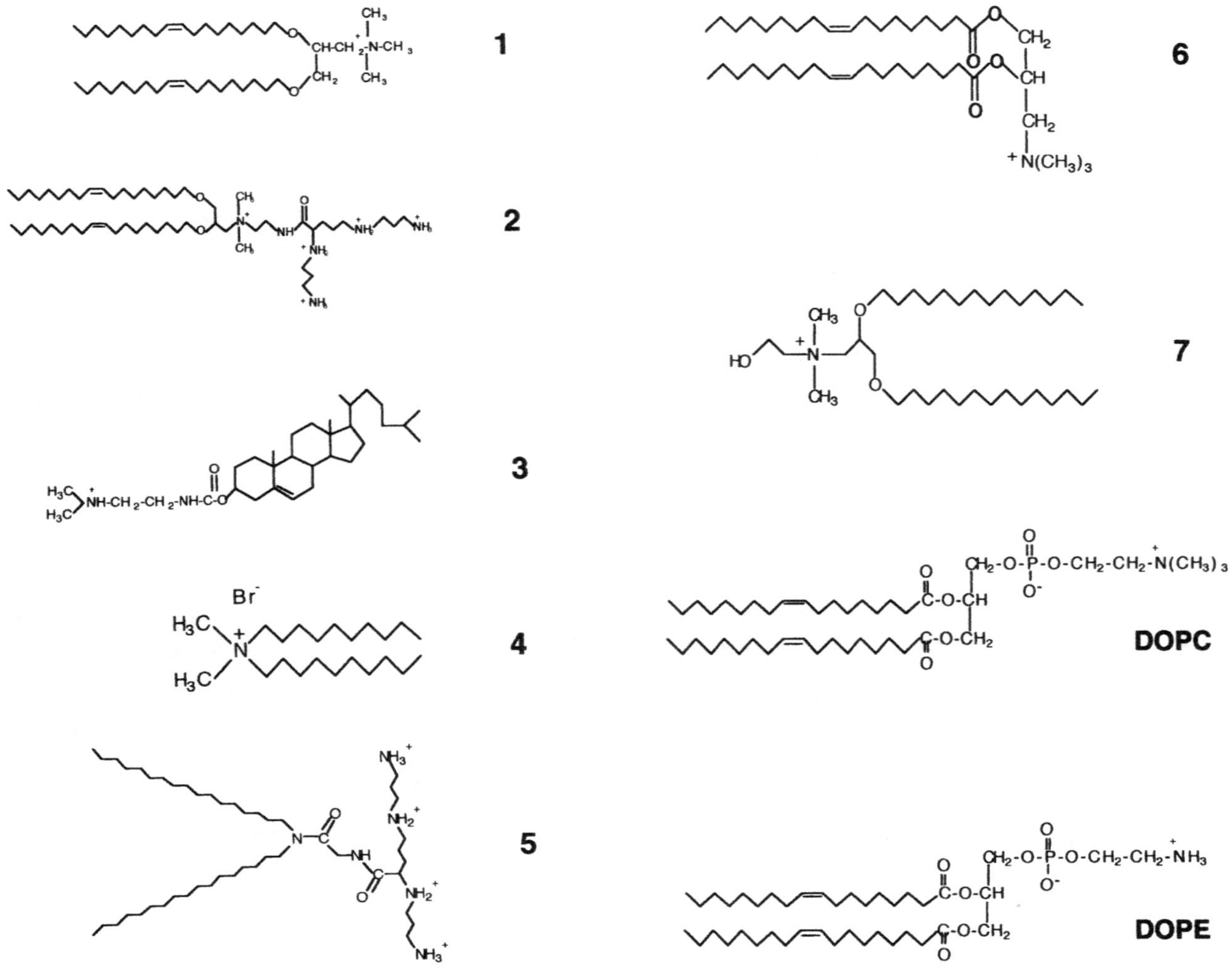

Figure 1 Structures of cationic lipids (1–7) and phospholipids (DOPC and DOPE). 1, DOTMA; 2, DOSPA; 3, DC-Chol; 4, DDAB; 5, DOGS; 6, DOTAP; 7, DMRIE. (Adapted from Reference 64.)

In addition to the optimal transfection activity achieved with a critical ratio of DNA to cationic liposomes, the formation of aggregates is prominent with several formulations of cationic liposome. Aggregation occurs by bridging between DNA molecules mediated by the complexed cationic liposomes. The aggregate size varies and can be several micrometers in diameter. These large aggregates are not expected to enter cells by endocytosis because the size of the endosome is limited to a few hundred nanometers in diameter.[36] Smaller DNA/liposome complexes may enter cells by endocytosis, as was shown by electron microscopy.[35] A model implicating endocytosis in the uptake of lipospermine-complexed DNA was proposed by Barthel et al.[33] The model sees the complex as a globule with a net cationic charge that undergoes a zipper-like mechanism of uptake by endocytosis (Figure 2). However, no direct data were provided to support the model. On the other hand, the model proposed by Zhou and Huang[35] provided direct electron microscopy evidence for endocytosis of lipopolylysine-complexed DNA (Figure 3). Lipospermine and lipopolylysine are cationic lipids with similar overall structures and components, and both are capable of complexing and efficiently condensing the DNA. Whether the model is applicable with other cationic liposome formulations is currently unclear.

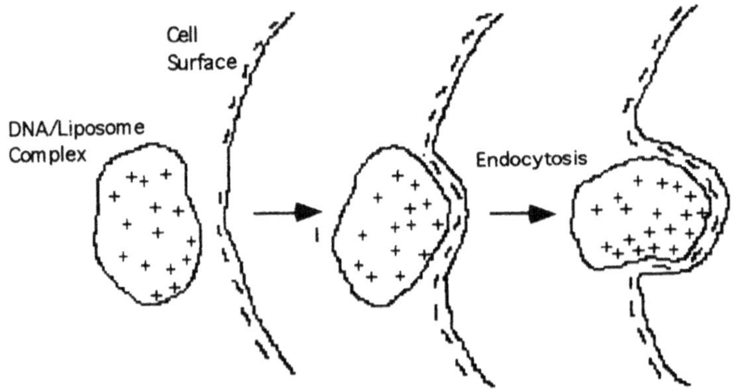

Figure 2 A model for DNA delivery by cationic liposomes (Modified from Barthel, F., Remy, J.-S., Loeffler, J.-P., and Behr, J.-P. (1993). *DNA Cell Biol.* 12, 553-560. With permission.)

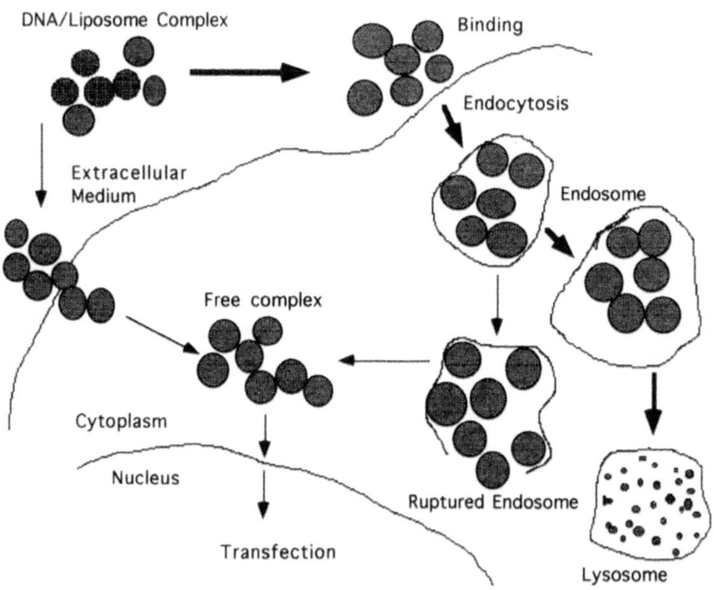

Figure 3 A Model for delivery of DNA by cationic liposomes. Thick arrows indicate the major pathway. (Modified from Zhou, X. and Huang, L. (1994) *Biochim. Biophys. Acta* 1189, 195–203. With permission.)

B. THE ROLE OF THE PHOSPHOLIPID IN DNA DELIVERY

The role of DOPE in cationic liposome formulations is two-fold. Some cationic lipids, such as cationic cholesterols, can not form vesicles without the phospholipid component.[25,26,31] This particular role for vesicle formation and stabilization can be performed by many types of lipids and is not unique to DOPE.[35] However, the second proposed role of the phospholipid is dependent on DOPE and in general can not be substituted by other lipids.[26,35] DOPE has a strong tendency to form inverted hexagonal (H_{II}) phases under physiological conditions.[37] The primary amine group of DOPE hydrogen bonds with neighboring phosphate group, resulting in a poor hydration of the ethanolamine head group.[38] The bulky oleoyl chains of the molecule further favor an aggregation state of inverted micelles.[38] In the presence of charged amphiphiles, such as a cationic lipid, the level of interfacial hydration is sufficiently increased such that stable bilayer liposomes (L_α phase) are produced. When the charge of the amphiphile is neutralized by binding with an opposite charge, such as DNA, it is highly likely that DOPE will revert to the inverted micellar structure (Figure 4). If a sufficient amount of DOPE is added to a biological membrane, it is also conceivable that the membrane would be destabilized by forming localized nonbilayer structures. This property of DOPE is important for the penetration of DNA/liposome complex across a biological membrane for transfection. Substitution of DOPE with the membrane-stabilizing phospholipid known as dioleoyl phosphatidylcholine (DOPC) (Figure 1) impaired the DNA delivery function of lipopolylysine liposomes, which again became dependent on scraping.[35] DOPC is an L_α phase-forming lipid which stabilizes a lipid bilayer as a result of the readily hydratable choline head group.[38]

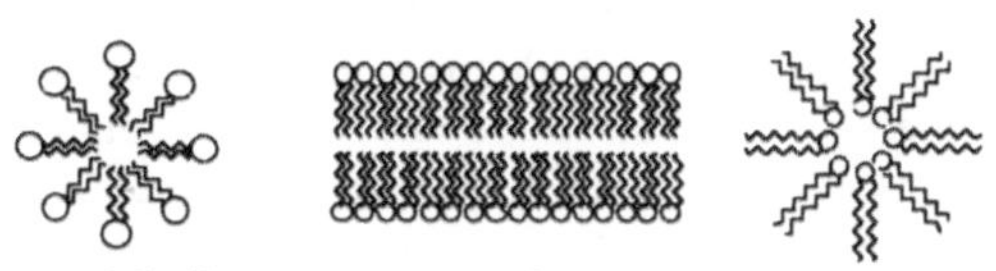

Figure 4 Different lipid phases structures.

The content of DOPE in the liposomes varies from 0–50 mol% depending on the preparation. Approximately 30–50 mol% of DOPE is needed for DNA delivery. Most attempts to replace DOPE with other lipids have failed to produce an efficient DNA carrier.[26,35] Only a few cationic liposomes can deliver DNA without DOPE (mentioned earlier). There is a direct relationship between macromolecule (DNA and protein) delivery and% DOPE content. We have found optimal delivery at 80% DOPE. However, available cationic liposome formulations do not exceed 50% DOPE because of the need for high cationic lipid content (50–70%) to complex with and neutralize the large quantity of negative charges of DNA.

In general, entry of macromolecules into cells (largely studied with DNA) mediated by cationic liposomes has been shown to take place essentially by adsorptive endocytosis with minor delivery activity directly through the plasma membrane.[17,35] The available models of macromolecule entry into cells predict the disruption of endosomal membrane (facilitated by DOPE) by cationic liposomes, an event that releases the macromolecules into the cytosol of transfected cells. The electron microscopy observations[35] showing release of DNA/lipopolylysine complexes into the cytosol of mammalian cells through a disrupted endosomal membrane lend strong support to the endocytosis model. Additonal support for this model comes from micrographs showing uptake of DNA complexes by coated pits at the plasma membrane surface of cells.[35] The mechanism by which DNA molecules enter into the nucleus from the cytosol is unknown.

III. DELIVERY OF MACROMOLECULES BY CATIONIC LIPOSOMES

DNA delivery was the motivation for the development of cationic liposomes.[23] The choice of DNA is justified by the stability of DNA in comparison to other macromolecules, including RNA and protein. Moreover, DNA, being the information source of both RNA and protein, can provide a constant and long-lasting supply of both macromolecules. Additionally, gene therapy by DNA transfer can potentially cure a genetic disease by permanent replacement of the defective gene with a normal gene. Permanent therapy of genetic diseases is not possible without DNA delivery. In practice, however, this ideal form of gene therapy is not within reach at this point and multiple doses of the gene in the form of DNA, RNA, or protein can provide a temporary relief for the patients.

The large size of DNA and the difficulties associated with its delivery has lent support to attempts for delivery of macromolecule other than DNA using cationic liposomes. In addition, the need for transient expression of a gene is best fulfilled by delivery of unstable gene products such as RNA and protein.

A. RNA DELIVERY BY CATIONIC LIPOSOMES

The first attempt to deliver RNA by cationic liposomes was reported by Ballas et al.[27] Tobacco mosaic virus RNA was successfully delivered into tobacco and petunia plant protoplasts using cationic liposomes and was assayed using a viral capsid protein quantitation assay. The cationic liposome composition used was atypical and contained cholesterol, phosphatidylcholine, and a quaternary ammonium detergent. The liposomes were complexed with RNA by simple mixing, similar to DNA complexing. Removal of the cationic detergent abolished the majority of RNA delivery activity. This is an expected result, since the cationic detergent is used for electrostatic interaction with RNA and cells. An interesting comparison was made between the delivery of RNA complexed to the cationic liposomes and the delivery of the same RNA encapsulated in the same cationic liposomes.[27] Complexed RNA showed higher delivery efficiency than the encapsulated RNA, was simpler and more reproducible, and showed no toxic effects.

Another delivery study of RNA complexed to cationic liposomes was reported a year later.[39] This study used Lipofectin™ to deliver luciferase RNA, synthesized *in vitro*, into a variety of mammalian cells as well as amphibian and insect cells. The study was aimed toward a comparison between the role of capped vs. uncapped RNA and the role of β-globin 5′ and 3′ untranslated sequences on translation in a reporter gene expression assay. The study showed a 1000-fold enhancement in gene expression when capped RNA coding for the untranslated sequences was used. The kinetics of RNA delivery was rapid with easily detectable reporter protein activity in less than 1 h after the delivery was initiated. Within 1 h of RNA delivery, 71% of the delivered RNA was shown to adhere to cells, 32% of which were RNase-resistant, which shows the protecting effect of cationic liposome complexing. The delivery using cationic liposomes is simple, efficient, and reproducible. Delivery is also quite versatile and successful with respect to the variety of target cells, including those normally refractive to transfection.

B. VIRAL DELIVERY BY CATIONIC LIPOSOMES

Lipofectin™ was also tested for the delivery of murine leukemia virus and packaged retroviral vectors to cells normally uninfectable by these retroviral particles due to the absence of specific receptors on the cell surface.[40] Simple addition of cationic liposomes to the virus incubated with target cells allowed efficient delivery of the viral genome into the cells. In contrast to cationic liposomes, the presence of polybrene, a cationic polymer used for DNA transfection, did not allow the infection of cells. In addition to high infection efficiency, the simplicity and reproducibility of the cationic liposome-mediated viral infection of nonpermissive cells makes this delivery strategy an available remedy to overcome resistance of a variety of cells to infection.

C. PROTEIN DELIVERY BY CATIONIC LIPOSOMES

Transient delivery of macromolecules such as proteins can be a challenge and/or toxic or injurious when the common delivery protocols are employed, including electroporation,[41] scrape-loading,[42] and liposome encapsulation.[22] The introduction of cationic liposome-mediated delivery of proteins by Debs et al. in 1990 has simplified and popularized protein delivery.[43] The report described the delivery of a purified recombinant glucocorticoid receptor fragment, a mammalian transcriptional regulator protein, to mammalian cells using Lipofectin™.[43] The protein was delivered with or without a *cis*-acting DNA element that codes for a reporter gene under the control of the element responsive to induction by the glucocorticoid receptor protein. High levels of gene expression were induced in this system. The fact that the protein used in the study was amplified in and purified from bacteria lends support to the use of recombinant proteins for studying the protein function and structure, particularly for those that do not undergo post-translational modifications and can be manipulated in bacteria. An unexpected high efficiency of protein delivery was observed when the cationic lipid component of Lipofectin (i.e., DOTMA) was used alone without DOPE.[43]

Recently, protein delivery by cationic liposomes was used in the field of molecular immunology through three different studies, of which all used cationic liposomes for delivery of antigens. The first was by Walker et al., who delivered the herpes simplex virus (HSV) glycoprotein B antigen.[44] The protein was recombinant and was delivered[44] to cells by mixing the protein with a cationic liposome formulation containing *N*-[1-(2,3-dioleoyloxy)propyl]-*N,N,N*-trimethylammonium methyl sulfate (DOTAP) and DOPE.[31] The delivery resulted in the presentation of a peptide fragment of the protein by the class I major histocompatibility (MHC) complex receptor and sensitized the cells to lysis by class I MHC-restricted, HSV-specific cytotoxic T lymphocytes (CTL). In addition, a memory CTL response was evoked specific for the HSV protein delivered by the liposomes. This memory response was stimulated in spleen cells from HSV-infected mice after the mice were injected subcutaneously with a mixture of DOTAP liposomes and the HSV protein. A similar injection strategy with complete Freund's adjuvant did not produce any memory immune response.[44] This novel strategy of immunization should help facilitate the induction of CTL response *in vivo* without introduction of the whole virus *in vivo*. Additionally, this simple delivery system may help in dissecting the class I MHC antigen presentation pathway.

The second study used lipopolylysine liposomes to deliver the anionic and antigenic protein ovalbumin which could readily form complexes with the cationic liposomes.[45] A mouse cell line was the target cell for delivery of the antigen. Successful uptake of the complexed antigen caused sensitization of the mouse cells to killing by an ovalbumin-specific CTL. Optimum sensitization to killing occurred at an ovalbumin concentration of 20 μg/ml. The delivery was more potent if the target cells were treated with phorbol myristate acetate, which may activate endocytosis leading to enhanced uptake of the complex.[45] Lipopolylysine delivery of antigen was compared favorably to the osmotic loading method.[45] Studying the antigen presentation pathway can be simplified using lipopolylysine-mediated antigen delivery.

The third study reported using DOTAP for the delivery of ovalbumin protein with subsequent presentation by the class I MHC pathway.[46] In this study, electroporation, osmotic loading, and cationic liposomes (DOTAP) were tested for ovalbumin delivery. Electroporation and liposomes were highly efficient delivery vehicles in comparison with osmotic loading. However, the toxicity associated with electroporation and the simplicity of cationic liposome delivery make liposomes the preferred delivery carrier.

Enzyme delivery by complexing to cationic liposomes (Lipofectin™) was reported by Lin et al.[47] The enzyme delivered was a purified prostatic acid phosphatase protein, and human prostate carcinoma cells were used as targets for the delivery. Enzymatic activity was demonstrated by a tartrate-sensitive acid phosphatase assay. Retention of enzymatic activity after delivery was maintained for at least 48 h. Immunofluorescence staining localized the delivered enzyme to the cytoplasm of cells. Delivery kinetics were as fast as RNA delivery, as 70% of the enzyme delivered was incorporated into cells within 1 h of incubation. This report took advantage of the acidic nature of the enzyme which is negatively charged at physiological pH, allowing efficient complexing with cationic liposomes. The data from this work set a precedence for future work on enzyme delivery which may help the studies in which enzymes are chemically and/or enzymatically manipulated *in vitro* before delivered to cells for functional test.

An interesting and novel application for protein delivery with cationic liposomes was introduced by Gao and Huang.[48] A well-known barrier to DNA delivery is the nuclear uptake of DNA after the DNA reaches the cytosol.[49] This problem was overcome by a cytoplasmic expression system borrowed from bacteriophage T7 which encodes for an enzyme (T7 RNA polymerase) that has been shown to transcribe DNA into RNA in the cytosol of mammalian cells.[48] Gao and Huang mixed a purified recombinant T7 RNA polymerase with DNA coding for a reporter gene under the control of T7-specific promoter and complexed the protein/DNA mix to cationic liposomes. The ternary complex was incubated with mammalian cells for transfection. Cationic liposomes containing 3β [*N*-(*N′,N′*-dimethylaminoethane)-carbamoyl] cholesterol (DC-Chol) and DOPE as well as Lipofectin™ were used successfully in this study. Reporter gene expression was shown to occur with faster kinetics than the expression of the reporter gene under the control of a nuclear promoter, which is nonresponsive to T7 polymerase. This was indicative of rapid cytoplasmic gene expression independent of nuclear transcription factors. The short half-life of this gene expression under the control of the T7 system was also consistent with the notion of cytoplasmic expression since nuclear gene expression can last at least twice as long as the T7 system did. The T7 system described can be a useful new tool for gene therapy applications where a fast, transient, but robust level of transgene expression is preferred.

IV. GENE THERAPY USING CATIONIC LIPOSOMES

One of the most important applications of gene transfer studies is gene therapy of human diseases.[1] In spite of the fact that gene therapy was born in an effort to cure inherited diseases, gene therapy is now targeted toward many diseases existing with or without available therapy. This wide range of diseases spans genetic diseases such as cystic fibrosis (CF), cancer diseases including benign and malignant tumors, and infectious diseases such as AIDS, among others.[1,7-9]

The idea behind exploiting cationic liposomes for gene therapy was born as an alternative to other gene therapy protocols that had potential toxicity, biohazard, and immunogenicity.[7-9] Cationic liposomes showed minimal toxicity effects in cell culture systems and were not expected to be immunogenic since they are composed of nonimmunogenic lipids.

A. ORGAN DISTRIBUTION, GENE EXPRESSION, AND TOXICITY STUDIES USING CATIONIC LIPOSOMES

The first report on the use of cationic liposomes for gene delivery in animals came from Brigham et al. in 1989.[50] Mice were used for testing the delivery of a reporter gene DNA complexed to Lipofectin™ to lung tissue. Delivery of complexed DNA intravenously or intratracheally, but not intraperitoneally, resulted in gene expression in the lung with no detectable activity in the liver or spleen. Gene expression was persistent for at least 1 week from the time of injection. The data presented an opportunity to use cationic liposomes as a DNA carrier for gene therapy since all previous work was performed on cultured cells.

Another *in vivo* study took advantage of Lipofectin™ to deliver a reporter gene to lungs of mice.[51] Aerosolization of complexed DNA was the route of delivery employed in this study, which allowed lung-specific targeting of DNA. Gene expression persisted for at least 21 days after administration of reporter gene. *In situ* immunostaining also detected the reporter gene expression in the majority of epithelial and alveolar lining cells in the lung. No injury was detectable with histological screening in spite of the extensive gene expression in the lung sections. These results lend more support to gene therapy with cationic lipid vehicles.

In addition to the previous two studies[50,51] and to set the stage for the first human cancer gene therapy clinical trial using cationic liposomes, 5 years after the first introduction of cationic liposomes as a mediator of gene delivery,[23] animal testing studies were initiated in mice by Nabel and co-workers, using DC-Chol cationic liposome.[52,53] The first objective of these studies was the evaluation of the toxicity and organ distribution of DNA/DC-Chol cationic liposome complex using different assays.[52] The route of complex delivery was by intravenous or direct intratumor injection. The majority of DNA was localized in the heart and lung tissues, analyzed by quantitative polymerase chain reaction (PCR) assays, 9–11 days after i.v. injection of the complex. Intratumor injection resulted in detection of DNA in the tumor injected. To assess the toxicity of the complex, histological examinations were performed in addition to assays examining tissue-specific serum enzymes, biochemical analysis, and electrocardiographic monitoring of treated mice. None of the toxicity tests performed showed any abnormalities caused by the inclusion of liposomes in the treatment.

The initial studies showing lack of toxicity with liposomes mandated further tests with other animals such as pigs and rabbits.[53] No detectable histological toxicity or serum biochemical abnormalities were seen. Pertinent to the long-term expression of the transgenes delivered were tests aimed at immunologic responses potentially causing autoimmune damage, potential gonadal localization of injected complex, and possible toxic effects if animals other than mice were used.[53] Histological testing included all major organs of the mouse, revealing no detectable immunopathological damage to the organs. PCR analysis also could not detect any gonadal localization of transfected genes, including testes and ovaries, but could easily detect the genes in lung, kidney, spleen, and liver. The data from all the biodistribution and toxicity studies support the use of DC-Chol cationic liposomes for DNA delivery in a human cancer gene therapy protocol.[18]

A surprisingly efficient DNA delivery by cationic liposomes was reported by Zhu et al. *in vivo* in 1993.[54] A single direct injection of Lipofectin™-complexed DNA coding for a reporter gene into mice, via the tail vein, led to transfection of almost all mouse tissue, including vascular endothelium, lung, spleen, heart, liver, kidney, bone marrow, and lymph nodes, detectable by immunohistoanalysis. Toxicity was absent from the treatment protocol and gene expression was efficiently extended to at least 9 weeks after the injection. These unexpected results delivered a new hope for gene therapy aimed toward targeting expression to all the major organs in the body. However, more reports in the future are required to confirm the study.

Another *in vivo* study described the transfection of splenic T lymphocytes by intraperitoneal injection of a reporter gene DNA complexed to three different types of cationic liposomes in mice.[55] In addition to lymphocytes, bone marrow-derived hematopoietic cells were also transfected. Gene expression lasted for at least 2 weeks with no detectable toxicity. This study introduces the opportunity to use gene therapy to combat diseases affecting the immune system.

Local transfection of tissues in an organ was shown recently by Roessler et al. in a study describing the injection of DNA, coding for a reporter gene, complexed to Lipofectin™ into the mouse brain.[55] Plasmid DNA expression lasted at least 3 weeks after transfection. The results were accumulated through histochemical analysis of tissue sections of the brains treated. This report was subsequent to the preliminary and similar report by Ono et al. who also described the same protocol, resulting in up to 9 days of gene expression post-injection of the complex into a neonatal mouse brain.[56]

One of the most recent reports described an *in vivo* transfection with cationic liposomes by demonstrating gene transfer into the lungs of rabbits.[57] The animals were injected with DNA, coding for a recombinant human α1-antitrypsin gene, complexed to Lipofectin™ intravenously or by aerosol. The human gene was expressed efficiently regardless of the injection route for at least 7 days post-injection. The intravenous injection route led to gene expression in the pulmonary endothelium, while the aerosol route led to alveolar epithelium expression. Expression in the airway epithelium was a result of both routes of injection. Radiolabeling of DNA injected allowed its detection in endothelial cells in the lung. These data set precedence for targeting DNA delivery to specific subsets of cells in the lung by a selective route of delivery.

Cationic liposome-mediated gene delivery was compared favorably to the biolistic particle delivery system in a report by Hui et al.[58] DC-Chol:DOPE liposomes were used to deliver either a reporter gene or an allogeneic class I MHC gene to tumor cells *in vivo*. A strong CTL response was generated against tumor cells expressing the MHC gene after the injection of DNA/liposome complex directly into mouse spleen. This strategy should lend support to a therapy protocol which employs a direct injection of the spleen of cancer victims in order to activate an immune response against tumor cells.

B. GENE THERAPY IN MICE BY GENE TRANSFER WITH CATIONIC LIPOSOMES

Cancer immunotherapy in an animal model (mouse) using cationic liposome-mediated gene transfer was the final step before the actual human gene therapy trial. Nabel and co-workers introduced DNA complexed to DC-Chol cationic liposomes into mice bearing malignant tumors (melanoma) by direct injection of the complex into the tumor lesions.[59] The delivered DNA codes for a foreign MHC class I antigen gene which is expected to stimulate tumor rejection by allospecific CTL if the gene is expressed. Expression of the gene induced a specific CTL response to the cells expressing the foreign gene in addition to a general immune response to other tumor cells not expressing the gene. As a result of the specific and general immune response, tumor growth was hindered and in some cases a complete tumor regression was observed.

Two other animal studies reported recently were aimed toward gene therapy of CF in a mouse model of the disease.[60,61] In the first report, a transgenic mouse with a disrupted (mutated) cystic fibrosis transmembrane conductance regulator (CFTR) gene was used to test the delivery of wild type CFTR gene via Lipofectin™.[60] The DNA/Lipofectin™ complex was delivered intratracheally, which allowed transfection of the epithelia of the airway and alveoli deep in the lungs. CFTR gene expression and correction of the ion conductance defects in the trachea were demonstrated by *in situ* hybridization and voltage clamping techniques.

The second report used CFTR mutant mice and human CFTR gene complexed to either Lipofectin™ or DOTAP cationic liposomes.[61] The complexes were delivered into the airways by nebulization, a treatment that resulted in 50% correction of the deficient ion transport involved in CF in some of the mutant mice. The human CFTR gene expression was detected by reverse transcriptase PCR. The success of these two reports paved the road for human CF gene therapy clinical trials using cationic liposomes.[20]

C. HUMAN GENE THERAPY TRIALS USING THERAPEUTIC DNA COMPLEXED TO CATIONIC LIPOSOMES

The significant results with animal studies prompted Nabel et al. to pursue a human cancer gene therapy clinical trial using DC-Chol cationic liposomes as a carrier for therapeutic genes.[19] Five patients with melanoma refractory to all available therapy participated in this study which aimed toward demonstrating gene transfer, gene expression, and safety.[19] All patients received injections of DNA/DC-Chol liposome complexes intratumorly. The DNA injected codes for a foreign MHC protein, HLA-B7. All patients were HLA-B7 negative. Three to seven days after injection tumor biopsies were tested, using PCR, for the presence of injected DNA which was detected in the tumor nodules but not in the serum. Gene expression was also detected using immunostaining for HLA-B7 protein. In addition, immune responses to HLA-B7 protein were detected in injected patients. No toxicity was detected in any of the five patients. A significant finding was the regression of two injected tumor nodules at different sites in one of the treated patients. Moreover, regression of uninjected tumor nodules at a distant site were also observed in the same patient after treatment.

CF was the second target disease for human gene therapy using cationic liposomes as a carrier for therapeutic DNA. However, the human clinical trial used DC-Chol cationic liposomes.[20] The DNA/liposome complex was delivered to the nasal epithelia of CF patients as part of the phase I clinical trial.[20] The trial is still in progress, and the results will be reported in the near future. As more of these clinical trials are approved, the efficacy and long term safety of the treatment should become more established.

V. CONCLUSIONS

It is evident from this introduction that the field of cationic liposome-mediated delivery has had tremendous progress since the first report of a cationic lipid formulation for gene delivery.[23] It is also evident that the progress was much more rapid in the applications of cationic liposomes than in basic research and formulation development. This is mainly due to the demanding field of gene delivery and therapy which is in favor of novel delivery systems devoid of undesirable side effects such as toxicity, biohazard, and immunogenicity. Preliminary studies with DC-Chol liposomes to address the issues of toxicity and immunogenicity[19,50,51] clearly established the safety of cationic lipid carriers for gene therapy. Yet to be demonstrated is the efficacy of these carriers in different disease models and in repeated tests in animals and humans. These tests should be imminent since the preliminary success with cationic liposomes has attracted the attention of many clinicians interested in using the liposomes in human clinical trials around the world.

Another important field yet to benefit from cationic liposome-mediated gene delivery is biotechnology. This field has produced a large number of transgenic animals for a variety of applications which have proven to be extremely useful in basic research as well as in biomedical research.[62,63] Delivery of engineered genes into embryonic cells of animals in order to produce a transgenic animal is challenging and is certainly in need of improved DNA delivery vehicles. Cationic liposomes may have future application in this area, provided that delivered DNA is stably expressed in the transgenic tissue. Stable expression of a transgene delivered by cationic liposomes is a nascent area of study and is quickly developing. One can envision cationic liposome-mediated delivery of protein co-factors which may control the integration of a co-delivered transgene, which codes for a stable expression DNA element. This type of stable delivery system is being explored in our laboratory as part of our effort to control gene expression by co-delivery of protein factors with DNA complexed to cationic liposomes.

REFERENCES

1. **Wolff, J. A.**, Ed. (1994) *Gene Therapeutics*, Birkhauser Press, Boston.
2. **Wolff, J. A., Malone, R. W., Williams, P., Chong, W., Acsadi, G., Jani, A., and Felgner, P. L.,** (1990) *Science* 247, 1465-1468.
3. **Ulmer, J. B., Donnelly, J. J., Parker, S. E., Rhodes, G. H., Felgner, P. L., Dwarki, V. J., Gromkowski, S. H., Deck, R. R., DeWitt, C. M., Friedman, A., Hawe, L. A., Leander, K. R., Martinez, D., Perry, H. C., Shiver, J. W., Montgomery, D. L., and Liu, M. A.,** (1993) *Science* 259, 1745-1749.
4. **Yang, N.-S., Burkholder, J., Roberts, B., Martinell, B., and McCabe, D.,** (1990) *Proc. Natl. Acad. Sci. U.S.A.* 87, 9568-9572.
5. **Berkner, K. L.,** (1988) *BioTech* 6, 616-630.
6. **Eglitis, M. A. and Anderson, W. F.,** (1988) *BioTech* 6, 608-614.
7. **Anderson, W. F.,** (1992) *Science* 256, 808-813.
8. **Miller, A. D.,** (1992) *Nature* 357, 455-460.
9. **Mulligan, R. C.,** (1993) *Science* 260, 926-932.
10. **Temin, H. M.,** (1990) *Hum. Gene Ther.* 1, 111-123.

11. **Findeis, M. A., Merwin, J. R., Spitalny, G. L., and Chiou, H. C.,** (1993) *Trends Biotechnol.* 11, 202-205.
12. **Hug, P. and Sleight, R. G.,** (1991) *Biochim. Biophys. Acta* 1097, 1-17.
13. **Smith, J. G., Walzem, R. L., and German, J. B.,** (1993) *Biochim. Biophys. Acta* 1154, 327-340.
14. **Singhal, A. and Huang, L.,** (1994) in *Gene Therapeutics* (Wolff, J. A., Ed.) pp.118-142, Birkhauser, Boston.
15. **Xiang, G. and Huang, L.,** (1993) *J. Liposome Res.* 3,17-30.
16. **Farhood, H., Gao, X., Kyonghee, S., Yang, Y.-Y., Lazo, J. S., Huang, L., Barsoum, J., Bottega, R., and Epand, R. M.,** (1994) in *Gene Therapy for Neoplastic Diseases* (Huber, B. E. and Lazo, J. S., Eds.) pp.23-35, *Ann. N. Y. Acad. Sci.*, New York.
17. **Legendre, J.-Y. and Szoka, F. C., Jr.,** (1992) *Pharm. Res.* 9, 1235-1242.
18. **Nabel, G. J., Chang, A., Nabel, E. G., Plautz, G., Fox, B. A., Huang, L., and Shu, S.,** (1992) *Hum. Gene Ther.* 3, 399-410.
19. **Nabel, G. J., Nabel, E. G., Yang, Z.-Y., Fox, B. A., Plautz, G. E., Gao, X., Huang, L., Shu, S., Gordon, D., and Chang, A. E.,** (1993) *Proc. Natl. Acad. Sci. U.S.A.* 90, 11307-11311.
20. **Caplen, N. J., Gao, X., Hayes, P., Elaswarapu, R., Fisher, G., Kinrade, E., Chakera, A., Schorr, J., Hughes, B., Dorin, J. R., Porteous, D. J., Alton, E. W. F. W., Geddes, D. M., Coutelle, C., Williamson, R., Huang, L., and Gilchrist, C.,** (1994) *Gene Ther.* 1, 139-147.
21. **Gregoriadis, G., Ed.,** (1984) *Liposome Technology,* 3 volumes, CRC Press, Boca Raton, FL.
22. **Collins, D. and Huang, L.,** (1987) *Cancer Res.* 47, 735-739.
23. **Felgner, P. L., Gadek, T. R., Holm, M., Roman, R., Chan, H. W., Wenz, M., Northrop, J. P., Ringold, G. M., and Danielsen, M.,** (1987) *Proc. Natl. Acad. Sci. U.S.A.* 84, 7413-7417.
24. **Leventis, R. and Silvius, J. R.,** (1990) *Biochim. Biophys. Acta* 1023, 124-132.
25. **Gao, X. and Huang, L.,** (1991) *Biochem. Biophys. Res. Commun.* 179, 280-285.
26. **Farhood, H., Bottega, R., Epand, R. M., and Huang, L.,** (1992) *Biochim. Biophys. Acta* 1111, 239-246.
27. **Ballas, N., Zakai, N., Sela, I., and Loyter, A.,** (1988) *Biochim. Biophys. Acta* 939, 8-18.
28. **Pinnaduage, P., Schmitt, L., and Huang, L.,** (1989) *Biochim. Biophys. Acta* 985, 33-37.
29. **Ito, A., Miyazoe, R., Mitoma, J.-Y., Akao, T., Osaki, T., and Kunitake, T.,** (1990) *Biochem. Inter.* 22, 235-241.
30. **Rose, J. K., Buonocore, L., and Whitt, M. A.,** (1991) *BioTech* 10, 520-525.
31. **Stamatatos, L., Leventis, R., Zuckermann, M. J., and Silvius, J. R.,** (1988) *Biochemistry* 27, 3917-3925.
32. **Felgner, J. H., Kumar, R., Sridhar, C. N., Wheeler, C. J., Tsai, Y. J., Border, R., Ramsey, P., Martin, M., and Felgner, P. L.,** (1994) *J. Biol. Chem.* 269, 2550-2561.
33. **Barthel, F., Remy, J.-S., Loeffler, J.-P., and Behr, J.-P.,** (1993) *DNA Cell Biol.* 12, 553-560.
34. **Zhou, X., Klibanov, A. L., and Huang, L.,** (1991) *Biochim. Biophys. Acta* 1065, 8-14.
35. **Zhou, X. and Huang, L.,** (1994) *Biochim. Biophys. Acta* 1189, 195-203.
36. **Helenius, A., Mellman, I., Wall, D., and Hubbard, A.,** (1983) *TIBS* 8, 245-250.
37. **Gruner, S. M.,** (1985) *Annu. Rev. Biophys. Biophys. Chem.* 14, 211-238.
38. **Litzinger, D. and Huang, L.,** (1992) *Biochim. Biophys. Acta* 1113, 201-227.
39. **Malone, R. W., Felgner, P. L., and Verma, I. M.,** (1989) *Proc. Natl. Acad. Sci. U.S.A.* 86, 6077-6081.
40. **Innes, C. L., Smith, P. B.,** Langenbach, R., Tindall, K. R., and Boone, L. R., (1990) *J. Virol.* 64, 957-961.
41. **Shigekawa, K. and Dower, W. J.,** (1988) *BioTech* 6, 742-751.
42. **McNeil, P. L., Murphy, R. F., Lanni, F., and Taylor, D. L.,** (1984) *J. Cell Biol.* 98, 1556-1564.
43. **Debs, R. J., Freedman, L. P., Edmunds, S., Gaensler, K. L., Duzgunes, N., and Yamamoto, K. R.,** (1990) *J. Biol. Chem.* 265, 10189-10192.
44. **Walker, C., Selby, M., Erickson, A., Cataldo, D., Valensi, J.-P., and Nest, G. V.,** (1992) *Proc. Natl. Acad. Sci. U.S.A.* 89, 7915-7918.
45. **Nair, S., Zhou, X., Huang, L., and Rouse, B. T.,** (1992) *J. Immunol. Methods* 152, 237-243.
46. **Chen, W., Carbone, F. R., and McCluskey, J.,** (1993) *J. Immunol. Methods* 160, 49-57.
47. **Lin, M.-F., DaVolio, J., and Garcia, R.,** (1993) *Biochem. Biophys. Res. Commun.* 192, 413-419.
48. **Gao, X. and Huang, L.,** (1993) *Nucleic Acids Res.* 21, 2867-2872.
49. **Capecchi, M. R.,** (1980) *Cell* 22, 479-488.
50. **Brigham, K. L., Meyrick, B., Christman, B., Magnuson, M., King, G., and Berry, L. C., Jr.,** (1989) *Am. J. Med. Sci.* 298, 278-281.
51. **Stribling, R., Brunette, E., Liggitt, D., Gaensler, K., and Debs, R.,** (1992) *Proc. Natl. Acad. Sci. U.S.A.* 89, 11277-11281.
52. **Stewart, M. J., Plautz, G. E., Buono, L. D., Yang, Z. Y., Xu, L., Gao, X., Huang, L., Nabel, E., and Nabel, G. J.,** (1992) *Hum. Gene Ther.* 3, 267-275.
53. **Nabel, E. G., Gordon, D., Yang, Z.-Y., Xu, L., San, H., Plautz, G. E., Wu, B.-Y., Gao, X., Huang, L., and Nabel, G. J.,** (1992) *Hum. Gene Ther.* 3, 649-656.
54. **Zhu, N., Liggitt, D., Liu, Y., and Debs, R.,** (1993) *Science* 261, 209-211.
55. **Roessler, B. J. and Davidson, B. L.,** (1994) *Neurosci. Lett.* 167, 5-10.
56. **Ono, T., Fujino, Y., Tsuchiya, T., and Tsuda, M.,** (1990) *Neurosci. Lett.* 117, 259-263.
57. **Canonico, A. E., Conary, J. T., Meyrick, B. O., Brigham, K. L.,** (1994) *Am. J. Respir. Cell Mol. Biol.* 10, 24-29.
58. **Hui, K. M., Sabapathy, Tr. K., Oei, A. A., Singhal, A., and Huang, L.,** (1994) *J. Liposome Res.* 4, 1075-1090.

59. **Plautz, G. E., Yang, Z.-Y., Wu, B.-Y., Gao, X., Huang, L., and Nabel, G. J.,** (1993) *Proc. Natl. Acad. Sci. U.S.A.* 90, 4645-4649.
60. **Hyde, S. C., Gill, D. R., Higgins, C. F., and Trezise, A. E. O.,** (1993) *Nature* 362, 250-255.
61. **Alton, E. W. F. W., Middleton, N. J., Caplen, N. J., Smith, S. N., Steel, D. M., Munkonge, F. M., Jeffery, P. K., Geddes, D. M., Hart, S. L., Williamson, R., Fasold, K. I., Miller, A. D., Dickinson, P., Stevenson, B. J., McLachlan, G., Dorin, J. R., and Porteous, D. J.,** (1993) *Nature Genetics* 5, 135-142.
62. **Jaenish, R.,** (1988) *Science* 240, 1468-1474.
63. **Hanahan, D.,** (1989) *Science* 246, 1265-1275.
64. **Sorgi, F. L. and Huang, L.,** Drug delivery applications of liposomes containing non-bilayer forming phospholipids. In *Structural and Biological Roles of Lipids Forming Non- Lamellar Structures*. Epand, R. M., Ed., JAI Press, Greenwich, CT. In press.

Chapter 4

Advances in the Design and Application of Cytofectin Formulations

Philip L. Felgner, Yali J. Tsai, and Jiin H. Felgner

CONTENTS

I. INTRODUCTION

When cationic liposomes and DNA are mixed together in an aqueous environment, the two macromolecular systems associate ionically, and the lipids and DNA reorganize in close association with each other.[1,2] These lipid-DNA complexes, when properly formulated, can facilitate DNA, RNA, and synthetic oligonucleotide uptake into living cells *in vitro* and *in vivo*.[3-19] The cationic lipid molecules comprising this type of formulation have been broadly termed "cytofectins", and the cationic lipid-polynucleotide complexes that form after mixing cytofectin liposomes with DNA are termed "cytisomes" (Figure 1).* The cytisome contains packaged, condensed DNA (or RNA), much as viruses capture and condense their genome; and the lipid envelop surrounding the cytisomes confers the property of endosomal escape, in much the same way that the viral envelop confers this quality to the virus.

The transfection activity of cytofectin formulations depends greatly on the cationic lipid chemical species, on the presence of particular neutral co-lipids in the formulation, on the lipid-to-DNA ratio, the DNA concentration, and on the ionic properties of the suspending vehicle. Evidence suggests that these formulation variables yield lipid-DNA complexes, i.e., "cytisomes" with different physical and structural properties, and that these qualities determine the activity of the final formulation. We are working toward a better understanding of how formulation variables control the physical and structural properties of the complexes, in order to rationally improve the activity of cationic lipid molecules and cytofectin formulations.

This review summarizes progress in the development of improved cytofectin formulations since their original description in 1987.[3] Approaches toward the rational design of these formulations are described, and a theoretical model to explain how cationic lipids may organize around the DNA strand is presented. This model takes into account the molecular dimensions of individual lipid molecules, known macromolecular organizations for lipids of this kind, the dimensions of the DNA strand, and the net charge of the system after the two macromolecular species (DNA and cationic liposomes) are mixed.

* The origin of the term "cytofectin" may require some explanation. The original cationic lipid found to have DNA delivery activity, DOTMA (Ref), was referred to as "Lipofectin". Since Lipofectin carries a trademark referring specifically to the particular reagent consisting of DOTMA, the term "lipofectins" could not be used to broadly define all classes of cationic lipid molecules capable of delivering functional DNA into cells. The term "cytofectins", which is not trademarked, was defined for this purpose, i.e., to broadly describe the class of cationic lipid molecules that have the property of interacting with polynucleic acids (i.e., DNA, mRNA and oligonucleotides) and facilitating their entry into cells. Similarly, the term "Lipofection" which has been used to generically define the cationic lipid-mediated transfection procedure, also falls under the trademark restriction and should therefore not be used for this purpose. For this reason, the term "cytofection" is preferred.

0-8493-4013-6/96/$0.00+$.50

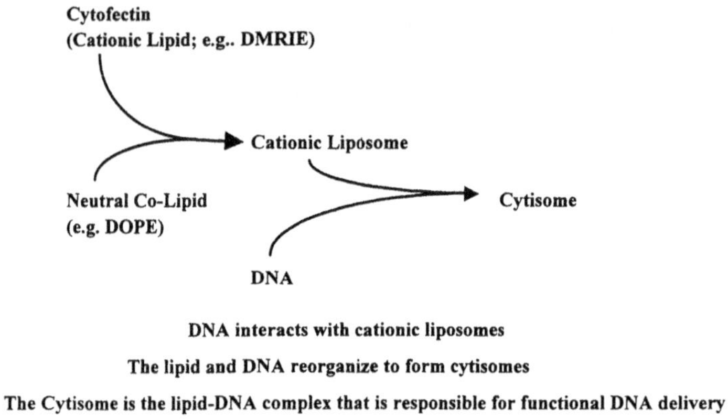

Figure 1 Pathway to functional assembly of cytisomes.

II. DEFINITION OF THE CYTISOME

The chemical design of DOTMA was based on theoretical considerations, elaborated by Israelachvili,[20] which emphasize that the geometry of amphipathic lipid molecules is critically important in the formation of macromolecular structures in solution, such as emulsions, micelles, and liposomes. These considerations emphasize that amphipathic molecules that are "wedge"-shaped, i.e., containing a polar head group with a larger cross-sectional area than the aliphatic tail, tend to adopt a micellar configuration in solution. "Brick"-shaped or rectangular amphipathic molecules that have a similar cross-sectional area of the polar head group and aliphatic region result in bilayer (i.e., liposome) formation. A further consideration in the design of DOTMA was that the lipid should have a prominent aliphatic region to avoid the high transfer rates from lipid vesicles that are problematic with molecules such as stearylamine.

There have been many reports concerning the characterization and use of positively charged lipid vesicles,[21-23] but prior to the description of DOTMA, the molecules used to render these vesicles cationic, from the standpoint of liposome preparation, were less than ideal. The two agents most frequently used for this purpose were stearylamine (SA)[22] and dioctadecyldimethylammonium bromide (DDA).[23] Neither of these lipids spontaneously forms physically and biologically stable liposomes without the addition of other liposome-forming (usually phospholipid) components. Based on previous background available on transfer rates of fatty acids out of phospholipid vesicles,[24,25] one could predict that phospholipid vesicles comprised of SA would lose their positive charge within seconds after exposure to another membrane population devoid of SA. The observation that positively charged SA liposomes are cleared biologically at a rate very similar to neutral lipid vesicles supported this hypothesis.[26] Furthermore, SA has been shown to have specific inhibitory effects related to protein kinase-C, which can lead to SA-mediated tissue culture cell death.[27,28] The tendency for DDA to form unusual lipid structures, such as inverted micelles, when it is used alone or in combination with other phospholipids, makes its use difficult to control.[29] Our interest, therefore, was to prepare a chemically stable, positively charged lipid, which would form physically stable liposomes, with the expectation that these vesicles would have more predictable physical properties.

Based on these considerations, the molecules like those shown in Figure 2 were synthesized. The parallel orientation of the aliphatic chains in this molecule would be expected to favor bilayer, rather than micelle, formation. The polar head group bears a quaternary amine so that vesicles comprised of DOTMA will be positively charged. Further, the ether linkages afford greater chemical stability in aqueous solutions than the comparable ester derivatives.[31] As expected, suspensions of DOTMA, alone or in combination with other phospholipids, in aqueous buffers resulted in the formation of multilamellar liposomes (MLV) which could be sonicated to form small unilamellar vesicles (SUV; 0.03 micron diameter as determined by quasielastic laser light scattering). Multilayer structures in preparations of MLV were apparent by freeze/fracture electron microscopy as were small vesicles in SUV preparations. These vesicles were capable of entrapping fluorescent dextran and had a typical liposome appearance as judged by freeze/fracture electron microscopy.[41]

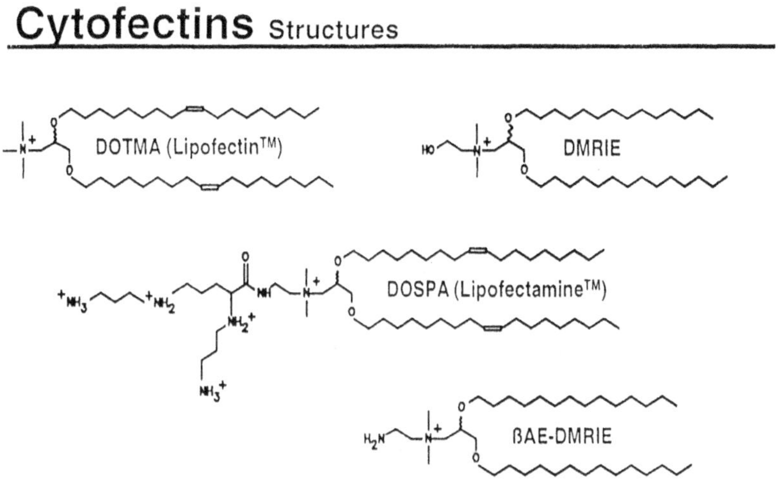

Figure 2 Cytofectin structures.

Since their original description, liposomes have been discussed as vehicles that could be used as carriers of pharmaceutically active agents and as possible gene delivery systems . The pioneering studies of Fraley, Papahajopaulos and Szoka provided many of the technical and theoretical guidlines pertaining to the use of lipid vesicles for polynucleotide delivery.[32-36] A fundamental problem associated with the use of negatively charged or neutral liposomes for DNA delivery was that the internal diameter of the vesicle is usually smaller than the longest dimension of a typical DNA molecule — a situation that leads to low encapsulation efficiency. This made the development of such delivery systems inefficent and inconvenient. We anticipated that by utilizing positively charged vesicles, the practical problem of encapsulating DNA into liposomes with high efficiency would be mitigated. As expected, liposome/DNA complexes formed spontaneously after mixing, and when a sufficient amount of lipid was used the resulting complexes could entrap 100% of the input DNA. The sucrose density gradient profile in Figure 3 is one of the first experiments done with these cationic liposomes, showing this property of 100% capture. DNA (pSV2CAT plasmid) applied to the top of a 5–20% sucrose density gradient migrates to the bottom of the gradient after centrifugation (Beckman airfuge, 90 min). However, if the DNA is mixed with Lipofectin™ prior to application on the gradient, all of the DNA remained on the top. Incorporation of radioactive lipid into Lipofectin™ indicated that all the lipid material was also on the top of the gradient, whether mixed with DNA or not. Surprisingly, the complexes produced with this very simple procedure gave rise to relatively homogeneous suspensions that were not grossly aggregated and did not necessarily settle out of solution. The complexes could be sized in the range of about 100–800 nm, depending on the details of how the samples were mixed. And it was a further surpising that this simple system was effective in cultured cells at delivering transcriptionally active nucleic acid, without the use of any additional specific ligand to encourage cell uptake, endosomal escape, or nuclear entry.

This information supports the scheme shown in Figure 1 which defines an entity referred to as the "cytisome". DNA interacts with cationic liposomes, and the lipid and DNA reorganize to form the cytisome. The structure and physical properties of the cytisome are determined by the composition of the liposomes and by the way in which the DNA and liposomes are mixed; and the physical properties of the resulting cytisome determines how it will interact with biological samples, and whether it will deliver functional DNA into the appropriate transcriptionally active intracellular compartment.

III. CYTISOME FORMULATION ASPECTS

Several laboratories have attempted to identify new cationic lipid molecules that have improved potency relative to the original lipofectin reagent (DOTMA). Accurate potency ranking of different cationic lipid species is complicated because of the large number of formulation dependent variables that affect transfection potency. Factors that have been shown to affect functional cytofectin-mediated nucleic acid delivery in cultured cells are listed in Table 1.

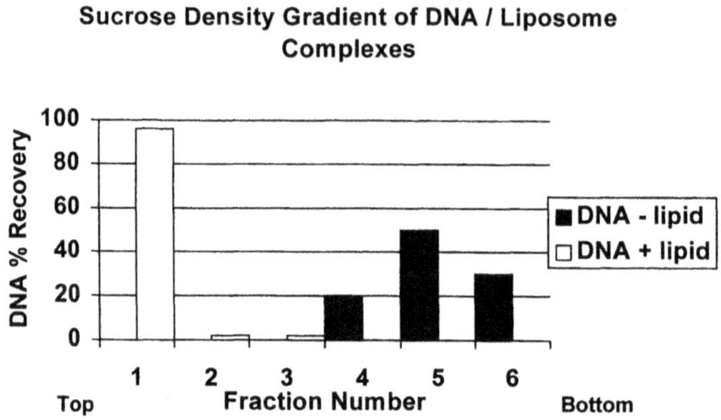

Figure 3 Demonstration of spontaneous cytisome formation.

Table 1 Factors Affecting Gene Delivery

Cytofectin – DNA Formulation Aspects
Cationic lipid species
Cationic lipid/DNA ratio
DNA (and cationic lipid) dose concentration
Amount and identity of co-lipid
Vehicle composition
Cationic liposome size
Temperature during cytisome assembly
Cell type
Extracellular matrix elements
DNA purity

With respect to the cationic lipid species dependency, several published studies have compared the relative *in vitro* transfection activities of different cationic lipid molecules. Drawing general conclusions about the relative potencies of different cationic lipid molecules from these studies is difficult, however, because the relative potency ranking is cell type dependent. For example, dioctadecyldimethlyammonium bromide (DDAB) is quite effective in baby hamster kidney (BHK)[8] cells but is much less active than the lipofectin reagent (DOTMA) in 3T3 cells.[16] In addition to the cell type-dependent differences, the precise optimization conditions can vary substantially for different reagents. At low cationic lipid and DNA concentrations, transfection activity increases in a dose dependent manner. However, at higher cationic lipid levels lipid-mediated toxicity leads to a dose-dependent decline in transfection activity — and furthermore, at high DNA levels, aggregation of the cationic lipid DNA complexes can result in a dose-dependent decline in transfection activity. Therefore, firm conclusions about the relative *in vitro* activities of different reagents can not be made definitively until thorough optimization studies are completed.

In order to address this optimization variable, we have developed and routinely implemented a 96-well microtiter plate *in vitro* transfection assay, which can be used to evaluate and qualify the activity of reagents by scanning 64 different assay conditions. Cells are plated into a 96-well microtiter plate. In separate mixing plates, serial dilutions of cationic liposomes and DNA are prepared, and the variously diluted samples are added into the wells of the plate containing cells. In this way the dose concentrations and of both DNA (0.2–20 μg/ml) and cationic lipid (1.3–170 μ*M*) can be independently varied in this assay over a 100-fold range, and the cationic lipid/DNA ratio is varied over a 10,000-fold range. The assay can be run using any convenient reporter gene, such as beta-galactosidase, luciferase, chloramphenicol acetyl transferase, or any gene with an ELISA endpoint. An example of the output from one of these assays is shown in Figure 4.

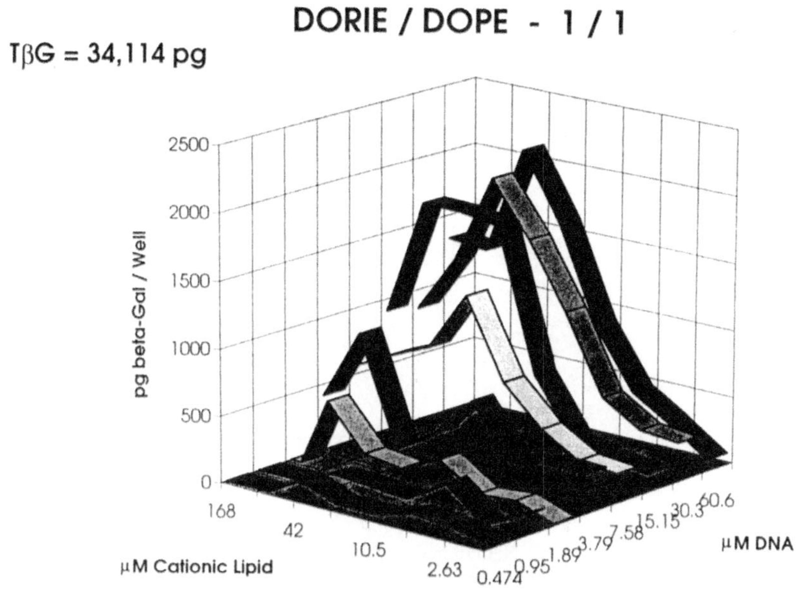

Figure 4 Output from representative transfection assay.

This assay has allowed us to identify the following general characteristics of cytofectin-mediated transfection. At low DNA concentrations, transfection activity increases in proportion to the amount of DNA added, and a plateau is usually observed at high DNA concentrations. At low cationic lipid concentrations the activity also increases in proportion to the amount of cationic lipid added, but higher cationic lipid concentrations lead to toxicity, cell death, and a corresponding decline in the *in vitro* transfection signal; therefore, the cationic lipid dose-response curves have a bell shape. There is an optimal cationic lipid/DNA ratio for each reagent studied and the optimization profile obtained in this assay varies depending on the cationic lipid species and on the formulation examined. For monovalent cationic lipids such as DOTMA, this optimum usually occurs at or near the point of charge neutralization, when the molar equivalents of positive charge contributed by the lipid are equal to the negative charge contributed by the nucleic acid. For polyvalent cationic lipids such as DOSPA (the Lipofectamine reagent)[16] and DOGS (Transfectam)[11] optimal activity occurs under conditions in which the amount of positive charge contributed by the cationic lipid exceeds the molar equivalents of negative charge contributed by the DNA. Each cationic lipid species has an optimal neutral lipid requirement and this optimum can vary depending on the cell type examined. For example, DMRIE (1,2-dimyristyooxypropyl-3-dimethyl-hydroxyetheyl ammonium bromide) requires DOPE for optimal activity, and DOPC will not substitute. In COS.7 monkey kidney cells the oprimal DOPE content is 50 mol%, while in mouse muscle C2C12 cells the optimum is 75 mol%. DOSPA also requires DOPE for optimal activity; however, unlike DMRIE, DOPC can substitute for the DOPE requirement in DOSPA formulations. Cholesterol can substitute for DOPE in DMRIE formulations, but not in DOSPA formulations. The mechanism of action to explain these formulation-dependent differences is not understood. However, the cytisomes that result when liposomes comprised of different lipid species are mixed with DNA would be expected to confer different physical properties on the complex. And the physical properties of the cytisome may influence the intracellular distribution of endocytosed complexes and the level of transcriptionally active nucleic acid that enters the nucleus.

The physical properties of cytisomes can also be influenced by controlling the conditions in which the complexes are formed. The composition of the suspending vehicle, the temperature, the rate and order of addition of the cationic liposomes and DNA, and the size of the initial liposomes, all combine to determine the physical properties and transfection potency of the final complexes. The most conveniently measured parameter characterizing the physical properties of cytisomes is the particle diameter determined by laser light scattering. Cytisomes tend to aggregate at the point of charge neutralization when the amount of positive charge contributed by the cationic lipid is equal to the amount of negative charge contributed by the DNA (Figure 5). Aggregation is greater in physiological saline solutions than

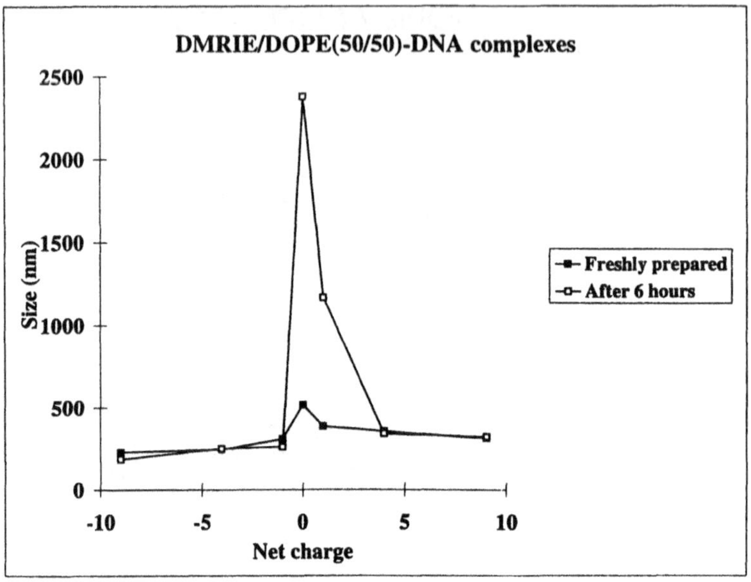

Figure 5 DMRIE/DOPE (50/50) DNA Complexes.

in low ionic strength media. Polyvalent anionic salts such as phosphate, citrate, or EDTA can greatly increase aggregation of the complexes. Excessive aggregation of the complexes can cause a decrease in transfection potency. The effect of the order and rate of addition is illustrated by the example shown in Figure 6. DOSPA/DOPE multilamellar liposomes have the particle size distribution shown in Figure 6, ranging between 140–540 nm in diameter. The addition of DNA leads to a slight shift toward lower particle diameter. This "condensing" effect in which the complexes exhibit a smaller diameter than the original cationic liposomes has been observed consistently with other cytofectin preparations. Most notably, however, when the complexes are mixed by slow addition with constant stirring, very small cytisomes ranging in size from 100–140 nm can be produced. When the slow addition method is used, the order of addition is particularly important. If the final preparation has an excess of negative charge equivalents contributed by the DNA, then the cationic lipid must be added into the DNA solution. Conversely, if the final preparation has an excess of positive charge contributed by the cationic lipid, then the DNA should be added slowly into the lipid solution. Following this procedure insures that the complexes will avoid passing through the charge neutral condition in which aggregation is the greatest.

Polyanionic extracellular matrix elements, such as chondroitin sulfate, or heparin sulfate can also adversely affect the DNA delivery potency of cytofectin formulations. The inhibitory effect of polyanionic macromolecules may be mediated by binding to positively charged cytisomes and competing with negatively charged cell surface components that serve as receptors for the cytisomes. Alternatively, the polyanions may alter the physical properties of the cytisomes and adversely affect their ability to deliver transcriptionally active nucleic acid into the nucleus. Polycations also inhibit cytofectin-mediated transfection, presumably by binding to negatively charged cell surface components that are cytisome receptors. DNA purity is another factor that can influence cytofectin-mediated transfection activity. High levels of endotoxin contamination have been shown to inhibit cytofectin-mediated transfection.[37] Contaminating RNA can compete with DNA during cytisome assembly and reduce the amount of functional cytisome produced.

Thus, there are numerous variables that can influence the *in vitro* transfection activity of cytofectin formulations, making it difficult to rank potency of different formulations or cationic lipid chemical species on an absolute scale and obtain definitive structure activity relationships. However, by maintaining a common cell type, and by using an internal standard, such as the Lipofectin reagent to normalize data, it is possible to generate data like those shown in Figure 7 which rank the transfection activity of several cationic lipids screened in COS.7 cells. These data show approximately a 20-fold enhancement in transfection activity of DMRIE and DOSPA relative to the original cytofectin reagent DOTMA. Although the magnitude of these differences may vary somewhat among different cell types, this figure conveys

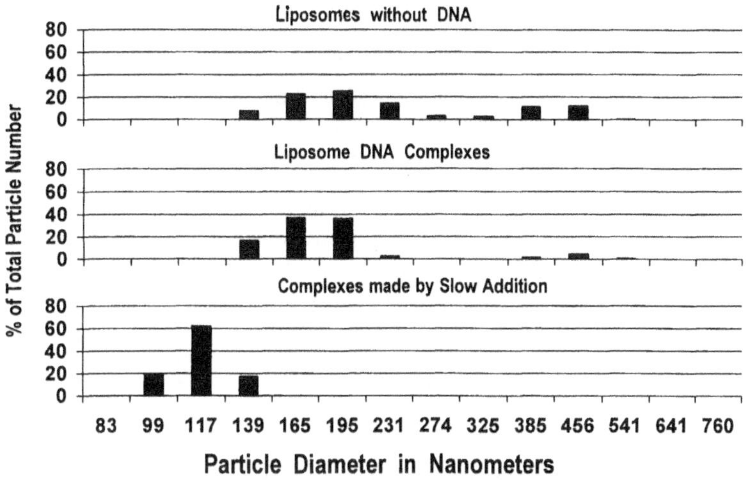

Figure 6 Particle size of DOSPA/DOPE–DNA complexes.

a reasonably accurate message that the reagents in development today are about 1 order of magnitude (i.e., not 2 or more orders of magnitude) more active then Lipofectin. In addition to the results shown in Figure 7, a homologous series of monovalent cationic lipid molecules were compared using the *in vitro* transfection assay, and the following general conclusions were made. Increasing the length of the saturated alkyl chains led to a progressive decline in transfection activity, and unsaturation improved transfection activity. This indicated that fluidity is a beneficial physical property for the cytofectins in this series. However, the saturated C14 compound (DMRIE) was more active then the homologous unsaturated C18 compound (DORIE). Differential scanning calorimetry showed that the C14 analog is in the fluid phase at 37°C, consistent with the conclusion that fluidity is a beneficial physical property for cytofectins. Sonicated cationic lipid vesicles are generally less active than the larger vortexed multilamellar liposomes. This presumably reflects differences in the way that the small highly curved sonicated vesicles interact with DNA, compared to larger vortexed vesicles that have more flexible bilayers.

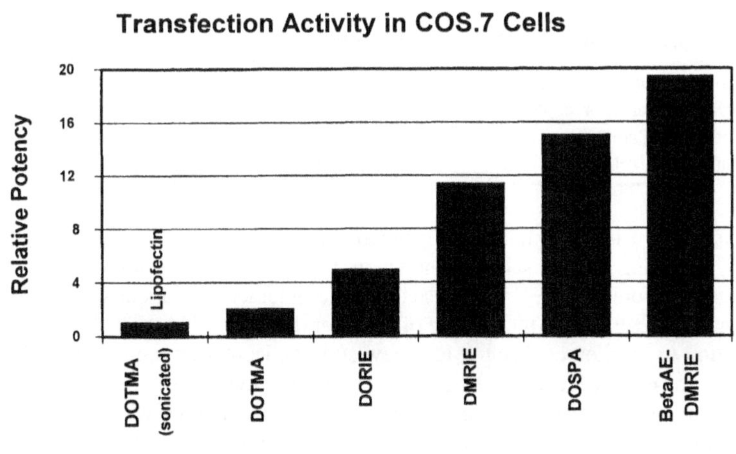

Figure 7 Progress toward more active cytofectins.

Most cytofectins require a neutral co-lipid for optimal activity and the most widely used co-lipid for this purpose is DOPE. The reason for the neutral lipid requirement is unclear; however, pure DOPE can adopt the inverted hexagonal phase which may be responsible for its ability to activate the cytofectins. Consistent with this conclusion is the observation that sequential methylation of the primary amino group on DOPE produces a series of neutral co-lipid analogs with progressively less tendency to form the inverted hexagonal phase under physiological temperatures and correspondingly less *in vitro* transfection activity. Thus, there is a direct correlation between transfection-enhancing activity of DOPE analogs and their tendency to form the inverted phase. The inverted hexagonal phase may be important in the packing of cationic lipid around DNA to produce an optimally condensed cytisome. Alternatively, DOPE is known to induce membrane fusion, a property which, if present in cytisomes, may enable delivery of DNA directly across the plasma membrane or escape from the endosomal compartment. Incorporation of DOPE into cytisomes also introduces a titratable amino group into the system which is titratable between pH 7.5 and 9.0.[15] The significance of this titratable amine is not clear. In addition to DOPE, cholesterol has also been shown to enhance *in vitro* transfection activity of cytofectins. Although cholesterol has a small polar head group that might help to encourage highly curved bilayers, it does not contain a titratable moiety.

IV. BARRIERS TO FUNCTIONAL NUCLEIC ACID DELIVERY

In the design of synthetic delivery systems, scientists have looked at viruses as models of efficient nucleic acid delivery systems. The three intrinsic viral attributes that have attracted the most attention are the target cell surface receptor interaction, viral membrane fusion proteins, and nuclear targeting signals (Table 2). Viral receptors enable infectious particles to specifically attach to the surface of target cells, whereas fusion proteins enable escape of the viral genome from the endosomal compartment into the cytoplasm following receptor-mediated endocytosis. Viruses lacking a membrane fusion capability would reach the lysosomal compartment and be degraded by hydrolytic and oxidative activity located there. Upon exit of the viral nucleic acid from the endosome, viruses benefit from protein components that help to concentrate the nucleic acid in the vicinity of the nucleus. Finally, viruses take advantage of capsid proteins which package the genome into a small, highly condensed form, leading to a particle of minimum size. The condensed DNA is protected from exogenous nuclease activity, and the small particle is presumably easier to be taken up by cells than a large extended DNA structure would be. Using this knowledge as a model for the development of more effective synthetic vehicles for gene delivery systems would involve incorporating into the system the corresponding elements as shown on the right hand side of Table 2. Thus, synthetic vehicles that are as effective as viruses would be expected to condense the DNA into a small particle, and incorporate cell targeting, endosomal fusion, and nuclear targeting elements.

Table 2

How Do Viruses Deliver Nucleic Acid?	How to Enhance Synthetic Delivery Systems?
1. Condense the extended DNA structure	1. Control DNA condensation and produce monodisperse systems
2. Engage specific cell surface receptors	2. Utilize cell surface receptors and avoid nonspecific interactions
3. Escape the endosomal compartment	3. Escape from the endosomal compartment
4. Incorporate nuclear targeting	4. Concentrate nucleic acid into the nucleus

Our understanding of the factors affecting functional delivery of cytofectin/DNA complexes can be summarized by referring to the scheme shown in Figure 8. The cationic lipids (cytofectins) form positively charged liposomes.[1-3,9,12,17,40,41] These liposomes interact spontaneously with DNA, which under appropriate conditions can lead to the capture of 100% of the input DNA.[3] The lipid and DNA reorganize in close apposition to each other, leading to a new supramolecular structure referred to as a cytisome. When cytisomes are introduced into cell culture or *in vivo*, they interact directly with the target cell surface in a potentially productive fashion, or in a nonproductive manner with extracellular factors.[1,3,12,17] After attachment to the cell surface, cytisomes can either fuse directly with the plasma membrane or be endocytosed. The endocytosed cytisomes can either fuse with the membranes of early (high pH) or late (low pH) endosomes, or be delivered into the lysosome where the DNA is degraded.[15,17,41] Fusion of the cytisome with either the plasma membrane or the endosomal membranes leads to diffusion of the lipid into target cell membranes; this displaces the ionic interactions between the cationic lipid and DNA and the DNA is free to diffuse into the cytoplasm.[3,5,41] No lipid-facilitated uptake of DNA from the cytoplasm into the nucleus has been demonstrated, although exposure of cells to cationic lipid has been reported to result in the formation of a "nucleoplasmic reticulum" (personal communication, D. Friend), thus indicating that cationic lipid can effect the structure of the nucleus. The nuclear membrane may represent a barrier to the entry of DNA into nucleus. Since the nuclear membrane is absent during cell division, one might predict that transfection would be more efficient in rapidly dividing cells. In support of this suggestion, cell division has been reported to increase the level of cytofectin mediated gene transfer *in vitro* and *in vivo*.[39]

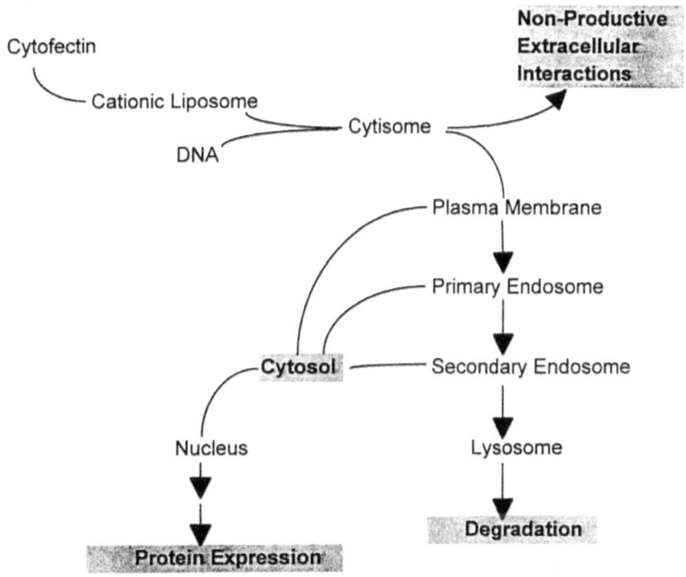

Figure 8 Cytisome uptake and functional DNA delivery.

V. STRUCTURAL MODEL FOR CYTISOME ASSEMBLY

Based on molecular packing considerations and on experimental results from several laboratories, it is possible to draft a working model describing the structure of cytofectin DNA complexes to describe how cationic lipids can organize around the DNA strand. This model takes into account the molecular dimensions of individual lipid molecules, known macromolecular organizations for lipids of this kind, the dimensions of the DNA strand, and the net charge of the system after the two macromolecular species (DNA and cationic liposomes) are mixed.

We predict that at least two types of structure will result when cationic liposomes and DNA are mixed. These two structures are referred to as filamentous lipoidal DNA and close packed lipoidal DNA (Figure 9). The filamentous structure consists of DNA ensheathed in a lipid bilayer along its entire length. The inner monolayer of this sheath is oriented with the polar head groups directed toward the DNA. This monolayer is highly curved, leading to an aqueous inner core diameter of ~5 nm. This dimension is similar to the internal aqueous core diameter of the inverted hexagonal phase, which is produced when pure DOPE is suspended in water. Thus it is physically possible for lipids of this kind to adopt a highly curved arrangement of this kind. A 5-nm aqueous core diameter is sufficiently large to accommodate the negatively charged DNA strand which has a ~2.5 nm diameter. The outer monolayer has its polar head group directed away from the central core. The total diameter of this ensheathed DNA structure is 10–15 nm. A mathematical description of the structure of filamentous lipoidal DNA is as follows:

Molecular Packing Hypothesis

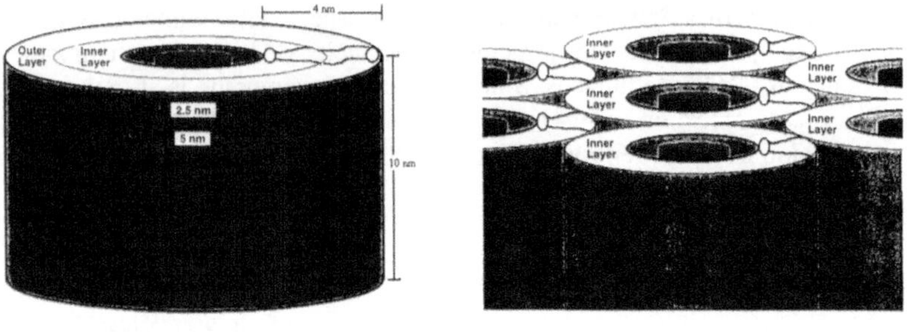

Filamentous Lipoidal DNA **Close-Packed Lipoidal DNA**

Figure 9 Molecular packing hypothesis of cytisome structure.

$$\text{\# of outer lipid molecules} = 2\pi R_O L/A_O$$

$$\text{\# of inner lipid molecules} = 2\pi R_I L/A_I$$

where R_O is the outer radius of the structure, and is equal to 5–10 nm.
R_I is the radius of the aqueous inner core, and is equal to 2–4 nm.
L is the length of the DNA and is equal to 1–1,000,000 nm. The DNA contains about 10 negative charges per nm. The diameter of the DNA strand is about 2.5 nm.
A_O is the average polar head group surface per molecule area occupied by the lipids in the outer lipid layer, and is equal to 0.45–2.0 nm^2
A_I is the average polar head group surface area per molecule occupied by the lipids in the inner lipid layer, and is 0.3–1.5 nm^2. Because of the high curvature of the inner monolayer, the average cross-sectional area per molecule at the ends of the acyl (or alkyl) chains is about 2× greater than A_I.

The second structure is referred to as close-packed lipoidal DNA. This structure also contains a highly curved lipid monolayer around the DNA strand with the lipid polar head group facing toward the DNA. However, unlike filamentous lipoidal DNA, the monolayer coated strands fold back and forth, and the DNA segments stack together into a hexagonal array. A monolayer of lipid with the opposite orientation surrounds each bunch of hexagonally packed DNA segments. Filamentous lipoidal DNA uses >2 times the lipid per unit of DNA length than the close packed lipoidal DNA. This is because close-packed DNA contains only one layer of lipid per unit DNA length, whereas filamentous DNA contains two lipid layers. A description of this theoretical model is included in this report. A mathematical description of close packed lipoidal DNA follows:

$$\text{\# of inner lipid molecules} = 2\pi R_I L/A_I$$

R_I is the radius of the aqueous inner core, and is equal to 2–4 nm.
L is the length of the DNA, and is equal to 1–1,000,000 nm. The DNA may be linear, branched, or circular.
A_I is the average polar head group surface area per molecule occupied by the lipids in this layer and is 0.25–1.5 nm^2. The average cross-sectional area per molecule at the ends of the acyl (or alkyl) chains is about 2× greater than A_I.

Based on these approximations, it is possible to predict the size of a cytisome consisting of close-packed lipoidal DNA with a plasmid 8000 nucleic acid base pairs in length. The structure shown in Figure 10 would be 50–100 nm in diameter. There is a growing body of experimental support for the models proposed here.[42]

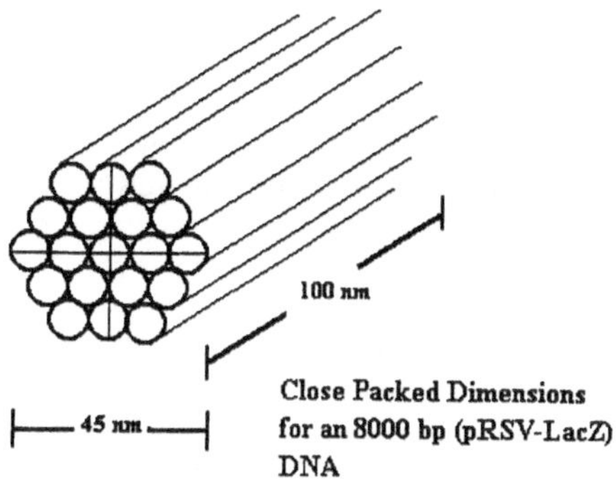

Figure 10 Close packed dimensions for an 8000 bp (pRSV-LacZ) DNA.

VI. *IN VIVO* STUDIES AND FUTURE PROSPECTS

The first cationic lipids shown to be useful for gene delivery were developed in the mid 1980s at Syntex Research in Palo Alto California.[1] These compounds were made commercially available as reagents to the molecular biology community as *in vitro* transfection reagents which are today widely applied. The development of the pharmaceutical gene therapy application of this technology began at Vical Inc. in 1988.[54,55] Today, numerous academic laboratories and biotechnology companies are developing technologies that allow direct transfer of plasmid DNA encoding specific proteins into different tissues of the body using similar cationic lipid gene transfer technology.

Vical is presently developing gene-based treatments for malignant melanoma as well as other solid tumors. Their first product, *Allovectin-7*, is a plasmid/lipid complex that is directly injected into tumors with the subsequent expression of HLA-B7, a type I major histocompatibility antigen (MHCI) on the cell surface of the tumors. The plasmid in *Allovectin-7* encodes a bicistronic mRNA that produces HLA-B7 (heavy chain) and beta-2 microglobulin (light chain) in approximately equimolar amounts. Since class I MHC expression requires both heavy and light chain coexpression and some tumors are deficient in beta-2 microglobulin, the bicistronic *Allovetin-7* should be more universally expressed in tumor cells than a plasmid containing only the HLA-B7 cDNA. The expression of the HLA-B7 protein by cancer cells is expected to stimulate the patient's immune system to recognize these transfected cells as "foreign" and to selectively destroy the tumor in a manner similar to a host vs. graft rejection which is initiated by "foreign" MHC molecules. Pre- clinical data with a prototype plasmid encoding only the heavy chain also suggest that the immune response generated against the primary tumor may also be effective in eliminating secondary tumors, or metastases. Under an investigator-sponsored IND, an initial phase I/II trial in five patients has been completed at the University of Michigan by Gary Nabel.[47,51,52] All patients treated with a prototype plasmid encoding only the heavy chain showed clear evidence of HLA-B7 protein expression, and no adverse reactions were observed. Vical is producing *Allovectin-7* under GMP conditions and supplying it to Dr. Nabel for further testing in cancer patients in a clinical protocol allowed by the NIH Recombinant Advisory Committee (RAC) and by the FDA. Vical has independently enlisted three additional trial sites to expand the full safety assessment and dosing optimization of *Allovectin-7*. An additional improvement in *Allovectin-7* is use of DMRIE/DOPE cationic lipid to permit higher DNA concentration to be formulated. These three clinical protocols for treating 15 patients each have been allowed by the RAC to proceed at the Mayo Clinic in patients with advanced colorectal carcinoma, at the University of Chicago in renal cell carcinoma, and at the Arizona Cancer Center in malignant melanoma. A Vical-sponsored IND application in support of all three protocols was allowed by the FDA in February 1994. Vical initiated these phase I/II trials in the second quarter of 1994.

Vical is currently planning a second gene therapy protocol for the direct injection of plasmid encoding interleukin-2, IL-2, for solid tumors. Interleukin-2 protein therapy has been approved by the FDA for systemic administration in patients with renal cell carcinoma. However, this therapy requires very high doses of IL-2 with a high incidence of toxicity. Vical is optimistic that the direct injection of plasmid/lipid complex into tumors, with the subsequent local expression of IL-2, will result in tumor reduction by stimulating the immune system but will avoid the high incidence of toxicity seen with systemic IL-2 delivery. Vical also intends to examine the feasibility of tumor vaccines as well as a therapy for B-cell lymphoma using gene based anti-idiotype antibody production.

In close collaboration with their partners at Merck & Co., Vical scientists provided the first unequivocal demonstration in relevant models that naked DNA gene transfer[44] may provide novel vaccine product opportunities. These data were published in March 1993 in an article in *Science*.[43] Merck & Co. expanded the collaboration in mid-1994 by acquiring exclusive option rights to license Vical's genetic immunization technology for a preventive vaccine against tuberculosis, exercising its options to Vical's technology for use in vaccines against hepatitis C and papilloma viruses (joining influenza and HIV as vaccine targets under license) and extending for an additional year its option rights for vaccines against hepatitis B and herpes simplex.

In the latter half of 1993, Vical concluded two additional agreements for specific indications in congenital disorders granting options to exclusive royalty-bearing licenses — the first with Genzyme Corp. in the field of cystic fibrosis, a frequent and debilitating congenital disorder, and the second with Baxter Healthcare Corp. in the field of hemophilia, a life-threatening blood clotting disorder. A clinical trial from Eric Sorscher at the University of Alabama, to treat the nasal mucosa of cytstic fibrosis patients with the CFTR gene formulated with DMRIE/DOPE, received approval from the recombinant advisory committee in 1994,[49,50] and the results of a similar clinical trial from England using DC-cholesterol were recently published.[45,46,48]

Avoiding the use of viral components in gene therapy eliminates a number of regulatory and safety concerns. It may further provide avenues for product manufacturing with cost advantages. Direct DNA administration could make Vical's approach comparable to the conventional drug business in that the appropriate DNA in a vial can be used for all patients with a certain indication. We anticipate that during the next 10 years the pharmaceutical development and application of DNA drugs will become as commonplace as recombinant protein drugs used today.

REFERENCES

1. **Felgner, P.L. and Ringold, G.M.,** (1989) Cationic liposome mediated transfection. *Nature* 337 (6205), 387-388.
2. **Felgner, P.L., Holm, M., and Chan, H.,** (1989) Cationic liposome mediated transfection. *Proc. West Pharmacol. Soc.* 32, 115-121.
3. **Felgner, P.L., Gadek, T.R., Holm, M., Roman, R., Chan, H.S., Wenz, M., Northrop, J.P., Ringold, G.M., and Danielsen, H.,** (1987) Lipofection: a highly efficient, lipid-mediated DNA-transfection procedure. *Proc. Natl. Acad. Sci. U.S.A.* 84 (21), 7413–7417.
4. **Felgner, P.L.,** (1990) Particulate systems and polymers for *in vitro* and *in vivo* delivery of polynucleotides. *Adv. Drug Delivery Rev.* 5, 167–187.
5. **Malone, R.W., Felgner, P.L., and Verma, I.M.,** (1989) Cationic liposome-mediated RNA transfection. *Proc. Natl. Acad. Sci. U.S.A.* 86 (16), 6077-6081.
6. **Chiang, M.-Y., Chan, H., Zounes, M.A., Freier, S.M., Lima, W.F. and Bennett, C.F.,** (1991) *J. Biol. Chem.* 266, 18171–18192.
7. **Bennett, C.F., Chiang, M.Y., Chan, H., Shoemaker, J.E., and Mirabelli, C.K.,** (1992) *Mol. Pharmacol.* 41 (6), 1023-1033.
8. **Rose, J.K., Buonocore, L., and Whitt, M.A.,** (1991) Biotechniques 10 (4), 520-525.
9. **Leventis, R. and Silvius, J.R.,** (1990) Interactions of mammalian cells with lipid dispersions containing novel metabolizable cationic amphiphiles. *Biochim. Biophys. Acta* 1023 (1), 124-132.
10. **Farhood, H., Bottega, R., Epand, R.M., and Huang, L.,** (1992) *Biochim. Biophys. Acta* 1111, 239–246.
11. **Behr, J., Demeneix, B., Loeffler, J., and Perez–Mutul, J.,** (1989) *Proc. Natl. Acad. Sci. U.S.A.* 86, 6982–6986.
12. **Legendre, J.-Y. and Szoka, F.C.,** (1992) Delivery of plasmid DNA into mammalian cell lines using pH-sensitive liposomes: comparison with cationic liposomes. *Pharm. Res.* 9 (10), 1235– 1242.
13. **Nabel, E.G., Plautz, G., and Nabel, G.J.,** (1990) *Science* 249 (4974), 1285-1288.
14. **Jiao, S., Acsadi, G., Jani, A., Felgner, P.L., and Wolff, J.A.,** (1992) *Exp. Neurol.* 115, 400–413.
15. **Felgner, J.H., Kumar, R., Sridhar, C.N., Wheeler, C.J., Tsai, Y. J., Border, R., Ramsey, P., Martin, M., and Felgner, P.L.,** (1994) Enhanced gene delivery and mechanism studies with a novel series of cationic lipid formulations. *J.Biol.Chem.* 269, 2550–2561.
16. **Hawley–Nelson, Ciccarone V., Gebeyehu, G., Jessee, J., and Felgner, P.L.,** (1993) LipofectaAMINE Reagent: a new, higher efficiency polycationic liposome transfection reagent. *Focus* 15, 73–79.
17. **Zhou, X. and Huang, L.,** (1994) DNA transfection mediated by cationic liposomes containing lipopolylysine: characterization and mechanism of action. *Biochim. Biophys. Acta* 1189, 195–203.
18. **Yaki, K., Noda, H., Kurono, M., and Ohishi, N.,** (1993) *Biochem. Biophys. Res. Commun.* 196, 1042–1048.
19. **Pinnaduwage, P., Schmitt, L., and Huang, L.,** (1989) *Biochim. Biophys. Acta* 985, 33–37.

20. **Israelachvili, J.N., Mitchell, D.J., and Ninham, B.W.**, (1977) Theory of self-assembly of lipid bilayers and vesicles. *Biochim. Biophys. Acta* 470, 185–201.
21. **Duzgunes, N.**, Membrane Fusion. *Subcell. Biochemistry* 11, 195–286 (1985).
22. **Martin, F.J. and McDonald, R.C.**, Lipid vesicle–cell interactions *J. Cell Biol.* 70, 494–505; 506–514; 515–526.
23. **Rupert, L.A.M., Hoekstra, D., and Engberts, J.B.F.N.**, (1985) Fusogenic behavior of didodecyldimethylammonium-bromide bilayer vesicles. *J. Am. Chem. Soc.* 107, 2628–2631.
24. **Nichols, J.W. and Pagano, R.E.**, (1981) Kinetics of soluble lipid monomer diffusion between vesicles. *Biochemistry* 20, 2783–2789.
25. **Doody, M.C., Pownal, H.G., Kao, Y.F., and Smith, L.C.**, (1980) Mechanism and kinetics of transfer of a fluorescent fatty acid between single-walled phosphatidylcholine vesicles. *Biochemistry* 19, 108–116.
26. **Steger, L.D. and Desnick, R.J.**, Enzyme therapy. VI. (1977) Comparative *in vivo* fates and effects on lysosomal integrity of enzyme entrapped in negatively and positively charged liposomes *Biocheim. Biophys. Acta* 464, 530–546.
27. **Yoshihara, E. and Nakae, T.**, (1986) Cytolytic activity liposomes containing stearylamine. *Biochim. Biophys. Acta* 854, 93–101.
28. **Hannun, Y.A. and Bell, R.M.**, (1989) Functions of sphingolipids and sphingolipid breakdown products in cellular regulation. *Science* 243, 500–507.
29. **Carmona–Ribeiro, A.M., Yoshida, L.S., and Chaimovich, H.**, (1985) Salt effects on the stability of dioctadecyldimethylammonium chloride and sodium dihexadecyl phosphate vesicles. *J. Phys. Chem.* 89, 2928–2933.
30. **Duzgunes, N. and Felgner, P.L.**, (1993) Intracellular delivery of nucleic acids and transcription factors by cationic liposomes. *Methods Enzymol.* 221, 303-6.
31. **Stamatatos, L., Leventis, R., Zuckermann, M.J., and Silvius, J.R.**, (1988) Interactions of cationic lipid vesicles with negatively charged phospholipid vesicles and biological membranes. *Biochemistry* 27, 3917–3925.
32. **Fraley, R., Subramani, S., Berg, P., and Papahadjopoulos, D.**, (1980) Introduction of liposome–encapsulated SV40 DNA into cells. *J. Biol. Chem.* 255, 10431–10435.
33. **Ostro, M.J., Lavelle, D., Paxton, W., Matthews, B., and Giacomoni, D.**, (1980) Parameters affecting the liposome-mediated insertion of RNA into eucaryotic cells *in vitro*. *Arch. Biochem. Biophys.* 201, 392-402.
34. **Fraley, R. and Papahadjopoulos, D.**, (1982) Liposomes: The development of a new carrier system for introducing nucleic acid into plant and animal cells. *Curr. Top. Microbiol. Immunol.* 96, 171–191.
35. **Fraley, R., Straubinger, R.M., Rule, G., Springer, E.L., and Papahadjopoulos, D.**, (1981) Liposome–mediated delivery of deoxyribonucleic acid to cells: enhanced efficiency of delivery related to lipid composition and incubation conditions. *Biochemistry* 20, 6978–6987.
36. **Straubinger, R.M. and Papahadjopoulos, D.**, (1983) Liposomes as carriers for intracellular delivery of nucleic acids. *Methods Enzymol.* 101, 512–527.
37. **Cotten, M., Baker, A., Saltik, Mediyha, S., Wagner, E., and Buschle, M.**, (1994) Lipopolysaccharide is a frequent contaminant of plasmid DNA preparations and can be toxic to primary cells in the presence of adenovirus. Gene Therapy 1, 1–8.
38. **Takeshita, S., Gal, D., Leclerc, G., Pickering, J.G., Riesen, R., Weir, I., and Isner, J.M.**, (1994) Increased gene expression after liposome–mediated arterial gene transfer associated with intimal smooth muscle cell proliferation. *In vitro* and *in vivo* findings in a rabbit model of vascular injury. *J. Clin. Invest.* 93, 652–661.
39. **Greber, U.F., Willetts, M., Webster, P., and Helenius, A.**, (1993) Stepwise dismantling of adenovirus 2 during entry into cells. *Cell* 75, 477–486.
40. **Gao, X. and Huang, L.**, (1993) Cytoplasmic expression of a reporter gene by co-delivery of T7 RNA polymerase and T7 promoter sequence with cationic liposomes. *Nucleic Acids Res.* 21, 2867-2872.
41. **Duzgunes, N., Goldstein, J.A., Friend, D.S., and Felgner, P.L.**, (1989) Fusion of liposomes containing a novel cationic lipid, N-[2,3-(dioleyloxy)propyl]-N,N,N-trimethylammonium: induction by multivalent anions and asymmetric fusion with acidic phospholipid vesicles. Biochemistry 28(23), 9179-9184.
42. **Sternberg, B., Sorgi, F.L., and Huang, L.**, (1994) New structures in complex formation between DNA and cationic liposomes visualized by freeze–fracture electron microscopy. *FEBS Lett.* 356, 361–366.
43. **Ulmer, J.B., Donnelly, J.J., Parker, S.E., Rhodes, G.H., Felgner, P.L., Dwarki, V.J., Gromkowski, S.H., Deck, R.R., DeWitt, C.M., Friedman, A., et al.**, (1993) Heterologous protection against influenza by injection of DNA encoding a viral protein. *Science* 259(5102), 1745-1749.
44. **Wolff, J.A., Malone, R.W., Williams, P., Chong, W., Acsadi, G., Jani, A., and Felgner, P.L.**, (1990) Direct gene transfer into mouse muscle *in vivo*. *Science* 247, 1465-1468.
45. **Hyde, S.C., Gill, D.R., Higgins, C.F., Trezise, A.E.O., MacVinish, L.J., Cuthbert, A.W., Ratcliff, R., Evans, M.J., and Colledge, W.H.**, (1993) Correction of the ion transport defect in cystic fibrosis transgenic mice by gene therapy. *Nature*, 362, 250–255.
46. **Alton, E.W.F.W., Middleton, P.G., Caplen, N.J., Smith, S.N., Steel, D.M., Munkonge, F.M., Jeffrey, P.K., Ged, D.M., Hart, S.L., Williamson, R., Fasold, K.I., Miller, A.D., Dickinson, P., Stevenson, B.J., McLachlan, G., Dorin, J.R., and Porteous, D.J.**, (1993) Non-invasive liposome–mediated gene delivery can correct the ion transport defect in cyctic fibrosis mutant mice. *Nature/Genetics* 5, 16–23.

47. **Nabel, G.J., Nabel, E.G., Yang, Z.-Y., Fox, B.A., Plautz, G.E., Gao, Z., Huang, L., Shu, S., Gordon, D., and Chang, A.E.,** (1993) Direct gene transfer with DNA-liposome complexes in melanoma: expression, biologic activity, and lack of toxicity in humans. *Proc. Natl. Acad. Sci. U.S.A.* 90, 11307–11311.
48. **Caplen, N.J., Alton, E.W.F.W., Middleton, P.G., Dorin, J.R., Stevenson, B.J., Gao, X., Durham, S.R., Jeffrey, P.K., Hodson, M.E., Coutelle, C., Huang, L., Porteous, D.J., Williamson, R., and Geddes, D.M.,** (1995) Liposome–mediated CFTR gene transfer to the nasal epithelium of patients with cystic fibrosis. *Nature Medicine* 1, 39–46.
49. **Sorscher, E.J. and Logan, J.,** (1994) Gene therapy for cyctic fibrosis using cationic liposome mediated gene transfer: a phase I trial of safety and efficacy in the nasal airway. *Hum. Gene Ther.* 5, 1259–1299.
50. **Logan, J., Bebok, Z., Walker, L., Felgner, P., Burgess, S., Shaw, W., Siegel, G., Frizzel, R., Dong, J., Matalon, S., Duvall, M., and Sorscher, E.J.,** Cationic lipids for CFTR gene transfer. *Gene Therapy* (in press).
51. **San, H., Yang, Z.Y., Pompili, V.J., Jaffe, M.L., Plautz, G.E., Xu, L., Felgner, J.H., Wheeler, C.J., Felgner, P.L., Gao, X., et al.,** (1993) Safety and short-term toxicity of a novel cationic lipid formulation for human gene therapy. *Hum. Gene Ther.* 4, 781-788.
52. **Nabel, G.J., Chang, A.E., Nabel, E.G., Plautz, G.E., Ensminger, W., Fox, B.A., Felgner, P., Shu, S., and Cho, K.,** (1994) Immunotherapy for cancer by direct gene transfer into tumors. *Hum. Gene Ther.* 5, 57-77.
53. **Jiao, S., Acsadi, G., Jani, A., Felgner, P.L., and Wolff, J.A.,** (1992) Persistence of plasmid DNA and expression in rat brain cells *in vivo*. *Exp. Neurol.* 115, 400-413.
54. **Felgner, P.L.,** (1993) Genes in a bottle [editorial] *Lab. Invest.* 68, 1-3.
55. **Felgner, P.L. and Rhodes, G.,** (1991) Gene therapeutics. *Nature* 349, 351-352.

Chapter 5

Gene Transfer to Plants by Lipofection and Electroporation

Mauro Maccarrone, Antonello Rossi, and Alessandro Finazzi Agrò

CONTENTS

I. INTRODUCTION

Much knowledge in plant biology at the molecular level resulted from the development of efficient methods for transferring foreign DNA into plant cells. Only a small portion of the introduced DNA becomes stably integrated into the plant chromosomes and is inherited as single Mendelian trait.[1] Nevertheless, the extrachromosomal DNA is transcriptionally active, thus forming the basis of extremely useful transient gene expression assays.[2] Detailed information has been gained concerning gene structure and regulation, elucidating the features of promoters, enhancers, silencers, terminators, organ/development-specific regulators and light-, temperature- or stress-responsive elements, as well as trans-acting factors. A great deal of interest exists in gene transfer and manipulation techniques for the predictable modification of plant genomes, in particular those of economically or nutritionally important plants. In recent years, a large number of gene transfer methods in plants have been developed.[3,4] The most efficient way of transferring genes to a wide variety of dicotyledonous, and some monocotyledonous, plants is the use of *Agrobacterium tumefaciens*.[5] Unfortunately, the *Agrobacterium* host range excludes important crops, especially the graminaceous monocotyledonous cereals.[6] One important exception is maize, which can receive DNA (T-DNA) from *Agrobacterium* by either agroinfection[7] or co-cultivation.[8] Plant systems that are not amenable to *Agrobacterium*-mediated transformation can be subjected to direct gene transfer procedures,[9,10] in which DNA uptake by protoplasts can be enhanced by polymers such as polyethylene glycol (PEG).[11,12] DNA transfer to intact cells or tissues can be achieved by viral infection,[13,14] macroinjection into floral tillers,[15] microinjection into proembryos,[16] UV laser micropuncture[17] or ultrasonication.[18,19] Furthermore, foreign genes can be taken up during incubation of dry embryos[20] or pollen grains[21] in DNA solutions, or they can be guided to the zygote through the pollen tube pathway.[22] Over the past years, particle bombardment (biolistics)[23-25] has evolved into a useful tool for molecular biologists, allowing direct gene transfer to a broad range of plant cells and tissues.[26] Each of the aforementioned techniques has its own spectrum of advantages and disadvantages, including efficiency, toxicity, technical difficulty, equipment needs and specificity. One major distinction between gene transfer techniques is whether they can be applied to walled cells or require naked cells (*i.e.*, protoplasts) as recipients. The main drawback of protoplast technology is the regeneration of whole plants,[27] although several stably transformed plant cells have been obtained from protoplasts.[2] Transfer of DNA into intact cells bypasses the regeneration problems, but stable transformation efficiencies can be very low.[2,3] An exciting perspective in gene transfer is the transformation of plant cell organelles, *e.g.*, chloroplasts[28,29] and mitochondria.[30] These various recipients can be targets of liposome-mediated transfection[31] or electroporation,[32] established methods for the production of transgenic plants.

II. LIPOFECTION

Liposomes have been used extensively to entrap a large variety of molecules, including macromolecules of biological interest, and to deliver them into cells by either fusion with the plasma membrane or endocytosis-like processes. Liposome-mediated gene transfer offers several advantages, e.g., protection

0-8493-4013-6/96/$0.00+$.50

of entrapped DNA from nuclease activity, targeting to specific cells and delivery of genes into a cell network through walls and plasmodesmata.[31] In plants, mostly protoplasts have been used for liposome-mediated delivery, other cells and tissues being occasionally employed. The main applications of liposome-mediated delivery to plants are summarized in Table 1.

Table 1 Liposome-Mediated Delivery to Plants

Recipient Cells	Application	Ref.
Tomato protoplasts	Uptake of fluorescent probes	33
Tobacco protoplasts	Transient transformation	34–36
	Stable transformation	37, 38
	Viral RNA-mediated infection	39–42
	Liposome-cell interaction	43
Petunia protoplasts	Viral RNA-madiated infection	42
Maize protoplasts	Stable transformation	44
Pea pollen grains	Stable transformation	45
Carrot protoplasts	Uptake of labeled DNA	46–48
	Uptake of fluorescent probes	49
Lentil protoplasts	Transient transformation	50
Catharanthus protoplasts	Uptake of fluorescent probes	51
Vinca protoplasts	Liposome-cell interaction	52

Most liposome compositions used for macromolecular transfer to plants were negatively charged vesicles, which interact with the cell membrane in the presence of cations or polycations.[53] More recently, highly efficient transfection of eukaryotic cells has been achieved using cationic liposomes, a process called lipofection.[54,55] Basically, anionic vesicles are loaded inside with the DNA molecules to be delivered and they either fuse to or are engulfed by the recipient cells (Figure 1A). On the other hand, cationic liposomes form a complex with DNA, which stays outside the vesicles; such a complex has a net positive charge and spontaneously interacts with the negatively charged target cell surface, increasing membrane permeability to the exogenous gene (Figure 1B). Stearylamine-based liposomes were used for the first cationic liposome-mediated delivery to plant protoplasts.[33] In fact, in 1978 Cassells reported the ability of lecithin/stearylamine vesicles to deliver fluorescein diacetate to tomato protoplasts, showing the migration of liposomes from the plasma membrane to the central vacuole within 15 h from mixing cells and vesicles (in a 1:1 ratio).[33] Later on, Cutler et al.[51] confirmed the ability of stearylamine-containing liposomes to deliver their content to naked plant cells and gave a quantitative estimate of the delivery process and its constraints. In 1987, Ahokas succeeded in transfecting pea pollen grains by means of positively charged, oligolamellar liposomes.[45] The cationic vesicles, prepared from a mixture of phosphatidylcholine, stearylamine and β-sitosterol (4:1.8:0.9), were found at a low frequency in living, germinated pollen grains and pollen tubes, yielding a putative, stable transfectant with pleiotropic effects on the phenotype.[45] Tobacco and petunia protoplasts have been lipofected by Ballas et al., who developed cationic liposomes composed of phosphatidylcholine, cholesterol and the quaternary ammonium detergent DEBDA (diisobutylcresoxyethoxyethyldimethylbenzylammonium, hydroxyl form).[42] These vesicles were able to transfect about 30% of the recipient cells under optimal conditions, RNA–liposome complexes resulting more efficient than liposomes bearing the RNA inside.[42] More recently, the liposome formulation originally developed by Felgner et al.[54] (termed Lipofectin) for mammalian cell transfection, has been used to efficiently transform maize[44] and tobacco[38] protoplasts. Antonelli and Stadler[44] obtained stable kanamycin-resistant secondary transformants (frequency = 6%) from naked maize cells which had been incubated for 6 h at 28°C with a Lipofectin-DNA mixture. Analogously, Sporlein and Koop[38] achieved transient expression and stable transformation of tobacco protoplasts treated with the Lipofectin cationic liposomes, the transformation frequencies being comparable to those found with other techniques.[56] Parameters affecting lipofection efficiency have been investigated by our group, using lentil (*Lens culinaris*) protoplasts as a model system and cationic vesicles made of dipalmitoylphosphatidylcholine and stearylamine (9:1).[50] Transmission electron microscopy showed that the addition of circular DNA plasmids of different lengths to the liposomes led to the formation of vesicle clusters around the DNA filament, the cluster dimension being dictated by the ratio DNA/lipid. Aggregates of average diameter = 2.5 μm were obtained when the ratio DNA/phospholipids was 20 μg/μmol, whereas the addition of more than 20 μg DNA per μmol phospholipids yielded aggregates too big to be engulfed

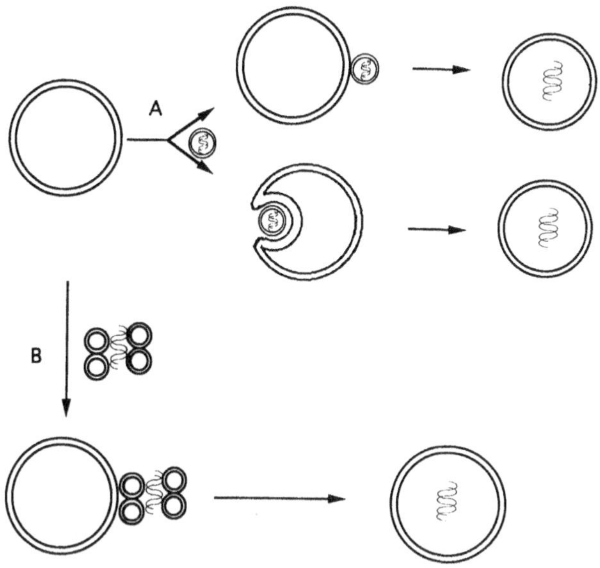

Figure 1 Scheme of the liposome-cell interactions. A) Negatively charged vesicles, loaded with DNA, either fuse to or are engulfed by the recipient cell. B) Positively charged liposomes form a complex with DNA molecules and spontaneously interact with the cell surface.

by the cell. Interestingly, the plasmids' length had no influence on their binding to the liposomes. Lentil protoplasts were successfully transfected by using plasmid pBI 221 (which carries the β-glucuronidase reporter gene), vehiculated by the stearylamine-containing liposomes.[50] The lipofection efficiency was com parable to that of the PEG procedure, but with the advantage of less severely affecting cell survival.[50] In the last years, new cationic lipid reagents have been synthesized which promise to be more efficient agents of lipofection.[57] This class of polynucleotide delivery reagents is referred to as cytofectins[58] and should broaden the potentiality of the lipofection strategy for plant engineering.

III. ELECTROPORATION

Electroporation has become a popular tool for introducing exogenous DNA into plant protoplasts. Many species have been transformed either transiently or permanently, shedding light on several features of gene regulation and expression.[59] The physicochemical properties of direct gene transfer in cells by electroporation and the many biochemical applications of cell electropermeabilization have been recently reviewed.[60,61] The utilization of electroporation for the genetic manipulation of plants has grown during the last decade.[62] Table 2 summarizes some useful information on the applications of electroporation to plant engineering. In Figure 2 a scheme is presented for the permeability induced by pulsed electric fields in cell plasma membranes, a process which can be interpreted according to different theories.[85]

Besides the wide utilization of electroporation for transient gene expression experiments, permanent transformation and viral RNA-mediated infection of plant protoplasts (Table 2), the technique has been increasingly used for antisense RNA-mediated inhibition of gene expression. Plants were the first multicellular organisms in which endogenous genes were successfully down-regulated by antisense RNAs.[86] Ecker and Davis succeeded in blocking the expression of target genes in naked carrot cells by electric field-mediated transfer of chimeric plant genes.[87] Analogously, Vaucheret et al. used electroporation to inhibit nitrite reductase activity in tobacco cells by expression of antisense RNA,[88] whereas Carneiro et al. modulated gene expression in tobacco by co-electroporating suppressor tRNAs.[89] The possibility of inhibiting enzyme activity in plants by electrotransferring inhibitory antibodies has been investigated by us.[79] Anti-lipoxygenase (EC 1.13.11.12) monoclonal antibodies were transferred into lentil protoplasts by delivering exponentially decaying pulses to the cells, with an average time constant of 1.5 ms. The transferred immunoglobulins retained their functional and structural integrity and were able to inhibit the intracellular target enzyme. A linear relationship was found between the inhibition of lipoxygenase activity and the amount of incorporated antibodies. Moreover, the lipoxygenase inhibition correlated well with the increase of cell viability, suggesting that the enzyme might be involved in the control of

Table 2 Electroporation-Mediated Delivery to Plants

Recipient cell	Application	Ref.
Tobacco protoplasts	Transient and stable transformation	63–65
	Viral RNA-mediated infection	66–69
	Gene rescue	70
Carrot protoplasts	Transient transformation	71–74
Maize protoplasts	Transient transformation	71, 75
Soybean protoplasts	Stable transformation	76, 77
Tomato protoplasts	Replication footprint analysis	78
Lentil protoplasts	Antibody transfer	79
Rice, wheat and sorghum protoplasts	Transient transformation	80
Beta protoplasts	Viral RNA-mediated infection	69
Solanum protoplasts	Transient transformation	81
Various monocot and dicot plant protoplasts	Transient and stable transformation	82–84

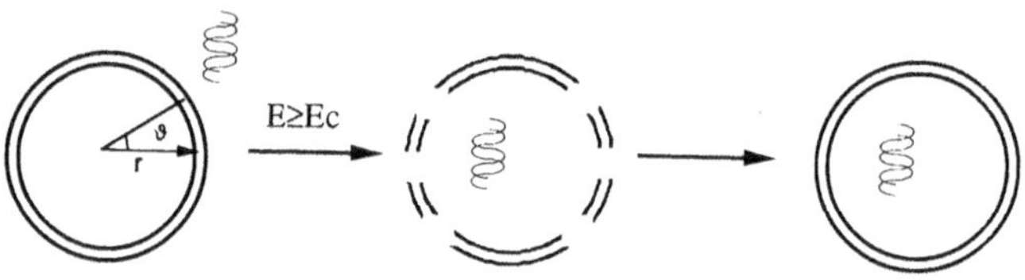

$$\Delta\Psi = f\ r\ E\ \cos\vartheta$$

Figure 2 Schematic representation of the effect of pulsed electric fields on cells at strenghts higher than the critical field (Ec) for electroporation. The induced membrane potential ($\Delta\Psi$) of the cell membrane at a given point is proportional to i) the cell shape, defined by a factor f (equal to 1.5 for a spherical cell), ii) the cell radius, r, iii) the external field (E) and iv) cos ϑ, where ϑ is the angle between the given point and the field direction. The Ec value is different from cell to cell.

membrane integrity. These results indicate that antibody electrotransfer could be instrumental for studying basic processes of plant physiology, being complementary, or alternative, to the RNA antisense technology.

A promising application of electropermeabilization is the direct introduction of foreign genes into intact walled plant cells, a technique referred to as electroinjection.[90] This method offers the advantage of bypassing the difficulties inherent to protoplast preparation and plant regeneration and allows the study of the expression of factor-regulated and tissue-specific chimeric genes. Electroinjection has been successfully used in tobacco,[90] sugar beet[91] and rice[92] cells. Recently, it has been extended to genetic engineering of plant tissues from different sources, and has been termed tissue electroporation.[93,94]

IV. COMBINED USE OF LIPOFECTION AND ELECTROPORATION

A classical example of the combination of the advantages of both liposome technology and electroporation has been described by Machy et al.[95] They prepared protein A-coated liposomes, which encapsulated a resistance gene and were able to selectively bind a defined cell type. Transfection of target cells bound to protein A-coated vesicles was achieved by means of electric fields. Gene transfer efficiency with electroporation of targeted liposomes was comparable to transfection efficiency of electrotransferred free plasmids, although in the latter case cell selectivity was not possible. The combined use of electroporation and targeted liposomes,[95] immunoliposomes[96] and chimerasomes[97] as cell-specific gene transfer strategy can be extended to plant cells.[4] For instance, tobacco protoplasts have been transfected by a combination of cationic liposomes (Lipofectin) and electroporation. Sporlein and Koop reported higher efficiencies of transformation after electroporation in the presence of Lipofectin, suggesting an additive effect of both treatments.[38] The combined use of cationic liposomes and electric fields for DNA delivery to plants has been investigated also by us, using lentil protoplasts as a model system.[98] Cationic liposomes made of dipalmitoylphosphatidylcholine and stearylamine (9:1)[50] were able to deliver plasmid DNAs to the cells with the same efficiency as electroporation at 500 μF and different field strengths. The combination of the two techniques yielded transient expression levels some 50% higher than those

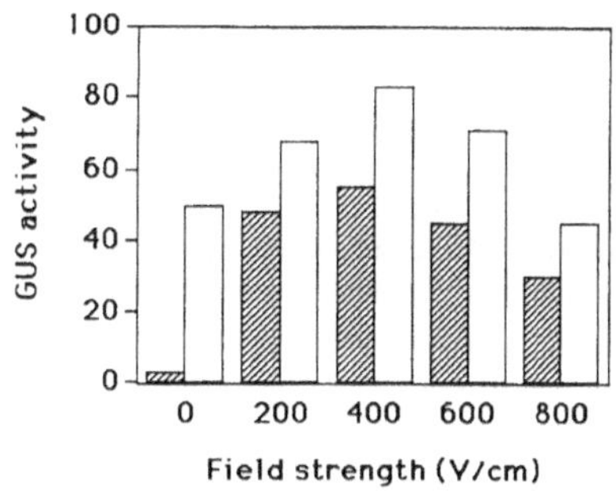

Figure 3 Effect of the combination of lipofection and electroporation on the transient expression of the β-glucuronidase (GUS) reporter gene. Lentil protoplasts ($1.0x10^6$) were electroporated at 500 μF and different field strengths, in the presence of 20 μg pBI 221 and 50 μg carrier DNA (hatched bars), or after 30 min of incubation, at room temperature, with pBI 221-loaded liposomes (20 μg pBI 221/μmol phospholipids; 1 μmol phospholipids/test), without carrier DNA (empty bars). After 24 h of culture, GUS activity was determined[99] and expressed as pmol 4-methylumbelliferone produced per min per mg protein.

recorded after either treatment alone (Figure 3). On the other hand, the combination of the two procedures reduced cell survival,[100] because of the physical damage due to the electric shock and the chemical toxicity of stearylamine.[50,98] Furthermore, it was found that the presence of carrier DNA, while doubling the transfection efficiency of electroporation alone, was detrimental for the interaction of liposomes with cells during electric shock, yielding a 4-fold decrease of the transient expression levels. This finding is explained by recalling that cationic liposomes tend to cluster to form aggregates too big to be engulfed by plant cells, when they are incubated with more than 20 μg DNA per μmol phospholipids.[50] Thus, carrier DNA should be avoided in procedures involving DNA/liposome complexes, because it might lead to the formation of useless macroaggregates. The mechanism underlying the improvement of gene transfer efficiency by combining the use of liposomes with electric fields is hard to interpret, because of the many uncertainties still existing on the mode of action of lipofection[57,58] and of electroporation[101] alone. Finally, it must be pointed out that lipofection can be also combined with other transfection methods, like the polyethylene glycol (PEG)[11,12] procedure. Such combinations, successful so far in prokaryotic systems,[102] might improve lipofection as a tool for genetic engineering of plants as well.

V. CONCLUDING REMARKS

A wide array of transformation techniques is available, but there is to date no universal, ideal, generally applicable method for gene transfer to plants. This is indeed the stumbling block towards an easy production of transgenic plants. Nevertheless, the above-mentioned features of liposome technology and electroporation suggest that important outcomes may be anticipated by the combination of these techniques for plant transformation. In particular, the potentialities of the co-electrotransfer of proteins and DNA molecules into recipient cells deserve attention, because they may elucidate basic processes like trans-activation of gene expression[103] and modulation of two (or more) cellular function at the same time.[104] In this respect, electric field-aided lipofection can become the procedure of choice, because of the ability of liposomes to vehiculate genes,[37,38] proteins[105,106] or both,[95-97] and the efficacy of electroporation in promoting cell/liposome interactions.[98] The possibility of electrotransferring monoclonal antibodies into plant cells,[79] coupled to gene transfer by lipofection, might open new avenues to genetic manipulation of plants, leading to crop improvement.[107] Exciting outcomes may be foreseen in this fast-growing, promising field of plant science.

REFERENCES

1. **Potrykus, I., Saul, M. W., Petruska, J., Paszkowski, J., and Shillito, R. D.,** Direct gene transfer to cells of a graminaceous monocot, *Mol. Gen. Genet.*, 119, 183, 1985.
2. **Steinbiss, H.-H. and Davidson, A.,** Transient gene expression of chimeric genes in cells and tissues of crops, in *Subcellular Biochemistry*, Vol. 17, Biswas, B. B. and Harris, J. R., Eds., Plenum Press, New York, 1991, 143.
3. **Potrykus, I.,** Gene transfer to plants: assessment and perspectives, *Physiol. Plant.*, 79, 125, 1990.
4. **Sawahel, W. A. and Cove, D. J.,** Gene transfer strategies in plants, *Biotech. Adv.*, 10, 393, 1992.

5. **Fraley, R. T., Horsch, R. B., Matzke, A., Chilton, M. D., and Sanders, P. R.,** *In vitro* transformation of *Petunia* cells by an improved method of co-cultivation with A. tumefaciens, *Plant Mol. Biol.*, 3, 371, 1984.
6. **De Cleene, M.,** The susceptibility of monocotyledons to *Agrobacterium tumefaciens, Phytopathol.* Z., 113, 81, 1985.
7. **Grimsley, N. H., Hohn, B., Hohn, T., and Walden, R.,** Agroinfection, an alternative route for viral infection of plants using the Ti plasmid, *Proc. Natl. Acad. Sci. U.S.A.*, 83, 3282, 1986.
8. **Shen, W.-H., Escudero, J., Schlappi, M., Ramos, C., Hohn, B., and Koukolikova-Nicola, Z.,** T-DNA transfer to maize cells: histochemical investigation of β-glucuronidase activity in maize tissues, *Proc. Natl. Acad. Sci. U.S.A.*, 90, 1488, 1993.
9. **Paszkowski, J., Shillito, R. D., Saul, M., Mandak, V., Hohn, T., Hohn, B., and Potrykus, I.,** Direct gene transfer to plants, *EMBO J.*, 3, 2717, 1984.
10. **Potrykus, I., Shillito, R. S., Saul, M. W., and Paszkowski, J.,** Direct gene transfer. State of the art and future potential, *Mol. Biol. Rep.*, 3, 117, 1985.
11. **Negrutiu, I., Shillito, R., Potrykus, I., Biasini, G., and Sala, F.,** Hybrid genes in the analysis of transformation conditions. I. Setting up a simple method for direct gene transfer in plant protoplasts, *Plant Mol. Biol.*, 8, 363, 1987.
12. **Shimamoto, K., Terada, R., Izawa, T., and Fujimoto, H.,** Fertile transgenic rice plants regenerated from transformed protoplasts, *Nature*, 338, 274, 1989.
13. **French, R., Janda, M., and Ahlquist, P.,** Bacterial gene inserted in an engineered RNA virus: efficient expression in monocotyledonous plant cells, *Science*, 231, 1294, 1986.
14. **Donson, J., Kearney, C. M., Hilf, M. E., and Dawson, W. O.,** Systemic expression of a bacterial gene by a tobacco mosaic virus-based vector, *Proc. Natl. Acad. Sci. U.S.A.*, 88, 7204, 1991.
15. **De la Pena, A., Lorz, H., and Schell, J.,** Transgenic rye plants obtained by injecting DNA into young floral tillers, *Nature*, 325, 274, 1987.
16. **Neuhaus, G., Spangenberg, G., Mittelsten Scheid, O., and Schweiger, H. G.,** Transgenic rapeseed plants obtained by microinjection of DNA into microspore-derived embryoids, *Theor. Appl. Genet.*, 75, 30, 1987.
17. **Weber, G., Monajembashi, S., Greulich, K. O., and Wolfrum, J.,** Injection of DNA into plant cells with a UV laser microbeam, *Naturwissenschaften*, 75, 35, 1988.
18. **Zhang, L.-J., Cheng, L.-M., Xu, N., Zhao, N.-M., Li, C.-G., Yuan, J., and Jia, S.-R.,** Efficient transformation of tobacco by ultrasonication, *Bio/Technology*, 9, 996, 1991.
19. **Joersbo, M. and Brunstedt, J.,** Sonication: a new method for gene transfer to plants, *Physiol. Plant.*, 85, 230, 1992.
20. **Topfer, R., Gronenborn, B., Schell, J., and Steinbiss, H.-H.,** Uptake and transient expression of chimeric genes in seed-derived embryos, *Plant Cell*, 1, 133, 1989.
21. **Ohta, Y.,** High efficiency genetic transformation of maize by a mixture of pollen and exogenous DNA, *Proc. Natl. Acad. Sci. U.S.A.*, 83, 715, 1986.
22. **Luo, Z. and Wu, R.,** A simple method for the transformation of rice via the pollen-tube pathway, *Plant Mol. Biol. Rep.*, 6, 165, 1988.
23. **Klein, T. M., Wolf, E. D., Wu, R., and Sanford, J. C.,** High-velocity microprojectiles for delivering nucleic acids into living cells, *Nature*, 327, 70, 1987.
24. **Klein, T. M., Harper, E. C., Svab, Z., Sanford, J. C., and Fromm, M. E.,** Stable genetic transformation of intact *Nicotiana* cells by the particle bombardment process, *Proc. Natl. Acad. Sci. U.S.A.*, 85, 8502, 1988.
25. **Sanford, J. C.,** The biolistic process, *Trends Biotechnol.*, 6, 299, 1988.
26. **Christou, P.,** Genetic transformation of crop plants using microprojectile bombardment, *Plant J.*, 2, 275, 1992.
27. **Steinbiss, H.-H. and Davidson, A.,** Genetic manipulation of plants: from tools to agronomical applications, *Sci. Prog.*, 73, 147, 1989.
28. **Blowers, A. D., Bogorad, L., Shark, K., and Sanford, J. C.,** Studies on Chlamydomonas chloroplast transformation: foreign DNA can be stably maintained in the chromosome, *Plant Cell*, 1, 123, 1989.
29. **O'Neill, C., Horvath, G. V., Horvath, E., Dix, P. J., and Medgyesy, P.,** Chloroplast transformation in plants: polyethylene glycol (PEG) treatment of protoplasts is an alternative to biolistic delivery systems, *Plant J.*, 3, 729, 1993.
30. **Johnston, S. A., Anziano, P. Q., Shark, K. B., Sanford, J. C., and Butow, R. A.,** Mitochondria transformation of yeast by bombardment with microprojectiles, *Science*, 240, 1538, 1988.
31. **Gad, A. E., Rosenberg, N., and Altman, A.,** Liposome-mediated gene delivery into plant cells, *Physiol. Plant.*, 79, 177, 1990.
32. **Saunders, J. A., Matthews, B. F., and Miller, P. D.,** Plant gene transfer using electrofusion and electroporation, in *Electroporation and Electrofusion in Cell Biology*, Neumann, E., Sowers, A. E., and Jordan, C. A., Eds., Plenum Publishing Corporation, New York, 1989, 343.
33. **Cassells, A. C.,** Uptake of charged lipid vesicles by isolated tomato protoplasts, *Nature*, 275, 760, 1978.
34. **Nagata, T.,** Liposome entrapment for delivery of genetic material to plant protoplasts, *Methods Enzymol.*, 149, 176, 1987.
35. **Rosenberg, N., Gad, A. E., Altman, A., Navot, N., and Czosnek, H.,** Liposome-mediated introduction of the chloramphenicol acetyl transferase (CAT) gene and its expression in tobacco protoplasts, *Plant Mol. Biol.*, 10, 185, 1988.
36. **Meyer, P., Walgenbach, E., Bussman, K., Hombrecher, G., and Saedler, H.,** Synchronized tobacco protoplasts are efficiently transformed by DNA, *Mol. Gen. Genet.*, 201, 513, 1985.

37. **Deshayes, A., Herrera-Estrella, L., and Caboche, M.,** Liposome-mediated transformation of tobacco mesophyll protoplasts by an *Escherichia coli* plasmid, *EMBO J.*, 4, 2731, 1985.
38. **Sporlein, B. and Koop, H.-U.,** Lipofectin: direct gene transfer to higher plants using cationic liposomes, *Theor. Appl. Genet.*, 83, 1, 1991.
39. **Nagata, T., Okada, K., Takebe, I., and Matsui, C.,** Delivery of tobacco mosaic virus RNA into plant protoplasts mediated by reverse-phase evaporation vesicles (liposomes), *Mol. Gen. Genet.*, 184, 161, 1981.
40. **Fraley, R. T., Dellaporta, S. L., and Papahadjopoulos, D.,** Liposome-mediated delivery of tobacco mosaic virus RNA into tobacco protoplasts: a sensitive assay for monitoring liposome-protoplast interactions, *Proc. Natl. Acad. Sci. U.S.A.*, 79, 1859, 1982.
41. **Nagata, T.,** Interaction of plant protoplast and liposome, *Methods Enzymol.*, 148, 34, 1987.
42. **Ballas, N., Zakai, N., Sela, I., and Loyter, A.,** Liposomes bearing a quaternary ammonium detergent as an efficient vehicle for functional transfer of TMV-RNA into plant protoplasts, *Biochim. Biophys. Acta*, 939, 8, 1988.
43. **Guerineau, J. F. and Tailliez, R.,** Interaction of liposomes with the plasmalemma of tobacco mesophyll protoplasts, *Cytobios*, 48, 39, 1986.
44. **Antonelli, N. M. and Stadler, J.,** Genomic DNA can be used with cationic methods for highly efficient transformation of maize protoplasts, *Theor. Appl. Genet.*, 80, 395, 1990.
45. **Ahokas, H.,** Transfection by DNA-associated liposomes evidenced at pea pollination, *Hereditas*, 106, 129, 1987.
46. **Uchimiya, H. and Harada, H.,** Transfer of liposome-sequestering plasmid DNA into *Daucus carota* protoplasts, *Plant Physiol.*, 68, 1027, 1981.
47. **Rollo, F., Galli, M. G., and Parisi, B.,** Liposome-mediated transfer of DNA to carrot protoplasts: a biochemical and autoradiographic analysis, *Plant Sci. Lett.*, 20, 347, 1981.
48. **Matthews, B. F. and Cress, D. E.,** Liposome mediated delivery of DNA to carrot protoplasts, *Planta*, 153, 90, 1981.
49. **Wang, C. Y., Hughes, K. W., and Huang, L.,** Improved cytoplasmatic delivery to plant protoplasts via pH-sensitive liposomes, *Plant Physiol.*, 82, 179, 1986.
50. **Maccarrone, M., Dini, L., Di Marzio, L., Di Giulio, A., Rossi, A., Mossa, G., and Finazzi Agrò, A.,** Interaction of DNA with cationic liposomes: ability of transfecting lentil protoplasts, *Biochem. Biophys. Res. Commun.*, 186, 1417, 1992.
51. **Cutler, A. J., Constabel, F., Kurz, W. G. W., and Shargool, P. D.,** Quantitation of the delivery of liposome contents into plant protoplasts, *Anal. Biochem.*, 139, 482, 1984.
52. **Fukunaga, Y., Nagata, T., Takebe, I., Kakehi, T., and Matsui, C.,** An ultrastructural study of the interaction of liposomes with plant protoplasts, *Exp. Cell Res.*, 144, 181, 1983.
53. **Gad, A. E., Elyashiv, G., and Rosenberg, N.,** The induction of large unilamellar vesicle fusion by cationic polypeptides: the effects of mannitol, size, charge density and hydrophobicity of the cationic polypeptides, *Biochim. Biophys. Acta*, 860, 314, 1986.
54. **Felgner, P. L., Gadek, T. R., Holm, M., Roman, R., Chan, H. W., Wenz, M., Northrop, J. P., Ringold, G. M., and Danielsen, M.,** Lipofection: a highly efficient, lipid-mediated DNA-transfection procedure, *Proc. Natl. Acad. Sci. U.S.A.*, 84, 7413, 1987.
55. **Felgner, P. L. and Ringold, G. M.,** Cationic liposome-mediated transfection, *Nature*, 337, 387, 1989.
56. **Tyagi, S., Sporlein, B., Tyagi, A. K., Herrmann, R. G., and Koop, H. U.,** PEG- and electroporation-induced transformation in *Nicotiana tabacum*: influence of genotype on transformation frequencies. *Theor. Appl. Genet.*, 78, 287, 1989.
57. **Gao, X. and Huang, L.,** Cationic liposomes and polymers for gene transfer, *J. Liposome Res.*, 3, 17, 1993.
58. **Felgner, P. L.,** Cationic lipid/polynucleotide condensates for *in vitro* and *in vivo* polynucleotide delivery — The cytofectins, *J. Liposome Res.*, 3, 3, 1993.
59. **Zachrisson, A. and Bornman, C. H.,** Electromanipulation of plant protoplasts, *Physiol. Plant.*, 67, 507, 1986.
60. **Neumann, E.,** Membrane electroporation and direct gene transfer, *Bioelectrochem. Bioenerget.*, 28, 247, 1992.
61. **Orlowski, S. and Mir, L. M.,** Cell electropermeabilization: a new tool for biochemical and pharmacological studies, *Biochim. Biophys. Acta*, 1154, 51, 1993.
62. **Van Wert, S. L. and Saunders, J. A.,** Electrofusion and electroporation of plants, *Plant Physiol.*, 99, 365, 1992.
63. **Shillito, R. D., Saul, M. W., Paszkowski, J., Muller, M., and Potrykus, I.,** High efficiency direct gene transfer to plants, *Bio/Technology*, 3, 1099, 1985.
64. **Riggs, C. D. and Bates, G. W.,** Stable transformation of tobacco by electroporation: evidence for plasmid concatenation, *Proc. Natl. Acad. Sci. U.S.A.*, 83, 5602, 1986.
65. **Schocher, R. J., Shillito, R. D., Saul, M. W., Paszkowski, J., and Potrykus, I.,** Co-transformation of unlinked foreign genes into plants by direct gene transfer, *Bio/Technology*, 4, 1093, 1986.
66. **Nishiguchi, M., Langridge, W. H. R., Szalay, A. A., and Zaitlin, M.,** Electroporation-mediated infection of tobacco leaf protoplasts with tobacco mosaic virus RNA and cucumber mosaic virus RNA, *Plant Cell Rep.*, 5, 57, 1986.
67. **Okada, K., Nagata, T., and Takebe, I.,** Introduction of functional RNA into plant protoplasts by electroporation, *Plant Cell Physiol.*, 27, 619, 1986.
68. **Hibi, T., Kano, H., Sugiura, M., Kazami, T., and Kimura, S.,** High efficiency electro-transfection of tobacco mesophyll protoplasts with TMV RNA, *J. Gen. Virol.*, 67, 2037, 1986.

69. **Watts, J. W., King, J. M., and Stacey, N. J.,** Inoculation of protoplasts with viruses by electroporation, *Virology,* 157, 40, 1987.
70. **Gallois, P., Lindsey, K., Malone, R., Kreis, M., and Jones, M. G. K.,** Gene rescue in plants by direct gene transfer of total genomic DNA into protoplasts, *Nucleic Acids Res.*, 20, 3977, 1992.
71. **Fromm, M., Taylor, L. P., and Walbot, V.,** Expression of genes transferred into monocot and dicot plant cells by electroporation, *Proc. Natl. Acad. Sci. U.S.A.*, 82, 5824, 1985.
72. **Langridge, W. H. R., Li, B. J., and Szalay, A. A.,** Electric field mediated stable transformation of carrot protoplasts with naked DNA, *Plant Cell Rep.*, 4, 355, 1985.
73. **Ecker, J. R. and Davis, R. W.,** Inhibition of gene expression in plant cells by expression of antisense RNA, *Proc. Natl. Acad. Sci. U.S.A.*, 83, 5372, 1986.
74. **Boston, R. S., Becwar, M. R., Ryan, R. D., Goldsbrough, P. B., Larkins, B. A., and Hodges, T. K.,** Expression from heterologous promoters in electroporated carrot protoplasts, *Plant Physiol.*, 83, 742, 1987.
75. **Fromm, M. E., Taylor, L. P., and Walbot, V.,** Stable transformation of maize after gene transfer by electroporation, *Nature*, 319, 791, 1986.
76. **Christou, P., Murphy, J. E., and Swain, W. F.,** Stable transformation of soybean by electroporation and root formation from transformed callus, *Proc. Natl. Acad. Sci. U.S.A.*, 84, 3962, 1987.
77. **Dhir, S. K., Dhir, S., Savka, M. A., Belanger, F., Kriz, A. L., Farrand, S. K., and Widholm, J. M.,** Regeneration of transgenic soybean *(Glycine max)* plants from electroporated protoplasts, *Plant Physiol.*, 99, 81, 1992.
78. **Smith, C. R., Tousignant, M. E., and Kaper, J. M.,** Replication footprint analysis of cucumber mosaic virus electroporated into tomato protoplasts, *Anal. Biochem.*, 200, 310, 1992.
79. **Maccarrone, M., Veldink, G. A., and Vliegenthart, J. F. G.,** Inhibition of lipoxygenase activity in lentil protoplasts by monoclonal antibodies introduced into the cells via electroporation, *Eur. J. Biochem.*, 205, 995, 1992.
80. **Ou-Lee, T.-M., Turgeon, R. and Wu, R.,** Expression of a foreign gene linked to either a plant-virus or a *Drosophila* promoter after electroporation of protoplasts of rice, wheat and sorghum, *Proc. Natl. Acad. Sci. U.S.A.*, 83, 6815, 1986.
81. **Jones, H., Ooms, G., and Jones, M. G. K.,** Transient gene expression in electroporated *Solanum* protoplasts, *Plant Mol. Biol.*, 13, 503, 1989.
82. **Hauptmann, R., Ozias-Akins, P., Vasil, V., Tabaeizadeh, Z., and Vasil, N.,** Expression of electroporated DNA in several mono- and dicotyledonous plant species, *Plant Physiol.*, 80, 102, 1986.
83. **Hauptmann, R. M., Vasil, V., Ozias-Akins, P., Tabaeizadeh, Z., Rogers, S. G., Fraley, R. T., Horsch, R. B., and Vasil, I. K.,** Evaluation of selectable markers for obtaining stable transformants in the Gramineae, *Plant Physiol.*, 86, 602, 1988.
84. **Joersbo, M. and Brunstedt, J.,** Direct gene transfer to plant protoplasts by electroporation by alternating, rectangular and exponentially decaying pulses, *Plant Cell Rep.*, 8, 701, 1990.
85. **Joersbo, M. and Brunstedt, J.,** Electroporation: mechanism and transient expression, stable transformation and biological effects in plant protoplasts, *Physiol. Plant.*, 81, 256, 1991.
86. **Mol, J. N. M., Van der Krol, A. R., Van Tunen, A. J., Van Blokland, R., De Lange, P., and Stuitje, A. R.,** Regulation of plant gene expression by antisense RNA, *FEBS Lett.*, 268, 427, 1990.
87. **Ecker, J. R. and Davis, R. W.,** Inhibition of gene expression in plant cells by expression of antisense RNA, *Proc. Natl. Acad. Sci. U.S.A.*, 83, 5372, 1986.
88. **Vaucheret, H., Kronenberger, J., Lepingle, A., Vilaine, F., Boutin, J.-P., and Caboche, M.,** Inhibition of tobacco nitrite reductase activity by expression of antisense RNA, *Plant J.*, 2, 559, 1992.
89. **Carneiro, V. T. C., Pelletier, G., and Small, I.,** Transfer RNA-mediated suppression of stop codons in protoplasts and transgenic plants, *Plant Mol. Biol.*, 22, 681, 1993.
90. **Morikawa, H., Iida, A., Matsui, C., Ikegami, M., and Yamada, Y.,** Gene transfer into intact plant cells by electroinjection through cell walls and membranes, *Gene*, 41, 121, 1986.
91. **Lindsey, K. and Jones, M. G. K.,** Transient gene expression in electroporated protoplasts and intact cells of sugar beet, *Plant Mol. Biol.*, 10, 43, 1987.
92. **Yang, J., Ge, K., Wang, Y., Wang, B., and Tan, C. C.,** Highly efficient transfer and stable integration of foreign DNA into partially digested rice cells using a pulsed electrophoretic drive, *Transgenic Res.*, 2, 245, 1993.
93. **Dekeyser, R. A., Claes, B., De Rycke, R. M. U., Habets, M. E., Van Montagu, M. C., and Caplan, A. B.,** Transient gene expression in intact and organized rice tissues, *Plant Cell*, 2, 591, 1990.
94. **D'Halluin, K., Bonne, E., Bossut, M., De Beuckeleer, M., and Leemans, J.,** Transgenic maize plants by tissue electroporation, *Plant Cell*, 4, 1495, 1992.
95. **Machy, P., Lewis, F., McMillan, L., and Jonak, Z. L.,** Gene transfer from targeted liposomes to specific lymphoid cells by electroporation, *Proc. Natl. Acad. Sci. U.S.A.*, 85, 8027, 1988.
96. **Wang, C.-Y. and Huang, L.,** Highly efficient DNA delivery mediated by pH-sensitive immunoliposomes, *Biochemistry*, 28, 9508, 1989.
97. **Gould-Fogerite, S., Mazurkiewicz, J. E., Raska, K., Jr., Voelkerding, K., Lehman, J. M., and Mannino, R. J.,** Chimerasome-mediated gene transfer *in vitro* and *in vivo*, *Gene*, 84, 429, 1989.
98. **Maccarrone, M., Di Marzio, L., Rossi, A., and Finazzi Agrò, A.,** Gene transfer to lentil protoplasts by lipofection and electroporation, *J. Liposome Res.*, 3, 707, 1993.

99. **Jefferson, R. A.,** Assaying chimeric genes in plants: the GUS gene fusion system, *Plant Mol. Biol. Rep.*, 5, 387, 1987.
100. **Widholm, J. M.,** The use of fluorescein diacetate and phenosafranine for determining viability of cultured plant cells, *Stain Technol.*, 47, 189, 1972.
101. **Sukharev, S. I., Klenchin, V. A., Serov, S. M., Chernomordik, L. V., and Chizmadzhev, Yu. A.,** Electroporation and electrophoretic DNA transfer into cells. The effect of DNA interaction with electropores. *Biophys. J.*, 63, 1320, 1992.
102. **Caso, J. L., Hardisson, C., and Suarez, J. E.,** Transfection of *Micromonospora* spp., *Appl. Environ. Microb.*, 53, 2544, 1987.
103. **Verhoef, K., Koken, S. E. C., and Berkhout, B.,** Electroporation of the HIV Tat trans-activator protein into cells, *Anal. Biochem.*, 210, 210, 1993.
104. **Maccarrone, M., Veldink, G. A., Finazzi Agrò, A., and Vliegenthart, J. F. G.,** Lentil root protoplasts: a transient expression system suitable for coelectroporation of monoclonal antibodies and plasmid molecules. *Biochim. Biophys. Acta.*, 1243, 136, 1994.
105. **Mossa, G., Di Giulio, A., Dini, L., and Finazzi Agrò, A.,** Interaction of dipalmitoylphosphatidylcholine/cholesterol vesicles with ascorbate oxidase, *Biochim. Biophys. Acta*, 986, 310, 1989.
106. **Douma, A. C., Veenhuis, M., Driessen, A. J. M., and Harder, W.,** Liposome-mediated introduction of proteins into protoplasts of the yeast *Hansenula polymorpha* as a possible tool to study peroxisome biogenesis, *Yeast*, 6, 99, 1990.
107. **Wilson, T. M. A.,** Strategies to protect crop plants against viruses: pathogen-derived resistance blossoms, *Proc. Natl. Acad. Sci. U.S.A.*, 90, 3134, 1993.

Chapter 6

The Contribution of Cationic Derivatives of Phospholipids and Cholesterol to the Bioadhesiveness of Liposomes and Their Transfection Performance

Luke S. S. Guo, Ramachandran Radhakrishnan, and Ming Man

CONTENTS

I. INTRODUCTION

Cationic liposomes are efficient vehicles for association with negatively charged cell surfaces for transfection.[1-3] The first such liposome tested, a combination of DOTMA and DOPE, was shown to be very effective in transfecting a variety of mammalian cells *in vitro*.[4,5] Several progenitor amphiphiles have appeared in the literature[6-10] for the same end use, but most of them are limited by toxicity to treated cells. Recently, the molecular interaction between positively charged lipid molecules and genetic material has been probed by physical techniques such as metal-shadowing electron microscopy.[11] In this study, cationic liposomes were shown to bind initially to DNA molecules to form clusters of aggregated vesicles along the nucleic acids. DNA-induced membrane fusion and liposome-induced DNA collapse were shown to occur as subsequent events at a critical liposome density. Thus, the major steps appeared to be the formation of transfecting DNA-liposome complexes with a total encapsulation of negatively charged DNA in cationic liposomes, followed by electrostatic interactions between components leading to lipid fusion and collapse of the genetic material. In an attempt to understand the structure–function relationships in the cationic lipids, a homologous series of hydroxyalkyl quaternary ammonium derivatives of DOTMA with different alkyl chain substitutions were prepared and tested.[12] Based on this investigation, Felgner et al. reported the order of transfection efficiency as dimyristyl > dioleyl > dipalmityl > distearyl. Cationic liposomes prepared from these lipids and 50 mol% of DOPE were more effective than 50 mol% of neutral lipid components. In general, LysoPE analogs were less effective.

Efficient and directed gene transfer to cells offers great promise for the *in vivo* generation of therapeutic proteins such as cytokines, correcting genetic deficiencies such as cystic fibrosis, and offering means for antibody responses to *in vivo* expressed vaccine antigens. In a clinical study designed to demonstrate gene delivery and consequent gene expression, HLA-B7 negative patients with advanced melanoma were injected with liposome complexes of the gene for the MHC protein, HLA-B7.[13] The treatment was well tolerated and evidence for the *in vivo* synthesis of recombinant protein was demonstrated in tumor biopsies by immunochemistry. Limited regression of tumor nodules was even observed in this study, indicating a promise for future gene therapies.

We have synthesized a series of cationic lipids by condensing phosphatidylethanolamine or cholesterol with basic amino acids such as lysine, arginine etc.[14] Liposomes prepared from these cationic lipids in admixture with neutral lipids were shown to bind to mucosal tissues strongly, presumably via electrostatic interactions. In an attempt to understand the impact of the charge density of the cationic lipids in the

0-8493-4013-6/96/$0.00+$.50

liposomes and the location of the cationic charge in the lipids at specific distances from the membrane interface, a series of suitably designed lipids was prepared and tested for the efficiency of bioadhesion to mucosal tissues *in vitro* and *in vivo*. In spite of extensive literature on the utility of the cationic lipid formulations to deliver macromolecules, the application of nontoxic lipids towards these applications remains limited. The preparation of such lipid derivatives, their use in bioadhesion, and limited applications in gene delivery will be reviewed here.

II. SYNTHETIC CATIONIC PHOSPHOLIPID AND CHOLESTEROL DERIVATIVES

In order to develop biocompatible cationic lipids for bioadhesion and cellular trasfection studies, we have synthesized a series of cationic molecules based on phospholipids and cholesterol as lipid backbones and amino acid moieties to provide charged amino groups (Figure 1). Because the linkages used, such as amide or ester, are cleavable *in vivo*, the synthetic molecules are expected to be degraded *in vivo* to their parent nutrient molecules. The lysinyl, argininyl, histidinyl, and ornithinyl derivatives of PE are designed to provide one net positive charge per molecule. The di-lysinyl and tri-lysinyl PE are designed to provide two and three net positive charges per molecule, respectively (Table 1). Synthetic cationic cholesterol derivatives were designed such that the charged amino groups can be spaced at different carbon chain lengths from the C3 oxygen of the cholesterol (Table 2).

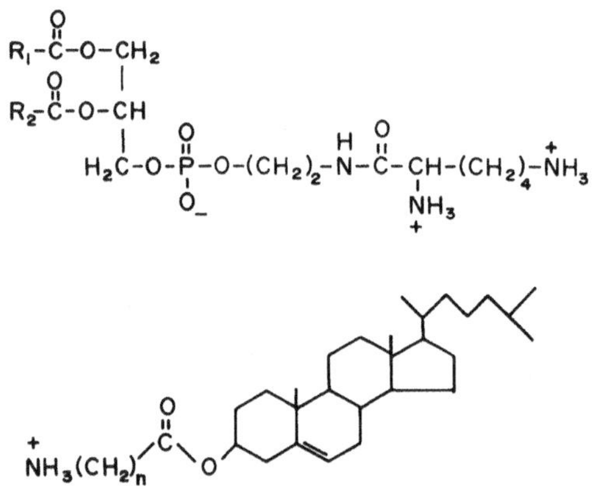

Figure 1 Chemical structure of synthetic lysinyl PE (top) and cholesteryl esters of amino acids of different carbon chain lengths (bottom).

Table 1 Synthetic Cationic Phospholipid Derivatives

Derivative	Amino Acid	Net Positive Charge
Lysinyl PE	L-Lysine	1
Argininyl PE	L-Arginine	1
Histidinyl PE	L-Histidine	1
Ornithinyl PE	L-Ornithine	1
Di-lysinyl PE	L-Lysinyl lysine	2
Tri-lysinyl PE	L-Lysinyl lysinyl lysine	3

Note: Phosphatidylethanolamine (PE) was obtained from egg yolk. In the case of lysinyl PE and lysinyl lysinyl PE, dipalmitoyl phosphatidylethanolamine (DPPE) was also used as a starting material to prepare the saturated derivatives.

When these cationic derivatives were formulated in liposomes and tested *in vitro* with a monkey kidney cell line, both the synthetic phospholipid and cholesterol derivatives were found to be noncytotoxic. Contrary to experience with these novel synthetic lipids, cell growth was significantly inhibited with liposomes containing the more traditional cationic surfactants such as stearylamine and CTAB.[14]

Table 2 Synthetic Cationic Cholesterol Derivatives

Derivative	Amino Acid	Number of Carbon Spacers from the C_3 Oxygen of Cholesterol
Cholesteryl glycine (C-2)	Glycine	2 carbons
Cholesteryl β-alanine (C-3)	β-Alanine	3 carbons
Cholesteryl aminobutyric acid (C-4)	4-Aminobutyric acid	4 carbons
Cholesteryl aminovaleric acid (C-5)	5-Aminovaleric acid	5 carbons
Cholesteryl aminocaproic acid (C-6)	6-Aminocaproic acid	6 carbons

III. LIPOSOMES PREPARED WITH THE CATIONIC PHOSPHOLIPID OR CHOLESTEROL DERIVATIVES

The cationic phospholipid or cholesterol derivatives at 10 to 40 mol% were routinely used in the liposome preparations. In fact, lysinyl PE or cholesteryl β-alanine by itself has been shown to form liposomes. The hydrated lipid suspensions were predominantly spheres, heterogeneous in size, and were birefringent under polarizing light microscopy. In order to ensure that discrete liposomal structures can be formed with the above cationic lipids, attempts were made to prepare small unilamellar liposomes (SUVs) containing PE, lysinyl PE, or lysinyl lysinyl PE (Table 3). Similar particle sizes and specific radioactivities (^{14}C-sucrose) associated with these liposomes confirm that they were in fact sealed vesicles with a defined inner aqueous volume.

Table 3 Mean Diameters and Encapsulation Volumes of SUVs Containing PE, Lysinyl PE or Lysinyl Lysinyl PE

Lipid Composition	Mean Diameter (nm)	Encapsulated ^{14}C-Sucrose (CPM/μmol PL)	Encapsulation Volume (μl/μmol PL)
PC/PE			
9:1	—	6,314	0.41
8:2	43	10,081	0.65
PC/lysinyl PE			
9:1	43	6,803	0.41
8:2	38	8,275	0.53
PC/lysinyl lysinyl PE			
9:1	42	7,856	0.52
8:2	44	8,036	0.50

Note: All values are means of duplicate assays. Small unilamellar liposomes were produced by sonication.

Assays of the charged amino groups using a membrane-impermeable reagent, trinitrobenzenesulfonic acid, to ascertain the topography showed that the distribution of the cationic lipids between the two bilayer leaflets of the SUVs can be manipulated. High concentrations of the cationic lipids containing one positive charge per molecule, such as lysinyl PE, or lower concentration of the cationic lipids containing a greater number of charges per molecule, such as lysinyl lysinyl PE, appeared to be preferentially located on the exterior surface or in the outer leaflet of the liposomes. For examples, liposomes containing 10 mol% of lysinyl PE had a similar distribution of the amino groups on the outer leaflets (58%) to noncharged liposomes containing 10 and 20 mol% of PE (54–58%, Table 4). Increasing the lysinyl PE concentration to 20 mol% or inclusion of 10 mol% of lysinyl lysinyl PE resulted in a much greater proportion of the amino group exposed to the exterior surface (76%). In liposomes containing 20 mol% of lysinyl lysinyl PE, up to 92% of the charged groups were found on the exterior surface. However, it is not known whether this kind of lipid asymmetry will be realized in larger, unstrained liposomes.

Table 4 Distribution of Amino Groups on the Surface of SUVs Containing PE, Lysinyl PE, or Lysinyl Lysinyl PE

Lipid Composition	Percent of Surface Amino Reactive Group
PC/PE	
9:1	58 (55–61)
8:2	54 (53–56)
PC/lysinyl PE	
9:1	58 (54–61)
8:2	76 (71–86)
PC/lysinyl lysinyl PE	
9:1	76 (73–79)
8:2	92 (89–95)

Note: Reactive amino groups were determined by a trinitrobenzene sulfonic acid assay. Means and ranges (in parentheses) from two determinations are given.

IV. ADHESION OF CATIONIC LIPOSOMES TO MUCOSAL TISSUES

Liposomes prepared with various types of cationic phospholipid or cholesterol derivatives were evaluated for bioadhesion using two different systems, the intact eye in unanesthetized rabbits[14] and dissected mucosal tissues. The unanesthetized rabbit eye is particularly useful because it is an intact tissue and the mucus layer on the ocular surface is protected by a thin film of liquid — the so called tearfilm. Foreign materials applied to the eyes are naturally washed away by two effective removal mechanisms, the blink motion of the eyelids and continuous tear flow, unless the exogenous materials interact specifically with the surface mucus layer.

Using the rabbit eye model, we found that inclusion of cationic lipid molecules into the liposomes significantly enhanced liposome retention compared to neutral or negatively charged liposomes (Table 5). Enhanced retention was observed with all six PE derivatives and four of the five cholesterol derivatives (see below). Among the four PE derivatives of a single amino acid, lysinyl PE was found to be the best adhesive molecule owing to the ε-amino group in lysine being presumably most basic. Therefore, the lysinyl and the lysinyl lysinyl PE were used as the most appropriate choice in our investigations.

Table 5 Retention of Liposomes of Various Surface Charges in the Rabbit Eye

Surface Charge	Lipid Composition	Percent of Retention 30 min	Percent of Retention 60 min
Neutral	PC/C (6:4)	5 ± 4	3 ± 2
Negative	PC/C/PG (3:4:3)	7 ± 1	4 ± 3
	PC/C/gangliosides (4:4:2)	5 ± 3	4 ± 2
Positive	PC/C/lysinyl PE (4:4:2)	30 ± 12	25 ± 12
	PC/C/lysinyl lysinyl PE (4:4:2)	47 ± 15	36 ± 15

Note: All liposomes were formed by thin film hydration and sonicated to form SUVs. Values are mean (±SD) from four rabbit eyes.

Similar results were also obtained with dissected mucosal tissues. For instance, when approximately equal concentrations of neutral, stearylamine and lysinyl lysinyl PE liposomes were incubated with dissected pieces of trachea, esophagus, stomach, small intestine and rectum tissue samples, both positively charged liposome preparations showed enhanced adhesion to most of the mucosal tissue types compared to neutral liposomes (Table 6). Lysinyl lysinyl PE liposomes showed almost twice the percent adhesion to the trachea, esophagus and small intestine relative to the stearylamine containing liposomes.

Mucin is a polyanionic glycoprotein which coats the surface of epithelial tissues. The enhanced adhesion of cationic liposomes to mucosal tissues is presumably via electrostatic interactions between the positively charged groups on the lipids and the negatively charged muco-glycoproteins. The fact that

Table 6 Adhesion of Neutral and Cationic Liposomes to Mucosal Tissues

Tissue	Percent of Adhesion		
	Neutral (PC:C/6:4)	Cationic A (PC:C:SA/4:4:2)	Cationic B (PC:C:Lysinyl Lysinyl PE/4:4:2)
Trachea (n = 3)	2.1 ± 0.8	6.6 ± 2.9	13.6 ± 4.1
Esophagus (n = 5)	2.2 ± 0.4	21.4 ± 7.3	38.4 ± 10.1
Stomach (n = 4)	1.2 ± 0.6	3.2 ± 1.9	5.2 ± 0.6
Small Intest. (n = 5)	7.7 ± 4.5	6.1 ± 2.4	14.0 ± 4.4
Rectum (n = 3)	2.3 ± 1.8	6.1 ± 3.3	8.5 ± 1.0

Note: Approximately equal concentrations of lipid and radioactive counts (^{125}I) of neutral, stearylamine (SA) and lysinyl lysinyl PE liposomes were incubated with rat mucosal tissue samples. Numbers (means ± SD) are relative % adhesion after normalization to 100 mg tissue weight for comparison within each tissue type.

the ocular surface was saturable with respect to applied cationic liposomes further suggests a specific adhesion of the liposomes to the surface of the mucus layer.[14]

V. SIGNIFICANCE OF LIPOSOME STRUCTURE ON BIOADHESIVENESS

Three different lines of studies with the cationic liposomes further established the significance of liposome structure in mucosal tissue adhesion. For example, increasing the amount of lysinyl PE from 0 to 40 mol% in liposomes increased the percent of retention by a factor of at least ten (Figure 2A). Moreover, a similar degree of adhesion was observed in liposomes of equivalent charge density whether they were prepared with molecules containing a single cationic group, such as lysinyl PE or multiple cationic groups, such as lysinyl lysinyl PE or tri-lysinyl PE. Therefore, the adhesion of cationic liposomes to mucosal tissue appears to be dependent on the charge density, but appears to be independent of the distribution of charged groups on the liposomes.

Poor ocular retention was observed with liposomes constructed from lysinyl PE or lysinyl lysinyl PE in which the fatty acyl chains are unsaturated (Figure 2B). The fluid and flexible chains in these liposomes may permit the polar head group to be conformationally mobile at the interface. Therefore, the charged groups presumably do not interact sufficiently with the mucus layer. On the other hand, inclusion of cholesterol at 40 mol% in the same liposomes sufficiently rigidified the lipid bilayer and resulted in a significant improvement of retention. Prolonged liposomal retention, however, can be achieved by using derivatives of saturated phospholipids, such as lysinyl DPPE, without the addition of cholesterol.

The other line of experiments used a set of synthetic cationic cholesterol derivatives with the charge being located at different distances from the 3-β hydroxyl group of the cholesterol. Liposomes composed of PC and cholesterol derivatives with at least three carbon residues (e.g., cholesteryl β-alanine) were quite effective in mucosal adhesion, whereas the short-chain derivatives were cleared by lacrimal flow (Figure 2C). This study clearly pointed toward the synthesis of "designer" cationic lipids that are capable of orienting away from bilayer interface and effectively interacting with negatively charged mucoglycoproteins.

VI. INTERACTION OF CATIONIC DETERGENTS AND PC IN LIPID BILAYERS

Stearylamine (SA) and benzyldimethylstearylammonium chloride (BDSA) are two commonly used cationic detergents with a single C-18 fatty acyl chain. Liposomes composed of these compounds where the positively charged amino groups would be expected to orient closely to the bilayer interface should have very few or no bioadhesive properties. Surprisingly, moderate ocular adhesion was observed with liposomes containing 20–30 mol% of SA or BDSA (data not shown).

In order to determine the orientation of these cationic molecules in lipid bilayers, we prepared SUVs, using hydrogenated soybean phosphatidylcholine (HSPC) alone or HSPC containing 20 mol% BDSA and characterized them by ^{31}P-NMR. As shown in Figure 3 (top), the HSPC liposomes revealed a sharp symmetrical spectrum due to free motion of the phospholipids in the plane of the liposomal bilayers.[15]

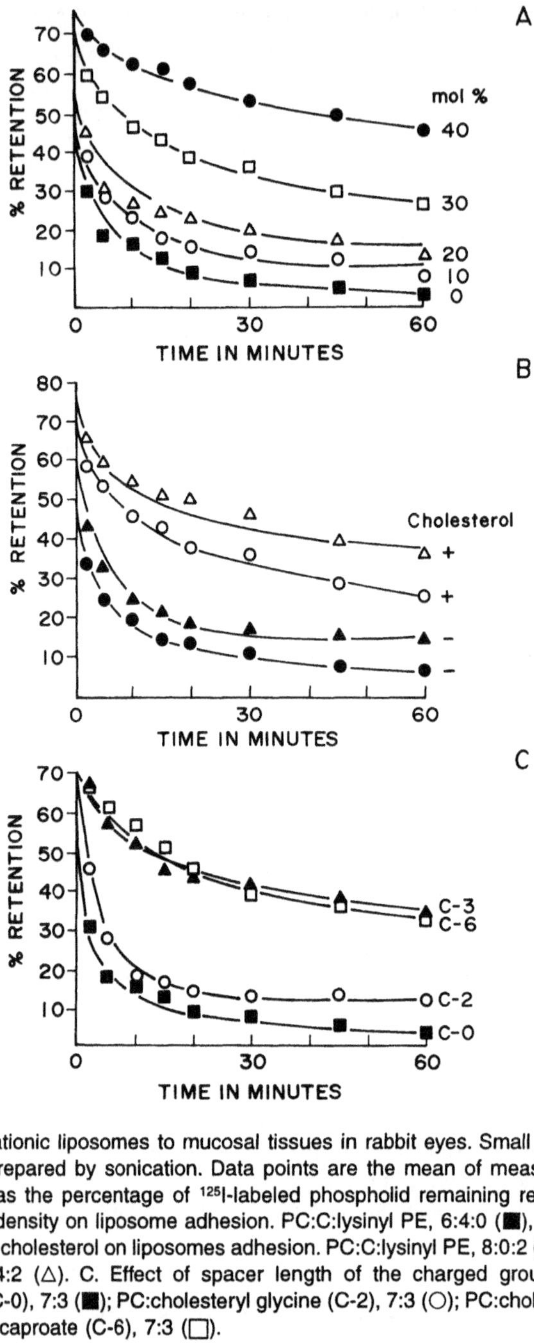

Figure 2 Adhesion of cationic liposomes to mucosal tissues in rabbit eyes. Small liposomes of the following molar proportions were prepared by sonication. Data points are the mean of measurements from four rabbit eyes and are expressed as the percentage of ^{125}I-labeled phospholid remaining relative to the initial instilled dose. A. Effect of charge density on liposome adhesion. PC:C:lysinyl PE, 6:4:0 (■), 5:4:1 (○), 4:4:2 (△), 3:4:3 (□), 2:4:4 (●). B. Effect of cholesterol on liposomes adhesion. PC:C:lysinyl PE, 8:0:2 (●), 4:4:2 (○); PC:C:lysinyl lysinyl PE, 8:0:2 (▲), 4:4:2 (△). C. Effect of spacer length of the charged group on liposome adhesion. PC:epicholesterylamine (C-0), 7:3 (■); PC:cholesteryl glycine (C-2), 7:3 (○); PC:cholesteryl β-alanine (C-3), 7:3 (▲); PC:cholesteryl aminocaproate (C-6), 7:3 (□).

The BDSA-containing liposomes, on the other hand, showed a much broader ^{31}P-spectrum (Figure 3, bottom). Incorporation of BDSA into the liposomes may result in an electrostatic interaction of the positively charged quaternary ammonium group on BDSA with the nearby negatively charged phosphate of the phospholipids and thus restrict the molecular motions of the PC in these liposomes. The electrostatic interactions between BDSA and PC may also free the positively charged choline head group on PC and make it extend away from the membrane surface for efficient interactions with the negatively charged cell surface.

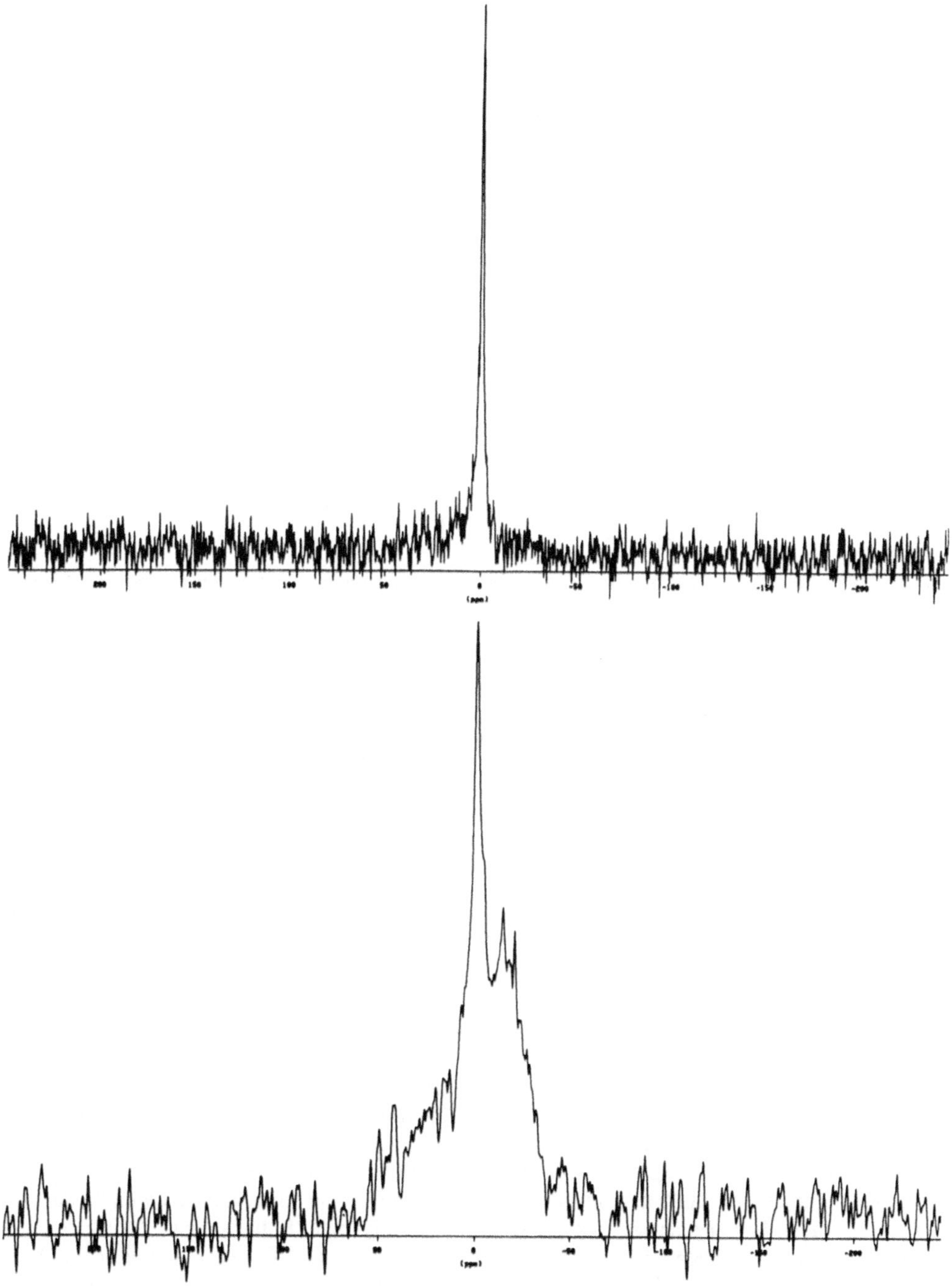

Figure 3 ^{31}P-NMR spectrum of small sonicated liposomes prepared with HSPC (top) and HSPC containing 20 mol% of a cationic detergent BDSA (bottom).

VII. CELLULAR TRANSFECTION WITH LIPOSOMES CONTAINING CATIONIC PHOSPHOLIPID AND CHOLESTEROL DERIVATIVES

Liposomes prepared from synthetic cationic lipids, such as DOTMA, DOTAP, DC-chol, etc., have been shown to form stable complexes with polyanionic materials like DNA, promote their entry into cells and thus enhance the expression of exogenous genetic material.[4,6,7] Because the electrostatic interactions between cationic phospholipid and cholesterol derivatives and muco-glycoproteins are the probable cause of adhesion observed in the eye, and these cationic lipids may have a similar interaction with DNA, possible applications of these lipid derivatives in cell transfection were studied.

The CAT gene expression has been studied in CV-1 cells with cationic liposomes composed of lysinyl PE and cholesteryl β-alanine.[16] As shown in Figure 4, they were at least as effective as cationic liposomes composed of DDAB:DOPE.

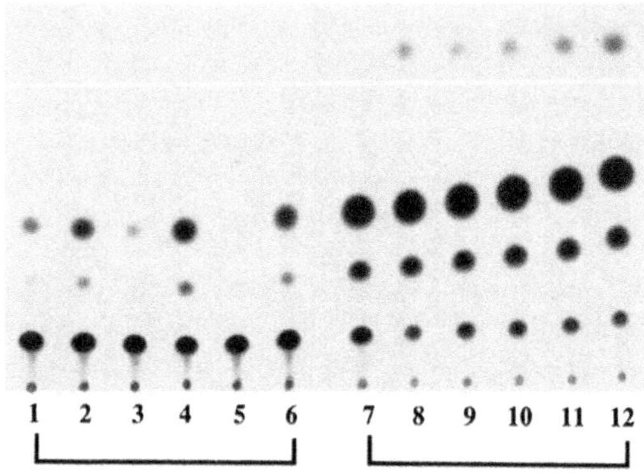

Figure 4 Expression of the CAT gene into CV-1 cells using cationic liposome-mediated transfection. All plasmid-liposome complexes were prepared at the same time. Complexes were added immediately or incubated at 37°C for 1 or 4 h before being transferred to cells in 100-mm dishes. Cells were harvested 48 h after trasfection, and cellular extracts prepared and assayed for CAT activity. 1. DDAB:DOPE/1:2, serum-free, complex added immediately; 2. DDAB:DOPE/1:2, 10% serum, complex added immediately; 3. DDAB:DOPE/1:2, serum-free, complex incubated 1 h; 4. DDAB:DOPE/1:2, 10% serum, complex incubated 1 h; 5. DDAB:DOPE/1:2, serum-free, complex incubated 4 h; 6. DDAB:DOPE/1:2, 10% serum, complex incubated 4 h; 7. Lysinyl PE:Chol β-alanine/6:4, serum free, complex added immediately; 8. Lysinyl PE:Chol β-alanine/6:4, 10% serum, complex added immediately; 9. Lysinyl PE:Chol β-alanine/6:4, serum free, complex incubated 1 h; 10. Lysinyl PE:Chol β-alanine/6:4, 10% serum, complex incubated 1 h; 11. Lysinyl PE:Chol β-alanine/6:4, serum free, complex incubated 4 h; 12. Lysinyl PE:Chol β-alanine/6:4, 10% serum, complex incubated 4 h.

VIII. DISCUSSION

Cationic liposomes prepared from a series of phosphatidylethanolamine derivatives or cholesterol derivatives with basic amino acids, such as lysine, arginine, histidine, ornithine, lysinyl lysine and lysinyl lysinyl lysine were designed to be biodegradable, have the capacity to provide desirable charge densities and interact efficiently with cells at variable distances from the membrane interface (Tables 1 and 2). As expected, these liposomes were shown to be noncytotoxic at least in cell culture[13] compared to cationic surfactants such as stearylamine and CTAB. The formation of sealed vesicles composed of such cationic lipids in combination with neutral lipids has also been demonstrated (Table 3) with the encapsulation of radiolabeled sucrose. Biochemical studies with these liposomes to assess the topographical distribution of the cationic lipids have revealed very interesting membrane asymmetry with PE derivatives composed of more than one cationic residue such as lysinyl lysinyl PE. For example, with 20 mol% of this lipid included in sonicated liposomes, predominant distribution of this cationic lipid could be seen in the outer leaflet (Table 4), offering the potential for total interaction with DNA or cells. Even though the asymmetric distribution in large vesicles may not be realized, the application of smaller vesicles with "coated" cationic lipid derivatives has enormous potential for bioadhesion and trasfection.

In vivo studies in rabbits have demonstrated the ocular retention of these cationic liposomes compared to neutral liposomes, specifically, lysinyl PE liposomes were particularly effective in prolonging the ocular residence times (Table 5). Thus, drug delivery for ocular diseases as well as delivery of DNA to correct genetic conditions such as retinitis pigmentosa may be feasible. The latter indication, at least in one form, has been shown to be associated with a single amino acid substitution in the primary sequence of an ocular protein.[17,18] Even though cationic liposomes were not equally effective with all mucosal tissue types (Table 6), their enhanced adhesion to trachea, esophagus and small intestine is sufficient to attest to their affinity for mucosal surfaces and their ability to act as carrier of negatively charged DNA and other macromolecules.

A closer investigation of the structure–activity relationship in bioadhesion revealed some interesting features. For example, an increase in the amount of liposome charge lead to an increase in adhesion (Figure 2A), and charge density of the liposome composition was the decisive factor in the extent of adhesion phenomenon. Second, rigid cationic liposomes had the maximal interaction with cell surfaces, indicating that in these liposomes the charged moieties may be more oriented toward the negatively charged mucosal surfaces for favorable electrostatic interaction. Third, the location of the charged group in cationic lipids needs to be arranged at a distance away from the membrane interface of the liposomes for optimal interaction with mucosal cell surfaces (Figure 2C).

We found that cationic liposomes, composed of a C-18 chain amphiphile, BDSA, exhibited a moderate ocular retention, even though the cationic charge of this molecule is believed to be located close to the liposomal membrance interface. A systematic study of ^{31}P-NMR spectra of HSPC liposomes and HSPC/BDSA liposomes revealed differences (Figure 3). The results imply that the positively charged amino groups of BDSA are capable of interacting with the phosphate of neighboring PC molecules and constraining their motion. Consequently, this frees the quaternary ammonium group of choline in PC molecule to interact with cell surfaces. The resulting adhesion, even though moderate, points toward the exercise of caution in engineering liposomes for bioadhesion and/or gene delivery.

In limited studies with cationic liposomes composed of lysinyl PE and cholesteryl β-alanine, we have demonstrated that these liposomes are quite effective in CAT gene delivery and compared favorably to the DDAB:DOPE liposomes which are currently used in experimental investigations.[9] In addition, the cationic lipids used in the present investigation may have the advantage of being biodegradable, with lower cytotoxic effects. Further work with similar synthetic lipids for the preparation of liposomes and their applications for gene delivery is in progress.

ABBREVIATIONS

DOTMA, N[1-(2,3-dioleyoxy) propyl]-*N,N,N*-trimethylammonium; DOPE, dioleoyl phosphatidylethanolamine; DOTPA, 1,2-dioleoyloxy-3-(trimethylammonio) propane; CTAB, cetyltrimethylammonium bromide; DC-chol, 3β(*N*-(*N*′,*N*′-dimethylaminoethane) carbamoyl cholesterol; DDAB, dimethyl dioctadecylammonium bromide.

REFERENCES

1. **Felgner, P. L. and Ringold, G. M.,** Cationic liposome-mediated transfection. *Nature,* 337, 387, 1989.
2. **Hug, P. and Sleight, R. G.,** Liposomes for the transformation of eukaryotic cells. *Biochim. Biophys. Acta,* 1097, 1, 1991.
3. **Behr, J. P.,** Gene transfer with synthetic cationic amphiphiles: prospects for gene therapy. *Bioconjugat. Chem.,* 5, 382, 1994.
4. **Felgner, P. L., Gadek, T. R., Holm, M., Roman, R., Chan, H. W., Wenz, M., Northrop, J. P., Ringold, G. M., and Danielsen, M.,** Lipofectin: a highly efficient, lipid-mediated DNA-transfection procedure. *Proc. Natl. Acad. Sci. U.S.A.,* 84, 7413, 1987.
5. **Bringham, K., Meyrick, B., Christman, B., Magnuson, M., King, G., and Berry, L.,** *In vivo* transfection of murine lungs with a functioning prokaryotic gene using a liposome vehicle. *Am. J. Med. Sci.,* 298, 278, 1989.
6. **Pinnaduwage, P., Schmitt, L., and Huang, L.,** Use of a quaternary ammonium detergent in liposome mediated DNA transfection of mouse L-cells. *Biochim. Biophys. Acta,* 985, 33, 1989.
7. **Leventis, R. and Silvius, J. R.,** Interactions of mammalian cells with lipid dispersions containing novel metabolizable cationic amphiphiles. Biochim. Biophys. *Acta,* 1023, 124, 1990.
8. **Gao, X. and Huang, L.,** A novel cationic liposome reagent for efficient transfection of mammalian cells. *Biochem. Biophys. Res. Commun.,* 179, 280, 1991.
9. **Rose, J. K., Buonocore, L., and Whitt, M. A.,** A new cationic liposome reagents mediating nearly quantitative transfection of animal cells. *Biotechniques,* 10, 520, 1991.
10. **Ito, A., Miyazoe, R., Mitoma, J., Akao, T., Osaki, T., and Kunitake, T.,** Synthetic cationic amphiphiles for liposome-mediated DNA transfection. *Biochem. Internatl.,* 22, 235, 1990.
11. **Gershon, H., Ghirlando, R., Guttman, S. B., and Minsky, A.,** Mode of formation and structural features of DNA-cationic liposome complexes used for transfection. *Biochemistry,* 32, 7143, 1993.

12. **Felgner, J. H., Kumar, R., Sridhar, C. N., Wheeler, C. J., Tsai, Y. J., Border, R., Ramsey, P., Martin, M., and Felgner, P. L.,** Enhanced gene delivery and mechanism studies with a novel series of cationic lipid formulations. *J. Biol. Chem.,* 269, 2550, 1994.
13. **Nabel, G. J., Nabel, E. G., Yang, Z., Fox, B. A., Plautz, G. E., Gao, X., Huang, L., Shu, S., Gordon, D., and Chang, A. E.,** Direct gene transfer with DNA-liposome complexes in melanoma: Expression, biologic activity, and lack of toxicity in humans. *Proc. Natl. Acad. Sci. U.S.A.,* 90, 11307, 1993.
14. **Guo, L. S. S., Radhakrishnan, R., and Redemann, C. T.,** Adhesion of positively charged liposomes to mucosal tissues. *J. Liposome Res.,* 1, 319, 1989-1990.
15. **Cullis, P. R. and De Kruijff, B.,** Lipid polymorphism and the functional role of lipids in biologic membranes. *Biochim. Biophys. Acta,* 559, 399, 1979.
16. **Guo, L. S. S., Radhakrishnan, R., Redemann, C. T., Brunette, E. N., and Debs, R. J.,** Cationic liposomes containing noncytotoxic phospholipid and cholesterol derivatives. *J. Liposome Res.,* 3, 51, 1993.
17. **Naash, M. I., Hollyfield, J. G., Al-Ubaidi, M. R., and Baehr, W.,** Simulation of human autosomal dominant retinitis pigmentosa in transgenic mice expressing a mutated murine opsim gene. *Proc. Natl. Acad. Sci. U.S.A.,* 90, 5499, 1993.
18. **Drrja, T. P., McGee, T. L., Reichal, E., Hahn, L. B., Cowley, G. S., Yandell, D. W., Sandberg, M. A., and Berson, E. L.,** A point mutation of the rhodopain gene in one form of retinitis pigmentosa. *Nature,* 343, 364, 1990.

Chapter 7

Liposomes in Cosmetics: How and Why?

Guy Vanlerberghe

CONTENTS

I. INTRODUCTION

The efficiency of many cosmetic preparations is closely related to the interplay between the physicochemical properties of lipids and related compounds, on one side, and the biophysics of the skin and its appendages, on the other side. Local, rather than systemic effects are the ultimate goal of the cosmetician. The extent to which this is obtained will be evaluated by scientific experimentation as well as by the final user's appraisal.

As consumer goods, cosmetics must be prepared on a reproducible commercial scale, and their shelf-lifetime in various environments shall not be less than 2 years.

By and large, emulsions fulfill the latter requirements, but in a distinctive manner according to the nature of their continuous phase. They are effective as carriers as well as agents through their basic components, i.e., oil and water. Whereas the latter is generally acknowledged as a plasticizer of keratin, the former imparts lubricity to the skin surface. Quite remarkable is the fact that hygroscopic, water-soluble substances contained within the corneum cells cannot be extracted by water as long as these cells are not damaged. Such is no longer the case when their membranes are disrupted by detergents, as has been shown by J.D. Middleton.[1]

These results demonstrate the importance of a suitable environment for the conservation of moisturizing factors, whether they be naturally present in the stratum corneum or delivered topically. Oil in water emulsions, stabilized by hydrophilic surfactants, are obviously the least effective in this respect.

0-8493-4013-6/96/$0.00+$.50

On the other hand, aqueous dispersions of lipophilic compounds, self-organized in closed units, appear as optimal carriers as well as structuring agents. A.D. Bangham discovered that phospholipids, dispersed in water, form closed vesicles selectively permeable to ions.

Whereas the potential applicability of these physicochemical systems in biology and medicine was soon realized, their advantage in skin care was only perceived some years later. Reduction to practice owes much to the development of improved materials and appropriate technology.

II. MEMBRANE COMPONENTS

Among the many factors that contribute to the efficacy of liposomes in cosmetics, the chemical structure of their membrane components is of cardinal importance. It must be acknowledged that much of the aura around these little objects has been generated through their relationship with Life and Nature. This should not obscure the basic physicochemical principles which sustain their existence. In this respect, quite illuminating is the structural diversity of the molecules which self-assemble in vesicles.

Listing all the compounds that have been mentioned for the preparation of liposomes would not be very productive. Our review will be limited to those which have gained a commercial significance.

A. BASIC MEMBRANE COMPONENTS

1. Glycerophospholipids

Lecithins are the common representatives of this class of lipids. They are complex mixtures, of animal or vegetable origin, in which polar lipids, insoluble in acetone, are the most abundant. Fat-free lecithins are now currently available in several grades containing 70 to 98% phosphatidyl-choline (PC). Egg and soybean are the mean sources from which they are extracted. Soybean is distinguished by its high content of unsaturated fatty acids, mainly linoleic acid (55%). Egg lecithin contains much less of this polyunsaturated fatty acid. Partially or totally hydrogenated lecithins are also produced commercially. They are characterized by their iodine number and their main phase transition temperature, which may be above 50°C.

2. Nonionic Amphiphiles

In 1975, we found that single-chain polyglycerol monoalkyl ethers (Formula I in Figure 1), whose synthesis and properties had been presented earlier,[2] form closed lamellar structures. The flexibility of design of this family of nonionics is only limited by the rather small number of appropriate long-chain alcohols available from natural or synthetic sources.[3]

A much wider range of molecular structures is obtained when two hydrocarbon chains are linked to a glyceryl group[4] (Formula II, Figure 1).

Oligomeric hydrophobic blocks can also be introduced into structures represented by Formula III (Figure 1).

3. Sphingolipids

These compounds are derivatives of sphingosine. This term currently designates long-chain 2-amino-1,3 diols, irrespective of the number of carbon atoms, unsaturation and stereoisomerism. More precisely, it should be restricted to trans-4-sphingenine, sphinganine being the saturated homolog. Only the D-erythro form or 2S,3R has been found in nature, and the 18-carbon sphingenine is predominant in most mammalian spingholipids.

Phytosphingosine is identified as 4-D hydroxy sphinganine.

Ceramides are *N*-acyl sphingosines. They are characterized by high melting points and weak hydrophilicity. Accordingly, in general, they are not able to form fluid mesophases by themselves. Such is no longer the case, when the ceramide moiety is linked to phosphocholine as in sphingomyelin, or to carbohydrates as in glycosphingolipids, e.g: cerebrosides, or to ionic carbohydrates as in gangliosides and sulfatides. The *N*-acyl group in ceramides and their derivatives is long-chained, occasionally α-hydroxylated.

The complex and varied chemical structure of sphingolipids sets a serious problem in quality control. Preference goes now, in cosmetology, to materials obtained from plants or by synthesis.

Liposomes in which sphingolipids are the building blocks have been termed "sphingosomes".[5]

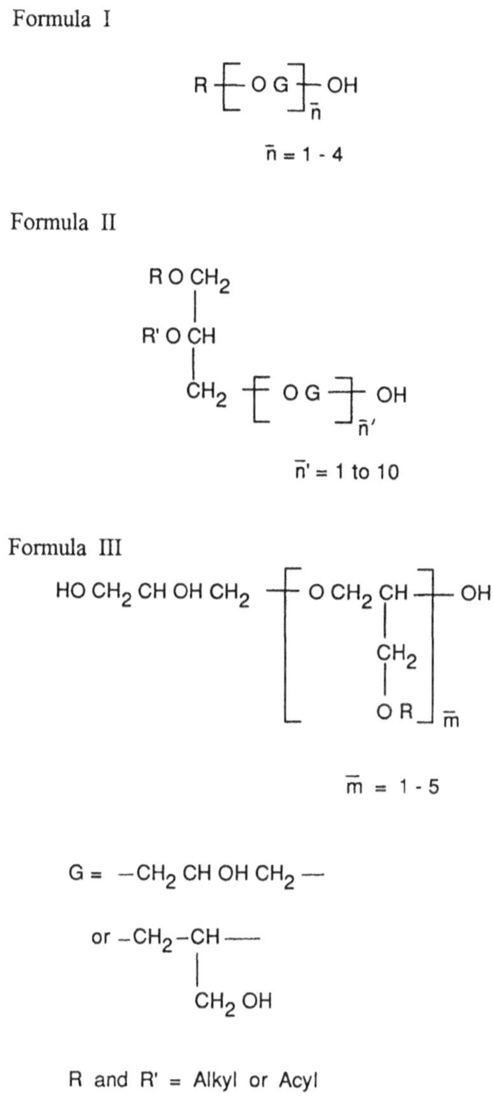

Figure 1 Chemical structure of nonionic polyglycerol derivatives forming vesicles.

B. STABILIZERS

It will be conceded that the aforementioned compounds, although they form vesicles by themselves in appropriate conditions, cannot fulfill the many requirements imposed on the formulator. For instance, long-chain amphiphiles, below their transition temperature, are prone to revert to the gel state. This process is abolished in the presence of several stabilizers.

Among them, sterols, mainly cholesterol, are most important. It has often been postulated that its activity is related to the formation of hydrogen bonds between the carbonyl groups of phospholipids and the 3-β-OH group of sterols. The results we have obtained with polyglycerol ethers show that this mechanism is not a general prerequisite to this molecular association. In the present case, cholesterol is incorporated up to an equimolecular ratio. Small-angle X-ray diffraction indicates that it disrupts interdigitation of the hydrocarbon chains in the Lβ phase. Bilayer fluidity of nonionic vesicles is also affected by cholesterol to an extent that depends on the polar head: polyoxyethylene ethers are less responsive in this respect than their polyglycerol counterparts, as evidenced by differential polarized phase fluorometry.[6]

It has been found recently[7] that polyisoprenic derivatives, such as phytanetriol (3,7,11,15-tetramethyl-1,2,3-trihydroxyhexadecane), can be substituted for cholesterol. However their behavior is different, as can be seen in Figure 2.

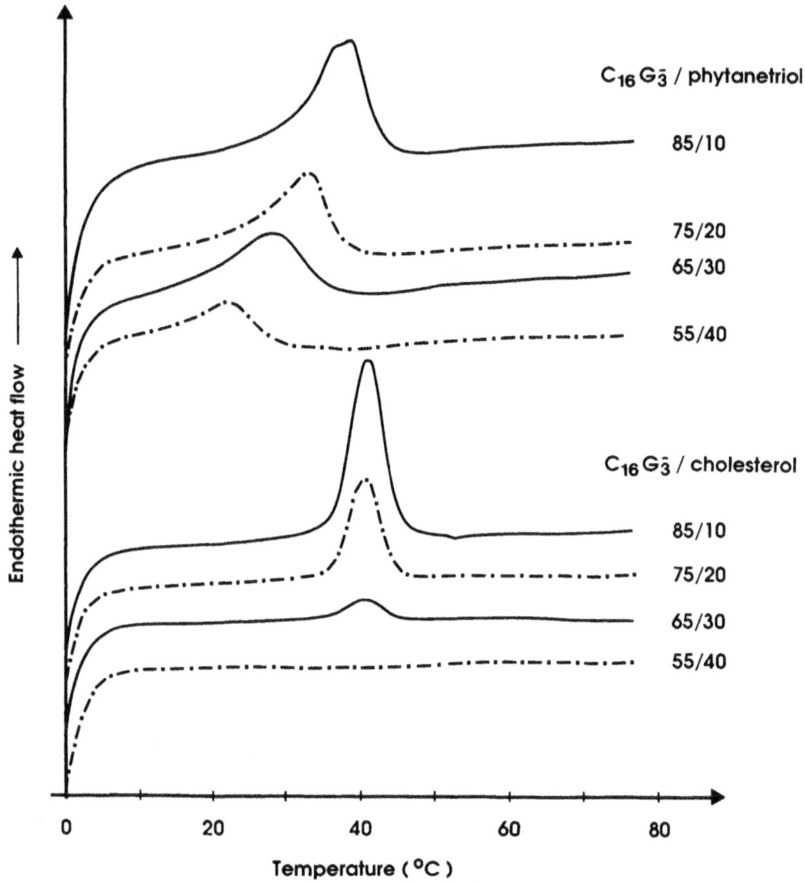

Figure 2 Differential scanning calorimetry of several mixtures of $C_{16}G_3$ (compound of Formula I) with increasing amounts of cholesterol and phytane triol.

Aqueous dispersions of vesicles must be stabilized against flocculation. Ionic amphiphiles, such as dicetyl phosphate, phosphatidic acid, cholesterol sulfate, cholesterol phosphate, lipoaminoacids and long-chain alkyl-ammonium salts are efficient for this purpose at concentrations less than 10% by weight of the total lipids. Concomitantly, they increase the thickness of the aqueous layers between the polar heads.

C. MIXED SURFACTANTS

It is well known that single-chained surfactants, predominantly hydrophilic, form micelles and several mesophases in water. As early as in 1946, Dervichian[8] showed that these compounds can associate with hydrophobic partners in molecular complexes, displaying "limited swelling" in water just as lecithins do. Ufasomes, obtained from partially neutralized unsaturated fatty acids provide an illustrative example of this phenomenon. More recently, vesicles have been prepared from *N*-acyl sarcosinates, partially ionized or compounded with cholesterol.[9] Polyoxyethylene monoethers, can also be used in similar combinations.[10]

In this approach, the choice of components is highly dependent upon their skin tolerance and the stability of the vesicles prepared from them.

III. PROPERTIES OF VESICLES RELATED TO COSMETIC APPLICATIONS

A. STABILITY

It is obvious that any discussion concerning the performance of vesicles should seriously take into account the stability of the system under study. This requirement is not easily fulfilled, since it takes on chemical as well as physical aspects. Stability *"in vivo"* is still more uncertain.

1. Chemical Stability

This problem is common to carboxylic esters, whether they are glycero-phospholipids or polyol esters. For practical purposes, at least in a medium term, it is solved satisfactorily when vesicles are maintained at pH 6.5.[11] Fatty acids, produced by hydrolysis, have no detrimental effects as long as they are present in a small percentage.[12]

The formation of mixed micelles with lysolecithins occurs at high concentrations which are not significant in the present context.

Oxidability is the main concern with unsaturated lipids. Antioxidants are generally recommended by the suppliers of these materials. For instance, according to Röding,[12] tocopherol acetate at 0.2% stabilizes 10% dispersions of soybean liposomes.

2. Physicochemical Stability

a. Phase Transitions

The role of the chemical structure of membrane components has already been touched on. It is important, in the present context, to have in mind that molecular associations depend on rather weak forces; as a consequence, molecular competition and exchanger in bilayers are possible. For instance, cholesterol is expelled from the $L\alpha$ phase by long-chain alcohols.[13] In the case of nonionic vesicles made from polyoxyethylenated alcohols, Hofland et al.[14] have observed that, when the temperature is raised to 80°C for 5 min, unilamellar vesicles revert to a multilamellar organization whether cholesterol has been added or not, and whatever the length of the polyoxyethylene groups. A possible mechanism for this transition is demixing into a surfactant rich- and a surfactant-poor phase when the cloud point is passed.

b. Solvents

Many solvents are surface active. In this respect, ethanol and isopropanol are typical. According to Lautenschläger, the former may be introduced into liposome dispersions at concentrations of 10 to 20%, and propylene glycol at 5 to 10%.[15] This phospholipid mixture forming the membrane of such liposomes is described as containing a charge carrier.[16]

The aforementioned solvents act as preservatives. Occasionally, they are used to solubilize ingredients that are insoluble in water as well as in lipids.

c. Surfactants

The lytic activity of detergents is well documented. Accordingly, one suspects that the stability of vesicles will be affected by them. This general opinion must be qualified, due regard being given to concentration and structural effects.

Light-scattering measurements have shown that the interaction of surfactants with vesicles proceeds by different steps evidenced by several breakpoints[17,18] which provide a basis to comparison. From his investigations on the release of radioactive glutamic acid, Oleniacz[19] concluded that neither an anionic surfactant (alkyl benzene sulfonate) nor an amphoteric surfactant (Miranol C 2M), disrupts the liposome structure. A cationic (Arquad 2C-75) exerted a marked effect. A nonionic (Pluronic L-64) was found to be mild in this respect.

More systematic studies suggest a different view. According to Urbaneja et al.[20] electrostatic forces do not determine the solubilization of vesicles; hydrophobic interactions are much more important. The detergent concentration that causes the onset of solubilization decreases as the alkyl chain length increases. It increases linearly with the critical micellar concentration (c.m.c.) when the latter is $>6.0^{-3}$ *M*. It is notable that chain length is not an absolute determinant of solubilization. For instance, De la Maza and Parra Juez[21] have found that among the C_{10}, C_{12}, C_{14} alkyl sulfates, the C_{12} is the most effective to form mixed aggregates.

The strong correlation between the ability to modify the permeability of liposomes and the partition coefficient of the surfactant between water and the bilayer is another important indication.[22] The same observation was made with polyethoxylated nonionics.[23] It is noteworthy that, when the size of the polyoxyethylene chain varies to a large extent, an optimal solubilization is observed at an intermediate degree of polymerization. For cetyl ethers, this occurs when the latter is 15.[24]

The influence of vesicle structure on solubilization by detergents is no less important. For instance, the octyl glucoside/lipid ratio required for the transition of nonionic surfactant vesicles to mixed micelles is three times higher than it is for egg phosphatidyl choline SUV.[25] This different behavior may be ascribed to molecular cohesion and bilayer rigidity.

B. VESICLES AS CARRIERS

The problems related to the utilization of vesicles as carriers in cosmetics are not fundamentally different from those encountered in other fields of application.

A large part of cosmetic ingredients are water soluble, and as such, they are candidates to encapsulation in the aqueous compartment of vesicles. Up to now, no general rules have been established concerning their permeation. One can only remark that water solubility is not a safeguard against leakage: molecules comprising small hydrocarbon groups, mainly aromatic, are prone to diffuse through vesicle membranes. Much less so are hydrophilic polymers, among which polysaccharides and proteins are representative.

Lipid bilayers constitute another locus of solubilization in vesicles. Guest molecules that are accommodated therein share a common amphiphilic character. Such is the case with retinoids. Retinol, retinal and retinoic acid are incorporated in membranes, modifying their structure to an extent that depends mainly on their polar head, as has been evidenced by calorimetry,[26-27] NMR,[28] and ESR.[29] From these studies, it appears that retinol is more soluble in membrane lipids than all-trans retinoic acid. Results obtained by differential scanning calorimetry (DSC) suggest that phase separation begins, respectively, at molar fractions of 0.3 and 0.1.

Carotenoids behave in a different manner. The solubility limit of β-carotene in the $L\beta$ and $L\alpha$ phases of dipalmitoyl phosphatidyl choline (DPPC) has been found to be 0.05 to 0.075 molar[30] and 0.10 ± 0.02 in egg lecithin sonicated liposomes.[31] This molecule is located in the hydrophobic core of the lipid lamellae, the polyisoprenic chains being oriented almost parallel to the lamellar plane, whereas polar carotenoids such as zeaxanthin and astaxanthin span bilayers,[32] and their solubility therein is controlled by the length of their lipophilic segment.[33]

α-Tocopherol is commonly used in cosmetics as an antioxidant. Thanks to its long polyisoprenic chain, it is embedded in lipid membranes, the fluidity of which is altered substantially. Conversely, this condition modulates its activity and provides an interesting example of free radical reactions in an organized medium. The regulative effects of tocopherols on membrane fluidity are similar to those of cholesterol. In the former, the respective roles of the chromane ring and of the polyisoprenic side chain have not yet been totally discriminated.[34,35]

α-Tocopherol appears to be not homogeneously distributed, but rather located in fluid zones of membranes. In pure saturated phosphatidylcholines, the upper solubility limit is about 40% molar according to calorimetric data.[36] The permeability to aqueous solutes is mostly reduced by it in liposomes prepared from phospholipids containing relatively high proportions of arachidonic acid.[37] Although the foregoing results were obtained with rather high, nonphysiological proportions of guest molecules, they show the solubilizing potential of vesicle bilayers, which must be considered on a qualitative as well on a quantitative basis.

This is mostly important in mixed systems, in which active molecules, located in different media, are destined to interact cooperatively. For instance, vitamin E and vitamin C are associated to protect membranes against peroxidation.[38] Interfacial binding can also take place when liposomes are exposed to aromatic compounds bearing proton-donor groups. Such is the case of phenols, which are liable to form complexes with phospholipids. This process has been observed when propyl gallate is added to DPPC liposomes.[39] In the same way, flavonoids are anchored to the polar head of phospholipids. The resulting compounds have been described as being moderately soluble in fats, insoluble in water and as forming structures which resemble vesicles.[40]

When vesicles are prepared from nonionic amphiphiles, similar phenomena will occur, mostly with polyoxyethylene derivatives. The latter interact strongly with phenolic preservatives, inactivating them.

Obviously, several mechanisms of interfacial binding are possible, as is seen in the solubilization of 2-phenyl ethanol by didodecyl ammonium bromide (DDAB) vesicles which, more strongly than micelles, incorporate this solute.[41]

C. RHEOLOGICAL BEHAVIOR

Flow properties, to a large extent, determine the preparation, the stability and the application of cosmetics. Therefore, rheological measurements are useful in quality control, provided that preliminary studies have clearly delineated the rheological profile of the product to be tested. Figure 3 shows the variation of relative viscosity with shear stress for several dispersions of nonionic surfactant vesicles containing ionic additives. Shear-thinning appears when the volume fraction of the dispersed phase exceeds 0.3; detailed analysis of the data brings to evidence the influence of size and polydispersity.[42]

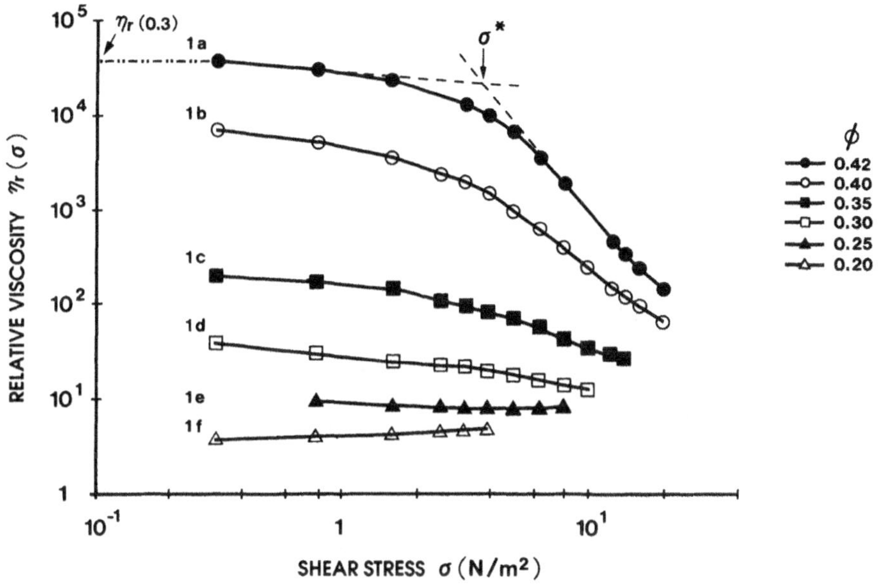

Figure 3 Rheological behavior of nonionic surfactant vesicles stabilized by dihexadecyl phosphate (zeta potential = –50 mv). Several dispersions with decreasing volume fractions have been prepared from the same stock solution.

D. FATE OF VESICULAR STRUCTURES IN THE SKIN

The maintenance of lipid vesicles in biological fluids, tissues and organs is one of the most problematic issues concerning their application *in vivo.* The skin is no exception in this respect; furthermore, its organization varies across its different layers. The stratum corneum, which is the target organ of cosmetics, is the front barrier to particles and hydrophilic substances. Therefore, it is improbable that it will passively allow the penetration of lipid vesicles.

This hypothesis has been checked experimentally. Among the available techniques, freeze fracture electron microscopy is most appropriate to visualize this phenomenon. Junginger et al.[43] were able to show the presence of vesicular structures in the uppermost, loosely packed layers of corneocytes as well as in the stratum compactum, when the apical side of stratum corneum was incubated with a dispersion of niosomes, the membrane of which contained 60% dodecyl-polyoxyethylene-ether and 40% cholesterol.

The behavior of phospholipid vesicles depends highly on their composition. Liposomes containing a high percentage of phosphatidylcholine were found to have a strong effect on human statum corneum; intercellular lipid layers showed flattened spherical structures. The conclusion of these authors is that lipid components of vesicles, when they have appropriate physicochemical properties, are able to interact molecularly with skin lipids. They suggested that vesicular structures may form again *in situ.* Destabilization of egg lecithin liposomes on the skin after topical treatment has been evidenced by R. Schubert et al.[44] by means of $\gamma\gamma$ angular correlation spectroscopy.

The aforementioned vesicular structures were obtained from substances of which some are alien to the lipids of stratum corneum. It is interesting to examine what occurs when these are made of endogeneous lipid mixtures. Such an approach has been followed by Abraham et al.;[45] they found that stratum corneum lipid-based liposomes are ultimately transformed into lamellar sheets after the addition of calcium chloride. For the moment, this process remains to be seen as an analogic model of the fusion of lamellar disks observed *in vivo*.

Since there are several indications that intact lipid vesicles do not penetrate substantially through the epidermis, their utility in topical applications has been questioned. Such an objection must be confronted with the efficacy of these systems in realistic conditions.

E. FUNCTIONAL ASPECTS IN BIOLOGICAL ENVIRONMENT

Having at our disposal vesicular systems displaying stability and notable loading capacity, we tested them on human whole skin.[46]

Three water-soluble substances were encapsulated: ^{14}C-labeled sodium pyroglutamate (PCNa) which is a moisturizer, 2-dihydroxy-acetone and tartaric dialdehyde which are skin-tanning agents. The results were the following: PCNa was concentrated in the upper layers of the skin that were taken off by stripping. A remarkable resistance to washing was observed when this solute was encapsulated in niosomes. Only a W/O emulsion, to a lesser degree, performed in a similar way. When an aqueous solution of dihydroxy-acetone and tartaric dialdehyde was enclosed in these vesicles, it produced an intense tanning resistant to washing, without discoloration. It appears therefore that vesicular structures (or materials), after deposition onto the skin, still protect aqueous solutes from lixiviation, whether these are located in the uppermost cell layers or in the lower ones whereto they have migrated.

It is very difficult to admit that such a process is mediated by the components of the lipid membrane which are nonionic and not liable to associated hydrophobically with an electrolyte like PCNa. One might evoke a penetration enhancement effect caused by them.

It will be remarked that empty niosomes had no plasticizing effect on the stratum corneum. Furthermore, when fluorescein was encapsulated, a histological study showed that penetration of the marker was limited to the stratum corneum. Therefore, we are led to conclude that the topological properties of these systems are involved in their efficiency. This does not mean that the closed structure of vesicles is indiscriminately maintained throughout the whole epidermis.

The fact that transport phenomena are affected by the chemical structure of the membrane, as has been shown in studies with paramagnetic probes,[47] is not inconsistent with this interpretation since physicochemical or biological stability, fluidity and permeability are altered concomitantly. The data reported by Patel and Moghimi[48] further confirm the present view. These authors observed that methotrexate entrapped in liposomes, applied topically, accumulates within the skin 3- to 4-fold more than the free drug applied with empty liposomes.

Finally, it must be stressed that the follicular route can contribute to percutaneous absorption.[49] Experiments with carboxy-fluorescein have shown an 8-fold enhancement of fluorescence in hamster ear sebaceous glands when liposomal preparations were compared to aqueous solutions and to the nonliposomal phospholipid mixtures.[50]

The incorporation of lipophilic substances in vesicular membranes has already been discussed from a physicochemical point of view. It is interesting to examine their behaviour *in vivo*. Masini et al.[51] observed a partial dissociation between tretinoin and DPPC related to a significantly reduced absorption and higher retention in epidermis, mainly stratum corneum. This study clearly shows the potential of liposome formulations to obtain local effects and to decrease systemic absorption.

A modest but significant protection against UV-mediated epidermal cell damage was found with hydrogenated phospholipid liposomes containing α-tocopherol.[52]

F. BIOLOGICAL EFFECTS OF VESICLE MEMBRANE COMPONENTS

1. Skin Tolerance

One of the most striking aspects of lipid composition in the epidermis is the disappearance, during differentiation and cornification, of amphiphiles forming liquid crystals in water, namely, phospholipids and glucosyl-ceramides. Correlatively, nonionic, less polar lipids become predominant and form the protective barrier of the stratum corneum.

The cosmetic implications of these phenomena have been set forth by Jass and Elias.[53] With these authors, one can logically conclude that "the ideal skin care cosmetic should avoid ingredients that have a disruptive effect on the stratum corneum." Soaps and detergents are typically such ingredients.

As regards vesicles, molecular associations of surfactants forming micelles with hydrophobic amphiphiles are a case in point. The fact that the hemolytic activity of surfactant is inhibited by addition of cholesterol should not convey the suggestion that such molecular complexes are mild to the skin. For instance, Blank and Gould[54] found that when the pH of sodium laurate solutions drops below 8.5, their penetration into excised human skin increases dramatically. This effect has been ascribed to the formation of lipid-soluble acid soaps.

Likewise, associations of well-tolerated ionic surfactants with long-chain fatty acids turn out to be irritant in occlusive patch tests. Similar adverse effects have been observed with some lipid-soluble nonionic surfactants. Florence et al.[55] have reviewed the interactions of polyoxyethylene alkyl and aryl ethers with membranes and other biological systems: "In many cases C12 hydrocarbon chain compounds appear to exert maximal effects and in all series a parabolic relationship between membrane activity and lipophilicity is observed."

French et al.[56] more recently have discussed the mechanisms of nonionic surfactant effects on skin permeability. They reported that the rate of penetration of polyoxyethylene ethers is generally quite low, but some possess a greater rate of permeation than their anionic counterparts. In a series of lauryl ethers, molecules of low HLB penetrated the skin at a higher rate than more hydrophilic homologs. Unsaturation of the hydrocarbon chain is another important factor: polyoxyethylene oleyl ethers enhance skin penetration of drugs probably by disordering the intercellular lipid lamellae. Cell toxicity evaluated on cultured human keratinocytes was found to be higher for polyoxyethylene-5 and polyoxyethylene-10 oleyl ethers compared to esters and not correlated to drug penetration enhancement.[57]

Obviously, the former is dependent on many factors. Hofland et al.[58] tried to relate it to physicochemical characteristics of nonionic surfactants. They concluded that neither HLB nor c.m.c. appeared to have an effect on cytotoxicity. In the present context, it should be noted that "lipophobicity," defined by Shinoda et al.[59] as the saturation concentration of singly dispersed surfactant in oil saturated with water is unduly overlooked.

The substantial differences in the properties of nonionics are observed according to the chemical structure of their polar head groups: one $-OCH_2CHOHCH_2-$ unit corresponds to approximately 5.6 $-OCH_2CH_2-$ units, in lipophobicity or associativity, according to these authors, and to 2–3 of these in polarity determined by chromatography.[2]

Another remarkable fact is the disappearance of lytic effects when several alkyl chains are joined to form a hydrophobic group. For instance, compounds represented by Formula II (Figure 1) are devoid of hemolytic activity at concentrations as high as 10%.[4]

In polyol esters, associativity is promoted by -OH and -COO- groups, and since these products are generally obtained as complex mixtures containing poly-substituted derivatives, physicochemical characteristics are brought together to limit adverse biological effects. Biodegradability is a further asset to this end in this case. The fact that molecules self assembling in membranes are generally safer than other amphiphiles is probably not fortuitous. Furthermore, these are indications that the vesicular state is favorable to skin tolerance, as we were able to show by comparison of vesicles and emulsions formed from nonionics at the same concentration. A rationale for this behavior might be found in reduced surface activity of vesicles comparatively to other metastable molecular assemblies, as found recently by Qui and MacDonald.[60]

2. Skin Hydration

This property is highly valuated in cosmetics for suppleness and smoothness of the skin. Artmann et al.,[61] using the corneometer which measures the capacitive resistance of the epidermis to a maximum depth of 80 μm, observed an increase in skin humidity that peaked at 30 min after application of a liposome preparation containing 80% phosphatidylcholine. They noted that other phospholipid compositions were inefficient.

A short-term effect of a liposomal gel was also observed by Gehring and Gloor.[62] However, according to these authors, no influence on the water-binding capacity could be detected after repeated application. From the experiments of Röding and Ghyczy,[63] the conclusion can be drawn that phospholipids are accumulated in the epidermis when liposomes were applied onto the skin twice a day for several weeks. Correlatively, skin humidity reaches a maximum after 7 days, and still remains elevated 2 days after the last application.

Using attenuated total reflectance Fourier transform infrared spectroscopy, Boddi et al.[64] found that liposomes promote the penetration of deuterium oxide into the skin but at the same time impair its capability to accumulate H_2O coming from inside. They ascribed this effect to an increased conformational disorder of lipid chains brought about by phospholipids. This study was carried out under occlusive conditions, using a lecithin mixture that had been found to be inefficient in Artmann's experiments. From the aforementioned investigations, it appears that the membrane lipids of liposomes can modify the water balance in the skin, according to the structure of their polar head. Resistance to water loss is obviously not an attribute of them, and is all the less so as their hydrocarbon chains are more loosely packed.

The main asset of vesicles for the alleviation of the "dry skin" syndrome is to be found in the emollient properties of their lipid components. Specific physiologic effects must be taken into account when polyunsaturated fatty acids are incorporated in lipid structures.

IV. COSMETIC FORMULATIONS BASED ON VESICLES

Since vesicles are formed and stabilized by a delicate balance of intermolecular forces, they cannot generally be considered as simple ingredients in the formulation of cosmetic products. The first problem to be addressed lies in their instability; it is circumvented by thickening the continuous dispersing phase. Polymer coatings can also be used to protect vesicles. In this approach much attention must be paid to the selection and to the dosage of the coating agent. As it has been discussed previously, the stability of vesicles is further impaired by surfactants.

However, cleansing compositions such as shampoos and shower gels containing stabilized liposome suspensions have been described.[65] According to the inventors, a sufficient amount of surfactant can be introduced into them to obtain detergency. Alkyl sulfosuccinates, acyl-amino-polyglycol ether sulfates, amino oxydes, phosphate esters and *N*-acyl amino acid salts, and derivatives of hydrolyzed proteins are mentioned as stabilizers against the effect of lytic surfactants.

In skin care products, various oils can be durably dispersed in an aqueous phase containing lipid vesicles. This composite physicochemical system does not require the addition of hydrophilic emulsifiers to be stabilized, provided that appropriate lipids are used to form the vesicles: natural lecithins are unsuitable for this purpose.

V. VESICLES AS A TOOL IN RESEARCH AND DEVELOPMENT OF NEW COSMETIC MATERIALS

The sensitivity of liposomes to surfactants and some of their similarities with biological membranes suggest they might be used to predict irritation. In fact, several authors have found significant correlations between the destabilization of liposomes and irritancy *in vivo*.[66,67]

In order to mimic more closely the composition of stratum corneum, lipid mixtures containing ceramides, cholesterol, free fatty acids and cholesterol sulfate have been used in liposomes to study the partition of sodium dodecyl sulfate into lipid bilayers.[68]

New materials produced by advanced technologies have recently been introduced into cosmetics; inorganic pigments are a typical example in this respect. Vesicles provide suitable environments to perform controlled precipitation reactions.[69] Their aqueous core is obviously a site available for this purpose: we have found that polymerization or precipitation can be achieved in vesicular dispersions and yield small-sized particles whether the reactants are encapsulated or not.[70] Melanin-like pigments are obtained by this process. This example shows that vesicles, nowadays, find their way upstream as well as downstream in cosmetics.

VI. CONCLUSION

Lipid vesicles have now reached their maturity in cosmetics. Nearly 20 years of effort in research and development have been necessary to reach this state. From the observations we have reported in this chapter, it appears there is still room for innovation in this field. The availability of new building blocks will permit the construction of supramolecular architectures with extended functions. In a less flamboyant perspective, substantial improvements in the safety, the efficacy, and pleasurability of cosmetics have already been and will be further gained.

REFERENCES

1. **Middleton, J. D.**, Development of a skin cream designed to reduce dry and flaky skin, *J. Soc. Cosmet. Chem.*, 25, 519, 1974.
2. **Vanlerberghe, G., Handjani-Vila, R. M., Berthelot, C., and Sebag, H.**, Synthèse et activité de surface comparée d'une série de nouveaux dérivés non-ioniques, in *Proc. Int. Congr. on Surface active agents*, Zürich, 1972, Carl Hauser Verlag, Munich, 1, 139, 1973.
3. **Vanlerberghe, G., Handjani-Vila, R. M., and Ribier, A.**, "Les Niosomes", une nouvelle famille de vésicules à base d'amphiphiles non-ioniques, in *Physicochimie des Composés Amphiphiles*, Bordeaux-Lac, Perron, R., Ed., *Colloq. Nat. CNRS* 938, 303, 1979.
4. **Sebag, H. and Vanlerberghe, G.**, Biomimetic surface-active agents, *Preprints 13th Congr. Fed. Soc. Cosm. Chem.*, 1, 111, 1984.
5. **Brunke, R. A. and Charlet, E.**, Production and identification of liposomes, sphingosomes and nanoparticles, *S. O. F. W.*, 117 (14), 514, 1991
6. **Ribier, A., Handjani-Vila, R. M., Bardez, B., and Valeur, B.**, Bilayer fluidity of non-ionic vesicles. An investigation by differential polarized phase fluorimetry, *Colloids Surf.*, 10, 155, 1984.
7. **Léty, A., Morançais, J. L., and Vanlerberghe, G.**, L'OREAL Int. Pat. Appl. WO 93/15708.
8. **Dervichian, D. G.**, Swelling and molecular organization in colloidal electrolytes, *Trans. Faraday Soc.*, 42B, 180, 1946.
9. **Wallach, D. F. H., Mathur, R., Redziniak, G. J. M., and Tranchant, J. F.**, Some properties of N-acyl sarcosinate lipid vesicles, *J. Soc. Cosmet. Chem.*, 43, 113, 1992.
10. **Mathur, R.**, Int. Pat. Appl., WO 93/05767.
11. **Gritt, M., De Smidt, J. H., Struijke, A., and Crommelin, D. J. A.**, Hydrolysis of phosphatidyl choline in aqueous liposome dispersions, *Int. J. Pharm.*, 50, 1, 1989.
12. **Röding, J.**, Stability, physical properties and characterization of liposomes in liquid and semi-solid preparations, *Parf. Kosm.*, 71, 80, 1990.
13. L'OREAL Research, Aulnay-sous-Bois, France, unpublished data, 1993.
14. **Hofland, H. E. J., Bouwstra, T. A., Gooris, G. S., Spies, F., Talsma, H., and Junginger, H. E.**, Nonionic surfactant vesicles: a study of vesicle formation, characterization and stability, *J. Coll. Interf. Sci.*, 161, 366, 1993.
15. **Lautenschläger, H.**, Liposomes in dermatological preparations, *Cosmet. Toiletr.*, 105, 63, 1990.
16. **Hager, J., Dürr, M., and Lünebach, E.**, EP. Appl. 0470 437 A1.
17. **Edwards, K. and Almgren, M.**, Solubilization of lecithin vesicles by $C_{12}E_8$, *J. Coll. Interf. Sci.*, 147, 1, 1991.
18. **Ollivon, M., Eidelman, O., Blumenthal, R., and Walter, A.**, Micelle-vesicle transition of egg phosphatidyl choline and octyl glucoside, *Biochemistry*, 27, 1695, 1988.
19. **Oleniacz, W. S.**, Lever Br. Cy U.S. Patent 3 957 971, 1976.
20. **Urbaneja, M. A., Alonso, A., Gonzalez, J. M., Goni, F. M., Partearroyo, M. A., Tribout, M., and Paredes, S.**, Detergent solubilization of phospholipid vesicles, *Biochem. J.*, 270, 305, 1990.
21. **De la Maza, A. and Parra Juez, J. L.**, Disintegration of liposomes by anionic surfactants and formation of mixed micelles, *Langmuir*, 9 (3), 870, 1993.
22. **De la Maza, A., Parra, J. L., Garcia, M. T., Ribosa, I., and Sanchez Leal, J.**, The alteration of small unilamellar liposomes by amphoteric alkylbetaine surfactants, *Colloids Surf.*, 61, 281, 1991.
23. **De la Maza, A., Parra, J. L., Garcia, M. T., Ribosa, I., and Sanchez Leal, J.**, *J. Coll. Interf. Sci.*, 148 (2), 310, 1992.
24. **Jin-Gu Kim and Jong-Duk Kim**, *J. Biochem.*, 110, 436, 1991.
25. **Lesieur, S., Madelmont, C. G., Paternostre, M. T., Mozeau, J. M., Handjani-Vila, R. M., and Ollivon, M.**, *Chem. Phys. Lipids*, 56, 109, 1990.
26. **Gurrieri, S. and Castelli, F.**, Thermotropic behaviour of dipalmitoyl-phosphatidyl choline liposomes containing retinoids, *Thermochim. Acta*, 122, 117, 1987.
27. **Castelli, F. and Gurrieri, S.**, Differential scanning calorimetric investigation on dipalmitoyl phosphatidyl choline — Retinoic and aqueous dispersions, *Bull. Mol. Biol. Med.*, 13, 1, 1988.
28. **De Boeck, H. and Zidovetzki, R.**, NMR study of the interaction of retinoids with phospholipid bilayers, *Biochim. Biophys. Acta*, 946, 244, 1988.
29. **Wassal, S. R., Phelps, T. M., Albrecht, M. R., Langsford, C. A., and Stillwell, W.**, Electron spin resonance study of the interactions of retinoids with a phospholipid model membrane, *Biochim. Biophys. Acta*, 939, 393, 1988.
30. **Holev, V. D. and Kafalieva, D. N.**, Miscibility of β-carotene and zeaxanthin with dipalmitoyl phosphatidyl choline in multilamellar vesicles: a calorimetric and spectroscopic study, *Photochem. Photobiophys.*, 11, 257, 1986.
31. **Holev, V. D. and Kafalieva, D. N.**, *C.R. Acad. Bulg. Sci.*, 38, 755, 1985.
32. **Subczynski, W. K., Markowska, E., Gruszecki, W., and Sielewiesuk, J.**, Effects of polar carotenoids on dimyristoyl phosphatidyl choline membranes: a spin-label study, *Biochim. Biophys. Acta*, 1105, 97, 1992.
33. **Milon, A., Wolff, G., Ourisson, G., and Nakatani, Y.**, Incorporation of zeaxanthin, astaxanthin, and their C_{50} homologues into dimyristoyl phosphatidyl choline vesicles, *Helv. Chim. Acta*, 69, 12, 1986.

34. **Fukuzawa, K., Chida, H., and Suziki, A.**, Fluorescence depolarization studies of phase transition and fluidity in lecithin liposomes containing α-tocopherol, *J. Nutr. Sci. Vitaminol.*, 26, 427, 1980.
35. **Urano, S., Yano, K., and Matsuo, M.**, Membrane stabilizing effect of vitamin E: effect of α-tocopherol and its model compounds on fluidity of lecithin liposomes, *Biochem. Biophys. Res. Commun.*, 150, 469, 1988.
36. **Ortiz, A., Aranda, F. J., and Gomez-Fernandez, J. C.**, A differential scanning calorimetry study of the interaction of α-tocopherol with mixtures of phospholipids, *Biochim. Biophys. Acta*, 898, 214, 1987.
37. **Diplock, A. T., Lucy, J. A., Verrinder, M., and Zieleniewski, A.**, α-Tocopherol and the permeability to glucose and chromate of unsaturated liposomes, *FEBS Letters*, 82, 341, 1977.
38. **Leung, H. W., Vang, M. J., and Mavis, R. D.**, The cooperative interaction between vitamin E and vitamin C in suppression of peroxidation of membrane phospholipids, *Biochim. Biophys. Acta*, 664, 266, 1981.
39. **Ishinaga, M., Okita, Y., and Ito, A.**, Calorimetric study on a dipalmitoyl phosphatidyl choline–propyl gallate mixture, *J. Biochem.*, 100, 1663, 1986.
40. **Bombardelli, E. and Spelta, M.**, Phospholipid-polyphenol complexes: a new concept in skin care ingredients, *Cosmet. Toiletr.*, 106, 69, 1991.
41. **Kondo, Y., Abe, M., Ogino, K., Uchiyama, H., Scamehorn, J. F., Tucker, E. E., and Christian, S. D.**, Solubilization of 2-phenylethanol in surfactant vesicles and micelles, *Langmuir*, 9, 899, 1993.
42. **Morinet, F., Morançais, J. L., Léty, A., and Vanlerberghe, G.**, Comportement rhéologique de dispersions aqueuses de vésicules lipidiques, *24e Colloque annuel du Groupe Français de Rhéologie*, MRT, 20, 1989.
43. **Junginger, H. E., Hofland, H. E. J., and Bouwstra, J. A.**, Liposomes and niosomes: interactions with human skin, *Cosmet. Toilet.*, 106, 45, 1991.
44. **Schubert, R., Joos, M., Deicher, M., Magerle, R., and Lash, J.**, Destabilization of egg lecithin liposomes on the skin after topical application measured by perturbed γγ angular correlation spectroscopy (PAC) with ^{111}, in *Biochim. Biophys. Acta*, 1150, 162, 1993.
45. **Abraham, W., Wertz, Ph. W., Landmann, L., and Downing, D. T.**, Stratum corneum lipid liposomes: calcium-induced transformation into lamellar sheets, *J. Invest. Dermatol.*, 88, 212, 1987.
46. **Handjani-Vila, R. M., Ribier, A., Rondot, B., and Vanlerberghe, G.**, *Int. J. Cosmet. Sci.*, 1, 303, 1979.
47. L'Oreal Research, unpublished data.
48. **Patel, H. M. and Moghimi, M.**, Liposomes and the skin permeability barrier, in *Drug Targeting Delivery*, Vol. 2, *Liposomes in Drug Delivery*, 1993, 137.
49. **Illel, B., Schaefer, H., Wepierre, J., and Doucet, O.**, Follicles play an important role in percutaneous absorption, *J. Pharm. Sci.*, 80, 424, 1991.
50. **Lieb, L. M., Ramachandran, Ch., Egbaria, K., and Weiner, N.**, *J. Invest. Dermatol.*, 99, 108, 1992.
51. **Masini, V., Bonte, F., Meybeck, A., and Wepierre, J.**, Cutaneous bioavailability in hairless rats of tretinoin in liposomes or gel, *J. Pharm. Sci.*, 82, 17, 1993.
52. **Werninghaus, K., Handjani, R. M., and Gilchrest, B.A.**, Protective effect of alpha-tocopherol in carrier liposomes on ultraviolet-mediated human epidermal cell damage in vitro, *Photodermatol. Photoimmunol. Photomed.*, 8, 236, 1991.
53. **Jass, H. E. and Elias, P. M.**, The living stratum corneum: implications for cosmetic formulation, *Cosmet. Toiletr.*, 106, 47, 1991.
54. **Blank, I. H. and Gould, A. B.**, Penetration of anionic surfactants into skin, *J. Invest. Dermatol.*, 37, 485, 1961.
55. **Florence, A. T., Tucker, I. G., and Walters, K. A.**, Interactions of non-ionic polyoxyethylene alkyl and aryl ethers with membranes and other biological systems, in *Structure/Performance Relationships in Surfactants*, M. J. Rosen, Ed., *ACS. Symp. Ser.* 253, 1984, 189.
56. **French, E. J., Pouton, C. W., and Walters, K. A.**, Mechanisms and prediction of non-ionic surfactant effects on skin permeability, *Drugs and Pharm. Sci.*, 59, 113, 1993.
57. **Kadir, R., Tiemessen, H. L. G. M., Ponec, M., Junginger, H. E., and Bodde, H. E.**, Oleyl surfactants as skin penetration enhancers. Effects on human stratum corneum permeability and in vitro toxicity to cultured human skin cells, *Drugs. Pharm. Sci.*, 59, 215, 1993.
58. **Hofland, H. E. J., Bouwsta, J. A., Verhof, J. C., Buckton, G., Chowdry, B. Z., Ponec, M., and Junginger, H. E.**, Safety aspects of non-ionic surfactants vesicles: a toxicity study related to the physicochemical characteristics of non-ionic surfactants, *J. Pharm. Pharmacol.*, 44, 287, 1992.
59. **Shinoda, K., Fukuda, M., and Carlsson, A.**, Characteristic solution properties of mono-, di,- and triglyceryl alkyl ethers: lipophobicity of hydrophilic groups, *Langmuir*, 6, 334, 1990.
60. **Qiu, R. and MacDonald, R. C.**, A metastable state of high surface activity produced by sonication of phospholipids, *Biochim. Biophys. Acta*, 1191, 343, 1994.
61. **Artmann, C., Röding, J., Ghyczy, M., and Pratzel, H. G.**, Influence of various liposome preparations on skin humidity, *Parf. Kosm.*, 71 (5), 326, 1990.
62. **Gehring, W. and Gloor, M.**, Effect of a liposomal gel on hydration and water binding capacity of stratum corneum, *Dermatol. Monatsschr.*, 179, 201, 1993.
63. **Röding, J. and Ghyczy, M.**, Control of skin humidity with liposomes. Stabilization of skin care oils and lipophilic active substances with liposomes, *S.O.F.W.*, 117(10), 372.

64. **Bodde, H. E., Pecatolo, L. A. R. M., Subnel, M. T. A., and De Haan, F. H. N.,** Monitoring *in vivo* skin hydration by liposomes using infrared spectroscopy in conjunction with tape stripping, in *Liposome Dermatics,* O. Braun-Falco, H. C. Korting, and H. I. Maibach, Eds., Springer Verlag, 1992, 137.
65. **Hart, G. L. and Charaf, U.** Johnson & Son Inc., Stabilized liposome process and compositions containing them, WO 92/04010.
66. **Charaf, U. K. and Hart, G. L.,** Phospholipid liposomes/surfactant interactions as predictors of skin irritation, *J. Soc. Cosmet. Chem.,* 42, 71, 1991.
67. **Redziniak, G.,** Phospholipids is cosmetics: in vitro technique based on phospholipids membrane to predict the in vivo effect of surfactants, in *Phospholipids: Biochem. Pharm. Anal. Consid.,* Proc. Int. Colloq. Lecithin, Hanin Israel and Pepeu Gianacarlo, Eds., Plenum, New York, 1990.
68. **Downing, D. T., Abraham, W., Wegner, B. K., Willman, K. W., and Marshall, J. L.,** *Arch. Dermatol. Res.,* 285, 151, 1993.
69. **Meldrum, F. C., Heywood, B. R., and Mann, S.,** Influence of membrane composition on the intravesicular precipitation of nanophase gold particles. *J. Coll. Interf. Sci.,* 161, 66, 1993.
70. **Morançais, J. L., Léty, A., and Vanlerberghe, G.,** L'Oreal, Fr. Pat. 2,677,897, 1993.

Chapter 8

Transport of Liposome-Entrapped Molecules into the Skin as Studied by Electron Paramagnetic Resonance Imaging Methods

Marjeta Šentjurc and Violeta Gabrijelčič

CONTENTS

I. INTRODUCTION

A. INTRODUCTORY REMARKS

Over the last decades, liposomes have been intensively studied as novel carriers for dermal and transdermal administration of drugs and skin care substances. For many drugs, when applied entrapped in liposomes, their exhibited effects were increased several times as compared to the conventional formulations. The results show higher concentrations of the delivered drug in different skin layers[1] or increased pharmacological effects[2] when the drug is applied on the skin entrapped in liposomes as compared to the application of the drug in conventional formulations. Although enhanced delivery of liposome-entrapped molecules is well documented, there is no detailed explanation of the mechanisms by which the liposomes facilitate the transport of the entrapped molecules into the skin. Do the liposomes disintegrate on the surface of the skin and the released molecules alone penetrate into the skin or do

0-8493-4013-6/96/$0.00+$.50

drugs penetrate together with some building domains of liposomes? Do even the liposomes penetrate intact through the stratum corneum? These are the basic questions in this field. Results of many investigations support the supposition that some liposomes can penetrate into the skin,[3,4] although a general opinion is that such large particles as liposomes can not cross the very narrow intracellular routes in the stratum corneum intact.[5,6] Some authors speculate that probably the transfolicular route is the path for liposome transport into the skin.[7]

Recent developments in one-dimensional electron paramagnetic resonance imaging (1D-EPRI)[8] and nitroxide reduction kinetic imaging methods[9,10] made possible continuous following of the evolution of concentration profiles into the skin, separately for liposome-entrapped molecules and for the released, free molecules, as well as to measure the decay rate of liposomes during their penetration into the skin.[10] The transport processes can be studied directly by entrapping paramagnetic molecules, which serve as markers for imaging the transport into the skin, into the liposomes. Owing to the reducing capability of skin, it is possible to distinguish the paramagnetic molecules that have been released after disintegration of liposomes and can therefore be reduced in the skin from those which are protected from reduction. Hydrophilic or lipophilic nitroxide radicals may be chosen for this purpose. They reflect and imitate the behaviours of drugs entrapped into the liposomes. The methods are convenient for the investigation of transport processes of spin-labeled molecules in skin and also for the investigation of metabolic activity of skin.[11,12] Other methods do not permit simultaneous following, on the same sample, of the spacial and time evolution of penetrating molecules in skin and, in addition, obtaining data on the stability of liposomes in contact with skin and on that portion of molecules which penetrate into the skin protected from reduction. Therefore EPR methods are specially helpful in the investigation of the mechanisms of facilitated transport into the skin of molecules applied to the skin entrapped in liposomes.

In this review the introductory notes briefly present the possible routes of transport of molecules into the skin to better understand the liposome–skin interactions. In the second part the electron paramagnetic resonance imaging methods are described and in the third part their application in the investigation of liposome–skin interactions is presented. The influence of liposome size and composition on their interaction with skin surface and on the transport of liposome-entrapped charged hydrophilic molecules into the skin have been investigated by these methods, and the results are overviewed in this work.

In order to understand better the mechanism of transport into the skin of molecules applied to the skin entrapped in liposomes, some investigations were performed on a model system consisting of lecithin and water which, at certain ratios, form lamellar phases similar to those in the stratum corneum.[13] Besides, the influence of varying salt concentrations on liposome dispersion and on their transport into a model system was examined in order to see how the difference in osmolality between this system and liposome dispersion affects the transport of liposomes and their shrinkage in an environment with greater osmolality, and how the elastic properties of liposome membranes influence the shrinkage.[14] This investigation gives an additional insight into the mechanism of liposome–skin interaction which is discussed in the fourth part of the review.

The last part presents the influence of hydrophilic polymers used as the thickening agents in final liposomal formulations which make liposomes convenient for dermal application, on the transport of liposomes and liposome-entrapped molecules into the skin. The type and concentration of hydrogel-forming polymer on liposome stability and on the rate of penetration of liposome entrapped molecules into the skin was investigated by EPR imaging methods.

B. TRANSPORT OF MOLECULES INTO THE SKIN

The skin has been recognised as an important drug delivery route. Through transdermal delivery of systemically acting drugs unpredicted gastrointestinal adverse effects and circumventing hepatic first-pass effects often associated with oral dosing may be avoided. Besides, through dermal delivery of locally acting drugs, unwanted systemic effects may be avoided.

However, intact skin is a barrier difficult to overcome.[15] It is designed to protect the body from the environment and to prevent water loss from the body.[16] Therefore, drugs permeate, if at all, only slowly through the intact stratum corneum, the upper skin layer.[17] Additionally, drug transport through the skin is difficult to control owing to the marked variations in the permeability of human skin depending on site, age and general condition.[18] Although skin permeabilities do vary considerably owing to these biological differences, the delivery of the drug has to be controlled primarily by the drug delivery system, and not by the skin itself.

Anyhow, in determining the amount and rate at which the penetrants permeate into the skin, it is important to know the possible penetration routes. There are two main routes for the drugs to permeate from the skin surface down into the viable tissue:

1. The appendageal route via the sebaceous glands, sweat pores, hair follicles.
2. The transepidermal route:
   - intracellular: directly crossing the skin lipids and the internal space of the corneocytes, or
   - intercellular: through the lipid matrix of stratum corneum.

Some authors favor the appendageal shunt routes as the principal mode of entry, whereas others consider this route to be unimportant.[19]

In general, it is difficult to predict which of the routes will be followed by the drug, since numerous factors are involved. Since stratum corneum is a dead tissue layer consisting of corneocytes embedded in a lipid matrix, diffusion is a passive process where only physicochemical laws are important. There is no active transport. The parameters that determine the route taken by a drug include its charge, size and partition coefficient. The pharmacokinetics of molecules which penetrate through the skin is determined by physicochemical parameters determining drug–skin, vehicle–skin and drug–vehicle interactions.

Whereas some (polar) drugs may in principle penetrate the stratum corneum transcellularly,[20] most drugs are thought to penetrate the stratum corneum mainly via the non tortuous intercellular route that involves two basic domains available for diffusion: the lipid bilayers and the interstitial hydrophilic layers.[18] However, in any case the intercellular lipids play an important role in the stratum corneum barrier function.[13]

As we already mentioned, the delivery of the drug has to be controlled primarily by the drug delivery system and not by the skin itself. Over the last few years different vesicles have been studied for possible use as novel carriers in topical administration of drugs. Liposomes have been found to increase the permeability of the skin for various compounds. They appear to be very promising for use as dermal and transdermal drug carriers.[21-29]

II. ELECTRON PARAMAGNETIC RESONANCE METHODS

A. BASIC PRINCIPLES

Electron paramagnetic resonance (EPR or ESR) is a spectroscopic method used in observation of paramagnetic centers. Paramagnetic centers can be native in the sample (free radicals, transition metal ions, etc.) or they can be artificially introduced into the sample (spin labels or spin probes); in the latter case they serve as markers providing data on their surroundings. Since paramagnetic centers are atoms, ions or molecules with one or more nonpaired electrons, they exhibit a magnetic moment different from zero. In an external magnetic field an interaction occurs between the magnetic moment of electrons and the magnetic field that splits the electrons into populations with different energies (E) which are expressed by the equation:

$$E = g\beta B m_s \qquad (1)$$

g = spectroscopic splitting tensor
β = Bohr magneton (9.273×10^{24} Am^2)
B = magnetic field density
m_s = eigenvalues of the components of the electron spin S along the magnetic field direction.

In the presence of electromagnetic waves with energy $h\nu$ (h = Plank's constant, ν = microwave frequency, typically 9.5 GHz) which is equal to the difference between the energy levels, a transition of electrons is induced between these levels. As a consequence electromagnetic waves are absorbed and this is detected as an EPR spectrum. From the position of absorption lines the data on the nature of the paramagnetic center can be obtained; the intensity of the spectrum is proportional to the concentration of paramagnetic centers in the sample; the line-shape of the spectrum provides data on the structure and dynamics of the surroundings of the paramagnetic centers.

A more detailed description of the basic principles of electron paramagnetic resonance and its use in biology and medicine is given elsewhere.[30]

Paramagnetic centers artificially introduced into the sample as spin probes or spin labels are usually nitroxide radicals. These are stable radicals with a piperidine or pyrrolidine ring. They are stable within a defined temperature and pH range, but in biological systems they may be metabolized to nonparamagnetic products; consequently, the intensity of EPR spectra decreases, providing additional data on some metabolic reactions in the system.

B. ONE- AND TWO-DIMENSIONAL EPR IMAGING

1. One-Dimensional Imaging

This method is used to measure concentration distribution of paramagnetic centers in samples in one dimension and is particularly useful in measuring the transport of paramagnetic molecules into the samples. Thus, the evolution of concentration distribution profiles can be followed in time. For this purpose a magnetic field gradient is applied in the direction of laboratory magnetic field B_0; gradients up to 1 T/m were found to be close to the optimum by some authors.[31] Due to this magnetic field gradient (G), paramagnetic centers which are spatially resolved fulfil the resonant condition at different laboratory magnetic field density (Figure 1). This is presented by the equation:

$$h\nu = g\beta(B_0+Gx) \tag{2}$$

B_0 = laboratory magnetic field density
G = magnetic field gradient in the direction of laboratory magnetic field dB/dx
x = the distance to the place where $B = B_0$

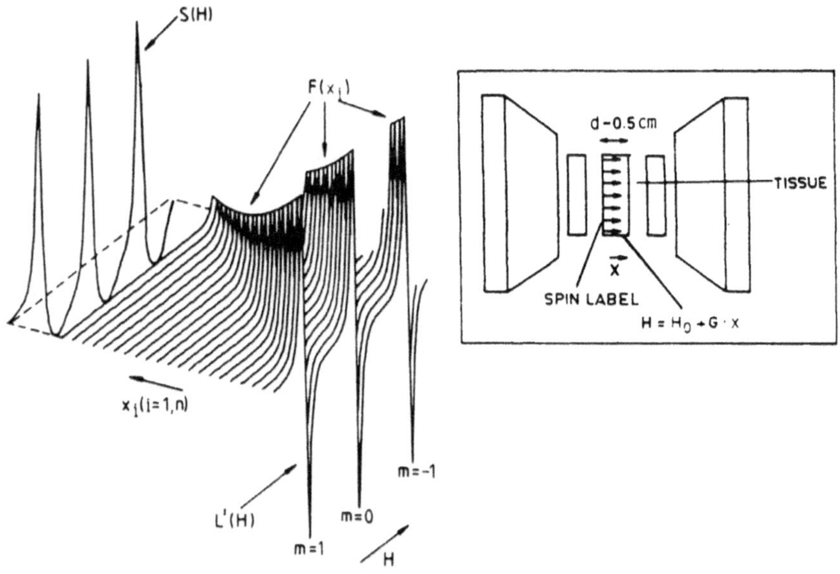

Figure 1 A graphical reconstruction of the magnetic field gradient EPR spectrum S(H) of a sample within which the diffusing molecules are distributed corresponding to F(x,t). The individual spectra of spatially separated molecules L′(H) are shifted stepwise in the magnetic field to illustrate the resulting spectrum S(H), calculated as their superposition by Equation 4 ($H = B/\mu\mu_o$, μ and μ_o are constants). The insert shows the geometry of the sample, magnetic field and magnetic field gradient for 1D-EPRI. The liposome formulation is applied at the left edge and then diffuses in the direction indicated by the arrows. (From Demsar, F., Cevc, P., and Schara, M., *J. Magn. Reson.*, 69, 258, 1986. With permission.)

During the measurement, magnetic field density B_0 is varied. Without gradient the resonant condition is fulfilled for all paramagnetic centers at the same time. Due to magnetic field gradient, the centers which are spatially distributed in the sample fulfil resonant condition at different values of B_0. Thus, spatially separated samples of paramagnetic probes can be detected as separate peaks. Resolution of two centers is determined by the relation:[32]

$$\Delta z = w_m/kG \tag{3}$$

Δz = resolution defined as a distance between two volume elements that can be detected separately,
w_m = line width of the EPR absorption line at magnetic field gradient $G = 0$,
k = 1.73 for Lorentzian line shape,
G = 1.18 for Gaussian line shape.

The method was first introduced by Karthe and Wehrsdorfer[32] and was later used for different purposes by different authors.[31,33-35] In biological systems it was used for measuring the diffusion of charged and neutral spin probes into liver tissue[36] and for measuring the diffusion of spin-labeled local anaesthetic lidocain in various ointment bases and into the skin.[37] A compilation of the relevant literature was given by Eaton and Eaton.[38]

If there is a concentration distribution of paramagnetic substance in the sample C(x,t) (for example, due to the transport of this substance into the sample) the observed spectrum S(B) is a superposition of the local Lorentzian spectra L_m from thin slices perpendicular to the direction of magnetic field gradient (x), which is expressed by the equation:

$$S(B) = \sum_i F(x_i, t) \sum_{m=1}^{-1} L_m'(B(x_i)) \tag{4}$$

where $B(x_i) = B_o + Gx_i$; L_m' are the first derivatives of the Lorentzian line shape representing the three hyperfine components of the paramagnetic probe; $F(x_i,t) = C(x_i,t)/C(0,0)$, where $C(x_i,t)$ is concentration distribution of the paramagnetic substance in the direction x, and C(0,0) is the concentration of the paramagnetic centre on the interface between the skin and the thread (surface concentration) in the time when the skin is put in contact with the thread.

The line shape depends on the concentration distribution of the paramagnetic probe in the sample. If the nitroxide molecules diffuse into the sample the concentration distribution profile can be calculated according to the model described by Demsar et al. The concentration distribution for the paramagnetic probe C(x,t) can be described by the one-dimensional diffusion equation:

$$\frac{\partial C(x,t)}{\partial t} = \frac{\partial}{\partial x} D \frac{\partial C(x,t)}{\partial x} \tag{5}$$

where D is the diffusion constant, x is the distance from the surface which is held at constant concentration of the paramagnetic probe C(0,0), and t is the time from the start of the diffusion.

If in the first approximation the model for a semifinite sample with constant surface concentration C(0,0) is taken into account, the solution of this equation with the boundary conditions: $C(0,t) = C(0,0)$ at $x = 0$ and $C(x,0) = 0$ at $x > 0$, is the distribution function:

$$F(x,t) = \frac{C(x,t)}{C(0,0)} = 1 - \mathrm{erf}\left(\frac{x}{2\sqrt{Dt}}\right) \tag{6}$$

where erf is the error function.

F(x,t) is plotted in Figure 2a, for different times t, and the corresponding calculated EPR spectra are presented in Figures 2b and c. As is evident from Figure 2, EPR spectra are asymmetric owing to the distribution of the paramagnetic probe in the sample. The parameter $l = I_1/I_2$ describes the asymmetry of the spectrum and is strongly diffusion dependent. It can be used in the determination of the diffusion constant by comparing the experimental and simulated spectra. The procedure can be used when the diffusion is slow compared to the time of sample handling and measurement, as is usually the case in biological samples.

The method has a broad application potential in dermatological research, to measure the transport of different spin-labeled molecules into the skin[11,36,39] or to measure the transport of such molecules in ointments.[37]

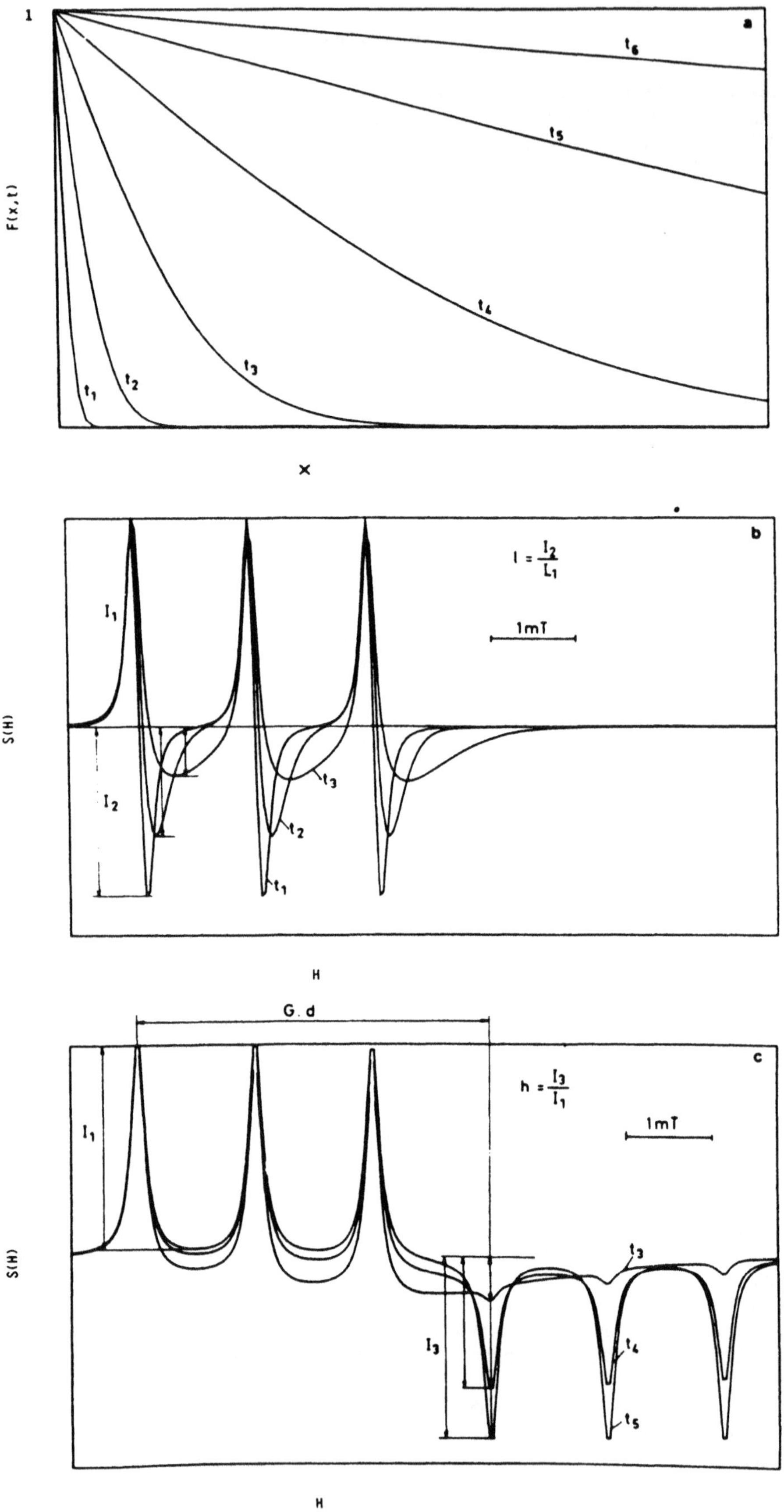
1
a
F(x,t)
t1
t2
t3
t4
t5
t6
x
b
I = I2/L1
1mT
I1
I2
S(H)
t1
t2
t3
H
G.d
c
h = I3/I1
1mT
I1
I3
S(H)
t3
t4
t5
H

2. Two-Dimensional EPRI and EPR *In Vivo*

By EPR it is also possible to get two- or three-dimensional images of concentration distribution of paramagnetic centers in the sample.[11,12,40,41] In comparison with other techniques suitable for measurement of concentration distribution in the sample (magnetic resonance tomography — MRI and ultrasound) EPR has some significant advantages such as high spatial resolution and measurement of some biochemical and biophysical characteristics of the sample.[11,12] With a lipid-soluble spin probe penetrating into the skin the point distinction resolution of 25 µm was obtained, although for the spectral line width of 0.1 mT and a field gradient of 5 T/m a theoretical resolution of 7 µm can be achieved.[11]

The main disadvantage of EPR imaging methods using conventional X-band (9.5 GHz) EPR spectrometers for biological samples is the high dielectric absorption due to the high concentration of water in the sample. Only thin tissue slices no more than 1 mm thick can be measured. Recently, spectrometers have been developed that operate at lower microwave frequencies (1–2 GHz, L-band). By these spectrometers, samples about 10 times thicker can be measured. However, it should be taken into account that by lowering the frequency the sensitivity of spectrometer decreases appreciably and only high concentrations of paramagnetic centers (about 1 m*M*) can be measured.

By using this type of spectrometers and specially designed surface detectors (surface coils), biological samples of 5 mm thickness can easily be imaged. The use of surface coils make possible the *in vivo* imaging of biological tissues.[42-44] The penetration of microwaves into the tissue varies considerably with the type of tissue investigated and is mainly a function of tissue water content. The water content in human epidermis is within the same range as that of total skin (65–70%), but in the horny layer the water content is only 10–30%.[11] Based on these data it is estimated that by X-band EPR spectrometer using surface coils, the imaging of the specimens with a total thickness of 3.5 mm is possible and with the L-band spectrometer the thickness of 35 mm can be measured; this suggests that X- and L-band EPR spectrometers can be used for imaging of skin and subcutaneous structures.[11] The limitation is the low concentration of paramagnetic centres in skin tissues.

Although 2D-EPRI provides more detailed information on the distribution of the penetrating paramagnetic substances into tissue, it allows measurement of paramagnetic center distributions in a selected tissue plane, and the analysis of spectral changes, the rapid information on liberation, penetration and accumulation of a paramagnetic compound in skin can be obtained already by 1D-EPRI. This approach is particularly suitable for measuring of fast penetration kinetics into the skin and is used in our investigations of the transport of spin-labeled molecules applied to the skin entrapped in liposomes.

3. Measurements of Liposome Transport into the Skin by 1D-EPRI

For this purpose, paramagnetic probes should be incorporated into the liposomes. Usually hydrophilic spin probe ASL, *N*-(1-oxyl-2,2,6,6-tetramethyl-4-piperidinyl)-*N*-dimethyl-*N*-hydroxyethylammonium iodide (insert in Figure 3) was used. It is easily soluble in water and, owing to its charge, does not penetrate the liposome membrane. ASL was incorporated into the liposomes during liposome preparation. Uncaptured ASL was removed by dialysis at 4°C for 24 h. The final concentration of ASL entrapped in liposomes was 0.01 mol/l.

However, some other hydrophilic probes entrapped in the liposome internal aqueous volume or lipophilic probes incorporated into the liposome membrane may also be used.

For measuring the transport, in our experiments a slice of pig ear skin approximately 1 mm thick and 1 cm long was placed in the tissue cell. A thread (1 cm long) soaked with liposome dispersion was put in contact with the narrow surface of the skin and both were put into the microwave cavity with the large plane parallel to the magnetic field gradient. The geometry of the sample in the magnetic field is shown in Figure 1. The EPR spectra obtained in the magnetic field gradient were measured at various intervals after the skin had been put in contact with the liposome dispersion. The change in the shape of the spectra reflected the concentration profile of the spin probe diffusing into the skin.

However, this method does not offer a possiblity of distinguishing the penetration of spin labeled molecules entrapped in liposomes from those released after the disintegration of the liposomes. Another

Figure 2 (a) Distribution function F(x,t) for spin-labeled molecules in tissue, calculated at different times (t) of diffusion. (b) and (c) The calculated EPR spectra line shapes for a tissue slice exposed to a constant concentration source of the diffusing nitroxide spin probe molecules. The spectra are scaled for better presentation, so that the first peak heights are equal. The peak size parameters *l* and *h* are defined. (From Demsar, F., Cevc, P., and Schara, M., *J. Magn. Reson.*, 69, 258, 1986. With permission.)

disadvantage of this method is that the reducing agents in the skin transform the paramagnetic spin probe, released from disintegrated liposomes, to the diamagnetic EPR nonvisible form. As a consequence, the EPR spectra intensity decreases with time. An additional term has to be included into the diffusion equation (5) to take into account the nitroxide reduction. For the first order reaction this is expressed by:

$$\frac{\partial C(x,t)}{\partial t} = \frac{\partial}{\partial x} D \frac{\partial C(x,t)}{\partial x} - kC(x,t) \tag{7}$$

where k is the reduction constant for the first order reaction.

On the other hand, this property, although a disadvantage for 1D-EPRI, has been explored in order to get additional information on the transport of liposome entrapped molecules into the skin. A method has been developed that gives data on the transport of liposome-entrapped and -released paramagnetic probes separately as well as the data on disintegration of liposomes. We named this the nitroxide reduction kinetic imaging method.

C. NITROXIDE REDUCTION KINETIC IMAGING

1. Reduction of Nitroxides by Biological Systems

By different redox reactions in biological systems, nitroxides can be reduced to hydroxylamines or can be oxidized to oxoamonium ions: neither is detectable by EPR. The reduction usually predominates and can be enzymatic or nonenzymatic. The enzymatic reduction can be direct by metabolic reactions catalysed by enzymes or indirect by organic radicals that are formed in metabolic reactions.[45] In living systems, the main sites of reduction are situated in mitochondria on the level of ubiquinone,[46,47] but can also be some other oxyreductases of the respiratory chain. Reduction also was observed in microsomes,[48] besides, some slow reduction was observed in cytosol.[49] Nonenzymatic reduction is usually produced by some reducing agents in the sample, usually by ascorbate and sometimes by thiols or free radicals formed at lipid peroxidation.[45] The pathway and rate of reduction depend strongly on the structure, charge and solubility characteristics of the nitroxide, and also on the concentration of oxygen in the sample, which is important especially, owing to potential reoxidation of hydroxylamines to nitroxides.[50]

2. Nitroxide Reduction in Skin

Our measurements show the reduction of spin probe ASL to be faster on the surface of the stratum corneum than in viable skin layers.[9] It has been reported[51,52] that the main reducing agent of nitroxides in skin is tioredoxin reductase, an enzyme bound to the external membrane of keratinocytes. The system tioredoxin/tioredoxin reductase is important for protection of skin against the damage due to free radicals produced by UV rays. Nitroxides are first reduced to hydroxylamines and later to secondary amines.[51] Immunohistochemically tioredoxin reductase was found in stratum germinativum, melanocytes, Langerhans cells, hair follicles and in sweat glands.

However, the validity of some of the major conclusions of this work have been challenged.[53,54] Namely, some reducing activities were stimulated by NADPH and could be inhibited by *N*-ethylmaleimide thiol group inhibitor and by Zn^{2+} ions, suggesting involvement of thiol-dependent processes. It was also demonstrated that the epidermis contains an effective antioxidant system that may play a significant role in preventing photobiologic damage of skin. The free radical scavenging activity of epidermis, together with this system and assessed by the reduction of some nitroxides, was considerably inhibited immediately after exposure to ultraviolet radiation.[53] The reduction of spin probe was also observed in membrane complexes extracted from the stratum corneum, suggesting the presence of free radical reducing substances in lipids of this tissues.[55]

Moreover, it has been reported that about 50% of the reduction in skin takes place in the extracellular space.[54] This is important for the reduction of charged nitroxides (ASL) that do not penetrate the cell membranes and are reduced primarily extracellularly. It has been suggested that ascorbate is important in the reduction of nitroxides in skin. It seems very probable that ascorbate is also present in extracellular space of the skin and represents one of the most important nonenzymatic nitroxide reducing agents.[56] This reaction is very fast and in many cases irreversible since formed dehydroascorbic acid in water hydrolyzes to 2,3-diketogulonic acid.

3. Measurements of Transport of Molecules by Nitroxide Reduction Kinetic Imaging

This method was developed to study the transport of liposome-entrapped nitroxides into the skin.[9,10] The ability of tissues to reduce the nitroxides to nonparamagnetic hydroxylamines was explored. Namely, from the rate of reduction of the spin probe molecules that had been released after disintegration of liposomes, it is possible to follow continuously the evolution of concentration profiles into the skin, separately for liposome-entrapped molecules and for the released, free molecules, as well as to measure the decay rate of liposomes during their penetration into the skin. This method identified the relationship between penetration of molecules released from and/or entrapped in liposomes and the reduction rates.

The experiment was designed in the same way as for 1D-EPRI (Section II.B.3). The kinetics of reduction of the spin probe ASL was determined by measuring the EPR spectra intensity decrease with time for the whole sample (thread and skin). The amplitude of the first hyperfine component of the EPR spectra was measured, as the line shape did not change with time significantly.

The reduction kinetics and evolution of the concentration profiles of the spatially spread reacting molecules were calculated using a model which included the transport of intact liposomes and of released spin-labeled molecules into the skin, the decay rate of the liposomes, and the differences between stratum corneum and viable skin layers. The following system of coupled differential equations describing the diffusion-reaction coupled system was used:

$$\frac{\delta C(x,t)}{\delta t} = D^{(j)} \frac{\delta^2 C(x,t)}{\delta x^2} - k_1^{(j)} C(x,t)^{(R)} C(x,t) + k_2^{(j)} C(x,t)^{(L)} \quad (8)$$

$$\frac{\delta C(x,t)^{(R)}}{\delta t} = -k_1^{(j)} C(x,t)^{(R)} C(x,t) \quad (9)$$

$$\frac{\delta C(x,t)^{(L)}}{\delta t} = D^{(j)(L)} \frac{\delta^2 C(x,t)^{(L)}}{\delta x^2} - k_2^{(j)} C(x,t)^{(L)} \quad (10)$$

where $C(x,t)$ and $C(x,t)^{(L)}$ are the concentration distribution functions of ASL released after liposome disintegration (free ASL) and of ASL entrapped in liposomes (or another form protecting ASL from reduction-entrapped ASL), and $C(x,t)^{(R)}$ is the concentration distribution of endogenous reducing agent. Index j denotes different compartments across the sample (thread, stratum corneum and viable skin layers), $k_1^{(j)}$ and $k_2^{(j)}$ are the rate constants for reduction of ASL and for liposome decay, and $D^{(j)}$ and $D^{(j)(L)}$ are the corresponding diffusion coefficients for free ASL and for entrapped ASL, respectively. This system of equations was solved numerically using the algorithm described,[10,57] and integrated using the Runge-Kutta method. From the best fit with the experimental reduction kinetic curves the concentration distribution functions $C(x,t)$ for free and entrapped ASL molecules in the skin were obtained.

It should be stressed that the reduction kinetic curves could be adequately fitted by different combinations of parameters k_1, k_2, D and $D^{(L)}$. In the application of reduction kinetic imaging method it is therefore essential to combine both 1D-EPRI measurements and reduction kinetic measurements. The concentration distribution function $C(x,t)$, calculated from the best fit of the reduction kinetic curve, should be used to calculate the line shape of 1D-EPRI spectra. The results are satisfactory when the calculated 1D-EPRI spectra fit the experimental spectra well.

An example: transport of egg yolk lecithin/cholesterol liposomes-entrapped spin probe into the skin

To match the lineshapes of the experimental spectra with those calculated from the reduction kinetic imaging, the difference between stratum corneum and viable skin layers should be taken into account. To the different concentrations of reducing agents,[9] the different diffusion rates of liposomes and free ASL should be used, as well as the different liposome decay rate in stratum corneum and viable skin layers.[10] Besides, it was assumed that there is no reduction of ASL and no decay of liposomes in the thread ($C(x,t)^{(R)}$, k_1 and k_2 are 0); the diffusion of liposomes in the thread and in the skin differs from the diffusion of free ASL. Through this approach the calculated spectra, corresponding to the spectra measured 5 and 30 min after application of liposomes to the skin, show a good fit with the experimental one (the calculated difference in parameter *l* determined from the EPR spectra measured 5 and 30 min after application of liposomes to the skin, Δl for REV is 0.03, and is the same as the experimental value) (Figure 3).

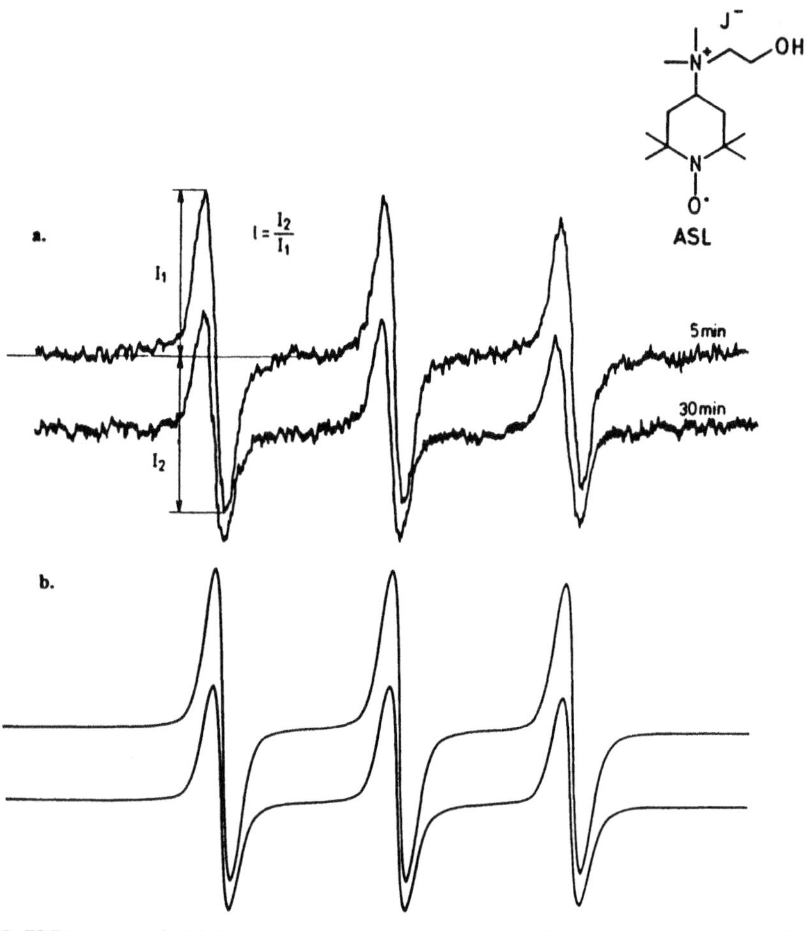

Figure 3 1D-EPRI spectra of ASL in the sample of pig ear skin 5 and 30 min after application of REV (EL:Ch = 7:3) with entrapped 0.01 mol/l ASL. (a) Experimental spectra, parameters used: magnetic field gradient 0.25 T/m, modulation amplitude 0.1 mT, microwave power 10 mW, sweep 10 mT. The spectra were measured at room temperature. (b) Calculated spectra: concentration distribution function C(x,t) of ASL obtained from the reduction kinetic experiment (Figure 4) was used. Spectra are a superposition of the spectra of released ASL (represented by a composite spectrum, Gaussian line shape 95%) and the entrapped ASL (Lorentzian line shape 70%, owing to the spin-spin interaction between ASL molecules). (The line shapes for different concentrations of ASL are evaluated for water solution in the concentration range from 10^{-4} to 10^{-2} mol/l). (From Gabrijelčič, V., Šentjurc, M., and Schara, M., *Int. J. Pharm.*, 102, 151, 1994. With permission.)

The best fits of the reduction kinetic curves, provided 1D-EPRI spectra calculated from the calculated concentration profiles match the best with the experimental 1D-EPRI spectra, are represented in Figure 4 as solid lines and the corresponding parameters are shown in Table 1. The set of parameters used in the calculation, which should not depend on the form in which the ASL is applied to the skin (rates of reduction and permeability constants for released or free ASL), were kept constant for all three examples. The concentration distribution profiles, separately for ASL entrapped in liposomes and for released ASL, are shown in Figure 5.

In Figure 6 the time development of the concentration profiles of the total amount of ASL (entrapped and released) for REV is presented. The observed development of the concentration profiles of spin

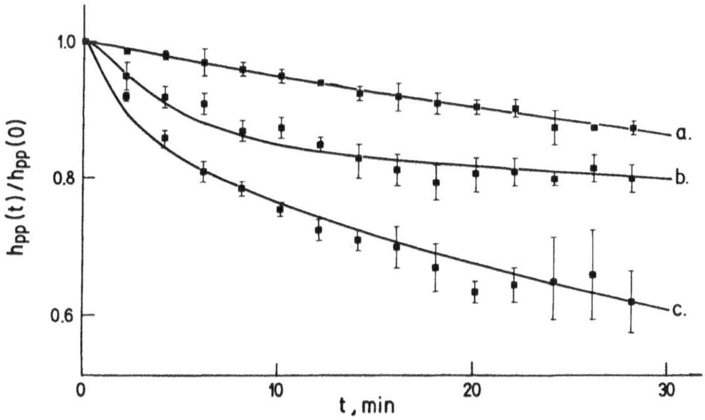

Figure 4 Kinetics of the EPR spectra amplitude decrease of ASL applied on the surface of the pig ear skin in: (a) water solution; (b) entrapped in MLV (EL: Chol = 7:3); (c) entrapped in REV (EL:Ch = 7:3). The dots are the measured amplitudes of m = 1 spectral component (mean values of three experiments, bars denote the standard error). Solid lines are the best fits of the experimental kinetic curves. The parameters used in the calculation are presented in Table1. They should also fulfill the condition that the calculated distribution function of ASL C(x,t), which is used for the EPR spectral line shape calculation, gives the best fit of the experimental 1D-EPRI spectra. Those parameters which should not depend on the form of preparation are kept the same for all three samples. (From Gabrijelčič, V., Šentjurc, M., and Schara, M., *Int. J. Pharm.*, 102, 151, 1994. With permission.)

Table 1 The parameters used in the calculation of the kinetics of the EPR spectra amplitude decrease (Equations 8,9,10), by which the best fits of the experimental curves (Figure 4) were obtained

	Compartment/ Type of Preparation	ASL in H_2O	MLV with ASL	REV with ASL
Decay rate constant k_2, s^{-1}	Thread	—	0.0	0.0
	St. corneum	—	0.08	0.02
	V. skin	—	0.0002	0.0002
Diffusion coefficient D, 10^{-7} cm^2 s^{-1}	Thread	1	0.05	0.5
	St. corneum	0.15	1	100
	V. skin	1.5	0.1	10
Entrapped volume, Vn, ml ml^{-1}		—	0.02	0.08

Note: In calculation, the three-compartment model was used: thread, stratum corneum and viable skin: k_2 = rate constant for liposome decay, D = diffusion coefficient of ASL, Vn = entrapped volume of liposomes measured by EPR.[10] The rate constant of ASL reduction k_1 = 5 l mol^{-1} s^{-1} is taken to be the same in stratum corneum and in viable skin, whereas the concentration of reducing agent in stratum corneum = 10^{-2} mol/l, and in viable skin is 3×10^{-4} mol/l, as was calculated from the best fit with the reduction kinetic curve for ASL solution.

St. corneum = stratum corneum, V. skin = viable skin layers.

From Gabrijelčič, V., Šentjurc, M., and Schara, M., *Int. J. Pharm.*, 102, 151, 1994. With permission.

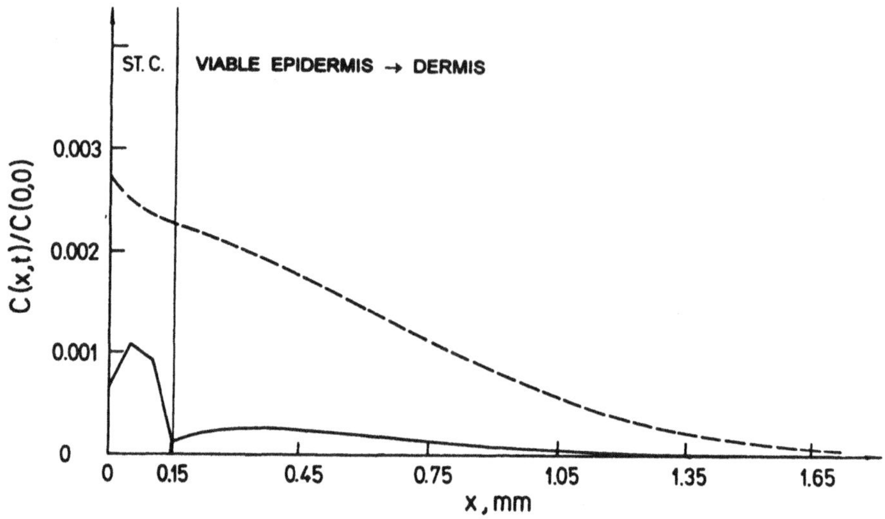

Figure 5 The distribution function of ASL in the pig ear skin 30 min after application of ASL entrapped in REV (EL:Ch = 7:3) separately for the released ——— and liposome-entrapped ASL - - - - -. The distributions were calculated from the described model (Equations 8,9,10), with the parameters presented in Table 1. The concentration of ASL in tissue C(x,t) is normalized to the total concentration of ASL (C(0,0)) in volume element used in the calculation and was calculated from the known concentration of ASL in liposomes (10^{-2} mol/l) and the internal volume of liposome dispersion. C(x,t)/C(0,0) = 0.001 corresponds to C(x,t), which for is REV equal to 8×10^{-7} mol/l. (From Gabrijelčič, V., Šentjurc, M., and Schara, M., *Int. J. Pharm.*, 102, 151, 1994. With permission.)

probe applied to the skin entrapped in liposomes is instructive. It directly reflects the role of a nonhomogeneous tissue structure such as epidermis. The results indicate that the concentration of reducing agent in stratum corneum is about 30 times higher than that in inner layers. The diffusion constants in the stratum corneum are higher than in the inner skin layers and are thousand times higher for the entrapped ASL than for the released or free ASL. Besides, Table 1 shows that the rate of liposome decay in the stratum corneum is two orders of magnitude larger than in inner layers of the skin. In the barrier, this rate is larger for MLV than for REV. Irrespective of the fast liposome decay in the stratum corneum it may be estimated (Figure 5) that after 30 min about 5% of the ASL molecules which were delivered into the skin remain in the form which protects them from reduction (entrapped ASL).

It should be mentioned that in general the reduction kinetic curves could be adequately fitted using a model where the skin is treated as a single homogeneous compartment in contact with the other compartment, the thread. However, when the distribution function C(x,t) obtained by this model was used to calculate the line shape of the 1D-EPRI spectra there was no agreement between the calculated and experimental 1D-EPR spectra; the calculated difference in l (Δl) was negligible, contrary to the experiments where $\Delta l = 0.03 \pm 0.01$ (mean value of three experiments) was measured.[9] These results clearly demonstrate that the nonhomogeneous structure of the epidermis is directly reflected in the reduction and diffusion processes in skin.

Since ASL is reduced fast when released from the liposomes, its penetration into the inner skin layers could not be measured. However, if in Equations 8, 9 and 10 the rate constant of ASL reduction is taken as 0, while the values of diffusion and liposome decay rates are taken as the same as in Table 1, an estimation for the penetration into the skin of some "stable" drugs entrapped in liposomes can be obtained.

4. Summary

To summarize the nitroxide reduction kinetic imaging experiment: The EPR spectra intensity decrease and the EPR spectra line-shape in the presence of the magnetic field gradient were measured simultaneously. The concentration distribution function evaluated from the kinetic imaging experiment C(x,t) was used to calculate the line-shape of the 1D-EPRI spectrum (Equation 4) which was then compared with the experimental one. Furthermore, the model used and the concentration distributions obtained were checked by two independent experiments. It should be stressed that the concentration of reducing

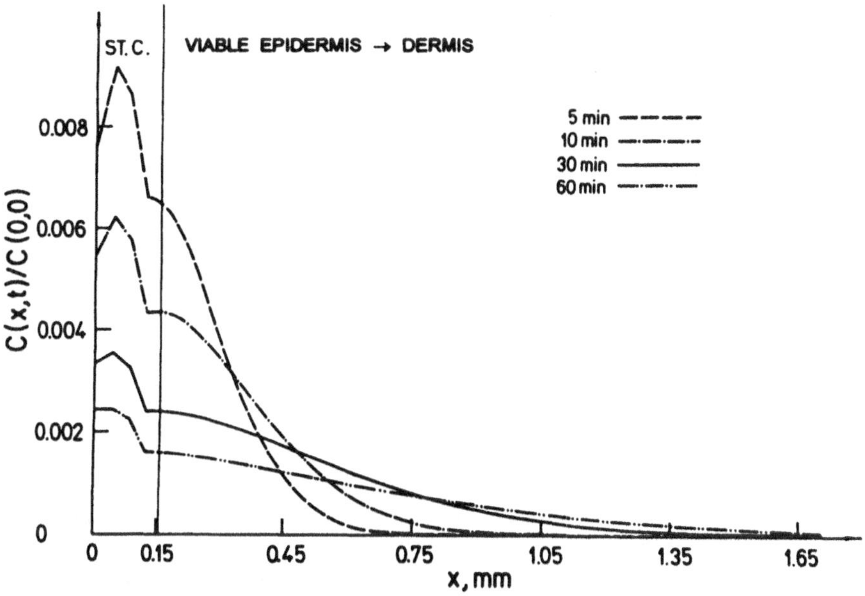

Figure 6 The time evolution of the distribution of the total amount of ASL in pig ear skin C(x,t)/C(0,0), calculated with the parameters presented in Table 1, 5 min – – – –, 10 min – · – · –, 30 min ———, and 60 min – ·· – ·· – after application of REV with entrapped ASL to the pig ear skin surface. (From Gabrijelčič, V., Šentjurc, M., and Schara, M., *Int. J. Pharm.*, 102, 151, 1994. With permission.)

agent, the nitroxide reduction constants and diffusion constants of released nitroxides in the stratum corneum and in viable skin were calculated from EPR spectra intensity decrease of free ASL solution applied to the skin. It needs to be kept the same for all the samples measured on the same skin. It should also be stressed that these constants vary appreciably for different skins and should be measured for each skin sample individually.

The use of nitroxide reduction kinetic imaging method provides data on both liposome decay in the stratum corneum and in viable skin layers and on the penetration and distribution of total and protected ASL delivered to the skin entrapped in liposomes.

III. INFLUENCE OF TYPE AND LIPID COMPOSITION OF LIPOSOMES ON THE TRANSPORT OF LIPOSOME-ENTRAPPED MOLECULES INTO THE SKIN

A. REVIEW OF RECENT RESULTS

The choice of lipids for liposome formulation depends on the effects we aim at in the application of liposomes to the skin. For example: the hydration of the horny layer is not just an important factor for the epidermal barrier function but is also responsible for the smoothness and the appearance of the skin and, hence, for the general well-being. The skin humidity is dependent on the environmental temperature and relative humidity, the extent of transepidermal water loss and the presence in the skin of components with water-binding capacity. Since phospholipids with high water-binding capacity such as phosphatidylcholine are excellent moisturizers, liposome formulations containing a high portion of phosphatidylcholine increase skin moisture whereas liposome formulations containing only 10% phosphatidylcholine and some portion of phosphatidylethanolamine, phosphatidylinositol and phosphatidic acid cause a reduction in the skin humidity.[58]

Besides, composition of liposome membranes determines the depth of liposome penetration into the skin. The appropriate choice of membrane components that assure adequate elastic properties to prevent liposome disintegration during their transport through the skin makes possible the transdermal drug delivery.[4]

However, many details concerning the efficacy of liposomes in topical drug delivery remain to be elucidated. Most of the studies presented in the literature are difficult to compare owing to differences in skin sources, in liposome lipid composition and liposome morphology. Conclusions are difficult to draw and are often questionable.

What is the influence of the type of liposomes, including their membrane composition, lamellarity, and the effect of their size and surface charge, on drug penetration into the skin?

Gehring et al.[59] investigated the effect of betamethasone incorporated in three different liposome formulations varying mainly in phosphatidylcholine content. They did not find any significant differences between these liposome formulations. On the other hand, some authors find, through different approaches, the delivery of the liposome-entrapped molecules into the skin to be better when they are entrapped into liposomes made of "liquid" phospholipids — phospholipids in the liquid crystalline state at room temperature.[39,60,61]

Weiner and Egbaria[62] found that both liposome composition and liposome preparation method influenced the penetration of interferon: liposomes made of "skin lipids" were found to deliver almost twice as much interferon as the liposomes made of egg lecithin and phosphatidylserine. Additionally, liposomes prepared by the dehydration-rehydration method were twice as effective as liposomes prepared by the reverse-phase evaporation method. It has been stated several times that in addition to the composition of the liposome membrane the method of liposome preparation is also an important factor in determining the efficacy of liposomes in drug delivery into the skin.[10,39,62]

B. 1D-EPRI AND NITROXIDE REDUCTION KINETIC IMAGING EXPERIMENTS

The influence of liposome type and composition was investigated more in detail by 1D-EPRI and reduction kinetic imaging.[10,39] The influence of the method of liposome preparation is presented in Figure 7, where the concentration profiles in the pig ear skin for REV, MLV and ASL solution are presented. They show that after 30 min free and MLV-entrapped ASL remain in the stratum corneum, whereas the concentration profile of ASL applied in REV develops deep into the skin, which correspond to the results of other authors.[4,6]

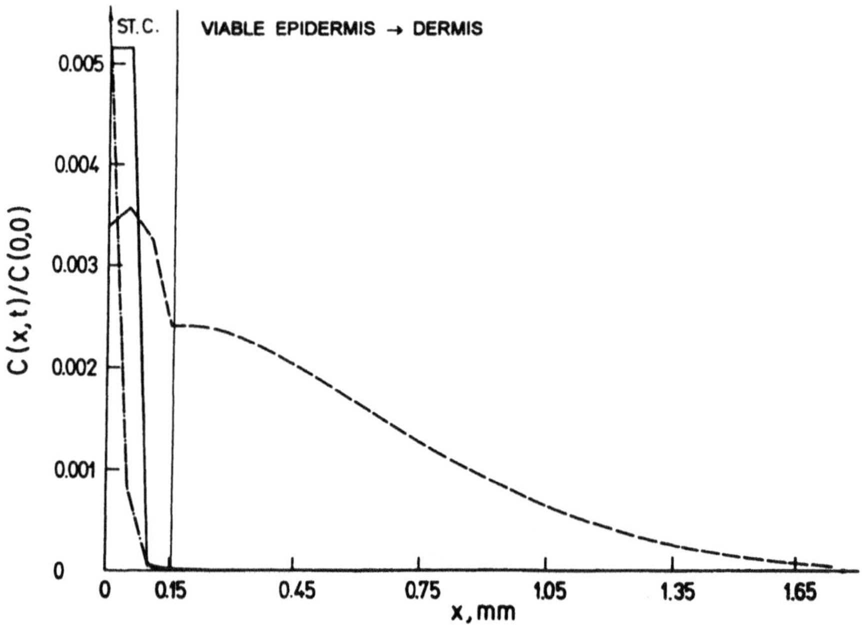

Figure 7 The total distribution function of ASL in the pig ear skin 30 min after application of ASL: in water solution ——— entrapped in REV (EL:Ch = 7:3) - - - - or entrapped in MLV (EL:Chol = 7:3) – · – · –. The distributions were calculated from the described model (Equations 8,9,10), with the parameters presented in Table 1. The concentration of ASL in tissue C(x,t) is normalized to the total concentration of ASL (C(0,0)) in volume element used in the calculation and was calculated from the known concentration of ASL in liposomes (10^{-2} mol/l) and the internal volume of liposome dispersion. C(x,t)/C(0,0) = 0.001 corresponds to C(x,t) which is for: REV, 8×10^{-7} mol/l; MLV, 2×10^{-7} mol/l, and ASL solution, 10^{-5} mol/l. (From Gabrijelčič, V., Šentjurc, M., and Schara, M., *Int. J. Pharm.*, 102, 151, 1994. With permission.)

The transport characteristics of charged hydrophilic molecules (ASL) applied to the skin entrapped in liposomes of different lipids and of different type, as obtained by the 1D-EPRI method are presented in Table 2. As may be concluded from Table 2, only ASL entrapped in small unilamellar vesicles (SUV) or reverse-phase evaporation vesicles (REV) of those phospholipids which exhibit phase transition from gel to liquid crystalline phase below room temperature (temperature of measurements) i.e., "fluid phospholipids" (egg or soya lecithin and phosphatidylserine), can diffuse into the skin, whereas free ASL or ASL entrapped in the multilamellar vesicles (MLV) or REV of phospholipids with phase transition above the temperature of measurement, i.e., "solid phospholipids" (dipalmitoylphosphatidylcholine, dipalmitoylphosphatidylglycerol with phase transition at 41°C) do not penetrate. No transport of ASL was observed if it had been applied to the skin in dispersion with empty REV. These results prove that the charged molecule ASL can be transmitted into the skin only when applied to the skin entrapped in "appropriate" liposomes.

Table 2 Diffusion Constant of ASL (D, cm^2/s) into Pig Ear Skin, from ASL Entrapped Liposomes of Various Lipid Composition and Size, and of Free ASL in Water and in Xanthan Gum Hydrogel (HG)

Lipid Composition of the Liposomes	Diffusion constant (10^{-6} cm^2/s)[a]
REV (EL) in water	3.1 ± 0.1 (3)
REV (EL: Ch = 1: 1) in water	2.4 ± 0.3 (3)
REV (EL: Ch = 1: 1) in HG[b]	1.8 ± 0.5 (4)
REV (EL: Ch = 7: 3) in HG[b]	0.6 ± 0.1 (2)
REV (EL: PS = 7: 3) in Tris[b]	2.5 ± 0.9 (5)
REV (EL: PS = 7: 3) in HG[b]	1.5 ± 0.1 (2)
REV (DPPC: DPPG = 4: 1) in HG[b]	0.0 ± 0.0 (6)
SUV (EL: Ch = 7: 3) in water[b]	1.2 ± 0.5 (5)
SUV (SL) in water	0.3 ± 0.1 (3)
MLV (EL: Ch = 6: 4) in HG[b]	0.0 ± 0.0 (3)
MLV (EL: PS = 7: 3) in Tris[b]	0.0 ± 0.0 (2)
MLV (SL) in HG[b]	0.0 ± 0.0 (3)
Free ASL (0.01 *M*) in water[b]	0.0 ± 0.0 (3)
Free ASL (0.01 *M*)in HG[b]	0.0 ± 0.0 (2)
REV (EL: Ch = 7: 3) + free ASL in water	0.0 ± 0.0 (2)
REV (EL:Ch = 7:3) + free ASL in HG[b]	0.0 ± 0.0 (2)

[a] Mean value ± standard deviation. In parentheses are the number of measurements on different skin samples. The measurements were take on various pig ear skins and at different time intervals after sacrifice of the animal.

[b] From Gabrijelčič et al.[39]

Based on data in Table 2, it may be concluded that the major factors influencing the penetration of liposomes are liposome type (MLV, SUV, REV), i.e., the method of liposome preparation and lipid composition. It seems that the fluidity of phospholipid components that mainly defines the elastic properties of the liposome membrane is of paramount importance.

The reduction kinetic imaging method was used to follow the transport of vesicles (the expression "vesicles" is used since they are not prepared from phospholipids but from nonionic surfactants) with different compositions of amphiphilic and hydrophobic molecules. Typical examples of concentration distribution in skin for three different vesicular systems are presented in Figure 8 (lipid composition is presented in the text to the Figure 8). The figure shows the influence of different composition of vesicles on the transport of hydrophilic substance applied to the skin entrapped in vesicles. If it is desirable that the substance penetrates deep into the skin, then vesicular system F should be applied; when it is desirable that the substance remains primarily in the epidermis, the dispersion B will be used; and if we want the vesicles to remain on the surface of the skin and in the stratum corneum, with prolonged release of the entrapped substance, the vesicular dispersion D should be applied (or any corresponding vesicular systems with similar concentration distribution characteristics). The studies performed on more than 20 vesicular systems (provided by L'Oréal), indicate that the main factor, with the strongest influence on the penetration of molecules applied to the skin entrapped in liposomes, is lipid composition: this factor determines their stability, flexibility and compatibility with skin lipids. The compounds which best mimic the lipid composition of skin fuse better with the stratum corneum and better delivery of entrapped substance into the skin can be obtained. The size of vesicles is of minor importance.

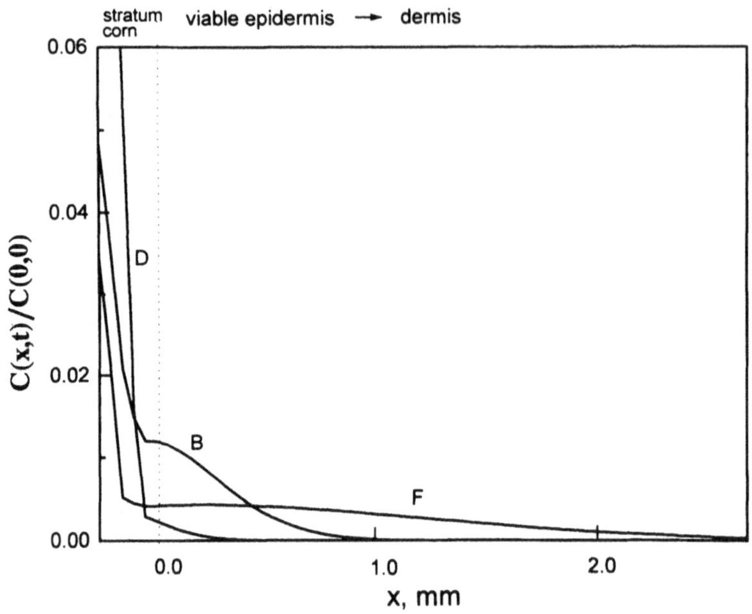

Figure 8 Total distribution of ASL in pig ear skin 30 min after application of ASL entrapped in vesicular systems to the skin. C(0,0) is initial concentration of ASL in the vesicular system and depends on the internal volume of vesticles. B: PEG isostearate/lipoaminosalt (90/10), C(0,0) = 9×10^{-4} mol/l; D: polyglyceryl dialkyl ether/dimyristyl phosphate (95/5), C(0,0) = 2.7×10^{-3} mol/l; F: polyglycerol mono, di and three alkyl mixture/cholesterol/dicetylphosphate (60/30/5), C(0,0) = 5×10^{-3}mol/l. (Vesicular systems are kindly delivered from L'Oréal Research Laboratories, Chevilly-Larue, France.)

C. CONCLUSIONS

Results of experiments performed up to now suggest that the major factor in the transport of molecules applied to the skin entrapped in liposomes is the lipid composition of the liposome membrane. Small additions of substances that increase membrane fluidity and the elasticity of the membranes are of prime importance. Liposomes that are at room temperature in the gel phase ("solid lipids") are not good enhancers of the transport of the entrapped molecules.

The type of liposomes, i.e., liposome preparation, also influences the transport of entrapped molecules. Some liposomes that are not able to transfer the entrapped molecules in MLV form, on the contrary facilitate transport if they are prepared as SUV or REV.

Liposomes can facilitate the transport into the skin only of these molecules that are entrapped in them. The transport is not enhanced when the molecules are applied to the skin in the dispersion of empty liposomes.

IV. MECHANISMS OF LIPOSOME–SKIN INTERACTIONS

A. RECENT STATE OF KNOWLEDGE

One of the most disputed issues regarding the topical use of liposomes is the question of the mechanisms lying behind the effects observed. Are the liposomes able to penetrate into the stratum corneum or even into the deeper layers of the skin? How deep do intact liposomes penetrate into the skin? These are the questions also discussed in this work.

It seemed unlikely that liposome particles, which are at least 20 nm in diameter, could penetrate the intact horny layer, either across the corneocytes or along the cross-linked intercellular lipid layers.[63] By small-angle X-ray scattering a fusion of liposomes with epidermal layers and intermixing of fluid liposome and rigid epidermal lipids, was shown. Besides, freeze-fracture electron microscopy revealed that the vesicles were able to introduce changes in the ultrastructure of the intercellular lipid lamellae of the stratum corneum.[6] However, in the intercellular lipid region between the two cells, vesicular structures were visualised that were comparable in their structure and size to vesicles applied to the skin.

Again, it is not clear yet by which mechanisms these structural changes are induced. There are at least two possible explanations:

1. Vesicular structures are due to a migration of intact vesicles through the intercellular lipid bilayers of the stratum corneum.
2. The surfactants of nonionogenic vesicles diffuse molecularly dispersed in the lipid bilayers of the horny layer, subsequently inducing an increase in curvature of the lipid bilayers, thereby forming vesicular structures.

Lasch and co-workers[5] tried to visualise intact liposomes in the human skin by fluoromicrography. Hydrophilic and lipophilic fluorophores were entrapped into the liposomes made of egg lecithin and cholesterol. Their conclusion was that intact liposomes are confined to the horny layer and do not penetrate deeper on. This technique does not provide an answer to the question: by which molecular mechanisms do liposomal phospholipids enhance the penetration of molecules into the skin? An interesting hypothesis is "fluidisation" of intercellular lipid domains in the stratum corneum by liposomal lipids favouring lipid intermixing. Also, perturbed angular correlation spectroscopy (PAC) studies of ^{111}In-DTPA entrapped into the liposomes suggest that egg lecithin liposomes disintegrate at the interface to the horny layer when water evaporates from the liposome dispersion on the skin surface.[63]

Other authors have offered different tentative explanations of enhanced skin penetration of molecules via liposomes. Phosphatidylcholine from the liposome envelope interacts with the lipid bilayers in the stratum corneum via lipid exchange. That process ends by fusion of liposome bilayers with intercellular bilayers of stratum corneum. Phosphatidylcholine has a great capacity for binding water molecules. When it is incorporated into the stratum corneum lipids it hydrates the horny layer, and this could be the basis of the enhanced permeation of the hydrophilic molecules. This hydration of the skin is temporary as is also the enhanced permeation. Fluidization of skin lipids also leads to an improved partition of the lipophilic and amphiphilic drugs into the lipid phases in deeper skin layers.[64] The "fluidization" effect of skin lipids by liposomes is not that pronounced if a fully hydrogenated, i.e., completely saturated phosphatidylcholine is employed. The enhanced penetration of molecules applied to the skin entrapped in liposomes can be controlled by the choice of the appropriate phospholipids for liposome preparation.[61]

Based on the statements described above, it may be concluded that liposomes act as skin penetration enhancers via fluidization of skin lipids. However, they are not just that. If the drug is added to empty liposomes, it is not entrapped/incorporated into the liposomes and there is no enhanced drug absorption into the skin. To enhance absorption, the drug must be entrapped in the vesicles.[39,62,65]

As determined by stripping technique after application of liposomes *in vitro*, the distribution of drug and liposome lipid label in the various skin strata revealed the data leading to the conclusions given below:

- The ratio of water and lipid soluble markers was found to be unaltered in various layers of the stratum corneum. It seems that liposome bilayers are able to carry their entrapped solutes into the skin, even to a depth beneath the stratum corneum.
- The ratio of labeled lipids of liposomes was essentially maintained unchanged through the skin strata. This again suggests a mixing of liposome lipids with the lipids of stratum corneum.[65]

On the other hand, electron microscopic observations using colloidal iron as an electrondense marker indicated that intact liposomes penetrated into the skin and deposited in the dermis where they acted as a slow release depot system.[3] On the basis of these results the following hypothetical model for liposome–skin interaction was proposed.

The multilamellar structure of liposomes protects the entrapped drug from premature clearance and provides a vehicle for sustained release of drugs. Multi- and unilamellar liposomes can be adsorbed to the skin surface intact before their penetration into the skin. Some liposomes can rupture on the surface of the skin, lipids intermix and better hydration of skin is achieved. After that the penetration of smaller vesicles is more probable. It is possible that the intradermally localized uni- and oligolamellar vesicles are derived from multilamellar liposomes, that have lost their outer bilayers during penetration.[3]

Cevc and Blume[4] suggest that gradients across the outer skin layers may result in driving forces which enforce a lipid flow into or through the intact skin surface, provided that lipids are applied on the skin in the form of special vesicles (transferosomes). The osmotic gradient created by the difference in the total water concentrations between the skin surface and the skin interior provides one possible source of such driving force. It is sufficiently strong to push these vesicles through the stratum corneum. On the contrary, the lipid concentration gradient does not contribute significantly to the lipid penetration into the dermis. Occlusion is therefore detrimental for the vesicle penetration into the skin. Conditio sine qua non for vesicle penetration through the skin is good elasticity of their membranes — the vesicles must be flexible enough to squeeze through the narrow, maze-like passages in the stratum corneum.

Experiments performed by Blume et al.[66] show that the interaction of phospholipids with stratum corneum model lipids leads to a mixing of lipid components. This could also occur *in vivo* and result in the formation of more hydrated lipid layers, which could lead to a structural rearrangement of the stratum corneum lipids in such a way to form stacked, more flattened vesicles similar to those found in the stratum granulosum. It is also possible that local lipid composites form in the layers, which are H_{II}-prone, forming intercellular attachments or inverted micellar intermediates, stages which are thought to be intermediate between lamellar and hexagonal or cubic phases. Once these structures are formed, other lipid vesicles could "squeeze" through these defects, reaching the lower layers of the skin.

This mixing process between liposome and stratum corneum lipids, either via monomer exchange or by direct fusion, is strongly dependent on the physical state of the constituents of the liposome membranes. Thus, it is understandable that different liposome formulations may vary in their ability to penetrate the skin and to deliver the drugs into or through the skin.

Of the possible routes of penetration of vesicles into the skin, transfollicular route should not be ignored. Namely, liposome formulations have been found very efficient for delivery of negatively charged marker that otherwise penetrates the stratum corneum only with great difficulty, through the follicular route. Targeted follicular delivery via the use of liposomes would certainly be advantageous for a number of follicular diseases.[7]

B. LYOTROPIC MODEL SYSTEMS

Experiments performed on model systems provide a means of obtaining more information on the processes taking place *in vivo*. The molecular basis for penetration enhancement, the mechanism of interaction of liposomes with the epidermal lipid barrier, and penetration properties of liposomes still need elucidation although they have been intensively discussed. To obtain some more information on these processes we studied the interactions of liposomes with lyotropic model systems of lecithin and water by 1D-EPRI and EPR kinetic imaging.[67] The results are summarized in Table 3.

Table 3 Transport of Free and Liposome Entrapped ASL into the Lyotropic Model Systems (Lecithin/Water) with Different Osmolality

Sample	Model System wt/wt	Transport yes/no	1D-EPRI Results $\Delta I(I_5-I_{25})$*
ASL in water	lecithin/water 60/40	no	0
Liposomes in water	lecithin/water 60/40	no	0
ASL in water	lecithin/water 50/50	yes	0.14 ± 0.02 (2)
Liposomes in water	lecithin/water 50/50	yes	0.06 ± 0.01 (4)
ASL in water	lecithin/NaAsc 50/50	to fast reduction for 1D-EPRI measurements	
Liposomes in water	lecithin/NaAsc 50/50	yes	0.17 ± 0.05 (2)
Liposomes in NaCl	lecithin/NaAsc 50/50	yes	0.03 ± 0.01 (4)
Liposomes in glucose	lecithin/glucose 50/50	yes	0.05 ± 0.01 (2)
Liposomes in glucose	lecithin/NaAsc 50/50	no	0
Liposomes in water	lecithin/glucose 50/50	yes	0.10 ± 0.03 (2)

Note: * Mean value ± standard deviation. The number of measurements is in parentheses.

The 1D-EPRI revealed that ASL applied in water solution or entrapped in liposomes could not penetrate into the model lyotropic system of 60% soya lecithin and 40% of water (the system was in lamellar phase[68]) and transport was observed into a system consisting of 50% soya lecithin and 50% water (coexistence of liquid and liquid crystal phase). The transport of free ASL into this system was faster than the transport of liposome-entrapped spin probe. Different explanations are possible:

1. Water-soluble spin probe ASL is small and can diffuse faster into this model system than the large vesicles.
2. Spin probe entrapped in liposomes should first be released: liposomes disintegrate when they come in contact with the lyotropic system and then the free probe diffuses into the system.

In order to get data on the stability of liposomes in contact with lecithin/water interfaces, instead of pure water, a solution of sodium ascorbate (NaAsc), which reduces the nitroxide spin probe to EPR nonvisible hydroxylamine (as explained in Section II.C.1) after it has been released from liposomes, was incorporated into the system. When ASL-entrapped liposomes prepared in water were applied to the surface of such a system an EPR spectra intensity decrease was observed due to the disintegration of liposomes at the interface. By 1D-EPRI the transport into the model system of lecithin/NaAsc was measured. The transport of free spin probe molecules into the system could not be measured because of the fast reduction of ASL by NaAsc. However, the transport of liposome entrapped ASL into the lecithin/NaAsc system was faster than the transport into the lecithin/water system, probably because of the differences in osmolalities of liposome dispersion and the system. When ASL entrapped liposomes prepared in isoosmolal sodium chloride solution were applied to the surface of lecithin/NaAsc system, transport was significantly slower as was also the reduction kinetics. It should be stressed that the rate of EPR spectra intensity decrease is directly related to the liposome decay which occurs owing to the interaction of liposomes with model system interface.

On the basis of these results it can be concluded that the physical state of the lecithin/water system and the difference in osmolality between liposome dispersion and model system are important for penetration of ASL applied entrapped in liposomes. This might also be important in the mechanism of liposome transport into the outer skin layer, the stratum corneum. These conclusions are also supported by the following results:

- There is no transport of ASL-entrapped liposomes prepared in glucose solution into isoosmolal model system of lecithin/NaAsc.
- The transport of ASL-entrapped liposomes prepared in glucose solution into isoosmolal model system of lecithin/glucose is slow.
- The transport of ASL-entrapped liposomes prepared in water into such lyotropic system is faster.

C. PROPOSED HYPOTHESES

With respect to our observations and hypotheses and results reported by other authors, the following mechanism of transport of molecules delivered to the skin entrapped in liposomes could be suggested:

At the interface with rigid lipid bilayers of stratum corneum a portion of liposomes disintegrates. Some liposomal lipids intermix with intercellular skin lipids. Skin lipid bilayers therefore become more fluid and hydrophilic;[61] their structure can change from lamellar to hexagonal or cubic phases, which enables some undisintegrated vesicles with suitable elastic properties to penetrate into the deeper layers of the skin. The difference in osmolality between liposome dispersion and skin seems to be important for this transport and defines the transdermal osmotic gradient, the driving force for liposome penetration through the stratum corneum.[4]

Based on the results obtained by the combination of 1D-EPRI and reduction kinetic imaging on the skin slices *in vitro,* it can be concluded that a portion of molecules applied to the skin in liposomes was not released from the liposomes but remained in the form which proteced them from reduction.[10] This supports the hypothesis that some liposomes can penetrate into the skin.[3,4] However, it is also possible that liposomes disintegrate on the surface of the skin, but ASL molecules, owing to their positive charge, remain associated with the liposome lipids that constitute some kind of envelope protecting the spin probe from reduction. This is supported by some other suggestions, namely, that after the destruction of liposomes on the surface of the skin a significant amount of water and entrapped substances remain associated with the liposome bilayer and are carried into the deeper layers of the skin.[6,65] It is not yet possible to distinguish between these two options, but we believe that by the application of different spin probes with different charge and structure, the distinction could be achieved. The experiments with some other spin probes are now in progress.

V. INFLUENCE OF HYDROGELS ON THE TRANSPORT OF LIPOSOME-ENTRAPPED MOLECULES INTO THE SKIN

Liposomes are usually applied to the skin in solution or in hydrogels since stable liposomal creams are difficult to formulate. For topical application of liposomes, hydrophilic polymers are suitable thickening agents since they make the formulations convenient for application. However, the type and concentration of hydrogel-forming polymer could affect the stability and rate of penetration of liposome-entrapped molecules into the skin. As found by some authors, the hydrogel formed from Hostacerin® traps the liposomes into a polymeric network and thus prevents an efficient carrier diffusion to and into the skin.[69]

We performed a more systematic study on hydrogels with different concentration of a hydrophillic polymer: carboxymethylcellulose or xanthan gum.[70] The investigations were performed on multilamellar liposomes composed from hydrogenated soya lecithin, cholesterol, and lipoaminosalt, with entrapped hydrophilic paramagnetic probe ASL, which were mixed up into the hydrogel. Electron paramagnetic resonance (EPR) together with one-dimensional EPR imaging (1D-EPRI) and EPR kinetic imaging[10] were used to follow the stability of liposomes in hydrogel formulations and the influence of hydrophilic polymers on the transport of liposome entrapped molecules into the skin.

By 1D-EPRI the variations in the asymmetry parameter *l* (defined previously) were followed. Figure 9 presents the difference in parameter *l* (Δl), 5 and 25 min after application of various liposome formulations to the skin. The results show that ASL applied to the skin entrapped in liposomes can penetrate into the skin irrespective of the formulation used; the same decrease (within the order of experimental error) in the parameter *l* with time is observed for liposome-entrapped ASL in water, in xanthan (0.5 and 1%) or in CMC (1%) hydrogels, irrespective of different viscosity of the formulations (values are given in text to the Figure 9).

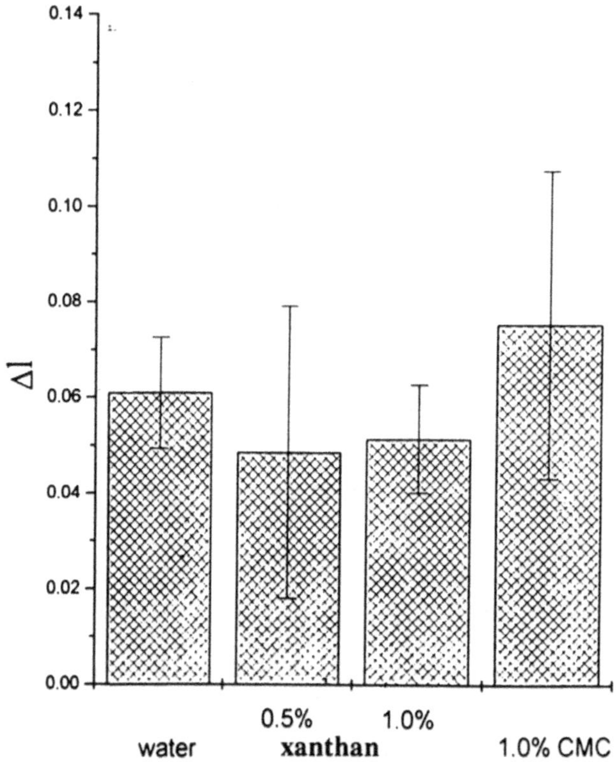

Figure 9 Difference in parameter *l*, 5 and 25 minutes after application of different liposome formulation to the skin surface (Δl). Bars denote standard deviations for three independent measurements. At sheer velocity of 120 s^{-1} viscosity of 0.5% xanthan hydrogel is 0.142 Pa s; for 1% xanthan hydrogel it is 0.325 Pa s and for 1% CMC it is 0.327 Pa s.

By EPR kinetic imaging experiment the rate of liposome decay and penetration rates of ASL (released from liposomes and entrapped in the liposomes) were calculated and are presented in Table 4 for liposomes in water and xanthan hydrogels (0.5 and 1%).

Table 4 Parameters for ASL Applied to the Pig Ear Skin Entrapped in HSL:Ch:LAS Vesicles in Solution and in 1% Xanthan Hydrogel Used in the Calculation of the Kinetics of the EPR Spectra Amplitude Decrease, by Which the Best Fits of the Experimental Curves Were Obtained

	Compartments/ Type of Preparation	ASL in H_2O	MLV with ASL in H_2O	MLV with ASL in Xanthan Hydrogel
	Thread	—	0.0	0.0
Decay rate constant k_2, s^{-1}	St. corneum	—	0.015	0.0075
	V. skin	—	0.00002	0.00002
	Thread	6.0	2.0	0.9
Diffusion coefficient	St. corneum	0.5	50.0	50.0
D, 10^{-7} cm^2 s^{-1}	V. skin	0.8	5.0	5.0
Entrapped volumes Vn, ml ml^{-1}		—	0.16	0.16

Note: For calculation the three-compartment model was used: thread, stratum corneum and viable skin. The rate constant of ASL reduction $k_1 = 130$ l $mol^{-1}s^{-1}$, is taken to be the same in the stratum corneum and in viable skin, while the concentration of reducing agent in stratum corneum is 10^{-2} mol/l and in viable skin 5×10^{-4} mol/l, as was calculated from the best fit with the reduction kinetic curve for ASL solution.

It is interesting to note that the concentration of polymer does not significantly influence the transport of liposome-entrapped molecules into the skin, in spite of increased viscosity at higher concentrations of hydrophilic polymer. Also, these results show that the transport into the skin was not prevented by polymers.

When the liposomes were applied to the skin in hydrogels instead of water, the penetration rate in thread was decreased for about factor of two as is evident from Table 4. This could be expected with respect to the restrictions caused by the polymeric network of hydrogels.[71] In skin layers the penetration rates were not affected by hydrogels.

It can be concluded that the polymers examined do not prevent efficient transport of liposome-entrapped substances into the skin. These results and the stability studies, which were performed simultaneously, support the use of hydrogels in liposome dermatics. Nevertheless, it should be taken into account that some hydrophilic polymers may influence the physical stability of liposomes; consequently each liposome preparation should be tested for stability in hydrogels.

ABBREVIATIONS

ASL	*N*-(1-oxyl-2,2,6,6-tetramethyl-4-piperidinyl)-*N*-dimethyl-*N*-hydroxyethylammonium iodide
Ch	Cholesterol
CMC	Carboxymethylcellulose
DPPG	Dipalmitoylphosphatidylglycerol
DPPC	Dipalmitoylphosphatidylcholine
EL	Egg yolk lecithin
EPR	Electron paramagnetic resonance
HG	Hydrogel
HSL	Hydrogenated soya lecithin
LAS	Lipoamino salt
MLV	Multilamellar vesicles
NaASC	Sodium ascorbate
NaCl	Sodium chloride
PS	Phosphatidylserine
REV	Reverse-phase evaporation vesicles
SL	Soya bean lecithin
SUV	Small unilamellar vesicles
Vn	Entrapped volume
1D-EPRI	One-dimensional EPR imaging

REFERENCES

1. **Wohlrab, W., Lasch, J., Taube, K. M., and Wozniak, K. D.,** Hautpermeation von liposomal inkorporiertem Hydrocortison, *Pharmazie*, 44, 333, 1989.
2. **Gesetz, A. and Mezei, M.,** Topical anaesthesia of the skin by liposome-encapsulated tetracaine, *Anesth. Analg.*, 67, 1079, 1988.
3. **Foldvari, M., Geszetes, A., and Mezei, M.,** Dermal drug delivery by liposome encapsulation: clinical and electron microscopic studies, *J. Microencapsulation*, 7, 479, 1990.
4. **Cevc, G. and Blume, G.,** Lipid vesicles penetrate into intact skin owing to the transdermal osmotic gradients and hydration force, *Biochim. Biophys. Acta*, 1104, 226, 1992.
5. **Lasch, J., Laub, R., and Wholrab, W.,** How deep do intact liposomes penetrate into human skin? *J. Contr. Rel.*, 18, 55, 1991.
6. **Hofland, H. E. J., Bouwstra, J. A., Ponec, M., Bodde, H. E., Spies, F., Verhoef, J. C., and Junginger, H. E.,** Interactions of non-ionic surfactant vesicles with cultured keratinocytes and human skin in vitro: a survey of toxicological aspects and ultrastructural changes in stratum corneum, *J. Contr. Rel.*, 16, 155, 1991.
7. **Lieb, L. M., Chandrasekharan, R., Egbaria, K., and Weiner, N.,** Topical delivery enhancement with multilamellar liposomes into pilosebaceous units. I. In vitro evaluation using fluorescent techniques with the hamster ear model, *J. Invest. Dermatol.*, 99, 108, 1992.
8. **Demsar, F., Cevc, P., and Schara, M.,** Diffusion of spin probes in tissues measured by field-gradient EPR, *J. Magn. Reson.*, 69, 258, 1986.
9. **Gabrijelčič, V., Šentjurc, M., and Schara, M.,** Liposome entrapped molecules penetration into the skin measured by nitroxide reduction kinetic imaging, *Period. Biol.*, 93, 245, 1991.
10. **Gabrijelčič, V., Šentjurc, M., and Schara, M.,** The measurement of liposome entrapped molecules penetration into the skin: A 1D-EPR and EPR kinetic imaging study, *Int. J. Pharm.*, 102, 151, 1994.
11. **Fuchs, J., Freislebens, J., Groth, N., Herrling, T., Zimmer, G., Milbrandt, R., and Packer, L.,** One- and two-dimensional electron paramagnetic resonance imaging in skin, *Free Rad. Res. Commun.*, 15, 245, 1991.
12. **Fuchs, J., Groth, N., Herring, T., Milbradt, R., Zimmer, G., and Packer, L.,** Electron paramagnetic resonance imaging in skin: biophysical and biochemical microscopy, *J. Invest. Dermatol.*, 98, 713, 1992.
13. **Williams, M. L. and Elias, P. M.,** The extracellular matrix of stratum corneum: role of lipids in normal and pathological function, *CRC Crit. Rev. Ther. Drug Carr. Syst.*, 3, 95, 1989.
14. **Ghosh, P. and Singh, U. N.,** Liposome shrinkage and swelling under osmotic-diffusional stress: evaluation of kinetic parameters from spectrophotometric measurements, *Biochim. Biophys. Acta.*, 1110, 88, 1992.
15. **Landmann, L.,** The epidermal permeability barrier. Comparison between in vivo and in vitro lipid structures, *Eur. J. Cell Biol.*, 33, 258, 1994.
16. **Potts, R. O. and Francoeur, M. L.,** The influence of stratum corneum morphology on water permeability, *J. Invest. Dermatol.*, 96, 495, 1991.
17. **Shah, V. P., Flynn, G. L., Guy, R. H., Maibach, H. I., Schaefer, H., Skelly, J. P., Wester, R. C., and Yacobi, A.,** In vivo percutaneous penetration/absorption, *Int. J. Pharm.*, 74, 1, 1991.
18. **Wiechers, J. W.,** The barrier function of the skin in relation to percutaneous absorption of drugs, *Pharm. Weekbla. Sci. Ed.*, 11, 185, 1989.
19. **Hofland, H. E. J.,** in Vesicles as Transdermal Drug Delivery Systems, Ph.D. thesis, University of Leiden, Leiden, Netherlands, 1992.
20. **Barry, B. W.,** Mode of action of penetration enhancers in human skin, *J. Controlled Rel.* 6, 85, 1987.
21. **Korting, H. C., Blecher, P., Schäfer-Korting, M., and Wendel, A.,** Topical liposome drugs to come: what the patent literature tells us, *J. Am. Acad. Derm.* 25, 1068, 1991.
22. **Krowczynski, L. and Stozek, T.,** Liposomen als Wirkstoffträger in der percutanen Therapie, *Pharmazie*, 39, 627, 1984.
23. **Lasch, J. and Wholrab, W.,** Liposome-bound cortisol: a new approach to cutaneous therapy, *Biomed. Biochim. Acta*, 45, 1295, 1986.
24. **Mezei, M.,** Liposomes in the topical application of drugs: a review, in *Liposomes as Drug Carriers*, Gregoriadis, G., Ed., John Wiley & Sons, New York, 1988, 663.
25. **Raab, W.,** Liposomen - eine neue Form dermatologischer Wirkstoffträger, *Ärtz Kosmetol.*, 18, 213, 1988.
26. **Schäfer-Korting, M., Korting, H. C., and Braun-Falco, O.,** Liposome preparations: a step forward in topical drug therapy for skin disease? *J. Am. Acad. Dermatol.*, 21, 1271, 1989.
27. **Škalko, N., Ćajkovac, M., and Jalšenjak, I.,** Liposomes with clindamycin hydrochloride in the therapy of acne vulgaris, *Int. J. Pharm.*, 85, 97, 1992.
28. **Touitou, E., Shaco-Ezra, N., Dayan, N., Jushynski, M., Rafaeloff, R., and Azoury, R.,** Dyphylline liposomes for delivery to the skin, *J. Pharm. Sci.*, 81, 131, 1992.
29. **Gehring, W., Ghyczy, M., Gloor, M., Scheer, T., and Röding, J.,** Enhancement of the penetration of dithranol and increase of effect of dithrano on the skin by liposomes, *Drug Res.*, 42, 983, 1992.
30. **Berliner, L. J.,** *Spin Labeling. Theory and Applications*, Vol. 1 and 2, Academic Press, New York, 1976 and 1979.

31. **Ohno, K.,** A method of EPR imaging: application to spatial distributions of hydrogen atoms trapped in sulphuric acid ices, *Jpn. J. Appl. Phys.*, 20, L179, 1981.
32. **Karthe, W. and Wehrsdorfer, E.,** The measurements of inhomogeneous distribution of paramagnetic centers by means of EPR, *J. Magn. Reson.*, 33, 107, 1979.
33. **Ohno, K.,** Application of ESR imaging to a continuous flow method for study of kinetics of short-lived radicals, *J. Magn. Reson.*, 49, 56, 1982.
34. **Ebert, B., Hanke, T., and Klimes, N.,** Application of ESR zeumatography, *Stud. Biophys.*, 130, 161, 1984.
35. **Herrling, T., Klimes, N., Karthe, W., Eweret, U., and Ewert B.,** EPR zeumatography using modulated magnetic field gradients, *J. Magn. Reson.*, 49, 203, 1982.
36. **Demsar, F., Swartz, H. M., and Schara, M.,** Use of field-gradient EPR to measure diffusion of nitroxides in tissues, *Magn. Reson. Med. Biol.*, 1, 17, 1988.
37. **Kristl, J., Pečar, S., Korbar-Smid, J., Demsar, F., and Schara, M.,** Drug diffusion: a field gradient electron paramagnetic resonance study, *Drug Development and Industrial Pharmacy*, 15, 1423, 1989.
38. **Eaton, G. R. and Eaton, E. E.,** EPR imaging: progress and prospects, *Bull. Magn. Reson.*, 10, 22, 1988.
39. **Gabrijelčič, V., Šentjurc, M., and Kristl, J.,** Evaluation of liposomes as drug carriers into the skin measured by one-dimensional EPR imaging, *Int. J. Pharm.*, 62, 75, 1990.
40. **Demsar, F., Walczak, T., Morse P. D., II, Bacic, G., Zolnai, Z., and Swartz, H. M.,** Detection of distribution and diffusion of oxygen by fast scan EPR imaging, *J. Magn. Reson.*, 76, 224, 1988.
41. **Berliner, L. J. and Fujii, H.,** Magnetic resonance imaging of biological specimens by electron paramagnetic resonance of nitroxide spin labels, *Science*, 227, 517, 1985.
42. **Swartz, H. M. and Walczak, T.,** In vivo EPR: prospects for the '90s, *Physica Medica*, 9, 41, 1993.
43. **Berliner, L. J. and Wan, X. M.,** In vivo pharmacokinetics by electron paramagnetic resonance spectroscopy, *Magn. Reson. Med.*, 9, 430, 1989.
44. **Eaton, G. R., Eaton, S. S., and Ohno, K., Eds.,** *EPR Imaging and in Vivo EPT*, CRC Press, Boca Raton, FL, 1991.
45. **Raukman, E. J., Rosen, G. M., and Griffeth, L. K.,** Enzymatic reactions of spin labels, in *Spin Labeling in Pharmacology*, Holtzman, J. L., Ed., Academic Press, New York, 1984, 175.
46. **Šentjurc, M., Morse, P. D., II, and Swartz, H. M.,** Influence of metabolic inhibitors on the nitroxide reduction in the cells, *Period. Biol.*, 88, 202, 1986.
47. **Chen, K., Morse, P. D., II, and Swartz, H. M.,** Kinetics of enzyme-mediated reduction of lipid soluble nitroxide spin labels by living cells, *Biochim. Biophys. Acta*, 943, 477, 1988.
48. **Iannone, A., Tomasi, A., Vannini, V., and Swartz, H. M.,** Metabolism of nitroxide spin labels in subcellular fractions of rat liver. I. Reduction by microsomes, *Biochim. Biophys. Acta*, 1034, 285, 1990.
49. **Iannone, A., Tomasi, A., Vannini, V., and Swartz, H. M.,** Metabolism of nitroxide spin labels in subcellular fractions of rat liver. II. Reduction in the cytosol, *Biochim. Biophys. Acta*, 1034, 290, 1990.
50. **Swartz, H. M., Šentjurc, M., and Morse, P. D., II,** Cellular metabolism of water-soluble nitroxides: effect on rate of reduction of cell/nitroxide ratio, oxygen concentrations, and permeability of nitroxides, *Biochim. Biophys. Acta*, 888, 82, 1986.
51. **Schallreuter, K. U. and Wood, J. M.,** The role of thioredoxin reductase in the reduction of free radicals at the surface of the epidermis, *Biochem. Biophys. Res. Commun.*, 136, 630, 1986.
52. **Schallreuter, K. U. and Wood, J. M.,** Free radical reduction in the human epidermis, *Free Rad. Biol. Med.*, 6, 519, 1989.
53. **Fuchs J., Huflejt, M. E., Rothfuss, L. M., Wilson, D. S., Carcamo, G., and Packer, L.,** Impairment of enzymic and nonenzymic antioxidants in skin by UV irradiation, *J. Invest. Dermatol.*, 93, 769, 1989.
54. **Fuchs, J., Mehlhorn, R. J., and Packer, L.,** Free radical reduction mechanisms in mouse epidermis skin homogenates, *J. Invest. Dermatol.*, 93, 633, 1989.
55. **Rehfeld, S. J., Plachy, W. Z., Hou, S. Y., and Elias, P. M.,** Localization of lipid microdomains and thermal phenomena in murine *Stratum corneum* and isolated membrane complexes: an electron spin resonance study, *J. Invest. Dermatol.*, 95, 217, 1990.
56. **Eriksson, U. G., Brasch, R. C., and Tozer, T. N.,** Nonenzymatic bioreduction in rat liver and kidney of nitroxyl spin labels, potential contrast agents in magnetic resonance imaging, *Drug Metab. Dispos.*, 15, 155, 1987.
57. **Schara, M.,** Molecular transport in membranes, in *Biophysics of Membrane Transport. School Proceedings II*, Kuczera, J., and Przestalski, S., Eds., Agricultural University of Warsaw, Warsaw, Poland, 1990, 117.
58. **Röding, J. and Ghyczy, M.,** Control of skin humidity with liposomes: stabilization of skin care oils and lipophilic active substances with liposomes, *S.Ö.F.W.*, 10, 372, 1991.
59. **Gehring, W., Ghyczy, M., Gloor, M., Hetzer, Ch., and Röding, J.,** Significance of empty liposomes alone and as drug carriers in dermatology, *Drug Res.*, 40, 1368, 1990.
60. **Ghyczy, M. and Niemann, R.,** Skin and liposomes, *RPR Nattermann Phospholipid GmbH Sci. Publ. No.1*, 17, 1992.
61. **Hofland, H. E. J., Bouwstra, J. A., Bodde, H. E., Spies, F., and Junginger, H.E.,** Interactions between liposomes and human stratum corneum in vitro: freeze fracture electron microscopical visualisation and small angle X-ray scattering studies, in: *Vesicles as Transdermal Drug Delivery Systems*, Ph.D. thesis, Hofland, H. E. J., University of Leiden, Leiden, Netherlands, 1992, 113.

62. **Weiner, N. and Egbaria, K.,** Topical delivery of liposomally encapsulated interferon evaluated by in vitro diffusion studies and in a cutaneous herpes guinea pig model, in *Liposome Dermatics*, Braun-Falco, O., Korting, H. C., and Maibach, H. I., Eds., Springer Verlag, Berlin, 1992, 242.
63. **Schubert, R., Joos, M., Deicher, M., Magerle, R., and Lasch, J.,** Destabilization of egg lecithin liposomes on the skin after topical application measured by perturbed angular correlation spectroscopy (PAC) with ^{111}in, *Biochim. Biophys. Acta*, 1150, 162, 1992.
64. **Artmann, C., Röding, J., Ghyczy, M., and Pratzel, H. G.,** Liposomes from soya phospholipids as percutaneous drug carriers. 1st communication: Qualitative in vivo investigations with antibody-loaded liposomes, *Drug Res.*, 40, 1363, 1990.
65. **Egbaria, K. and Weiner, N.,** Topical delivery of liposomally encapsulated ingredients evaluated by in vitro diffusion studies, in *Liposome Dermatics*, Braun-Falco O., Korting, H. C., and Maibach, H. I., Eds., Springer Verlag, Berlin, 1992, 172.
66. **Blume, A., Jansen, M., Ghyczy, M., and Gareiss, J.,** Interaction of phospholipid liposomes with lipid model mixtures for stratum corneum lipids, *Int. J. Pharm.*, 99, 219, 1993.
67. **Gabrijelčič, V., Šentjurc, M., and Schara, M.,** Interaction of liposomes with lecithin/water systems, *Acta Pharm.*, 43, 139, 1993.
68. **Chapman, D.,** Physical chemistry of phospholipids, in *Form and Function of Phospholipids*, Ansell, G. B., Hawthorne, J. N., and Dawson, R. M. C., Eds., Elsevier Scientific Publishing, Amsterdam, 1973, 117.
69. **Cevc, G.,** Rationale for the production and dermal application of lipid vesicles, in *Liposome Dermatics*, Braun-Falco, O., Korting, H. C., and Maibach, H. I., Eds., Springer Verlag, Berlin, 1992, 82.
70. **Gabrijelčič, V. and Šentjurc, M.,** Influence of hydrogels on liposome stability and on the transport of liposome entrapped substances into the skin, *Int. J. Pharm.*, accepted for publication, 1994.
71. **Kristl, J., Pečar, S., Šmid-Korbar, J., and Schara, M.,** Molecular motion of drugs in hydrocolloids measured by electron paramagnetic resonance, *Pharm. Res.*, 8, 505, 1991.

Chapter 9

Some Large-Scale, Nonmedical Applications of Nonphospholipid Liposomes

D. F. H. Wallach and Rajiv Mathur

CONTENTS

I. INTRODUCTION

Liposomes were described — more than 30 years before the term "liposome" was coined[1] by X-ray diffraction studies[2-4] defining the structure of particles consisting of organized lipid bilayers enclosing a water space. However, only in the past few years has there been effort or success in harnessing liposome technology for large-scale applications.

Until recently, liposome technology has been concerned mostly with vesicles composed of phospholipids, phosphatidylcholine predominantly. While it is true that such phospholipid liposomes, PL, are suitable for certain pharmaceutical applications, the advancement of PL technology has been beset by many obstacles: because phospholipids are biologically very active molecules, they tend to be unstable once extracted from their *in vivo* environment. Because of their lability and structure, they are expensive to purify or synthesize. In addition, manufacture of PL is complicated and costly to scale up.

For these reasons there has been increasing interest in nonphospholipid liposomes, NPL, composed of "membrane mimetic" amphiphiles.[5] These comprise a group of molecules that have a hydrophilic head group attached to a hydrophobic "tail". They include long-chain fatty acids and derivatives, long-chain alcohols and derivatives, long-chain amines and amides, and polyol lipids, sphingo- and glycerolipids. In the following we will describe the background and applications of this type of liposome, in particular the paucilamellar NPL manufactured by Micro Vesicular Systems, Inc.

Micro Vesicular Systems, Inc.'s NPL system was developed to allow large-scale production, at the same time avoiding the high raw material and production costs, vesicle instability and poor loading capacity that have plagued PL technology. The NPL are paucilamellar lipid spherules that can be formed from appropriate single-tailed amphiphiles, as well as from certain phospholipids. Each NPL has one or several membranes surrounding a large, unstructured core. Each membrane is composed of amphiphile

0-8493-4013-6/96/$0.00+$.50

molecules in bilayer array. The membranes of the NPL vesicles are stabilized by a number of interactions, including the hydrophobic effect[6] driving apolar moieties out of water, van der Waals attractions between ordered amphiphile residues and hydrogen-bonding of amphiphile head groups to the water at the hydrophilic surfaces of each bilayer. The central core accounts for most of the NPL volume, providing a high carrying capacity for water-soluble and water-immiscible substances. The high capacity for water-immiscible substances represents a unique advantage of the NPL vesicle system described.

The methods to be discussed for the manufacture of NPL involve the formation of micellar aggregates of membrane-mimetic amphiphiles and the controlled coalescence of these aggregates into membrane-bounded vesicles.

II. BULK MANUFACTURE OF NONPHOSPHOLIPID LIPOSOMES

As far as we know, no methods have been devised for the large-scale production of any liposome, except for the manufacture of Niosomes by L'Oreal, France[7-10] and of the NPL made by Micro Vesicular Systems, Inc. The bulk manufacture of Niosomes proprietary to L'Oreal has not been described in detail and we will therefore focus this manuscript on the production of NPL using the techniques published by Micro Vesicular Systems.[11-40] These techniques were originally developed for the production of food animal vaccines and have been applied to nonmedical areas such as cosmetics, personal care products, paints and coatings, agrichemicals and other bulk products (at costs equivalent to making simple emulsions). However, the Micro Vesicular Systems NPL technology can also be scaled down (e.g., for biotechnology applications).

A. BILAYER MATERIALS

1. Modular Membrane Design

NPL are engineered for particular applications. For each application, the NPL membranes are built from a series of membrane "modules", each module imparting a desired characteristic. The modular approach provides great breadth and flexibility in membrane design. It allows the combination of numerous single-tailed amphiphiles with each other (or even with phospholipids, sphingolipids, or both) to yield appropriate membrane hybrids.

2. Major Structural Modules

The major structural membrane amphiphiles account for more than 50% of NPL membrane lipid. They include fatty alcohols, fatty acids, ethoxylated fatty alcohols, fatty acids, and glycerol fatty acid monoesters, long-chain alkyl–glyceryl ethers, fatty acid glycol esters, fatty acid glycerol monoesters, fatty acid glycerol diesters, fatty acid diethanolamides, fatty acid *N,N*-dimethylacylamides, and "alkyd" amphiphiles (designed for the encapsulation of alkyd resin oils within NPL to create water-borne oil paints, as shown below. They copolymerize with the alkyd resins upon drying).

3. Membrane Modulating Molecules

The modular approach used in the design of NPL relies on the use of sterol, such as cholesterol. These "modulator" molecules intercalate between hydrocarbon chains in amphiphile bilayers, thereby allowing the intermixing of different acyl chains without phase segregation and broadening the range of temperature within which the crystalline → liquid-crystalline transition occurs. They interact with the polar termini of the chains, leaving them less free to change conformation than the more disordered long-chain segments. Phytosterols (e.g., sitosterol) and ergosterol impart similar effects and are used in place of cholesterol when there is a need to avoid cholesterol for cost or health reasons.

Additional flexibility can be achieved by use of ionogenic sterols such as cholesterol phthalate, cholesterol hemisuccinate, and cationic cholesterol derivatives, as well as by use of thio cholesterol as an attachment site for protein and peptides.

4. Ionogenic Modules

A number of ionogenic membrane amphiphiles can be used in NPL design. They can be used in conjunction with any of the nonionic modules listed previously, in proportions ranging from 0.05 to 10 mol%.

Of the anionic modules, the fatty acids and sarcosinates, as well as cholesterol phthalate or cholesterol hemisuccinate, must be used at bulk pH levels of less than 7 (membrane pH greater than 5). Diacylphosphates can be used over a broad pH range. The betaines are zwitterionic at pH levels above neutrality but become cationic below neutral pH.

Of the cationic molecules, the fatty amines are ionogenic up to about pH 10. Their applications are somewhat limited because of a high irritation potential. Single alkylated, and double alkylated quaternary amines, as well as quaternary amine derivatives of cholesterol, are cationic over the entire pH range. Cetyl pyridinium cation is ionized over most of the pH range and is widely used in oral hygiene products. Quaternary amines such as distearyl-dimethyl ammonium are extensively used in hair care preparations.

B. PROCESSES FOR NPL FORMATION

The generation of NPL by the Micro Vesicular Systems system is based on the known phase behavior of single-chain lipid amphiphiles[41-44] but can also be extended to some phospholipids and glycolipids. In the Micro Vesicular Systems approach, anhydrous, membrane-forming lipid (amphiphile, plus auxiliary molecules such as sterols, and ionogenic amphiphiles) are heated (if necessary) to yield a homogeneous liquid. In large-scale continuous flow applications, this liquid is then injected into a 3 to 15-fold excess of heated aqueous phase, through tangentially placed nozzles into a small, cylindrical mixing chamber.[12-20] One or two nozzles are used for the lipid phase and one or two alternating ones for the aqueous phase. The two liquids are driven by flow-controlled, positive displacement pumps. Depending on the hydraulic and viscous properties of the lipid phase, the injection velocity can range from 10 to 50 m/s through a 0.1 to 1.5-mm diameter nozzle. Initiation of hydration occurs at the nozzles within less than one millisecond. Photon correlation counting suggests precursor particle (micelle) size of less that 0.1 microns diameter. Cooling of the mixture from typical initial temperatures of 50°C–65°C to about 25° under conditions of high turbulence causes the micelles to fuse into paucilamellar NPL. NPL formed by this process are heat-stable under nearly all storage conditions. The process can also be conducted using motor-driven syringes.

C. NPL WITH A PREDOMINANTLY AQUEOUS CORE

Comparison of captured water volumes (calculated for spheres with 1 to 5 bilayer shells 6 nm thick and with interlamellar water computed as a bilayer component) and experimental data obtained with high- and low-capture volume NPL populations, with diameters recorded by photon correlation counting, indicate that most of the NPL have 2–3 bilayer shells. Capture of water-soluble solute parallels the entrapment of water into the vesicle core. By repeated washing of vesicles over dextran density gradients, we find an entrapment efficiency of 35–45% for water soluble proteins.[35]

1. "Humectant" NPL

Suspensions of paucilamellar NPL have three aqueous compartments. The first consists of the extravesicular phase and is immediately available after deposition on a surface such as the skin. The second consists of NPL-entrapped water and is released more slowly, depending on the osmotic activity of any solutes in the vesicle core. The third is bound, interlamellar water and is released very slowly. The second compartment, in particular, makes paucilamellar NPL effective skin humectants.

An example of humectant NPL made on a bulk scale is given in Table 1.

Table 1 Composition of (w/w; %) Lipid and Aqueous Phases for Humectant NPL

Lipid Phase		Aqueous Phase	
Glyceryl monostearate	20.8	Glycerine	11.2
Cholesterol	12.7	Diglycerine	11.2
Stearyl alcohol	8.6	Sodium lauryl sulfate	1.2
Polysorbate 60	8.6	Deionized water	~75.1
Mineral oil	49.1	Preservatives	As required

The ingredients of the lipid phase are combined in a steam-jacketed, stainless steel kettle to give a homogeneous solution at 70–73°C. The ingredients of the aqueous phase are similarly combined at 62–65°C. The two phases are then combined at a ratio of about 1:4.8 and at a rate of 5–7 l/min using a Micro Vesicular Systems high-capacity, continuous flow machine. Cooling yields a lotion with the properties given in Table 2.

Table 2 Composition (w/w; %) and Properties of the Humectant NPL Lotion

Composition		Properties	
Glyceryl monostearate	3.6	Vesicle diameter	0.270–0.570 microns
Cholesterol	2.2	Viscosity	50–350 cps
Stearyl alcohol	1.5	Specific gravity	0.970
Polysorbate 60	1.5	pH	6.5–7.5
Mineral oil	8.5		
Glycerine	9.8		
Diglycerine	9.8		
Sodium lauryl sulfate	1.0		
Deionized water	~62.1		
Preservatives	As required		

D. NPL WITH A WATER-IMMISCIBLE CORE

Liposome bilayers have a low capacity for most apolar molecules because partition of such substances into bilayers is between a structured and an isotropic phase rather than between two isotropic phases.[45] Thus, the partition coefficients of noble gases are 2- to 15-fold lower into a bilayer than into a bulk organic phase. Moreover, the uptake of apolar molecules declines with increasing chain density and factors that increase chain density of the bilayer (e.g., lowering temperature through the lipid transition temperature; incorporation of cholesterol above the lipid transition temperature) and the insertion of apolar molecules into the bilayer may disrupt the integrity of the bilayer. For these reasons there has been relatively little success in achieving stable incorporation of lipophilic substance (other than certain steroids) into PL bilayers at levels greater than 0.5% dry weight.

The NPL system described, in contrast, allows transport of apolar materials within the central core of the liposomes. Two processes allow loading of appropriate NPL cores with a wide variety of water-immiscible materials and solutions or suspensions of substances in water-immiscible carriers.

1. One-Step Process

In this process,[20,27] the water-immiscible, apolar substance is incorporated into the liquid lipid used to make vesicle membranes. During injection of the liquid lipid into the aqueous phase containing a small amount of "indifferent surfactant" — surfactant that does not form bilayers — the water-immiscible substance forms microdroplets (<0.1 micron diameter) stabilized at the surface by the indifferent surfactant around which the bilayers of membrane-forming surfactants coalesce. The captured apolar volume and encapsulation efficiency of this process are potentially very high and can range from traces to near 90%.

a. Lactic Acid-Loaded, Humectant NPL

The humectant properties of paucilamellar NPL can be augmented by alpha hydroxy acids such as lactic acid. In order to achieve a sufficiently high lactic acid concentration and, at the same time, avoid skin irritation, a humectant NPL with a nonaqueous lactic acid reservoir was developed. The alpha hydroxy acids also speed the desquamation of dead cells from the skin surface. The reservoir consists of lactic acid containing octyl hydroxystearate. At the pH of the formulation (pH 4; Tables 3 and 4) lactic acid partitions preferentially into the octyl hydroxystearate of the NPL core. As the pH rises, lactate is released into the aqueous phase.

Table 3 Composition (w/w; %) of Lipid and Aqueous Phases for Lactic Acid-Loaded, Humectant NPL

Lipid Phase		Aqueous Phase	
Glyceryl distearate	16.4	Deionized water	~100
Cholesterol	4.0	Preservatives	As required
POE (10) stearyl alcohol	8.1		
Octyl hydroxystearate	39.3		
Lactic acid (88%)	32.3		

Table 4 Composition (w/w; %) and Properties of a Lactic Acid, Humectant NPL Lotion

Composition		Properties	
Glyceryl distearate	7.7	Vesicle diameter	0.61 μm
Cholesterol	1.9	Specific gravity	0.92
POE (10) stearyl alcohol	3.8	pH	4.2
Octyl hydroxystearate	18.5		
Lactic acid (88%)	15.2		
Deionized water	~52.9		
Preservatives	As required		

The ingredients of the lipid phase (Table 3) are mixed to homogeneity in a steam-jacketed stainless steel kettle at 72°C. The lipid phase is then reacted at high velocity with water (ratio 1:1.1), preheated to 65°C, using Micro Vesicular Systems' rapid flow apparatus. Cooling to room temperature yields a lotion with the characteristics given in Table 4.

b. Retinol-Loaded NPL

The production of retinol-containing NPL is another example of the one-step process for encapsulation of water-insoluble materials for cosmetic use. The formulations for a retinol cream are given in Table 5.

Table 5 Composition (w/w; %) of Lipid and Aqueous Phases for Production of Retinol NPL

Lipid Phase		Aqueous Phase	
Glyceryl distearate	13.9	15 m*M* citrate buffer, pH 5.5	~100
Glyceryl monostearate	6.9	Preservatives	As required
Stearyl alcohol	12.5		
POE (10) stearyl alcohol	10.6		
POE (20) sorbitan monoleate	4.6		
(C_6–C_{12}) triglycerides[a]	46.2		
Retinol (45%)	5.1		

[a] $C_6 = 0.2\%$, $C_8 = 58.5\%$, $C_{10} = 40.8\%$, $C_{12} = 0.4\%$.

The ingredients of the lipid phase are mixed in a steam-jacketed stainless steel kettle at 70°C to give a homogeneous solution. The components of the aqueous phase are similarly combined at 65°C. The two phases are reacted at high velocity as before (ratio 1:3.6) using Micro Vesicular Systems' high-capacity rapid flow machine. Cooling to room temperature yields a cream with the characteristics given in Table 6.

Table 6 Composition (w/w; %) and Specifications of Retinol-Loaded NPL

Composition		Properties	
Glyceryl distearate	3.0	Mean vesicle diameter	0.52 μm
Glyceryl monostearate	1.5	pH	5.5
Stearyl alcohol	2.7	Specific gravity	0.98
POE (10) stearyl alcohol	2.3		
POE (20) sorbitan monoleate	1.0		
(C_6–C_{12}) triglycerides	10.0		
Retinol (45%)	1.1		
15 m*M* citrate buffer	~78.4		
Preservatives	As required		

c. Fungicide-Loaded NPL

This section describes the use of the one-step program for the encapsulation of 2-(thiocyanomethyl-thio) benzothiazole (TCMTB), a common agricultural fungicide for the pregermination treatment of cotton seed against fungal pathogens, avoiding the use of organic solvents.[30] The encapsulating agent, DMATO, is composed of *N,N*-dimethylamides of tall oil fatty acids, with the general formula

$$\begin{array}{l} R\text{-}\underset{\underset{\displaystyle O}{\|}}{C}\text{-}N(CH_3)_2 \end{array}$$

where R represents the alkyd chains derived from C_{12}–C_{18} straight-chain carboxylic acids such as oleic and linoleic acids. DMATO has been used in the formulation of products that act as penetrants, dispersants, plasticizers, and solvents. DMATO[46-49] has also been used as a special-purpose solvent for agricultural applications, including the delivery of TCMTB[46] and carbamate pesticides. DMATO has previously been used in organic solution or emulsified with surfactants, but we find that DMATO can form paucilamellar vesicles which can serve as delivery vehicles for water-insoluble fungicides such as TCMTB.[30] The head group of DMATO and other *N,N*-dimethlymides consists of two methyl residues on the amide nitrogen, with the oxygen on the neighboring carbon having the potential for hydrogen bonding. Molecular modeling suggests that the oxygen of the dimethylamide groups of adjacent DMATO molecules may be joined through hydrogen-bonding to a water molecule. This type of interaction between adjacent DMATO molecules would favor interdigitation of the hydrocarbon chains, rather than the end-to-end configuration found in lamellar structures composed of amphiphilic molecules.

Technical grade TCMTB (80% active) and DMATO was obtained from Buckman Laboratories International, Inc. (Memphis, TN), Busan^R 30 E.C., containing 30% TCMTB, and Nu-Flow ND, a combination product containing 23.5% 1,4,-dichloro-2,5-dimethoxybenzene (Chloroneb), another fungicide, plus 9.0% TCMTB was supplied by Wilbur-Ellis (Fresno, CA). NPL were formed by combining a lipophilic phase consisting of DMATO or DMATO/TCMTB with an aqueous phase containing 20% glycerine or propylene glycol in 1.5% sodium lauryl sulfate. The lipid and aqueous phase were combined at 55°C as described in the previous examples, yielding a fluid vesicle suspension.

One-pound batches of Chembred DES 119 or Acala SJ-2 cotton seed were treated using formulations prepared in 50-ml batches. Following treatment, the seeds were dried before bagging and storage. Greenhouse testing was conducted by the Plant Pathology Department, University California, Davis. Flats containing autoclaved Yolo loam sand (1:1) were infested with the fungal pathogens *Rhizoctonia solani* or *Pythium ultimum*. Each test involved six randomized seed treatments with the test formulations in randomized flats. Twenty seeds per flat were planted for each treatment. After 1 week, emergence and post emergence damping-off were recorded. Survival and damping-off was determined after 2 weeks.

Optical microscopy showed empty DMATO liposomes to consist of spherical vesicles with a clearly defined aqueous center. Particle sizes were determined using a Coulter N4SD submicron sizing instrument. DMATO vesicles containing TCMTB encapsulated within the central cavity were more than twice the diameter of empty DMATO vesicles (Table 7). After centrifugation of DMATO/TCMTB vesicles (3000 rpm, 30 min), no separation of free TCMTB occurred, indicating complete entrapment of this fungicide. No free TCMTB was observed when concentrated DMATO/TCMTB vesicles (30%) were either stored at elevated temperature (50°C) or diluted 1:10 with water. Stable mixtures were also obtained when DMATO/TCMTB liposomes were combined with powdered Chloroneb, a second fungicide, to give final active ingredient concentrations of 9.9% and 23.5%, respectively.

Table 7 Characteristics of DMATO/TCMTB NPL

	Vesicle Composition	w/w%	Mean Diameter (μm)
1.	DMATO	9.25	0.209
2.	DMATO + TCMTB	9.25 + 44.0	0.522

Table 8 shows the results of greenhouse testing of cotton seeds treated with four different formulations directed against *P. ultimum* and *R. solani*. Formula I is Nu-Flow ND, containing emulsified TCMTB with Chloroneb added as a wettable powder. Identical results were obtained with aqueous suspensions of DMATO/TCMTB vesicles in the presence of Chloroneb (Formula II). Formulations containing Chloroneb gave better protection against *Rhizoctonia* since Chloroneb acts primarily against this fungus.

Formula III is Busan 30 EC, which contains TCMTB and DMATO formulated with aromatic solvents (Tenneco 500-100) and ethylene glycol butyl ether. Formula IV, containing TCMTB/DMATO vesicles but no organic solvents, gave similar results against fungal pathogens to Busan 30 EC, showing that the efficacy of TCMTB is not modified by its encapsulation in DMATO vesicles. Greenhouse testing of cotton seeds which had been stored for 6 months following treatment with TCMTB/DMATO vesicles gave essentially the same results as the nonvesicular commercial formulations, confirming the stability of the TCMTB/DMATO vesicle formulations.

Table 8 Treatment of Cotton Seed with DMATO/TCMTB NPL

		Pythium Ultimum		Rhizoctonia	
Treatment	Rate[a] (fl oz/cwt)	Percent Emergence	Percent[b] Survival	Percent Emergence	Percent[b] Survival
Formula I	14.5	99.0	80.0[b]	83.9	31.9
Formula II	14.5	94.4	78.8	81.9	32.5
Formula III	4.35	85.6	59.9	0.6	0
Formula IV	3.04	83.8	32.5	3.8	0.6
Water	24.0	0	0	0	0

[a] Adjusted to give equivalent concentrations of TCMTB (and Chloroneb) per cwt seed.
[b] Duncan's multiple range values at 0.05 level of significance.

Delivery of TCMTB has previously required a combination of organic solvents and/or surfactants. Stability, ease of dilution, and impact on the environment are some among the factors that determine the final composition of the formulation. The fact that TCMTB can be delivered in stable DMATO vesicles that are freely dispersible in water and biodegradable after application avoids the use of organic solvents. The antifungal activity of TCMTB-laden NPL in aqueous dispersion demonstrate the commercial utility of NPL delivery systems.

d. Waterborne Alkyd Resin Paint

To create waterborne alkyd paints containing no or little organic solvent, we developed "alkyd" membrane-mimetic amphiphiles.[28] For this, unsaturated fatty acids are conjugated with polyols such as glycerol or pentaerythritol and further with an anhydride such as phthalic or trimellitic anhydrides, yielding anionic, "monomeric" alkyds (Table 9). Oils at low pH, these molecules form NPL at neutral pH and above. They can be used to efficiently encapsulate alkyd resins, to form waterborne oil paint. Preparations with 50% solids can be readily achieved. When painted out and dried, the "wall alkyds" copolymerize with the cargo alkyd resin to form normally "drying", clear, coherent films.

Table 9 Examples of Alkyd Membrane-Mimetic Surfactants[20]

C_{12}-C_{18} (0–4 unsaturations)	-O-CO	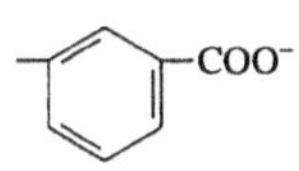
C_{12}–C_{18} (0–4 unsaturations)	-CO-O	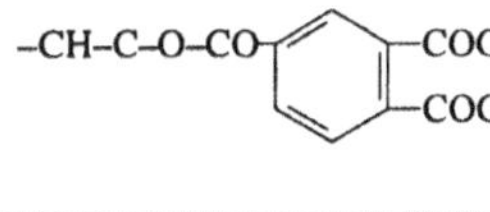

An example of an alkyd resin NPL system which has desirable polymerization characteristics, a particle size less than 1 micron, and shows little degradation at 50°C for 1 month is given in Table 10. The cargo resin is combined with the wall resin at 70°C and thoroughly mixed. The aqueous phase is then added at 65°C using a Micro Vesicular Systems motor-driven syringe mixer.

The final product, containing 53% (w/w) solids, is a thick, flowing, beige, water-miscible material (Table 11). Microscopic examination shows spherical vesicles less than 1 micron in size. No free resin is released upon centrifugation. The product produces a smooth, clear film on glass which sets to touch in 8 h and becomes hard in 16 h.

Table 10 Composition (w/w;%) of Lipid and Aqueous Phases Used to Form a NPL Alkyd Paint

Lipid Phase		Aqueous Phase	
Cargo resin	82.7	Ammonium lauryl sulfate (30%)	6.2
Wall resin	17.3	Ammonium hydroxide	1.6
		Deaerated, deionized water	92.2

Table 11 Composition (w/w; %) and Properties of Alkyd Resin-Laden NPL

Composition		Properties	
Cargo resin	53.0	Specific gravity	~0.92
Wall resin	10.0	pH	8.2
Ammonium lauryl sulfate (30%)	2.3	Solids	63.7
Ammonium hydroxide	0.6	Mean diameter	0.88 μm
Deionized, deaerated, water	34.1		

e. NPL Loaded with Diethyltoluamide (DEET)

DEET is currently the most widely used and most effective insect repellent. It is an oily liquid which is available in spray, emulsion and "stick" formulations. The disadvantages of current DEET formulation are short duration of action (2 h), unpleasant smell (spray applications), and in some individuals, penetration through the skin into the bloodstream, causing an unpleasant taste.

To reduce these disadvantages, a cream formulation consisting of NPL-encapsulated DEET was developed (Tables 12 and 13). Encapsulation of DEET dissolved in a short-chain triglyceride base provides for extended release of the insect repellent and reduces objectionable odor and skin penetration.

Table 12 Composition (w/w; %) of Lipid and Aqueous Phases Used to Make NPL Loaded with Diethyltoluamide (DEET)

Lipid Phase		Aqueous Phase	
Glyceryl Monostearate	8.3	Deionized water	~100
Glyceryl distearate	5.8		
Stearyl alcohol	5.3		
POE (10) stearyl alcohol	4.4		
Cholesterol	7.5		
(C_6–C_{12}) triglycerides[a]	29.6		
Diethyltoluamide	39.1		

[a] $C_6 = 0.2\%$, $C_8 = 58.5\%$, $C_{10} = 40.8\%$, $C_{12} = 0.4\%$.

Table 13 Composition (w/w; %) of NPL DEET Cream

Diethyltoluamide	25.0	Mean diameter	0.64 μm
Glyceryl monostearate	5.3	pH	5.0
Glyceryl distearate	3.7	Specific gravity	0.98
POE(10) stearyl alcohol	2.8		
Stearyl alcohol	3.4		
(C_6–C_{12}) triglycerides	18.9		
Cholesterol	4.8		
Deionized water	36.1		

The lipid phase is brought up to 72°C. as usual, the DEET being added last, and combined as before with aqueous phase at 65°C. Cooling to room temperature yields a cream with the composition given in Table 13. Formulations suitable for pump spray applications can be obtained by increasing the proportion of aqueous phase.

2. Two-Step Process

In the two-step process,[27] appropriately compositioned vesicles are loaded at ambient or near-ambient temperatures. The vesicles are mixed under low shear conditions with the water-immiscible cargo in the presence of indifferent surfactant. Microdroplets of the apolar material then enter the central vesicle space within seconds. This process is particularly useful for the loading of thermolabile materials (e.g., flavor or fragrance oils) and substances that interfere with micelle formation or fusion at higher temperatures.

a. NPL-Based Skin Cleaner

Many NPL formulations have an unexpected capability to solubilize and remove water-insoluble "dirt" from the skin and other surfaces.[20] This function may be enhanced by encapsulating "cleansing-oils" such as D-limonene or oxo-decyl acetate, Exate 1000.

The formulation for an oxo-decyl acetate-containing skin cleaner is given in Tables 13 and 14. "Empty" NPL are made from the component listed in Table 13 using the continuous flow apparatus under conditions already described. After 16 h, the PL are combined with oxo-decyl acetate (Table 15) at 25–32°C, using gentle stirring for 30–60 min, to provide a skin-cleaning lotion (mean diameter 0.2–0.5 microns; viscosity not less than 200 cps).

Table 14 Composition (w/w; %) of Lipid and Aqueous Phase for the Preparation of NPL to be Loaded with Exate 1000 (Oxo-Decyl Acetate)

First Lipid Phase		First Aqueous Phase	
POE (9) glyceryl monostearate	97.3	Deionized water	~99.1
Phyosterol	2.7	NaCl	0.9
		Preservatives	As required

Table 15 Loading of NPL with Oxo-Decyl Acetate

Second Lipid Phase	v/v (%)	Second "Aqueous" Phase	v/v (%)
Exate 1000	40.0	NPL from first step	60.0

REFERENCES

1. **Bangham, A.D., Standish, M.M., and Watkins, J.C.,** 1965, Diffusion of univalent ions across the lamellae of swollen phospholipids. *J. Mol. Biol.* 13, 238-252.
2. **Schmitt, F.O., Bear, R.S., and Clark, G.L.,** 1935, X-ray diffraction studies on nerve. *Radiology* 25, 131-151.
3. **Schmitt, F.O. and Bear, R.S.,** 1939, The ultrastructure of the nerve axon sheath. *Biol. Rev.* 14, 27-41.
4. **Bear, R.S., Palmer, K.J., and Schmitt, F.O.,** 1941, X-ray diffraction studies of nerve lipids. *J. Cell Comp. Physiol.* 17, 355-367.
5. **Fendler, J.,** 1982, *Membrane Mimetic Chemistry*, John Wiley & Sons, New York, 522.
6. **Tanford, C.,** 1980, *The Hydrophobic Effect*, Wiley, New York.
7. **Handjani-Vila, R.M., Ribier, A., and Vanlenberghe, G.,** 1985, *Les niosomes, in Les Liposomes*, Puisieux, F. and Delattre, J., Eds., Technique et Documentation, Lavoisier, Paris, 297-312.
8. **Handjani, R., Ribier, A., Vanlenberghe, G., and Handjani R.M.,** 1987, Preparation of more stable niosomes useful in preparation of cosmetic creams, pharmaceutical products, etc. and obtained with nonionic lipid phase and an aqueous phase. French Patent 2597346.
9. **Vanlenberghe, G. and Handjani-Vila, R.M.,** 1988, Aqueous dispersions of lipid spheres and compositions and contents of same. U.S. Patent 4,772,471.
10. **Griat, J., Handjani-Vila, R.M., Ribier, A., Vanlenberghe, G., and Zabotto, A.,** 1989, Cosmetic and pharmaceutical preparations containing niosomes and a water-soluble polyamide, and a process for preparing these compositions; encapsulated aqueous phase in spherule formed from a nonionic amphiphilic lipid; external phase of aqueous poly-B-alanine. U.S. Patent, 4,830,857.

11. **Wallach, D.F.H.,** 1989, Method of producing high-aqueous volume multilamellar vesicles. U.S. Patent 4,855,090.
12. **Yiournas, C. and Wallach, D.F.H.,** 1990, Method and apparatus for producing lipid vesicles. U.S. Patent 4,895,452.
13. **Wallach, D.F.H.,** 1990, Paucilamellar lipid vesicles. U.S. Patent 4,911,928.
14. **Wallach, D.F.H.,** 1990, Lipid vesicles formed of surfactants and steroids. U.S. Patent 4,197,951.
15. **Wallach, D.F.H.,** 1990, Encapsulated humectant. U.S. Patent 4,942,038.
16. **Tabibi, E. and Wallach, D.F.H.,** 1991, Theoretical consideration of lipid vesicle formation by Novamix. *INTER-PHEX-USA, Proc. 1991 Technical Program,* 61-68.
17. **Sakura, J.D., Mathur, R., Wallach, D.F.H., Schulteis, D.T., and Ostrom, J.K.,** 1991, The delivery of agricultural fungicides in paucilamellar amphiphile vesicles. In *Pesticide Formulations and Application Systems,* Vol. 12, ASTM STP 1146, B.N. Devisetty, D.G. Chasin, and P.D. Berger, Eds. American Society for Testing and Materials, Philadelphia, 155-162.
18. **Wallach, D.F.H.,** 1991, Protein coupling to lipid vesicles. U.S. Patent 5,000,960.
19. **Yiournas, C. and Wallach, D.F.H.,** 1991, Method and apparatus for producing lipid vesicles. U.S. Patent 5,013,497.
20. **Wallach, D.F.H.,** 1991, Removing oil from surfaces with liposomal cleaner. U.S. Patent 5,019,174.
21. **Wallach, D.F.H.,** 1991, Encapsulation of parasiticides. U.S. Patent 5,019,392.
22. **Wallach, D.F.H.,** 1991, Encapsulation of ionophore growth factors. U.S. Patent 5,023,086.
23. **Wallach, D.F.H.,** 1991, Paucilamellar lipid vesicles using charge localized, single chain, nonphospholipid surfactants. U.S. Patent 5,032,457.
24. **Wallach, D.F.H. and Philippot, J.R.,** 1992, New type of lipid vesicle: Novasome™, in *Liposome Technology,* 2nd ed., G. Gregoriadis, Ed., CRC Press, Boca Raton, FL, 141-151.
25. **Wallach, D.F.H.,** 1992, Reinforced paucilamellar lipid vesicles. U.S. Patent 5,104,736.
26. **Wallach, D.F.H.,** 1992, Paucilamellar lipid vesicles. U.S. Patent 5,147,723.
27. **Wallach, D.F.H.,** 1992, Method of making oil filled paucilamellar lipid vesicles. U.S. Patent 5,160,669.
28. **Wallach, D.F.H., Mathur, R., Chang, A.C., and Tabibi, E.,** 1992, Lipid vesicles having an alkyd as wall-forming material. U.S. Patent 5,164,191.
29. **Wallach, D.F.H., Mathur, R., Redziniak, G.J.M., and Tranchant, J.F.,** 1992, Some properties of N-acyl sarcosinate lipid vesicles, *J. Soc. Cosmet. Chem.* 43, 113-118.
30. **Wallach, D.F.H. and Mathur, R.,** 1993, Lipid vesicles having *N,N*-dimethylamide derivatives as their primary lipid. U.S. Patent 5,213,805.
31. **Wallach, D.F.H., Mathur, R., and Henderson, S.,** 1993, Gas and oxygen carrying lipid vesicles. U.S. Patent 5,219,538.
32. **Wallach, D.F.H., Mathur, R., and Albert, L.,** 1993, Lipid vesicle containing water-in-oil emulsions. U.S. Patent 5,256,422.
33. **Wallach, D.F.H.,** 1993, Hybrid paucilamellar lipid vesicles. U.S. Patent 5,234,767.
34. **Mathur, R. and Wallach, D.F.H.,** 1993, Blended lipid vesicles, U.S. Patent 5,234,915.
35. **Vandergriff, K., Wallach, D.F.H., and Winslow, R.M.,** 1994, Encapsulation of hemoglobin in nonphospholipid vesicles. *Art Cells, Blood Sub. and Immob. Biotech.,* 22, 849-854.
36. **Dowton, S.M., Hu, Z., Ramachandran, C., Wallach, D.F.H., and Wiener, N.,** 1993, The effect of liposomal composition on topical delivery of encapsulated cyclosporin. I. An *in vitro* study using hairless mouse skin. *STP Pharma Sci.* 3, 304-407.
37. **Dowton, S.M., Hu, Z., Ramachandran, C., Wallach, D.F.H., and Wiener, N.,** 1993, The effect of dosing volume on the disposition of cyclosporin-A in hairless mouse skin after topical application of a nonionic liposomal formulation: An *in vitro* diffusion study. *STP Pharma Sci.* 4, 145-149.
38. **Raychaudhuri, P., Katz, M., Wilkinson, D., Mathur, R., and Wallach, D.F.H., 1994,** Increased efficacy in psoriasis of topical anthralin encapsulated in nonphospholipid liposomes, *Arch. Dermatol.*, in press.
39. **Philippot, J. R. and Wallach, D.F.H.,** Modified lipids and nonphospholipid molecules as constituents of liposomes, in *Liposomes as Tools in Basic Research and Industry,* J. R. Philippot and F. Schuber, Eds., CRC Press, Boca Raton, FL, p. 44-54.
40. **Varanelli, C., Kumar, S. and Wallach, D.F.H.,** Nonphospholipid vesicles as experimental immunological adjuvants in *Liposomes as Tools in Basic Research and Industry.* J. R. Philippot and F. Schuber, Eds., CRC Press, Boca Raton, FL, p. 253-263.
41. **Mitchell, J., Tiddy, G.J.T., Waring, L., Bostock, T., and McDonald, M.P.,** 1983, Phase behavior of polyoxyethylene surfactants with water. *J. Chem. Soc. Faraday Trans.* 1, 79, 975-1000.
42. **Adam, C.D., Durrant, J.A., Lowry, M.R., and Tiddy, G.J.T.,** 1984. Gel and liquid-crystal phase structures of the trioxyethyleneglycol monohexadecyl ether/water system. *J. Chem. Soc. Faraday Trans.* 1, 80, 789-801.
43. **Randall, K. and Tiddy, G.J.T.,** 1984, Interaction of water and oxyethylene groups in lyotropic liquid-crystalline phases of polyoxyethylene n-dodecyl ether surfactants studied by ^2H nuclear magnetic resonance spectroscopy. *J. Chem. Soc. Faraday Trans. 1,* 80, 3339-3357.
44. **Carvell, M., Hall, D.Y., Lyle, I.G., and Tiddy, G.J.T.,** 1986, Surfactant water interactions in lamellar phases. *Faraday Discuss. Chem. Soc.* 81, 223-237.

45. **DeYoung, L.R. and Dill, K.A.,** 1988, Solute partitioning into lipid bilayer membranes. *Biochemistry* 27, 5281-5289.
46. **Pulido, M.L.**, 2-(Thiocyanomethylthio)benzothiazole: an effective treatment for cottonseed, *Proc. 1971 World Agric. Conf.*, October 11–15, 1971, 1-12.
47. **Lutey, R.W., King, V.M., and Cleghorn, M.Z.,** Mechanisms of action of dimethylamides as a penetrant/dispersant in cooling water systems, 50th Annu. Meeting Int. Water Conf., Pittsburgh, PA, October 23-25, 1989.
48. **Buckman, S.J., Pera, J.D., and Purcell, W.P.,** 1969, Pitch control in pulp and papermaking, U.S. Patent No. 3,274,050, 1966.
49. **Mod, R.R., Magne, F.C., and Skau, E.L.,** The plasticizing characteristics of some N,N-dimethylamides and ester amides of long-chain fatty acids, *J. Am. Oil Chem. Soc.*, 45, 385-387, 1968.

Chapter 10

Liposomal Formulations of Agrichemical Pesticides

Peter J. Quinn and Steven F. Perrett

CONTENTS

I. INTRODUCTION

Agricultural and horticultural practices increasingly rely on the application of a variety of pesticidal agents to protect crops and harvested products from attack by predatory organisms or combat weed infestations. Whilst there is a wide variety of pesticides available, differing in mode of action and biochemical potency, there is considerable scope for improvement in efficiency with which these agents are employed. The main impetus to achieve greater efficiency of agrichemical use is the increased awareness of the environmental impact of current and projected usage of these chemicals.

Some indication of the seriousness of environmental effects of agrichemical pesticides can be judged from initiatives taken by governments to draw up ever-tightening regulations to control the use of chemical products in agriculture. It is recognised that agriculture is the only economic sector that intentionally releases massive quantities of chemical substances directly into the environment which have the potential to jeopardise human health, wild life and soil quality. Some indication of the magnitude of the problem can be gauged from figures released on rates of fertilizer applications to arable land within countries of the European Union. The Netherlands heads the league table with rates of application of 342 kg/ha, followed by Belgium (313 kg/ha), Germany 294 kg/ha), Denmark (240 kg/ha) and the United Kingdom (120 kg/ha). Rates of pesticide application are believed to follow a similar pattern. It is not only the accumulation of bioactive residues in the environment that has undesirable effects since formulation aids such as surfactants can also reach significant levels (Parr, 1982; Ernst et al., 1982; Davis et al., 1982).

0-8493-4013-6/96/$0.00+$.50

There are two possible approaches to the problem of excessive pesticide usage. The first is to develop more specific and potent pesticides. This is a constant and increasingly expensive endeavour of the agrichemical industry. The second approach is to improve the efficiency with which the present generation of pesticidal agents is used by the development of more effective and specific formulation and delivery strategies. The general aims in the latter case are to apply pesticides evenly to the desired target, to protect it from chemical degradation for a designated period, to assist its relocation to the site of action and to ensure ultimately that residues are rapidly removed from the environment.

This chapter gives details of how phospholipids can be exploited to provide novel methods of pesticide formulation. The general principles involved in these strategies will be described and, in particular, the use of crude soya lecithin in formulation of pesticides will be exemplified.

II. CONVENTIONAL PESTICIDE FORMULATION STRATEGIES

The problem confronting the applicator is how to distribute defined quantities of highly biologically active chemical evenly over relatively large surface areas. The most common methods employ the use of dusts and granules, smokes and sprays. Dusts and granules are dry formulations; they are simple to apply and involve sprinkling the formulation from a container or ejecting it into an air stream. It is often hard to achieve an even distribution using this method. In this formulation the active ingredient is combined with an inert particulate material, such as a clay mineral like kaolinite or attapulgite, by either adsorption onto the surface from solution or simply milling the dry pesticide with the clay in a suitable mill. Particle sizes define dusts (3–30 μm) and granules (>250 μm). Smokes and fogs are other nonliquid forms of application which are suited to large areas. The chemical is dissolved in an oil with an appropriate flash point and injected into a hot gas stream. A dense fog is formed by condensation of the oil to produce an aerosol when it is discharged into the atmosphere. The most common form of application, however, is in the form of a liquid concentrate sprayed onto the target area. Reviews of these strategies can be found in Mathews (1982a–c) and Hassall (1990).

The universal solvent for spray applications of pesticides to arable land or crops is water. The majority of useful agrichemical pesticides, however, are relatively insoluble in water or are unstable and undergo rapid hydrolysis to inactive products. Table 1 is a list of the most commonly used pesticides that have low solubility in water and require formulation aids to produce water-based sprays. Formulations of concentrated bioactive agents must render the pesticide dispersible in water so that it will remain in a homogeneous suspension during the spraying operation. In order to form stable dispersions or emulsions, a variety of formulation aids are employed to prepare emulsifiable concentrates or wettable powders, including solvents, detergents and particulates. Kerosene, higher ketones and xylene are typical solvents used in the preparation of emulsifiable concentrates. Surfactants used to stabilize insoluble pesticides include sodium dodecyl benzene sulphonate (Santomerse 1), polyoxy-thioether (Sterox SK), the Tritons and Tweens (McWhorter, 1982; Falbe, 1987). As with dry granular formulations, wettable powders can consist of active agent adsorbed to a fine inert material that remains in colloidal or particulate suspension. The formulation will normally contain surfactant to prevent aggregation or coagulation of the particles in suspension. Typically, a water-dispersible powder consists of 50% active ingredient, 47% filler and 3% surfactant (Hassall, 1990).

In addition to these base formulation aids, other ingredients of wettable powders or emulsifiable concentrates are designed to improve the spray characteristics of the formulation, such as antifoaming agents or to facilitate the interaction of the spray with the target surface. Examples of such additives include wetting and sticking agents that assist the spread of spray drops over the waxy cuticular layer of plant foliage and in the retention at the target under weathering conditions, respectively. Wetting agents serve to reduce the surface tension of the spray drop, thereby reducing the tendency of the drops to bounce and increase the spreading of the drop on the leaf surface (Cowell, 1982). Surfactants, as well as affecting the surface tension of the spray drop, can alter spray drop characteristics (Fraser, 1958; Arnold, 1983) and augment the penetration or uptake of the active agent at the target (Smith et al., 1966; Baker and Hunt, 1988).

One particular problem encountered in the use of many formulations of solvents and detergents is that they are often phytotoxic, and concentrations in the spray need to be low to avoid damage to plant tissues. Since dilute tank mixes are required to reduce phytoxicity, this can hamper spray operations in wet conditions and they are unsuitable for aerial applications where spray volumes need to be low. Another difficulty with conventional formulations is that incompatibilities are frequently encountered in

Table 1 Pesticides That Show Relatively Low Solubility in Water and Are Suitable for Formulation With Soya Lecithin

Insecticides	Fungicides	Acaricides	Herbicides	Nematocides	Molluscicides
Chlorpyrifos	Benomyl	Amitraz	Atrazine	Aldicarb	Methiocarb
DementonSmethyl	Binapacryl	Azinophosmethyl	Aziprotryne	Dichloropropene	
Dicofol	Captan	Diazinon	Bensoylprop-ethyl		
Diflubenzuron	Carbendazim		Bromacil		
Dimethoate	Chlorothalonil		Carbetamide		
Iodofenphos	Cufraneb		Chlorbromuron		
Iprodione	Dichlorfluanid		Chloridazon		
Malathion	Imazalil		Chlorthaldimethyl		
Mephosfolan	Triadimenol		Diphenamid		
Permethrate	Triadimefon		Diuron		
Pirimicarb	Triforine		Flamprop		
Tetradifon	Dinocap		Lentacil		
	Etridiazole		Linuron		
	Maneb		Metamituron		
	Metalaxyl		Methbenzthiazuron		
	Propioconazole		Methazole		
			Metribuzin		
			Propham		
			Propyzamide		
			Simazine		
			Terbacil		
			Trietazine		

tank mixes of different pesticides as well as between pesticides and other agrichemical agents such as fertilizers, plant growth regulators, desiccants, etc. These incompatibilities are manifest as sludges or coagulants that serve to clog the spray machinery and result in uneven application to the target. Apart from incompatibilities amongst formulation mixes, many formulations do not form tank mixes that are stable for lengthy periods and they require constant agitation to prevent settling. In addition, high concentrations of dispersing agents, such as synthetic detergents, are required to produce stable emulsions, and these tend to be relatively expensive ingredients in the formulation.

III. NEW FORMULATION STRATEGIES

The use of vegetable oils as an alternative to water-based spray formulations has been explored. These include the use of cotton seed oil alone or as an oil-in-water emulsion for the application of a variety of insecticides and herbicides. Vegetable oils from other sources have been used in trials of the application of post-emergent herbicides. Claims that these formulations result in protection of the biologically active agent against degradation, leaching from the soil and weathering on exposed targets, potentiation of uptake into pests and plants, etc. have been made. The major disadvantage, however, is that conventional spraying equipment is unable to handle viscous oils, and high capital investment in special machinery is necessary.

More recently, surfactants derived from plant oils have been examined as alternatives to synthetic detergents which are derived largely from feedstocks supplied form the petrochemical industry. These surfactants are by-products of purification of the oil and they require little if anything in the way of chemical modification to adapt them for use in pesticide formulations.

Two types of patents cover the use of these natural surfactants and synthetic analogues of natural phospholipids in pesticide formulations. The first group (A. Nattermann GmbH) concern the use of phospholipids as general additives to existing formulations or in mixtures of pesticides with other formulation aids. These claim that addition of phospholipids to acaricidal, fungicidal and insecticidal formulations leads to improved efficacy. Because of the presence of other formulation aids, the form of phospholipids in the spray formulation is not likely to be in the characteristic bilayer and nonlamellar aggregates that are found in aqueous dispersions of these natural surfactants. The second group of patents (Acacia Chemical Co.) refer to the use of natural surfactants and, in particular the phospholipids of crude soya lecithin, as base formulation ingredients. In this strategy the detergent properties of these natural phospholipids are responsible for the formation of typical liposomal aggregates in water, and this is claimed to improve the performance of spray applications of pesticides. The viability in both strategies relies on the plentiful supply of various commercial grades of lecithin at market prices competitive with synthetic detergents.

IV. COMMERCIAL SOURCES OF PHOSPHOLIPIDS

Soya lecithin is one of the most plentiful supplies of surfactant phospholipids. Soya beans are processed by crushing to remove oil from the protein which is a major constituent of the seeds and used mainly in animal feedstuffs. Soya bean crushing yields an oil that contains about 15% by weight of phospholipids, and their removal is the primary step in purification of the oil. The extracted lecithin is considered to be a by-product of relatively low value. The extracted lecithin is used as an emulsifying agent in food processing and, in less significant quantities, as a base chemical in cosmetic manufacture, health foods and the like (Sartoretto, 1967; Scocca, 1976). To date, only limited use has been made of phospholipids or other natural surfactants in agriculture. Phospholipid sprays have been used to prevent mosquito larval development in stagnant pools through a reduction in surface tension from 70 to 30 $mN \cdot m^{-1}$ (Van Nieuwenberg, 1981). Aqueous formulations of phospholipids have also been used in the control of powdery mildew (Misato et al., 1977).

Lecithin is removed from the oil by treatment with super-heated steam. This creates a two-phase system consisting of oil (mainly triglycerides) and a concentrated emulsion of soya phospholipids and other surfactant materials. The soya lecithin is recovered usually by solvent extraction and subsequent removal of the solvent at high temperature. This product, with the consistency of a sludge, contains most of the phospholipid with some contaminating mono-, di and possibly triglycerides. The main phospholipid components of soya lecithin are phosphatidylcholines, phosphatidylethanolamines, phosphatidylserines, phosphatidylglycerols and lyso derivatives of the respective phospholipids. The chemical composition of a typical commercial grade of soya lecithin is presented in Table 2. The peroxide value varies considerably between grades of soya lecithin, and antioxidants are often added to protect against oxidation of unsaturated fatty acids during storage. The extent of oxidation of the lipid does not appear to cause significant alteration in surfactant properties, at least in samples stored for several years with no addition of antioxidant.

Table 2 Chemical Composition of a Typical Industrial Grade of Soya Lecithin

Agent	Value
Acid value	30 max
Saponification value	190–200
Free fatty acids	25 max
Peroxide value	25 meq/kg max
Moisture	1.0
Color (Garner 10% in CTC)	10 max
Acetone insolubles	62% min
Benzene insolubles	0.3% max
Phosphorus	3.2% max
Nitrogen	About 1%

V. SURFACTANT PROPERTIES OF PHOSPHOLIPIDS

The term "surfactant" covers a relatively diverse group of chemical compounds that have a common physical property, namely, their tendency to orient at interfaces between polar and less polar environments. This property arises because of the amphipathic character of the molecules, i.e., the nonpolar domain is separated discretely from the polar group. Several categories of surfactants are recognised according to the manner in which they interact with water which, in turn, depends on the amphipathic balance within the molecule and the chemistry of the polar group. As shown in the previous section, soya lecithin contains varying proportions of phosphatidylcholines, phosphatidylethanolamines, phosphatidylserines and phosphatidylglycerols, all of which are electrostatically charged and, with the exception of the choline phosphatides, carry net negative charges at neutral pH. Synthetic detergents commonly used in pesticide formulations are either neutral polyhydroxy compounds or cationic detergents.

Surfactant molecules differ in their surface activity according to their amphipathic balance. Detergent strength can therefore be related to relative solubility in water. Since solubility in water is inversely proportional to solubility in apolar solvents, this also provides a useful index of detergent strength. Another parameter that influences surface activity is the steric factors associated with the distribution of polar and nonpolar affinities within the surfactant molecule and these are much more difficult to quantitate. For example, the bulkiness and disposition of the respective polar and nonpolar domains can exert a considerable effect on detergent strength. An alkane, for instance, with two hydroxyl groups at one end of the hydrocarbon chain, will act as a rather strong detergent. By contrast, if the hydroxyls are located at either end of the chain a relatively weak detergent is formed. The physical extent of the two domains can also be exemplified by considering the detergent properties of a surfactant with four hydroxyl groups and two alkyl chains of equivalent length. Such a molecule might be expected to differ considerably from the example used above because of intermolecular interactions that would modify the behavior of the complex molecule in water.

In order to generalize the concept of detergent strength and to compare the surface activity of synthetic detergents with phospholipids, the relationship between detergent strength and critical micelle concentration is instructive. Critical micelle concentration is the concentration of molecules in free solution, usually water, which are in equilibrium with aggregates formed by the molecules. This relationship with detergent strength is illustrated schematically in Figure 1. This shows that the most powerful detergents in aqueous systems have critical micelle concentrations in the m*M* concentration range. Substances which have a high critical micelle concentration are those in which the polar affinity tends to dominate the character of the molecule, and this compensates for the energy required to expose the relatively small hydrophobic domain of the molecule to water. Molecules that possess a relatively extensive hydrophobic domain tend to have low critical micelle concentrations and they prefer to associate together to form stable aggregates in water. The aggregates are in equilibrium with molecules in free solution and there is a rapid exchange between the two forms.

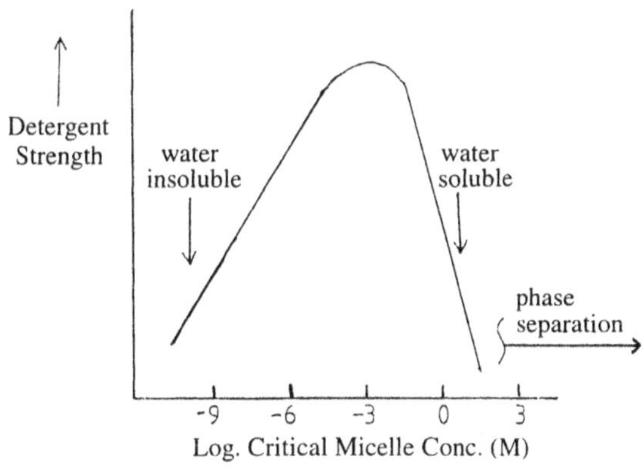

Figure 1 Relationship between surfactant strength and critical micelle concentration in water. The most powerful detergents lie in the m*M* range of critical micelle concentration.

The orientation of the molecules within aggregates is such that the hydrophobic domains assemble so that water is excluded, allowing the hydrophobic residues to interact through strong cohesive van der Waals forces. At the same time the polar head groups of the molecules are able to reside at the interface between the aqueous phase and the hydrophobic domain of the molecules where they are afforded maximum opportunity to interact with the surrounding water molecules. This is in contrast to oils, which tend to expose a minimum area of contact with an aqueous medium. When surfactants are added to an oil–water system, the molecules orient at the interfacial region serving to disperse the nonpolar component in the aqueous phase to create an emulsion.

The stability of emulsions are critical in pesticide tank mixes and is governed not only by the particular chemistry of the surfactant but also on the relative proportions of the polar and nonpolar domains and the presence of factors capable of modifying the polar interaction of the surfactant with water. The capacity of surfactants to form stable emulsions also depends on the way the polar interactions of the aggregates are modified by neighboring surfactant molecules. Modulation can arise through charge repression caused by the attraction of protons and counter ions to the charged interface, and the balance of these proximity effects, with the desire to exclude water molecules from the hydrophobic domain created by the surfactant assembly.

VI. STRUCTURE AND SIZE OF SURFACTANT AGGREGATES

Surprisingly little is known of the precise detail of the way strong surfactant molecules pack into micellar aggregates. The structures are usually depicted as spherical in shape with the polar head groups oriented at the surface and the hydrocarbon chains projecting into the interior. This can, at best, be regarded as schematic and is most likely quite different from the actual arrangement of molecules. It is obvious, for example, that there is not sufficient space within the interior to accommodate the bulk of the hydrocarbon residues contained in the chains. The problem can be likened to assembling a collection of match sticks so that there is maximum contact between the shafts of the matches while at the same time ensuring that the heads are separated by the greatest distance from one another. This task, it should be acknowledged, is slightly more difficult than one would expect in practice, where it is likely that the hydrocarbon chains possess a certain amount of flexibility and can bend to some extent to accommodate intermolecular packing in a more efficient manner. Thus the aggregate will have a configuration that maximises disorder in the system while obeying the physical constraints of packing the molecules into the available space. It may be envisaged that domains of hydrocarbon are exposed to water and incur a penalty in the form of decreased entropy of the system, but these would tend to be compensated by the additional hydration potential created at other sites within the assembly.

In contrast to the paucity of information about the structure of micelles formed by relatively strong detergents, there is a considerable body of data concerning the structure and phase behaviour of weak detergents. Phospholipid and glycolipid aggregates in aqueous systems have been characterised in some detail by diffraction, spectroscopic and thermal methods (Luzzati, 1968; Lindblom and Rilfors, 1992). Structural transitions in surfactant–water systems can be recognised as the critical micelle concentration of the surfactant decreases and the aggregation number increases. The aggregation number is the average number of molecules present in each aggregate. The relationship between aggregation number and critical micelle concentration is shown in Figure 2. It can be seen from this diagram that surfactants with low critical micelle concentrations can exist in a separate phase in which there is a fixed stoichiometry between surfactant and water and which coexists in equilibrium with bulk water. By contrast, as the critical micelle concentration increases there is a tendency for aggregation number to decrease because there is less need to protect the relatively small hydrophobic domain of the surfactant from exposure to water.

There are two features associated with the high aggregation number of weak surfactants like phospholipids that render them suitable for use as formulation agents in pesticide sprays. Firstly, they create structures such as bilayer and hexagonal phases in water that have relatively extensive hydrophobic domains into which pesticide molecules of low solubility in water may partition. Secondly, weak surfactants have a much greater tendency to stabilize water-in-oil emulsions than strong detergents which tend to micellize the system. Related to this feature is the tendency to adsorb to or coat finely milled particles and stabilize the particles in aqueous suspension.

VII. FORMATION AND STABILITY OF SOYA LECITHIN DISPERSIONS

Strong surfactants will readily form homogeneous micellar dispersions in water. Weak surfactants like phospholipids do not disperse readily in water, and this is one of the major problems that confronts the pesticide formulator. Whether the pesticide is formulated as an emulsifiable concentrate or wettable powder, it is essential that it disperses readily in a relatively large volume of water with a minimum of agitation. It is also desirable that the pesticide formulation flows readily from the container as residues from this source represent a considerable toxic hazard.

A. DISPERSION PROCEDURES

In general, the most efficient methods of dispersing phospholipids involve vigorous mechanical mixing or ultrasonic irradiation in dilute salt solutions, preferably at high temperatures. Such methods are

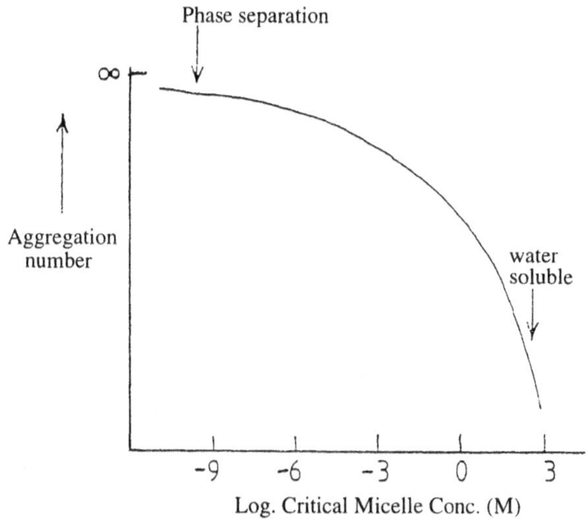

Figure 2 Relationship between critical micelle concentration of a surfactant and the aggregation number. Surfactants with low critical micelle concentrations tend to form a separate hydrated phase which can coexist with bulk water. There is a tendency of surfactants with high critical micelle concentrations to exist in free solution or as dimers or similar configurations.

obviously of no use in the practice of pesticide application in the field and alternative methods are required. One method is to use the product of steam-treated soya oil directly without further purification. The extract is already hydrated and can be simply mixed with the bioactive agent to form an emulsifiable concentrate. Solvents can be used to assist mixing and methanol, ethanol, butanol, propanol, isopropanol etc. are satisfactory for this purpose. An alternative method to improve dispersal properties of anhydrous soya lecithin preparations is to add an alcohol–water mixture containing the pesticide to the lipid. Lower alcohols mentioned above are also suitable solvents for this application. An example of this type of formulation comprises industrial grade soya lecithin, 10 parts by weight: water, 3 parts by volume: methanol, 3 parts by volume. This formulation is light brown in colour and has the consistency of a viscous oil. Upon addition to a large volume of water it disperses readily on gentle agitation to form a homogeneous opalescent dispersion.

B. DISPERSION STABILITY

A reliable method to assess the stability of dispersions is to monitor the rate at which the dispersion settles upon standing. Emulsions of soya lecithin produced by gentle shaking of a concentrated methanol–water formulation of the type described above in a concentration of 5 to 50 $g \cdot l^{-1}$ of dilute salt solution were found to be stable on standing for periods of longer than 1 year. Similar dispersions to which an anthraquinone dye was added as a marker were tested in accelerated sedimentation tests using centrifugation at 2000 g for 10 min and estimating the amount of lipid remaining in dispersion by quantitating the dye in the supernatant. Provided a small quantity of salt (~25 m*M*) was present in the aqueous medium, only abut 2% of the lipid was found to sediment under these fairly rigorous conditions. Similar stabilities were recorded with tap water, which presumably contains sufficient salts to ensure a stable dispersion.

In other tests, the pyrethroid insecticide permethrin (Wellcome) was formulated into an emulsifiable concentrate as described above and the stability of the dispersion examined in the centrifugation test. The results of analysis of the supernatant fraction are presented in Table 3. This shows that more than 90% of the permethrin and phospholipid remain in the supernatant after centrifugation; the remainder was recovered in the pellet. In treatments where the phospholipid was omitted, the permethrin was insoluble and precipitated rapidly to the bottom of the tube, leaving a clear supernatant.

Formulations of pesticides as wettable powders by dry mixing finely milled pure active agents with soya lecithin were found to produce readily dispersable concentrates that flow easily and form stable suspensions. A relatively high ratio of pesticide to soya lecithin can be achieved in such preparations before flocculation occurs.

Table 3 Percent of Soya Lecithin Formulation of Permethrin Remaining in Suspension After an Accelerated Sedimentation Test

Replicate Number	Soya Lecithin (% of total)	Permethrin (% of total)
1	90	90
2	98	100
3	95	95
4	92	90

Comparative tests have been performed to examine the performance of commercial formulations of representative pesticides with formulations in phospholipids of the type described for permethrin above. Formulations of permethrin as a wettable powder (Coopex; Wellcome), metalaxyl/mancozeb as a wettable powder (Fubol; Ciba Geigy) and propiconazole as an emulsifiable concentrate (Radar; ICI) were dispersed in water according to the manufacturers' instructions and subjected to sedimentation by centrifugation. All formulations were completely precipitated by this treatment. Formulations of pesticides in soya lecithin, at least where the ratio of phospholipid to pesticide is relatively large, form extremely stable dispersions in water.

VIII. SPRAY CHARACTERISTICS OF SOYA LECITHIN DISPERSIONS

There have been a number of reports that surfactants and emulsifying agents affect the spray characteristics, particularly the size of drops formed and the distribution of drop sizes. A particularly important drop index is the proportion of the drops with a diameter less than 100 μm. This is because drops of this size represent the upper limit of drop sizes that are subject to respiratory inhalation. When the proportion of such drops is high, operator hazards arise because of intoxication by breathing in the spray mist. Furthermore, sprays with these characteristics are prone to drift and instances of up to 1 km have been reported (Hock, 1987). It is obvious that in these circumstances action on non target areas is more difficult to control.

A. PRODUCTION OF SPRAYS

There are a variety of nozzles available to form an aqueous spray. The principal types of these are impact, cone, fan-jet, solid stream, air blast, vortical and spinning disc. Each produces a spray with different characteristics. The impact, cone, fan-jet and solid stream nozzles are those most often used. The fanjet nozzle is one of the most common and was selected as an example for studies presented here. Fan-jet nozzles use relatively low hydraulic pressures to force liquid through a small orifice. They produce a spray with a flat triangular shape; the width of the spray fan is dependent on the precise nozzle design. These nozzles are often used as multiples on a boom and are suited to spraying flat surfaces such as soil and walls; they are also used to delivery sprays from aircraft.

In the field, it has been shown that better coverage of a fixed dose of pesticide per unit area of crop is produced with smaller drops (Hislop and Baines, 1980). The greater coverage is in part related to a greater drop density (e.g., 1 liter over 1 hectare represents 2387 × 20 μm or 2.4 × 200 μm drops/cm^2) and in part to the increased tendency of larger drops to bounce off the target due to their higher momentum (Van Emden, 1989). However, if the average size of drops delivered from a conventional hydraulic nozzle is reduced, by increasing the pressure or decreasing the size of the nozzle orifice, the proportion of very small drops in the spray is also increased.

The range of drop sizes in an aqueous agrichemical spray is generally very broad, from 1 to 1000 μm in diameter. Aqueous sprays may be classified according to the predominant drop diameter. Thus: aerosols contain drops below 50 μm; mists, 50–100 μm; fine sprays, 100–200 μm; medium sprays, 200–400 μm; coarse sprays, above 400 μm (see Mathews, 1982b).

B. MEASUREMENT OF SPRAY CHARACTERISTICS

The size of the drops in a spray has traditionally been determined by collection of the drops on suitable surfaces, such as magnesium oxide-coated slides or water-sensitive paper; drop sizes are determined from the crater or stain left by the impact of the drop. Methods based on drop collection are useful in the field, but suffer from reduced collection efficiency of small drops (Arnold, 1983). Large drops may also shatter or bounce on impact. In the laboratory, methods based on the scattering of laser light (Swithenbank and Taylor, 1979; Weiner, 1979) are more useful since they allow measurements to be rapidly taken of the spray drops in flight, eliminating sampling problems.

One instrument used to measure spray drop size is a Malvern Instruments ST1800 (Malvern, Worcs.) laser diffraction particle analyser. When fitted with a lens of focal length 80 cm, a window for drop sizing in the range 15 to 1500 μm diameter is created. The spray is directed through a laser beam which results in a pattern of light diffraction that is characteristic of the size of the spray drops (Born and Wolf, 1959). The pattern falls on the detector assembly of the instrument and information on the size of the drops is derived by computer processing of the data (Crump and Seinfield, 1982; Heintzenberg and Welch, 1982).

In the studies described below, spray nozzles were placed in a holder positioned 30 cm from the laser beam and the spray was directed through the beam at 45° to the horizontal (to minimize the splash-back from a liquid collecting tray). Sampling was carried out with the beam passing through the spray fan transversely, i.e., through its shortest axis, at a distance of 30 cm. At this point, break-up of the spray liquid into drops has been shown to be complete (Arnold, 1983). For each measurement 300 separate scans were made (taking approximately 4 sec.) and a background level was recorded in a similar manner prior to each spray run. All nozzles were brass and obtained from Spraying Systems, Woodbridge Park, Guildford, U.K. The fan-jet nozzles comprised angles of 65°, 80°, and 110°; serial nos. 6503, 8003, 1103, respectively. Pressures of 200 kPa, 300 kPa and 400 kPa were employed; pressure was developed with compressed air.

The Malvern Particle Size Analyser provides two types of particle size distribution. Firstly, the volume median diameter (VMD), which is the drop diameter such that 50% of the total spray volume is of larger drops and, secondly, the number median diameter (NMD), the drop diameter such that 50% of the total number of spray drops is of larger diameter. The VMD is more relevant to the volume of pesticide delivered and is the distribution most often used. In the case of sprays from hydraulic nozzles, the NMD is low (typically 5–10 μm), and it is more useful to calculate the percentage of spray by volume below a particular drop diameter. A drop size of 100 μm was chosen as a convenient arbitrary value, as pointed out above, smaller drops are generally considered to be prone to drift (Arnold, 1983). The ratio of VMD to NMD is a purely mathematical concept which can be used to obtain a measure of the variability of drop sizes within a given spray. With a monodisperse spray, the ratio VMD/NMD is unit and a progressive increase in the value occurs with increasing polydispersity.

C. DROP SIZE DISTRIBUTION

As already indicated, the characteristics of the component drops of an aqueous agrichemical spray largely determine its behaviour in terms of crop coverage and potential to drift. Surfactants have been shown to affect spray drop characteristics (Fraser, 1958; Arnold, 1983; Quinn et al., 1986; Sundaram et al., 1987). The effect of an industrial grade of soya lecithin obtained from Lucas Meyer, Hamburg, Germany, has been examined using a Malvern Particle Size Analyser. An emulsifiable concentrate was prepared by stirring a mixture of the lecithin, methanol and water (6:2:2; by wt) for approximately 30 min at 60°C. When cooled the mixture was a homogeneous opaque gel. This gel readily disperses in water. The aqueous industrial-grade soya lecithin dispersions were prepared by dilution of this 60% concentrate with tap water, to give final lecithin concentrations of 1.25, 2.5, 5 and 7.5 g/l. Sprays were delivered through 110°, 80° and 65° fan-jet nozzles at pressures of 200, 300 and 400 kPa. The effect of soya lecithin on spray drop characteristics was found to be similar in all of the nozzle and pressure combinations used. The results obtained using an 80° fan-jet nozzle for volume medium diameter, % spray volume in drops <100 μm diameter and polydispersity of the spray drops is illustrated in Figures 3, 4 and 5, respectively.

The data presented in Figure 3 illustrates the effect of soya lecithin concentration on the VMD of aqueous sprays. The VMD is plotted as a function of lipid concentration and the inclusion of soya lecithin clearly increases the VMD of the sprays. The extent of this increase (approximately 30%) is independent of lecithin concentration in the range used. The proportion of the spray by volume in drops <100 μm diameter plotted as a function of lecithin concentration (Figure 4) shows that at all concentrations of lecithin there is a marked decrease the percentage of the drops less than 100 μm in diameter. Irrespective of the concentration of lecithin the decrease is approximately the same amount, i.e., about 70%.

The polydispersity of a spray is an index of the width of the size distribution such that a value of 1 indicates a monodisperse spray. Plots of polydispersity as a function of lecithin concentration (Figure 5) indicate that there is no consistent effect of the lecithin on polydispersity.

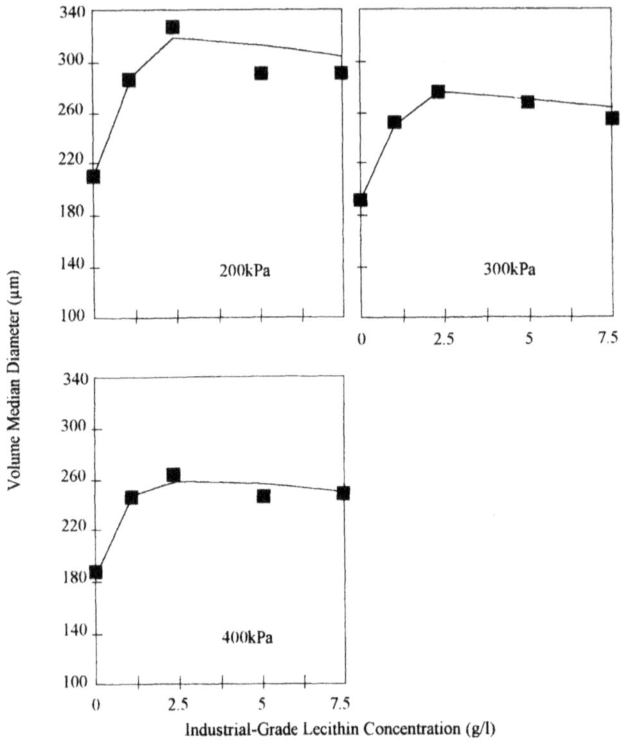

Figure 3 The effect of industrial-grade soya lecithin concentration on the volume median diameter (VMD) of aqueous spray drops. Soya lecithin was added to tap water as a 60% gel to give final concentrations as indicated. The sprays were delivered through an 80° fan-jet nozzle using the pressures indicated on the figures. The spray drop parameters were measured by laser diffraction using a Malvern ST1800 instrument as described in the text.

The effect of including industrial-grade soya lecithin in an aqueous dispersion is to increase the average size of the spray drops by around 30%. This is also reflected in a reduction in size of the small drop fraction. These effects are the reverse of that which would be expected from a decrease in surface tension, which generally leads to a decrease in drop size and or an increase in the size range of the drops (Sundaram et al., 1987; Arnold, 1983).

Additional studies of the effect of soya lecithin on drop characteristics have been performed using hollow cone nozzles at pressures of 500 and 1000 kPa. The results obtained with the use of such a nozzle at a pressure of 500 kPa are presented in Figure 6; qualitatively similar data was obtained for sprays produced by pressures of 1 MPa. It can be seen that soya lecithin, particularly at lower concentrations, causes an increase in volume medium diameter of the particles (Figure 6a), a reduction in the range of drop sizes (Figure 6b) and a decrease in the proportion of drops with a diameter less than 100 μm (Figure 6c). It was also found that sprays delivered at higher pressures produced similar but less pronounced effects on drop characteristics.

The effects of soya lecithin might be explained by considering the effects involved as the aqueous liquid is sheared into drops. A phospholipid monolayer will quickly form at the air/water interface on the drop surface, with the hydrophobic regions of the molecule oriented away from the drop center. The fact that the effects of lecithin are largely independent of concentration is consistent with its effect being confined to the air/water interface, since the total surface area of the drops is finite and will saturate when a monolayer of phospholipid is formed. This effect is illustrated schematically in Figure 7. The mechanism through which this monolayer might affect spray drop size is, however, unclear.

The feasibility of this model can be evaluated by estimating the amount of phospholipid required to cover the entire surface area of the spray drops. The area occupied by the head group of a phospholipid molecule is approximately 60 Å^2 (Hauser et al., 1981). If the total surface area of the spray drops is

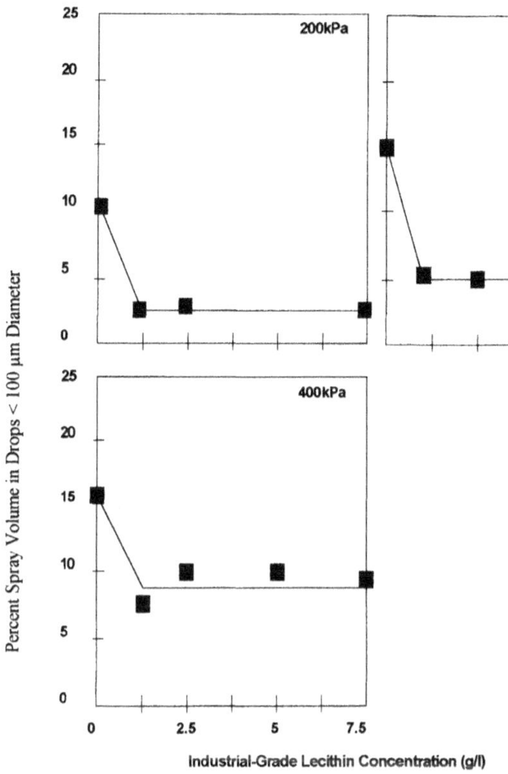

Figure 4 The effect of soya lecithin concentration on the percentage of the spray volume contained in drops less than 100 µm diameter. The conditions and the methods used were as described in the legend to Figure 3.

known it is, therefore, possible to calculate the aqueous concentration of phospholipid required to cover this area. Unfortunately, techniques based on laser diffraction do not record the number of spray drops produced from a given volume. However, a rough estimate of the total surface area of the spray produced from unit volume can be made from the VMD. Thus, taking the VMD of the sprays presented in Figure 3 as 260 µm, the volume of one drop is 9.2×10^{-6} cm^3, which corresponds to 1.09×10^9 drops per liter. The total drop surface area of 1 liter of a spray containing only 260 µm diameter drops is given by drop number multiplied by $4\pi r^2$: 2.12 Å^2 × 1.09×10^9 = 2.3^{22} Å^2. Thus $2.3^{22}/60 = 3.8 \times 10^{20}$ molecules of lipid would be needed to cover the total surface area. The amount of lipid can be estimated from Avogadro's number. Thus $3.8 \times 10^{20}/6.023 \times 10^{23} = 4 \times 10^{-4}$ moles of lipid. Taking an average molecular weight of 800 for the lipid gives a concentration of 0.32 g/l. This figure is a rough approximation, but it indicates that a soya lecithin concentration of 1.25 g/l (60% phospholipid), the lowest used in the study, should be capable of producing enough monolayer phospholipid to cover the entire surface area of spray drops in this size range.

The increase in spray drop size caused by industrial-grade soya lecithin is not an effect generally observed with other surfactants. These either have insignificant effects (Sparkes et al., 1988; Combellack and Mathews, 1981) or increase polydispersity and decrease the VMD (Arnold, 1983; Sundaram et al., 1987). It should be noted that the effects of soya lecithin are to cause an increase in VMD and therefore it has the opposite effect from that produced by strong surfactants. The presence of a monolayer of phospholipid on the drop surface has already been suggested but why this monolayer should exert effects which differ from those formed by typical detergents remains unclear. It may be associated with the greater hydrophobic character of the phospholipid monolayer which results from a difference in amphipathic balance compared to conventional synthetic detergents. Such a monolayer may also be effective in preventing the reduction of spray drop size through reducing the evaporation of water from the drop

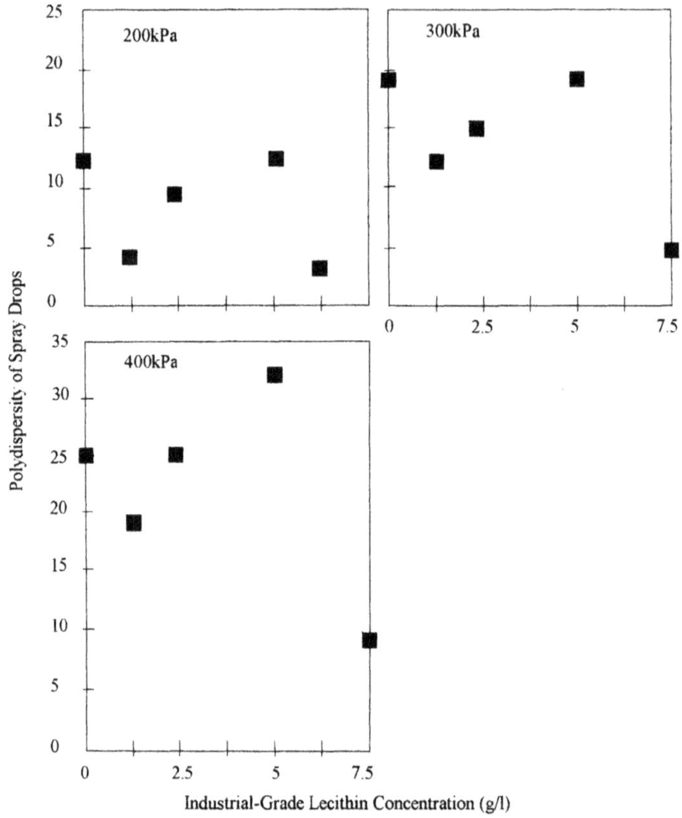

Figure 5 The effect of soya lecithin concentration on the polydispersity of aqueous spray drops. The conditions and the methods used were as described in the legend to Figure 3.

surface. Under field conditions in particular, the relatively large surface area to volume ratio of small drops means that losses of water by evaporation can result in a significant and rapid size reduction of very small drops. For example, it has been calculated that a 20°C and a relative humidity of 80% the lifetimes to extinction of drops of 50 μm and 200 μm diameter are 14 and 227 seconds, respectively (Mathews, 1982c). A condensed monolayer of phospholipid may well slow evaporation from drop surfaces and thereby reduce inflight changes in drop size.

D. SPRAY PATTERN

The pattern of distribution of spray is an important index of spray performance since it is a measure of the evenness with which the active ingredient is deposited on the target area. It is obviously important when potent pesticides are applied, since a uniform distribution ensures an effective biological action and avoids overdosing areas of the target. Spray patterns are determined in devices that measure the volume of spray delivered over segments of the area covered by the spray. Only two dimensions of fan-jet and hollow cone nozzles were examined in tests performed with soya lecithin dispersions.

The volume of spray recovered from a cross section of a 110° fan-jet nozzle delivering either water or a dispersion of soya lecithin (5 g/l) at 200 kPa is illustrated in Figure 8. It can be seen from the distribution of spray volume that the presence of soya lecithin creates a more even distribution over the target area compared with water. The pattern observed with water did not differ significantly from that obtained with Agral 90 and other synthetic surfactants commonly employed in pesticide formulations.

Although the differences in spray pattern recorded in Figure 8 do not appear dramatic, when the total volume delivered to the target is taken into account, considerable overdosing would occur towards the

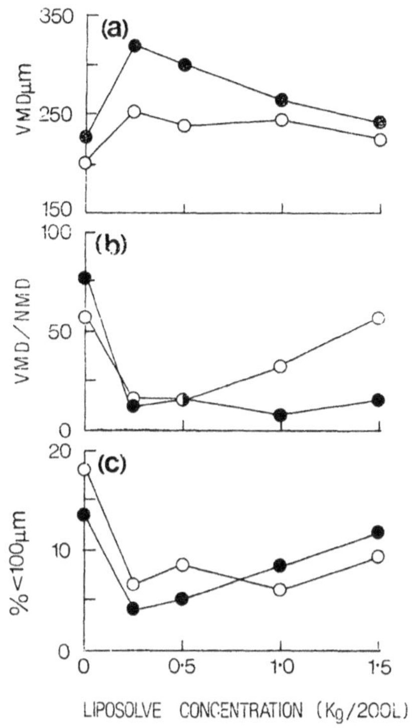

Figure 6 The effect of soya lecithin concentration on (a) volume medium diameter, (b) ratio of volume medium diameter to number medium diameter, (c) percentage by volume of particles less than 100 μm diameter in aqueous sprays delivered at 500 kPa pressure from hollow cone nozzles D6/25 (•) and D4/25 (○).

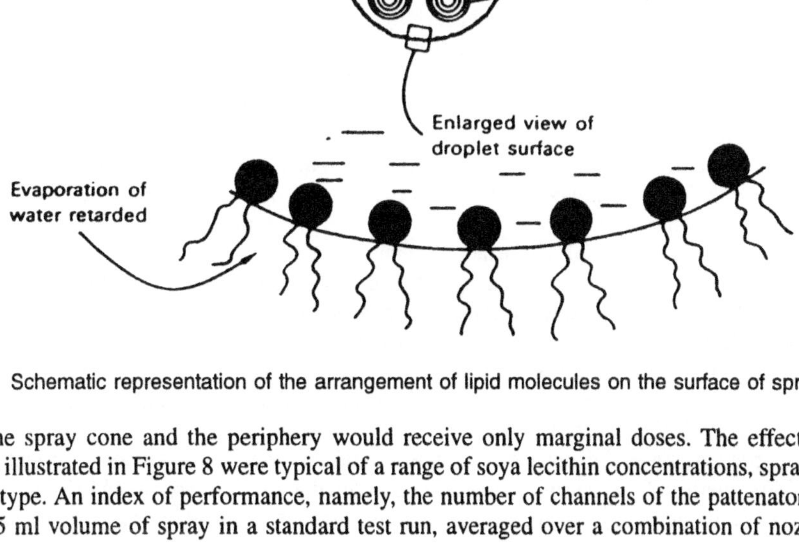

Figure 7 Schematic representation of the arrangement of lipid molecules on the surface of spray drops.

center of the spray cone and the periphery would receive only marginal doses. The effects on spray distribution illustrated in Figure 8 were typical of a range of soya lecithin concentrations, spray pressures and nozzle type. An index of performance, namely, the number of channels of the pattenator collecting more than 5 ml volume of spray in a standard test run, averaged over a combination of nozzles, spray

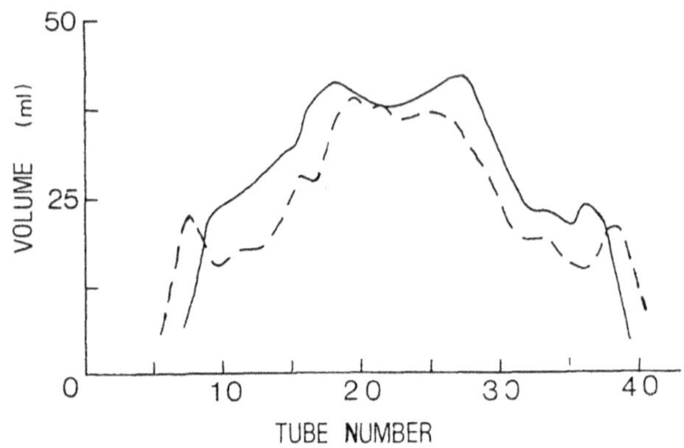

Figure 8 Volume distribution of spray at 200 kPa from a 110° fan-jet nozzle. Water (———); soya lecithin dispersion at a concentration of 1 kg/200 l (- - -).

Table 4 Indices of Spray Volume Distribution From Fan-Jet Nozzles

Lecithin Concentration (kg/200 l)	Fan-Jet Nozzle 110°	80°	65°
0	37	24	17
0.25	40	28	20
0.5	41	28	20
1.5	41	27	20

Note: Values represent the number of tubes containing more than 5 ml of spray volume under standard test conditions.

pressure and soya lecithin concentration is collated in Table 4. This shows that soya lecithin in the spray has a consistently beneficial effect on creating an even distribution of spray volume over the target area.

E. FORMATION OF BIOCOMPATIBLE SURFACES

In many applications of pesticides, the target is plant foliage, insects or microorganisms. One of the characteristic features of these surfaces is that they are hydrophobic and this is mainly due to the presence of waxes or wax-like coating materials. The use of soya lecithin in spray formulations has several important features in regard to improvement of efficiency of pesticide application.

As illustrated in Figure 7, drops formed from dispersions of soya lecithin become coated with a monomolecular film of phospholipid molecules which impart a hydrophobic character to the drop surface. Since the target is often of like character, there is a marked tendency for the drop to spread evenly over the target area. This is illustrated schematically in Figure 9.

The wetting ability of the spray drops can be assessed in quantitative terms by measuring their contact angle with a leaf surfaces (Holloway, 1970). This is the angle between a tangent to the contact point of the drop and the leaf surface as shown in Figure 10. The lower the contact angle of an aqueous drop, the greater the extent of wetting of the surface. Contact angles of pesticide formulations and soya lecithin dispersions have been determined by placing 20 μl drops onto sections of field bean *(Vicia faba)* leaf fixed with double-sided tape onto a flat glass tile. A sharp shadow of the drop was cast onto a screen using a slide projector. The leaf and drop outlines can then be traced and contact angles measured from these by drawing a tangent from the point of contact of one side of the drop to the leaf surface. This process was repeated for ten drops on sections from ten similar leaves taken from different plants at equal stages of growth (30 cm tall).

The mean contact angle of drops from samples containing increasing aqueous concentrations of the phospholipid/permethrin formulation (prepared as described above) are presented in Table 5. The contact angles decreased as the industrial-grade soya lecithin concentration increased. This effect is observed

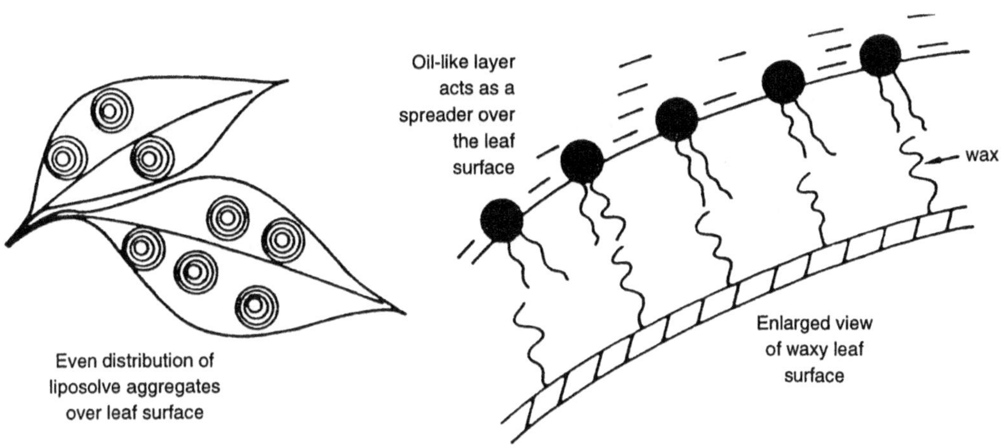

Figure 9 Schematic illustration of the origins of compatibility between spray drops coated with a monomolecular film of phospholipid molecules and waxy target surfaces.

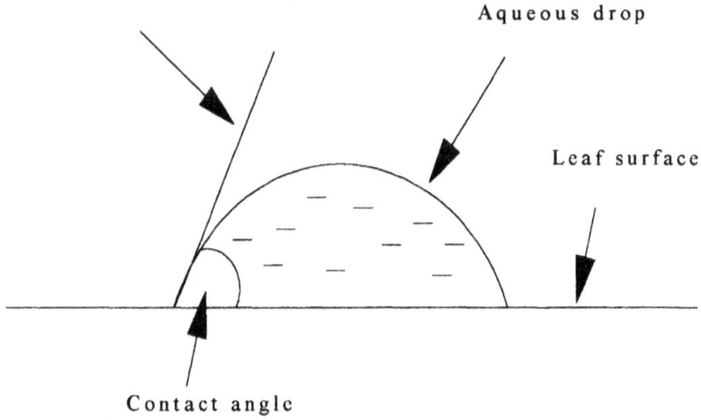

Figure 10 Determination of contact angle of drops on a planar surface.

Table 5 Effect of Soya Lecithin on Contact Angle of Aqueous Drops on the Surface of Field Bean Leaves

Lecithin Concentration (g/l)	Mean Contact Angle (degrees)
0	85
1.25	72
2.5	66
5.0	59
7.5	49

with most surfactants and is due to a reduction in surface tension, which causes the drops to spread over the leaf surface. The increased spreading of the drops, owing to the presence of soya lecithin, is also obvious from the diameter of drops formed on the leaf surface. It is clear from this type of data that soya lecithin is a very efficient wetting agent.

Related to biocompatibility effects is the fact that packaging pesticides in liposomes tends to extend rain fastness of the agent on the target surface. Sticking action can be demonstrated using dye-containing soya lecithin dispersions sprayed onto leaves and then dipped into aqueous solutions using standard test procedures. Measurements of dye retained on leaf tissue and removed in the washing solution when applied in soya lecithin of differing concentrations are presented in Table 6. It can be seen that soya lecithin acts as an effective sticking agent and prevents washoff of dye from the leaf surface. It is noteworthy that the effect is directly related to the concentration of soya lecithin in the spray in which the dye is applied. This effect of concentration indicates that soya lecithin may have particular utility in low volume applications of pesticides.

Table 6 Percentage Washoff of Residues From Washed Leaves Sprayed With Soya Lecithin Dispersions

Lecithin Concentration (g/l)	Range of % Washoff	Average
2.5	26–58	42
10	34–47	42
30	23–34	29
60	27–42	38
100	17–28	21

Similar results were obtained with commercial pesticide preparations including malathion and derris which were formulated according to manufacturers' instructions and to which varying amounts of soya lecithin were added. In this example, the effect is due to the additive effects of soya lecithin and not due to soya lecithin as a base formulation ingredient.

F. TOXICITY

A desired effect of pesticide formulation is that the biologically active agent will be taken up readily by the target organism. Packaging pesticides in soya lecithin can augment the action by virtue of creating a biocompatible surface, as outlined in the previous section. Thus, phospholipids have adjuvant effects on the uptake of hydrophobic molecules by creating biocompatible surfaces between pesticide vehicles and surfaces of living organisms.

Allied to this effect is the possibility of uptake of toxic chemical agents by operators in the course of spraying pesticides. These undesirable adjuvant effects can be counteracted to some extent by the reduced risk associated with the effect of soya lecithin on reducing the proportion of spray drops <100 µm in diameter.

IX. SOYA LECITHIN AS A FORMULATION AID

The above results show that the addition of phospholipid to water has significant effects on spray drop characteristics. It is necessary in an evaluation of soya lecithin performance to establish whether the addition of phospholipid to preformed aqueous dispersions of agrichemicals will also affect spray drop characteristics in a similar manner. The effect of adding soya lecithin to commercial pesticide formulations on spray characteristics was therefore examined.

The agrichemical formulations studied are representative examples of the primary means by which agrichemicals are dispersed in water (namely, emulsifiable concentrates, wettable powders and dispersible particulates). The agrichemicals were formulated in accordance with the manufacturers instructions. The first pesticide selected was Radar (ICI), which is an emulsifiable-concentrate of propiconazole. This was dispersed at a concentration of 2.5 ml per liter (0.5 l/ha). The second pesticide selected was Fubol (Ciba Geigy), which is a water-dispersible powder formulation of metalaxyl and mancozeb. This was dispersed at a concentration of 7.5 g/l (1.5 kg/ha). The final pesticide chosen was Avenge (American Cyanamid Co.), which formulated as a powder containing 63% (w/w) of the water-soluble methyl sulphate salt of difenzoquat. Avenge was dispersed with the wetting agent Agral 90 (alkyl phenol ethoxylate), respective final concentrations of each were 7.9 g/l (1.58 kg/ha) and 5 ml/l (1.0 l/ha). Avenge was also formulated without the inclusion of the wetting agent. Soya lecithin was added as a 60% concentrate in methanol:water described above, to give final lecithin concentrations of 1.25, 2.5, 5 and 7.5 g/l in the spray formulation. There was no obvious deterioration in the suspension stability of the agrichemical dispersions when soya lecithin was added. There was some flocculation of the water-dispersible powder (Fubol), but this was easily resuspended by gentle agitation. The nozzle and pressure combinations used and the measurement conditions were as previously described for studies performed on lecithin dispersions.

The spray drop parameters (VMD, the percentage of the spray drops below 100 μm diameter, and the polydispersity) were examined using a Malvern Particle Size Analyser as described above. The effect of phospholipid on spray drop characteristics was found to be broadly similar for all the nozzle and pressure combinations used. Representative examples are illustrated in Figures 11–13.

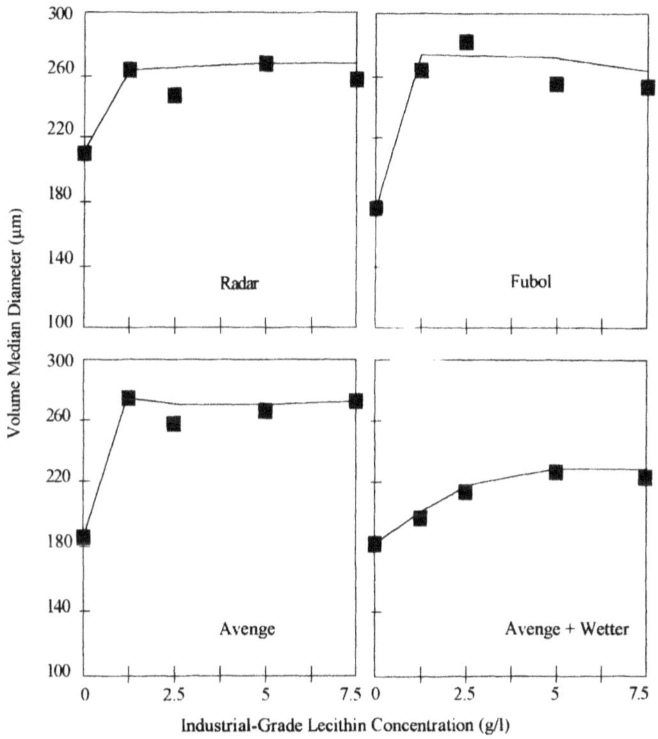

Figure 11 The effect of soya lecithin concentration on the volume median diameter (VMD) of spray drops of aqueous formulations of agrichemical formulations. The soya lecithin was added to the dispersions as a 60% (wt/vol) concentrate to give the final concentrations indicated. The sprays were delivered through an 80° fan-jet nozzle using 300 kPa pressure.

The VMD of typical agrichemical sprays as a function of soya lecithin concentration are plotted in Figure 11. The examples presented are: (a) Fubol (b) Radar (c) Avenge (d) Avenge + wetter (Agral 90). In each case the results were obtained using a 80° fanjet and 300 kPa pressure. The addition of soya lecithin increased the VMD of all agrichemical sprays. The increase was about 50% for the Avenge and Fubol dispersions at all lecithin concentrations. The increase in VMD of the Radar spray was lower (approximately 25%) at all lecithin concentrations. The increase in VMD of the Avenge plus wetter spray showed some dependence on lecithin concentration. At lower levels (<5 g/l), the value increases with concentration, until reaching a maximum of approximately 20% at 5 g/l. Higher lecithin concentrations result in no further increase in VMD.

The effect of the wetting agent, Agral 90, can be seen in isolation by comparison of the sprays containing Avenge both with and without wetting agent. The effect of wetting agent appears to counteract the effects of the phospholipid. The increase in VMD is lower and more phospholipid is required for maximum effect. This is probably due to the detergent effect of Agral 90 causing a disruption of the organisation of the phospholipid molecules in bilayer liposomes. It should be remembered that some

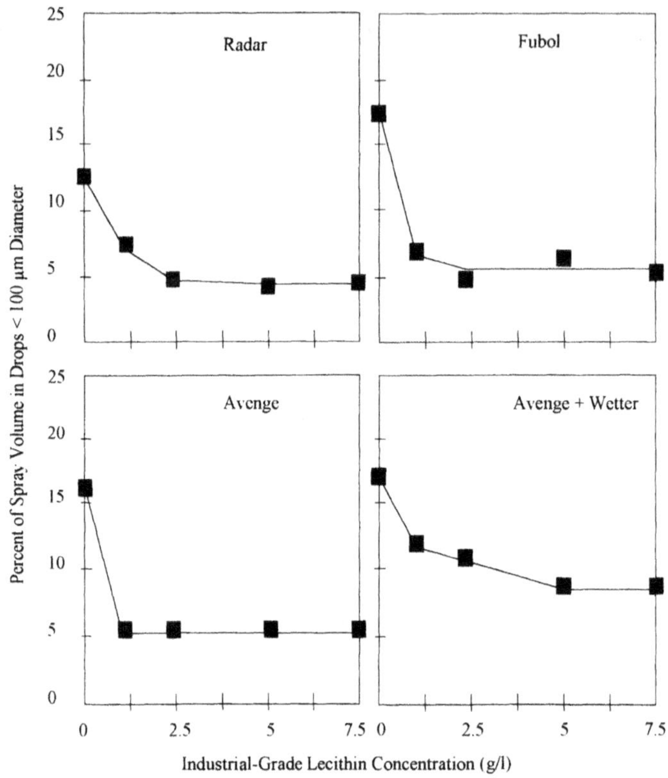

Figure 12 The effect of soya lecithin concentration on the percentage of the spray volume contained in drops <100 µm diameter of the aqueous agrichemical dispersions indicated. Measurement conditions were as described in the legend to Figure 11.

surfactant is also present in the Fubol and Radar sprays. However, since water-dispersible powder and emulsifiable concentrate formulations typically contain between 3 and 5% surfactant, the levels of surfactant present in these sprays would, therefore, be at least an order of magnitude lower than those containing Agral 90.

The effects of lecithin on the fraction of the spray volume in small drops (<100 µm diameter) as a function of soya lecithin concentration are presented in Figure 12. In the Avenge and Fubol sprays this size fraction decreases from approximately 17% to 5%. This reduction is independent of lecithin concentration in the range used. In the Radar sprays the small drop component was also reduced. The decrease in this fraction was slightly lower, in this case it decreased from 12% to 7% at a concentration of 1.25 g/l, then still further, to 5%, at higher lecithin levels. The effect of the addition of lecithin to the Avenge + wetter dispersions was also to decrease the small drop fraction. In this case, however, the decrease appears to be dependent on lecithin concentration, exhibiting progressive reductions from approximately 17% to 7%, as the level of lecithin is increased through the range from 0 to 7.5 g/l.

The effects of soya lecithin on polydispersity as a function of concentration in the agrichemical sprays are presented in Figure 13. Again, all formulations were delivered through an 80° fan-jet nozzle at 300 kPa pressure. The polydispersity decreased from approximately 25 to 10 in all formulations when lecithin was added. The size of the reduction showed some dependence on lecithin concentration in Radar and Fubol sprays, but approached the maximum at 2.5 g/l. The Avenge and Avenge + wetter sprays reached the maximum decrease shown in these systems at 1.25 g/l.

The effect of the agrichemical on the spray drop characteristics can be seen by a comparison of sprays of water only (Figures 3–5) with agrichemical sprays to which no lecithin was added. It can be seen that the agrichemicals alone do not significantly affect the parameters measured.

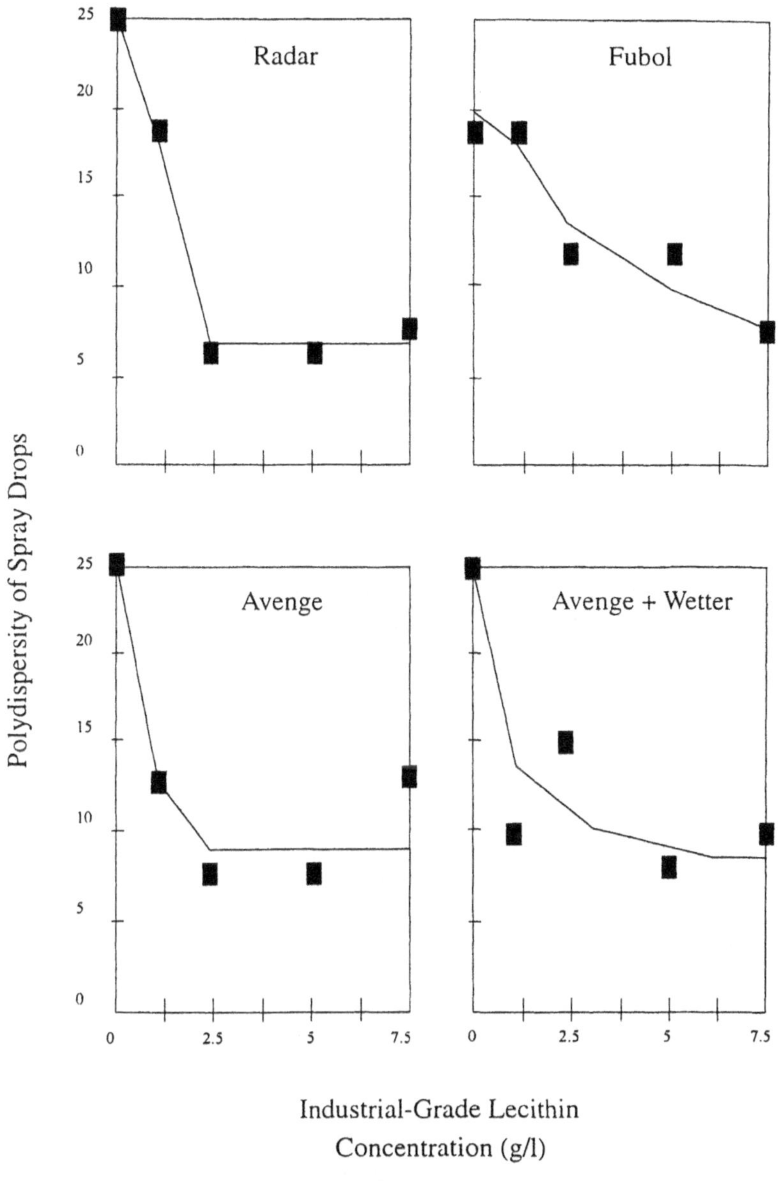

Figure 13 The effect of soya lecithin concentration on the polydispersity of spray drops of the aqueous agrichemical dispersions indicated on the figure. Measurement conditions are as described in the legend to Figure 11.

The effects of the addition of lecithin to agrichemical dispersions are to increase the size of the spray drops, to reduce the small drop fraction, and to narrow the drop size distribution of the spray. These effects are similar to those seen in the absence of agrichemical and are largely independent of the concentration of soya lecithin in the range examined. The effects were generally antagonised by the detergent Agral 90, consistent with a competitive interaction between the two amphipaths. This may occur at the air/water interface on the drop surfaces where both surfactants are likely act. This affect may be mediated through a disruption of the phospholipid monolayer by the detergent.

X. SOYA LECITHIN AS A BASE FORMULATION INGREDIENT

Small-scale laboratory trials have been performed with the pyrethroid insecticide, permethrin (Borroughs Wellcome), formulated directly with industrial-grade soya lecithin. The formulations were prepared by dissolving soya lecithin and pure permethrin (30:1, by wt) in solvent (chloroform), and evaporating the solvent to leave a dry mixture of the two components. An aqueous solution of 10 m*M* NaCl was added to the dry residue to give a final concentration of permethrin of 1 mg/ml. This is an effective concentration defined by the manufacturer. The mixture was dispersed by mechanical agitation and a stable suspension was formed (see Table 3).

A. STABILIZATION OF PESTICIDES

One of the major routes of inactivation of pesticides at the target site is photolysis by UV radiation from sunlight. The rapid photochemical decomposition of natural pyrethroid insecticides, for example, has led to the synthesis of more photo-stable analogues, such as permethrin (Elliot, 1977) which have improved residue stability but not necessarily the potency. The production of formulations which incorporate the less stable natural chemicals to replace synthetic analogues should be both commercially and environmentally attractive.

A strategy to achieve this objective is to use phospholipid bilayers as structures to package pesticide intimately with stabilizing agents such as antioxidants and UV screening agents. To exemplify this strategy, the effect of including the UV-absorbing compound Uvinul D49 (BASF) in the phospholipid/permethrin formulation on the rate of pesticide photochemical decomposition has been assessed. UV-absorbing compounds are often included in paints and plastics for protection from the effects of UV exposure which otherwise become discolored or brittle with age. As indicated above, compared to natural pyrethroids, permethrin is relatively photo-stable; however, it will decompose on exposure to UV light (Holmstead and Casida, 1977; Holmstead et al., 1978) and it can be used in photo-stability studies. Uvinul D49 was selected as one of its absorbance maxima is coincident with that of permethrin (278 nm). Uvinul D49 is soluble in nonpolar solvents so that the pesticide and the UV-absorber both partition into the hydrocarbon region of the liposome bilayer. The close association of the two compounds in the stacked leaflet arrangement of the multilamellar liposomes serves to maximise the photo-protective effect of the UV-absorber.

Permethrin–phospholipid dispersions were prepared as described above but containing additionally 0.5 mg/ml final concentration of Uvinul D49; this was dissolved in the chloroform with the pesticide and phospholipid at the first stage of preparation. Aliquots of formulation with and without Uvinul D49 were irradiated with a UV lamp and the amount of permethrin remaining measured by gas chromatography. The amount of permethrin remaining as a function of the time of UV exposure is presented in Figure 14. This shows that there is a rapid photolysis of permethrin in each dispersion. The decay of pesticide in the formulation which contained the UV-absorber, however, is markedly slower. The incorporation of UV absorber is clearly effective in slowing the photo-degradation of permethrin in the formulation. Formulation of permethrin with UV screening agents which do not have absorbance characteristics covering the absorption bands of permethrin are not effective in preventing photolysis, indicating a specific effect of the screening agent.

A photo-protective effect by UV absorbers on pyrethroids has been previously demonstrated (Miskus and Andrews, 1972; Abe et al., 1972). These studies, however, used mixtures of pesticide and UV absorber dissolved in organic solvent. It is acknowledged that the photo-protective effect demonstrated in the present studies may not prove to be unique to formulations based on phospholipids. These formulations, however, are aqueous dispersions and, as such, are more closely related to commercial formulations than photo-protected formulations in organic solvent. In this context, the phospholipid formulation represents a significant advance in the application of UV screening agents to stabilize photolabile pesticides.

A related effect of stabilizing pesticides is the ability of phospholipids to create hydrophobic domains into which hydrophobic pesticides may partition. This is likely to reduce hydrolysis of unstable pesticides. No studies have been performed, however, to assess this possible protective effect of phospholipids against hydrolytic degradation.

B. IMPROVEMENT OF EFFICACY

One index of efficacy is the persistence of pesticide residues at the target. Limited laboratory trials have been performed to compare the persistence of permethrin residues of commercial formulations and formulations of permethrin in soya lecithin.

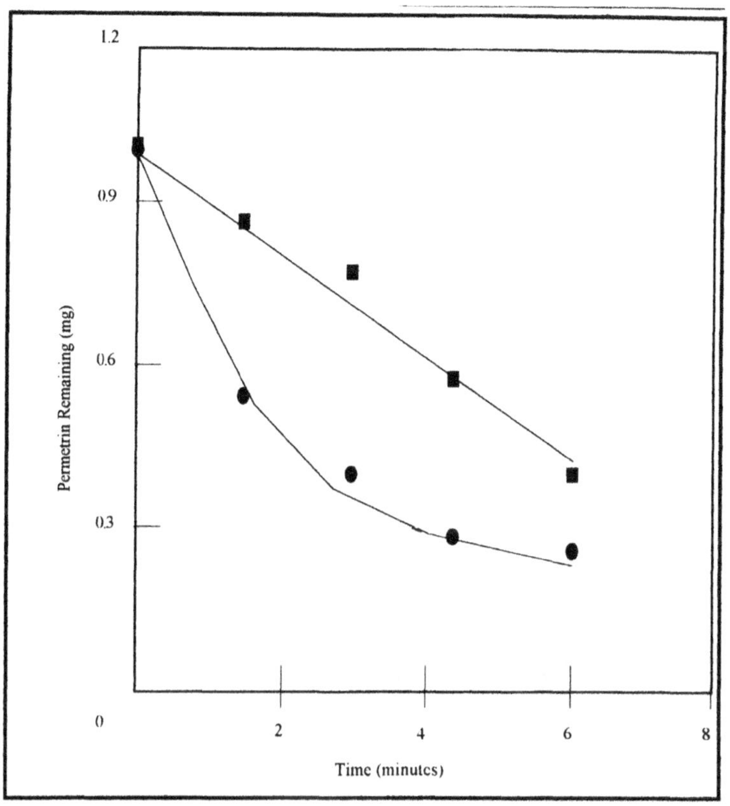

Figure 14 Protection of permethrin from photo-decomposition by the incorporation of the UV-absorbing chemical, Uvinul D49, into a phospholipid/permethrin formulation. Formulations were prepared with (■) and without (●) Uvinul D49. The concentrations of pesticide remaining have been expressed as the level which would be present in 1 ml of the original aqueous dispersion.

Permethrin is available commercially in water-dispersible powder and emulsifiable-concentrate formulations for aqueous dispersion. The water-dispersible powder formulation (Coopex) is reported by the manufacturer (Burroughs Wellcome) to be the most persistent of the permethrin formulations. Coopex contains 25% permethrin adsorbed onto the surface of an inert powder base. This formulation was compared with a soya lecithin–permethrin dispersion prepared as described above. The delivery efficiency and persistence of pesticide was studied by spraying pea plants with each of these formulations and then extracting and quantitating permethrin remaining on the plant leaves harvested at selected time intervals.

The plants used were 30 cm tall pea plants (var. Feltham First) grown in individual pots. These were arranged in 4 blocks of 30 plants: 2 of these blocks were each sprayed with 42 ml of soya lecithin–permithrin formulation, dispersed as described. The remaining two blocks were sprayed with similar volumes of the Coopex formulation. Both formulations were dispersed to a final aqueous volume of 1 mg/ml permethrin (the manufacturer's recommended dispersion rate for Coopex). The levels of permethrin recovered from the plant surfaces at intervals (together with corresponding rainfall data) over a 40-day time period are presented in Figure 15. The level of permethrin originally deposited on the leaf surface is two- or threefold greater for the soya lecithin formulation, in comparison to the Coopex formulation. The increased cover by the phospholipid formulation was visually obvious. The Coopex dried to blotched and uneven deposits of powder. The phospholipid formulation dried to an even waxy covering.

The permethrin level on the plants initially sprayed with the Coopex formulation exhibits a constant rate of decay to about half of its original level in the first few days. The rate is slower after this, and a further approximate halving of the permethrin level occurs over the remainder of the experimental period. Plants sprayed with the lecithin–permethrin dispersion also exhibit an early rapid loss of pesticide. The half-life of this phase is approximately 2 days. The rate of loss slows considerably after this period. A

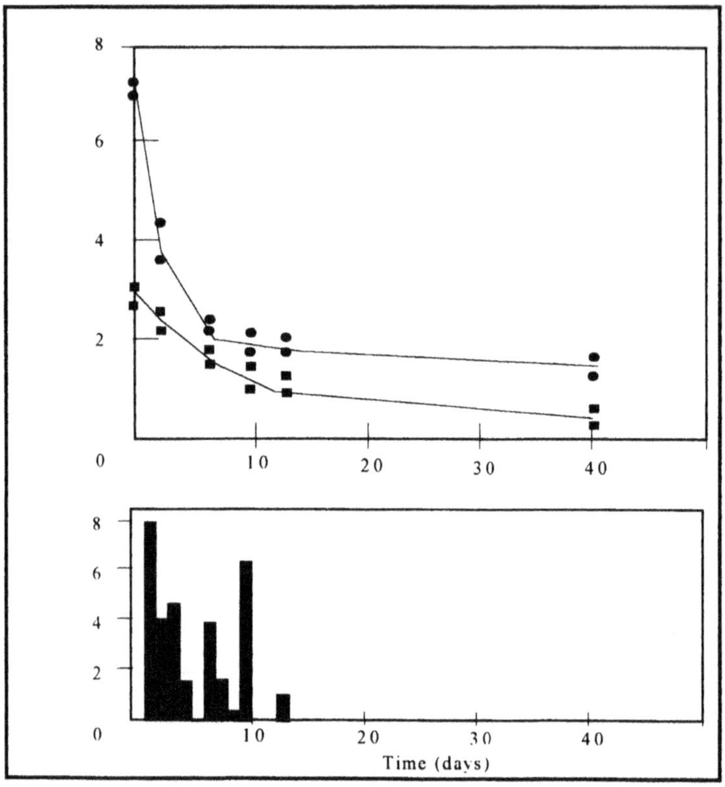

Figure 15 Residues of permethrin remaining on the surface of field peas with time after covering with aqueous sprays of soya lecithin (●) and a commercial water-dispersible powder (■) formulation of permethrin (Coopex). The aqueous dispersions contained 1 mg/ml pesticide and were prepared as described in the text. The plants were all at approximately the same stage of growth (30 cm tall) when originally treated. Pesticide was extracted from the plants by immersion in chloroform and permethrin was quantitated by gas chromatography. Rainfall data during the experimental period are also shown.

two-stage loss of permethrin, similar to that observed in this trial, from bean leaf surfaces sprayed with an emulsifiable concentrate formulation has been reported previously by Southwick et al. (1986). The first stage consisted of a rapid adsorption of permethrin into the leaf in the first 4 days. After this period, the leaves were found to be effectively saturated with pesticide and, thereafter, there was a second stage of permethrin loss, with a half-life of 13.9 days.

The principal mechanisms by which pesticides are lost from the surfaces of plants are through decomposition of the pesticide, absorption into the plant and leaching of the pesticide from the surface by rainfall. The rapid and then slower rates of permethrin loss recorded are unlikely to be due to pesticide decomposition, as this would be expected to proceed at a constant rate. The heavy rainfall recorded in the early stages of the trial may have caused the rapid early disappearance of the pesticide. Alternatively, the plant may have rapidly absorbed a saturating concentration of pesticide in the early stages, absorption being slow or negligible thereafter. Absorption may not be an undesirable feature, since the plant is likely to remain toxic to chewing insects and may be released again to the surface by diffusion as the residues in this layer are depleted via other routes. It is noteworthy that other surfactants have been shown to promote absorption of pesticides into plant tissues (Baker and Hunt, 1988).

From these studies, it should be noted that after 40 days, twice as much permethrin remained on the surfaces of plants treated with the lecithin–permethrin dispersion in comparison to the Coopex formulation. In addition more of the pesticide is initially deposited on the plants when a lecithin–permethrin formulation is used. It is, therefore, not unreasonable to expect that the combination of increased delivery and persistence of pesticide, evidenced by the lecithin formulation, should have the advantage of allowing less pesticide to be used for equivalent biological effect.

XI. FORMULATIONS UNDER FIELD CONDITIONS

Only limited field trails of soya lecithin in agrichemical spray formulations have been undertaken to date. One such trial is the use of soya lecithin as a formulation aid in the fungicide, Fubol (Ciba Geigy), used to control potato blight. The use of foliar fungicides against blight is variable depending on season and area. Nevertheless, some 85% of the crop area is usually treated at least once per season and seven or more sprays are not uncommon. The necessity for frequent spraying is largely due to the poor persistence of blight fungicides, and improvements in this characteristic would free resources required during the blight period for harvest of other crops, thereby allowing considerable financial savings. Because of the relatively low value of the product, savings in costs due to reduced application rates are considered as only of marginal benefit.

Some results of a limited field trial of the control of blight in potatoes using soya lecithin as an additive to Fubol are shown in Figure 16. Fubol 58 WP is a wettable powder formulation consisting of 10% (w/w) water soluble metalaxyl and 48% (w/w) mancozeb adsorbed to an inert support. Addition of soya lecithin to this formulation has no significant effect on stability of the suspension of the commercial formulation. Treatment of potato crops under field conditions were undertaken in a factorial trial. Plots of 24 m^2 were arranged in blocks with 12 treatments randomized within each block; each block was replicated 4 times. The trial areas were sprayed after full canopy had been reached according to seasonal conditions and the percent of the plot area infected by mildew during the third and fourth spray treatments is shown in Figure 16. It can be seen that application of Fubol at the recommended dose rate (1.5 $kg{\cdot}ha^{-1}$) effective control of mildew is achieved and soya lecithin does not markedly affect this performance. With lower dose rate of Fubol, however, the inclusion of soya lecithin in the formulation at a concentration equivalent to 1 $kg{\cdot}ha^{-1}$ markedly reduces the extent of mildew infection.

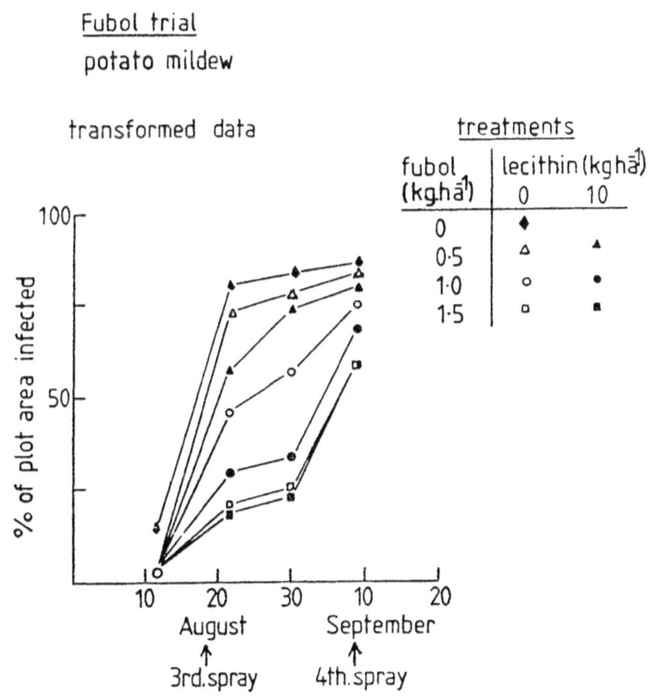

Figure 16 The percent of plot area of a potato crop infected by mildew during the terminal stage of maturity of the crop following spray treatments.

It may be concluded from this limited trial that inclusion of soya lecithin in a formulation of Fubol has a beneficial effect in control of mildew on potatoes, in particular, in permitting the use of lower application rates to achieve effective control of pest infestation. The precise mechanism of action of soya lecithin in achieving this effect is not known.

XII. CONCLUSIONS

This chapter describes the use of phospholipid liposomes as formulation ingredients in agrichemical pesticides. The particular feature of the phospholipid dispersion is not in its ability to encapsulate agrichemical pesticides or plant nutrients, but to act as a dispersal agent for pesticide molecules with sparing solubility in water. The character of weak surfactant activity typical of phospholipids is seen to impart certain desirable features to conventional formulations of pesticides as well as representing a useful base formulation material. Further trials are required to establish the utility of phospholipid dispersions in formulation of agrichemical pesticides. The area where greatest benefit is likely to arise is in the stabilization of pesticides, hitherto rejected because of susceptibility to hydrolysis and photochemical decomposition, thereby extending the range of effective agents in the armoury available for combating agricultural pests.

ACKNOWLEDGMENT

Spray performances were measured with the assistance of Adrian Arnold.

REFERENCES

Abe, Y., Tsuda, K., and Fujita, Y. (1972) *Botyu Kagaku* 37, 102.

Arnold, A. C. (1983) *Crop Protection* 2, 193.

Baker, E. A. and Hunt, G. M. (1988) in *Pesticide Formulations: Innovations and Developments* (Sher, H. B., Ed.), 8 pp., American Chemical Society, Washington, D.C.

Born, M. and Wolf, E. (1959) *Principles of Optics,* 394 pp., Pergamon Press, New York.

Combellack, J. H. and Mathews, G. A. (1981), *Weed Res.* 21, 77.

Cowell, R. D. (1982) in *Amphoteric Surfactants,* (Surfactant Science Series, Vol. 12: Bluestein, B. R. and Hilton, C. L., Eds.), 229–264 pp., Dekker, New York.

Crump, J. G. and Seinfield, J. H. (1982) *Aerosol Sci. Technol.* 1, 1.

Davis, D. G., Stolzenberg, R. L., and Stolzenberg, G. E. (1982) *Environ. Pollut.* 27, 197.

Elliot, M. (1977) *Am. Chem. Soc. Symp. Ser.* 42, 1.

Ernst, R., Ball, E. A., and Ardetti, J. (1982) *Am. J. Bot.* 69, 1340.

Falbe, J. (1987) *Surfactants in Consumer Products,* Springer Verlag, Berlin.

Farr, S. J., Kellaway, I. W., Parry-Jones, R., and Woolfrey, S. G. (1985) *J. Pharm. Pharmacol.* 37 Suppl., 61 p.

Fraser, R. P. (1958) in *Advances in Pest Control Research,* Vol. 2 (Metcalf, R. L., Ed.), pp. 1-106, Wiley Interscience, New York.

Hassall, K. A. (1990) *The Biochemistry and Uses of Pesticides,* p. 41, Macmillan, London.

Hauser, H., Pascher, I., Pearson, R. H., and Sundell, S. (1981) *Biochim. Biophys. Acta* 650, 21.

Heintzenberg, J. and Welch, R. M. (1982) *Appl. Optics* 21 (5), 822.

Hislop, E. C. and Baines, C. R. (1980) in *Spraying Systems for the 1980's,* Proc. Symp. held at Royal Holloway College, London, March 1980 (Walker, J. O., Ed.) BCPC Publications, Croydon.

Hock, W. K. (1987) *Am. Chem. Soc. Symp. Ser.* 336, 128.

Holloway, P. J. (1970) *Pestic. Sci.* 1, 156.

Holmstead, R. L. and Casida, J. E. (1977) in *Synthetic Pyrethroids, Am. Chem. Soc. Symp. Ser. no.* 42 (Elliot, M., Ed.), pp. 137-146.

Holmstead, R. L., Casida, J. E., Ruzo, L. O., and Fullmer, D. C. (1978) *J. Agric. Food Chem.* 26 (3), 590.

Lindblom, G. and Rilfors, L. (1992) *Adv. Colloid Interface Sci.,* 41, 101.

Luzzati, V. (1968) in *Biological Membranes* (Chapman, D., Ed.) p. 71, Academic Press, London.

Mathews, G. A. (1982a) *Pesticide Application Methods,* pp. 39-46, Longman, London.

Mathews, G. A. (1982b) *Pesticide Application Methods,* p. 57, Longman, London.

Mathews, G. A. (1982c) *Pesticide Application Methods,* p. 61, Longman, London.

McWhorter, C. G. (1982) in *Adjuvants for herbicides* (Hodgson, R. H., Ed.), pp. 10-25, Weed Science Society of America, Champaign, IL.

Misato, T., Homma, Y., and Ko, K. (1977) *Neth. J. Plant Pathol.* 83, 395.

Miskus, R. P. and Andrews, T. L. (1972) *J. Agric. Food Chem.* 20 (2), 313.

Parr, J. F. (1982) in *Adjuvants for Herbicides* (Hodgson, R. H., Ed.), pp. 93-113, Weed Science Society of America, Champaign, IL.

Quinn, P. J., Perrett, S. F., and Arnold, A. C. (1986) *Atomisation Spray Technol.,* 2, 235.

Sartoretto, P. (1967) in *Kirk-Othmer Encyclopaedia of Chemical Technology,* Vol. 12, 2nd ed., pp. 342-361, Wiley Interscience, New York.

Scocca, P. M. (1976) *J. Am. Oil Chem. Soc.* 53, 428.

Smith, L. W., Foy, C. L., and Bayer, D. E. (1966) *Weed Res.* 6, 233.

Southwick, L. M., Janes, J., Boethal, D. J., and Willis, G. H. (1986) *J. Entomol. Sci.* 21 (3), 248.

Sparkes, B. D., Sundaram, A., Kotlyar, L., Leung, J. W., and Curry, R. D. (1988) *J. Environ. Sci. Health Part B Pestic. Food Contam. Agric. Wastes* 23, 235.

Sundaram, A., Leung, J. W., and Curry, R. D. (1987) *J. Environ. Sci. Health Part B Pestic. Food Contam. Agric. Wastes* 22, 319.

Swithenbank, J. and Taylor, D. S. (1979) Size Distribution Measurement for Near Monodisperse Particles and Sprays, HIC 315, Department of Chemical Engineering and Fuel Technology, University of Sheffield, U.K. Feb.

Van Emden, H. F. (1989) *Pest Control,* 2nd ed., New Studies in Biology Series, p. 7, Edward Arnold, London.

Van Nieuwenberg, W. (1981) *J. Am. Oil Chem. Soc.* 58, 886.

Weiner, B. B. (1979) *J. Soc. Photo-Optical Instrumentation Eng.* Vol. 170.

Chapter 11

Liposomes and Cheesemaking

Patricia Dufour, Edith Laloy, Jean-Christophe Vuillemard, and Ronald E. Simard

CONTENTS

I. INTRODUCTION

The first fermented dairy product appeared some 8000 years ago. Since then, men quickly discovered how to provoke, control and develop cheese production at will. Cheesemaking has long been considered as an "art": each cheese is unique, and it is often difficult to reproduce cheese types from another place and time. Today, cheesemaking is an important industry, intent on increasing its profits, developing new markets and keeping consumers supplied with consistently superior products. It is not surprising then that many scientists have become involved in the study of cheese ripening.

During the past decade, cheese ripening technologies have received increasing attention. Interest has been focused mainly on the reduction of ripening period in order to reduce storage time (often several months) and therefore costs, and to enhance flavor.[1,2] Numerous strategies have been developed to reduce the ripening period: the addition of attenuated starters such as genetically modified, freeze-shocked or heat-shocked cells; the addition of enzymes such as proteinases, lipases, peptidases, β-galactosidase or mixtures of these; elevated ripening temperature and microencapsulated enzymes. All of these approaches were reviewed by Fox[3] and more recently by El Soda and El Soda and Pandian.[5]

Ever since 1965, when Bangham et al.[6] demonstrated that an aqueous dispersion of phospholipids forms closed structures that are relatively impermeable to entrapped ions, liposomes have been widely studied[7-9] as a model membrane and drug delivery system. Recently, considerable interest has been raised by the possibility of using this technology as an effective tool to accelerate cheese ripening by the entrapment of an enzyme.[10-12] Indeed, liposomes prevent early proteolysis of milk proteins during milk

0-8493-4013-6/96/$0.00+$.50

clotting by encapsulating proteases and by progressively releasing their enzymatic content during the ripening stage.

The aim of this chapter is to review the recent advances in the use of liposomes to reduce cheese ripening time. In the first part, emphasis is placed on the importance of the biochemical changes that occur during cheese ripening with consideration of the different methods available to shorten the ripening time. In the second part, the importance of liposome technology in cheesemaking is developed. Finally, future industrial applications of this technology are discussed.

II. CHEESEMAKING

A. CHEESE COMPOSITION AND MANUFACTURE

Cheese is known as the fresh or matured product obtained by draining off the liquid that results after the coagulation of milk, cream, skimmed milk or partly skimmed milk, buttermilk or a combination thereof. The typical composition of most cheeses is based on the following mixture: water, protein (more than 94% casein), fat, minerals (Ca, P,...) and trace elements and vitamins (vitamin A, riboflavin,...).[13,14] There is little or no concentration of lactose, because most of it passes into the whey or is converted into lactic acid. The proportion of all these components in cheese varies greatly and partially determines cheese specificity. Factors influencing the composition of cheese are the following:

- Milk quality: Milk from several species is used in cheese manufacture. Cow, sheep, goat or buffalo milk have very different protein and fat contents, and thus give very different cheeses. There are also significant differences in milk composition between breeds of cattle and these also influence cheese, as do seasonal, lactational and nutritional factors.
- The nature and the quantity of adjuncts: *starters* (lactic acid bacteria) are added to milk to convert lactose into lactic acid. The acidity of the curd has been shown to directly influence the calcium level in cheeses and the rate of syneresis.[15] Thus, residual lactose and moisture in the curd will vary with the activity and the quantity of starters. The essential property of a *rennet* is its milk-clotting activity. Rennet directly influences the degree of proteolysis of casein in the matrix, and indirectly affects the moisture retained in the curd.[14] Other additives are also used in cheesemaking, the most common and influential on cheese composition being calcium chloride, sodium chloride and in some cases whey proteins. Calcium chloride is added to obtain a satisfactory balance between soluble, colloidal and complexed calcium. Successful coagulation and the ratio of water to dry matter depend on this balance.[16] Denatured whey proteins could be added to increase the dry matter and reinforce the caseic matrix.
- Processing: production of the vast majority of cheese varieties can be subdivided into two well-defined stages, manufacture and ripening. Ripening does not have a major influence on the proportion of the major cheese components. However, the various basic manufacturing steps (acidification, coagulation, dehydration, shaping and salting) result in a dehydration process in which fat and caseins in milk are concentrated between six and twelve times, depending on the variety of cheese. Manufacture also determines ripening conditions and, as a result, the characteristic flavor and texture of cheese. Table 1 gives a summary of the different steps occurring in cheese manufacture, control parameters and their influence on cheese characteristics.

The result of this process is a fresh curd, or primary curd. Some cheeses are eaten at this stage, but most of them undergo a ripening period.

B. CHANGES IN CHEESE DURING RIPENING

During ripening (about 4 weeks and for some cheeses more than 2 years), complex biochemical changes convert a bland-tasting, rubbery-textured curd into a cheese with the right flavor and texture.[14,17,18] These changes are due to the action of:

1. Rennet;
2. Starter bacteria and their enzymes;
3. Surviving indigenous microflora of the milk;
4. Contamination microflora;
5. Indigenous milk enzymes.

Microorganisms described in (3) and (4) are defined as "secondary microflora" in Figure 1.

Table 1 Basic Steps of Cheese Manufacture and Their Influence on Cheese Characteristics

Step	Parameters	Influence
Milk treatment	Pasteurization	Microbiological quality of cheese
	Standardization	
	Homogenization	Composition of cheese
Addition of starter	Nature of the microorganisms	pH of the curd
	Temperature	Coagulation
	Time	Syneresis rate
		Texture and flavour
Addition of rennet and coagulation	Nature of the coagulant	Strength of the caseic matrix
	Method of application of the coagulant	Proteolysis rate
	Temperature	Moisture
	Acidity	
Breaking or cutting the coagulum	Size of the cubes	Syneresis rate
Stirring and cooking the curd	Temperature	Syneresis rate
	Rate of stirring	Growth of starters
	Time	
Whey removal	Volume of the whey	Syneresis rate
		Residual lactose
Texturing and milling (i.e., cheddarization)	Size of milled particle	Texture of the cheese
	Time	
Salting or brining the cheese	Size of curd cubes	Cell growth
	Concentration of salt	Syneresis rate
		Flavour of the cheese
Moulding to storage	Intensity of the pressing	Shape of the cheese
	Time	Texture of the cheese

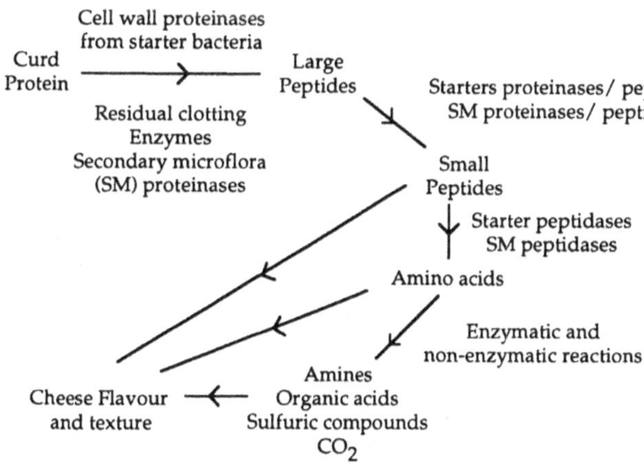

Figure 1 Schematic representation of protein breakdown.

The whole process of achieving flavor and/or texture is not yet well understood, but general mechanisms can be explained.

The main transformations of water-soluble substances, fat and proteins occurring in the curd during ripening are summarized in Figure 1. Primary changes involve protein and lipid modifications that occur during the first week following manufacture. Secondary changes correspond to later changes that lead to the desired texture and flavor.

1. Lipids and Ripening

Lipids influence cheese texture. For example, fat content is inversely proportional to firmness and elasticity. However, texture has not been shown to be influenced by lipolysis.[15]

Fat is also important for the detection and development of flavor. Studies on Cheddar cheese would seem to prove that one important action of fat is to dissolve and hold flavor components.[19] In some cheeses, lipolysis products have a marked and direct influence on flavor. Lipid hydrolysis results in free fatty acids which constitute flavor components (i.e., butyric acid in some soft Italian cheeses),[14] or the substrate for further specific reactions. Some of these reactions lead to the formation of methylketones in blue cheeses,[20] and in Cheddar cheese of some aliphatic and aromatic esters, and lactones. All these products contribute to the specific flavor of cheeses.

2. Proteins and Ripening

Firmness in cheeses is directly related to protein content: the increase of protein leads to a harder texture.[21] Chen et al.[22] showed that protein level is more significant than water, NaCl, fat content and pH in explaining differences in firmness among 11 different types of cheese.

The conversion of fresh cheese curd into mature cheese is largely determined by protein breakdown (Figure 1, Table 2). Early proteolysis by chymosin is recognized as a factor governing texture development in high moisture cheeses: α_{s1} casein degradation and softening of the curd seem to be correlated.[23] Creamer and Olsen[24] have suggested that proteolysis generates new ionic groups that compete for the available water in the system. This means that relatively low moisture cheeses, such as Cheddar, should become increasingly harder during ripening and more resistant to slight deformation.

Table 2 Relationships Between Protein Degradation and Cheese Flavor Formation

Products of Degradation	Method Degradation	Influence on Cheese Flavor	Ref.
Low molecular and hydrophobic peptides	Proteolysis (essentially of β-casein)	Bitterness	25, 35
Amino acids and small peptides	Proteolysis (bacterial enzymes)	Sweet flavor in Emmental and related cheeses	59
Ammonia	Deamination of amino acids	Typical flavor of Camembert	3
Alcohols	Deamination Transamination Decarboxylation of amino acids	Flavor characteristics of surface-ripened cheese	60
Sulphur compounds such as methane thiol and thioesters	Desulphurylation, demethiolation and unknown mechanisms	Relation with Cheddar and surface-ripened cheese flavors	3

For the development of an acceptable cheese flavor, a well-balanced breakdown of caseins into small peptides and amino acids is necessary.[25] These products act as flavor compounds or as precursors for later chemical reactions that will result in the formation of specific cheese flavors (Table 2).

C. METHODS USED FOR THE ACCELERATION OF CHEESE RIPENING

Since the beginning of the 1980s, the cheese industry has been looking for ways to reduce the cost of cheese production. This can be achieved by improving the control process during manufacture or by reducing ripening time.

Although all the details of cheese flavor and texture formation during ripening are not completely understood, the accumulated data have helped numerous investigators to formulate strategies on how to accelerate cheese ripening. Various methods for accelerating cheese ripening were described recently by El Soda.[4,5] Before summarizing the most interesting, it will be useful to mention the conditions required for an industrial application of an accelerated cheese ripening technology.

- Obviously, the cost of such a strategy must be lower than the cost of the traditional technology;
- The technology must not require specific expertise from the cheesemaker, and must not take too much time;
- Eventually, added products must be edible and recognized as safe for human consumption;
- The proposed technology must not induce modification of the cheese texture or flavor;
- Finally, the technology must be flexible enough to allow for regular adaptations when slight fluctuations occur in raw material quality or in the processing method.

1. Traditional Techniques

Since proteolysis and lipolysis are the main biochemical events that occur during maturation, the addition of free-form proteinases, peptidases and lipases was evaluated for its potential use.[5] Best results were obtained with "enzyme cocktails".[5] Some manufacturers have already proposed such an enzyme mixture, and Neutrase is probably the one most often used. But one of the major limitations is the method used for adding free enzymes. If free enzymes are added to the milk, whey is contaminated and some enzymes retained in the curd induce early proteolysis with possible off-flavor and low cheese yield. If added to the curd, enzyme distribution is not satisfactory.

2. Promising Techniques

Currently, the most promising techniques are probably the addition of attenuated bacterial cells,[5] genetically modified strains[3,4] and entrapped enzymes.[26] All of these methods are supposed to solve some of the problems related to the use of free enzymes and should accelerate research on cheese ripening.

Starter and nonstarter lactic acid bacteria in the attenuated form have been suggested as a simple way to increase the microflora enzymatic charge without over-production of lactic acid. Research in the field is focused on a satisfactory method of cell attenuation. Spray drying seems to be promising. Johnson and Etzel[27] showed that the consequent decrease in viability was obtained without significant impact on the activities of intracellular enzymes. Results of cheesemaking experiments with spray-dried cultures are not available at this time.

Genetically modified strains are essentially lactose negative.[3,28] First results are quite encouraging.[28] However, the use of genetically modified mutants in the food industry is a delicate subject and it may take some time before any application is allowed.

In the previous two techniques, cells are the natural components of an enzyme mixture thought to accelerate cheese maturation. It is quite difficult then to control the disponibility of enzymes. The entrapment of "cocktail enzymes" within milk fat capsules or liposomes could be a solution to this problem.

III. LIPOSOMES: APPLICATION IN CHEESEMAKING

A. INTEREST FOR LIPOSOMES IN CHEESEMAKING

Use of liposomes containing entrapped enzymes appears to be one of the most promising solutions for shortening the cheese ripening period. One of the basic steps in cheesemaking[29] (Figure 2) is to add lipid vesicles directly to milk.[30] Liposomes are retained in the curd matrix leading to a low whey contamination (less than 10%, compared to 95% in the case of free enzyme).[31] During the draining off of the whey from the curd, cheddaring, salting and pressing, liposomes remain mainly intact. The major changes occur during ripening when liposome breakdown results in the gradual release of enzymes.[31] Law and King[30] showed that the delayed and progressive enzyme leakage prolonged the half-life of liposomes is not satisfactorily distributed.

Preliminary experiments for reducing the ripening time of cheese were described by El Soda et al.[32] They obtained a good quality Cheddar by entrapping extracts of *Lactobacillus casei* and *L. helveticus* in small unilamellar vesicles (SUV). However, the reduction of ripening time was not significant. Despite this fact, numerous scientists continued to explore different ways to improve these inspiring first results. Indeed, this method offers several advantages: the milk fat proteins are protected against premature enzyme attack; liposomes are retained and distributed throughout the curd matrix and, consequently, the whey is free from exogenously added enzymes (whey represents 90% of the total weight of milk used in cheese manufacture and is recovered for further use);[33] the gradual release of the liposome content in the cheese matrix avoids the bitterness mentioned in the literature[31,34-36] when free enzymes are used; liposomes form very useful vesicles in the food industry because they are made up of natural biological materials. Moreover, the stability, permeability, surface net charge, size, sensitivity to pH or temperature of liposomes can be altered as a function of their preparation condition and could therefore be applied to different types of cheese.

For these reasons, the addition of an entrapped enzyme directly to milk is an interesting option.

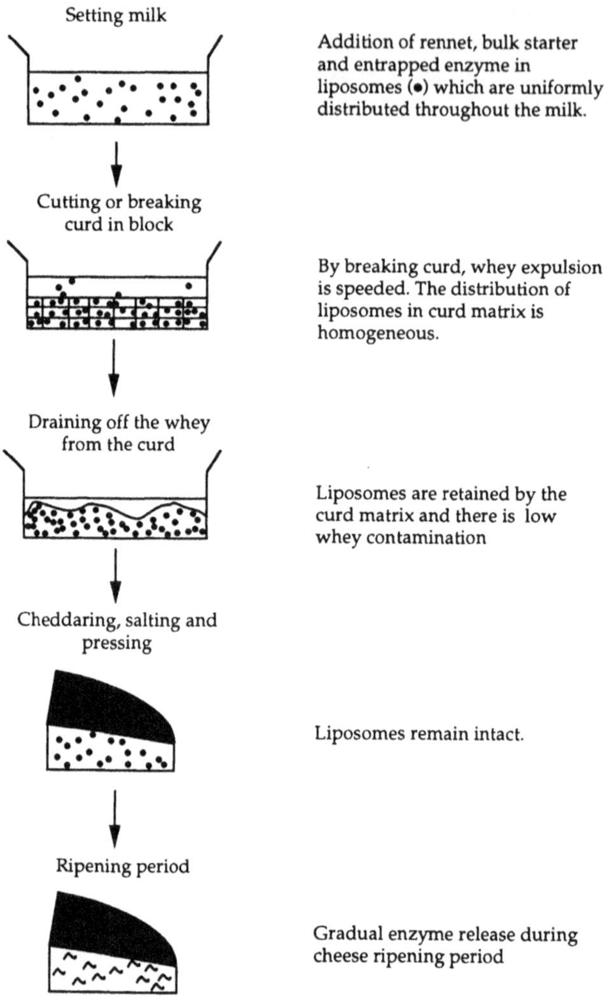

Figure 2 Basic steps in cheesemaking.

B. REDUCTION OF THE RIPENING PERIOD BY THE ADDITION OF LIPOSOMES TO CHEESE MILK

1. Entrapment Efficiency and Retention in Curd

Entrapment efficiency is defined as the percentage of the entrapped enzyme on the total amount of enzyme. The process, to be economic, must minimize the loss of enzyme during liposome preparation. Entrapment efficiency is essentially determined by two methods: measurement of the radioactivity of ^{14}C-labeled enzyme or detection of the entrapped enzyme activity.[10,12] The estimation of liposome retention in cheese curd is generally given by the total radioactivity detected in the cheese slurry or the fluorescence measurement of entrapped carboxyfluorescein (CF). Numerous studies have been reported in the literature and the results are summarized in Table 3.

Law and King[30] encapsulated a commercial neutral proteinase (Neutrase) in multilamellar vesicles (MLV). Although a comparison between cheeses containing Neutrase added directly to the curd or added through liposomes showed a much higher retention of active enzyme, in the latter case (17%), the efficiency of enzyme encapsulation in MLV stayed particularly low (1–2%). Piard et al.[37] compared the entrapment of a metallo-protease from *Micrococcus caseolyticus* (Rulactine) in three types of liposomes which were then added to Saint-Paulin type cheese. They found a higher encapsulation efficiency and retention rate for MLV (9% and 60%, respectively) than for REV (reverse phase evaporation vesicles: 5% and 40%, respectively) and SUV (3% and 36%, respectively). More recently, in order to study the evolution of proteolysis in Taleggio type cheese, Scolari et al.[38] compared cheese containing free added

Table 3 Liposome Entrapment Efficiency and Retention in Curd

Liposome Type	Enzyme	Entrapment Efficiency (%)	Retention in Cheese (%)	Size (nm)	Type of Cheese	Ref.
REV (+)	Neutrase	33	42	nd	St. Paulin	37
REV (–)		26	31	nd		
REV (0)		19	24	nd		
MLV (–)	Neutrase	17	60	nd	St. Paulin	5
REV (–)		30	3	nd		
MLV	Neutrase	1–2	17	250–2500	Cheddar	29
MLV (+)	Neutrase	4	nd	400–600	Cheddar	7
MLV (–)		8	nd			
MLV (0)		2	nd			
SUV (+)		nd	nd	nd		
SUV (–)		1	nd			
SUV (0)		1	nd			
DRV (+)		34	nd	200–400		
DRV (–)		26	nd			
DRV (0)		33	90			
REV (+)		5	nd	nd		
REV (–)		14	nd	nd		
REV (0)		10	nd	nd		
DRV (0)	Neutrase	22	42	nd	Tallegio	43
DRV (0)	*L. helveticus*	22	42	nd		
MLV(–)	Rulactine	9	60	500–5,000	St. Paulin	35
SUV (–)		3	36	50–200		
REV (–)		5	40	50–500		
MLV (–)	Corolase	11	nd	300–2,600	Gouda	27
SUV (–)		2	nd			
REV (–)		14	nd			
MLV	Trypsin	10	nd	1,300	Cheddar	39
MLV-MF		14	nd	170		
MF		12	nd	110		
DRV		13	nd	1,540		

Note: MLV, multilamellar vesicle; MLV-MF, Microfluidized multilamellar vesicle; MF, microfluidized vesicle, SUV, small unilamellar vesicle, REV, reverse phase vesicle; DRV, dehydrated-rehydrated vesicle; (+), Positively charged; (–), negatively charged; (0), neutral charged; nd, not determined.

or encapsulated proteolytic enzymes in DRV (dehydrated-rehydrated vesicles) liposomes. The encapsulation efficiency and the retention of the vesicles in curd were 22% and 42%, respectively. To date, the best results are attributed to Kirby et al.,[12] who reported an encapsulation efficiency and a curd retention efficiency of 33% and 90%, respectively, for DRV liposomes.

Although some of these studies were carried out on the same type of liposomes, results showed wide variations. These differences could not be explained solely by the different process of cheesemaking but also by differences in liposome characteristics.

a. Effect of Liposome Size

Depending on liposome size, their retention in curd could be altered. Kirby et al.[12] observed a liposomal retention in curd of 90% for DRV vesicles. This high level of retention was attributed to the low vesicle size of DRV (200–400 nm). However, in another study, a lower retention rate for SUV (50–200 nm) and REV (50–500 nm) as compared to MLV (500–5000 nm)[37] was found and was attributed to the smaller size of SUV and REV liposomes which caused them to be excluded from the curd matrix. These results show that retention of liposomes in curd is greatly affected by vesicle sizes. However, other factors such as the type of enzyme, phospholipids and number of bilayers certainly take a part in liposome interaction or exclusion from the matrix.

b. Effect of Liposome Net Charge

A comparative study on the influence of liposome net charge on Neutrase encapsulation efficiency was carried out by Kirby et al.[12] and Alkhalaf et al.[39] in two types of cheese, Cheddar and Saint-Paulin, respectively. Results (Table 3) showed that Neutrase could be entrapped in anionic, cationic or neutral liposomes. However, the highest encapsulation rate was observed for positively charged liposomes in both cases. This is due to the fact that the enzyme (Neutrase) is negatively charged at the encapsulation pH (6.5), reducing the repulsions with the phospholipids.

The rate of liposome retention in cheese was highest for positively charged (42%) rather than for negatively charged (31%) and neutral liposomes (24%).[39] The higher affinity of the positive liposomes for the cheese could be explained by electrostatic interaction with the casein micelles during the early steps of cheesemaking.

Entrapment efficiency and liposome retention in curd are greatly affected by the composition, size, surface net charge and type of liposome, as well as the cheese manufacturing process. The major drawback (Table 3) remains the low efficiency of encapsulation, which is not compatible with an industrial application. However, recent work carried out in our laboratory reported efficiencies of encapsulation of enzyme in liposomes as high as 90%.[40] Although no trials in cheese have been reported, this first result is promising.

2. Distribution in Curd

An interesting work[12] actually visualized the distribution of liposome entrapping carboxyfluorescein (CF) throughout the cheese by electron microscopy. Results showed that horseradish peroxidase entrapped in liposome was grouped in tight clusters which were confined between the fat globules and the casein matrix and not uniformly distributed in cheese as had been thought. Even if liposomes were not in intimate contact with the casein matrix, trials showed that α-casein was degraded when enzyme was released.[41] Authors[4,12] have suggested that retention and enzyme distribution can be increased by targetting liposomes in specific locations in curd matrix.

3. Behavior of Liposomes in Cheese and Their Effect on Cheese Ripening

a. Liposome Stability in Cheese Curd

During cheese manufacture (Table 1), different parameters (temperature, pH, osmotic pressure variation, phospholipases) can greatly affect liposome stability and consequently the release of the encapsulated material. In order to reduce maturation time and prevent bitterness and texture defect, enzyme release must be progressive. Studies to this effect have been carried out to determine enzyme release kinetics which can be measured by either the indirect or direct method. The first is proteolysis assessment, while the second one is a spectroscopic. Four types of spectroscopic probes[42] are currently being used to determine the kinetics of liposome leakage, including measurement by nuclear magnetic resonance (NMR); analysis by electron spin resonance (ESR); measurement by gamma ray perturbed angular correlation (PAC) and determination of the degree of self-quenching of the fluorescent intensity of carboxyfluorescein (CF) in a fluorimeter. Liposome stability in milk is generally determined by CF as a marker which shows two distinctly different spectroscopic characteristics when entrapped in and released from liposomes.

Although the addition of negatively charged phospholipids has resulted in a population of stable liposomes in cheese,[12,39] as a general rule, positively charged vesicles are more resistant to high temperature, low pH and NaCl than negatively charged vesicles.[39] However, negatively charged REV liposomes are more stable in cheese than negatively charged MLV.[39] In the case of REV liposomes, proteolysis was progressive during ripening, while MLV gave a too high proteolysis level, which leads to crumbly structure and a high level of bitterness.[39] The stability of liposomes can be enhanced by supplementing their composition with cholesterol (less than 1%).[12] But it seems that this type of liposome is too stable and stays intact for too long a time.[43]

The study of the effect of cheese ripening parameters revealed that more enzyme was liberated at neutral pH. Increasing NaCl concentration from 1% to 4% resulted in an increase of proteinase leakage, and the release of Neutrase was about 2.5 times higher at 40°C than at 20°C.[39,44]

Scolari et al.[38] reported that liposome leakage in curd depends on liposome content. In this respect, vesicles containing extracts from *Lactobacillus helveticus* and Neutrase were compared. After 7 days, the percentage of release was 51% for the former, while the latter was 19%. Destabilization seemed to be due to an alteration of the surface tension in the liposome instigated by *L. helveticus* cell-free extract.

To date, information on this aspect is not sufficient and no answers have been found. Numerous hypotheses exist,[30] one being that indigeneous phospholipases in cheese are responsible for liposome membrane disruption.

As a controlled-release system for the acceleration of cheese ripening, El Soda et al.[45] evaluated temperature-sensitive liposomes made of dipalmitoyl phosphatidylcholine. Although the enzyme was released at a specific temperature, the system presents many problems, mainly the partial or total denaturation of the enzyme due to the elevated temperature needed for the liposome preparation (45°C). Also, heating the cheese to release the enzyme was another limitation.

b. Protein Breakdown

Hydrolysis of protein results in a higher concentration of amino acids, peptides and polypeptides in curd. A list of criteria for evaluating accelerated ripening of cheese has been established.[46] To assess proteolytic activities in cheeses, different forms of nitrogen: total nitrogen (TN), soluble nitrogen (SN) and nonprotein nitrogen (NPN) are determined.

Law and King[30] showed that entrapped Neutrase does not cause detectable casein breakdown in milk up to the pressing step (Figure 2).

During the early ripening stage, the SN/TN and NPN/TN ratios of cheeses containing liposomes were closer to those obtained from control cheese.[10,38,47] Behavior of MLV liposomes was very different from the REV type. While the proteinase released from the REV liposomes generated a sharp increase in the SN/TN and NPN/TN ratios to a level close to that found for free enzyme-treated cheeses after 45 days of ripening,[10] MLV started proteolysis earlier with a rapid increase up to 30 days (SN represented more than 40% of TN).[10,12,30,47,48]

These observations were in accordance with electrophoretic analysis of MLV- and REV-added cheeses.[10] In the case of MLV liposomes, intensity of the band corresponding to the β-casein decreased after 48 h and new bands appeared. This β-casein hydrolysis rapidly increased until 30 days of ripening, whereas no changes occured in the control. Appearances of new bands were more progressive for REV, indicating a more gradual enzyme release.

Attempts carried out with SUV liposomes showed no alteration of the β-casein band after 7 days of ripening.[37] By incorporating cholesterol in liposomes, β-casein breakdown can be significantly delayed.[43] Moreover, protein breakdown is not affected by liposome charge. In fact, neutral and positively charged liposomes showed the same proteolysis pattern but with a lower rate than the negatively charged vesicles.[39]

c. Cheese Texture and Flavor

It is widely accepted that flavored cheese can be produced in a shorter time by adding an entrapped enzyme in liposomes. The type and amount of enzyme are critical parameters in the production of a desirable flavor and texture in cheese. Cheese texture analysis shows clearly the net advantage of entrapped Neutrase as compared to free Neutrase.[10] Studies with Saint-Paulin cheese show that the ripening of liposome-added cheese was complete after 15 days and was found equivalent to 45-day-old control cheeses.[10] The texture of cheese containing free Neutrase was found to be crumbly, very rigid and brittle. On the other hand, cheeses containing encapsulated enzyme in REV liposomes exhibited rheological characteristics similar to those of the control cheeses; except for MLV, crumbly texture was found.[10,47] All these results were confirmed by an informal taste panel assessment. Sensory analysis showed no significant difference between REV-treated cheese and the control product after 10 and 20 days ($P > 0.05$). However, the difference became significant after 30 days ($P < 0.05$) and highly significant after 45 days ($P < 0.01$) of ripening.[10]

Enzyme entrapment in liposome results in a significant decrease in bitterness.[10,12,48] Low levels of bitterness are probably due to the gradual release of entrapped enzyme from vesicles during ripening time.[48] However, these results were unsatisfactory as liposome technology can not completely eliminate the bitter defect due to the type of enzyme used.

To prevent bitter defect, entrapment of both endopeptidase and exopeptidase is proposed. The latter enzymes, which release low molecular weight peptides and amino acids, will enhance degradation of bitter peptides.[10,12,31,48-50] Attempts were carried out by preparing cheeses in which endopeptidase and exopeptidase were added in liposomal form to the milk. Compared to the control cheese, mature cheese was obtained in half the time, without flavor or texture defect, indicating the great potential for use of this enzyme cocktail.[43]

The flavor development of the Gouda cheese type made from ultrafiltered (UF) milk has been considerably enhanced by entrapping a blend of proteinase (Corolase PN) and cell-free extract of freeze-shocked *L. helveticus* cells. This mixture constitutes an alternative to overcome the resistance of UF cheese to proteolysis during ripening.[48]

IV. CONCLUSION

The basic concept of enzyme entrapment into liposome, the high retention and the delayed enzyme release into cheese seem to be of interest to reduce the maturation time. However, despite the fact that all the mentioned studies have attempted to understand liposome behavior in cheese, many questions still exist. The studies in the literature show great variations due mainly to differences in the size, type and net charge of liposomes in cheesemaking. Other parameters, such as the interaction between the liposome and the cheese components, and also the interactions between entrapped material and membrane, are important. Indeed, recent work has shown that a major part of the enzyme was not entrapped inside the liposome but fixed on the liposome surface, probably by hydrophobic interaction.[40] These interactions should be studied in the future. In order to render liposome ripening methods more attractive, it would be interesting to control liposome disruption, as experienced by El Soda et al.[45] with a temperature-sensitive liposome system.

V. PERSPECTIVES

A. ENTRAPMENT OF OTHER SUBSTANCES

The application of liposome technology in cheesemaking should not be restricted just to the acceleration of cheese ripening. Another reason for using liposomes is the prevention of microbial spoilage. Many varieties of washed curd cheeses, including Gouda, Edam and Emmental, are susceptible to spoilage by spore-forming bacteria, which produce undesirable flavor, texture and swelling due to butyric acid fermentation. This problem could be controlled by adding encapsulated bacteriocin to prevent the growth of contaminating microflora.[41,51,52] Similarly, liposomes could be used in food products for other purposes such as antioxidant systems for food protection[55] or nutritional improvement.[54,55]

B. INDUSTRIAL POTENTIAL

Liposome technology offers the possibility of preparing a wide range of vesicles (varying in size, net charge, stability…) and in some cases on a large scale.[56]

By using microfluidization, Koide and Karel[57] and Larivière et al.[47] obtained multilamellar liposomes without using any organic solvent, but despite the attractive advantage of large-scale production, this method is restricted by low encapsulation efficiency (14%, Table 3).

A new and very simple liposome preparation technique was recently proposed by Perret et al.[58] This technique consists of a three-phase mixture (phospholipids, ethanol and buffer), called a proliposome mixture, that can be converted into multilamellar vesicles by a two-stage dilution procedure. This process did not involve the use of an organic solvent, nor high energy requiring steps such as sonication, and it has possibilities for large-scale liposome production. Initial attempts on the encapsulation of model enzyme (chymotrypsin) appear to be very promising.[40,44] Thus, the proliposome approach for food application has much potential.

REFERENCES

1. **Magee, E. L., Jr. and Olson, N. F.**, Microencapsulation of cheese ripening systems: formation of microcapsules, *J. Dairy Sci.*, 64, 600, 1981.
2. **Ardö, Y. and Larson, P. O.**, Lindmark Mansson, H., and Hedenberg, A., Studies of peptidolysis during early maturation and its influence on low-fat cheese quality, *Milchwissenschaft*, 44(8), 485, 1989.
3. **Fox, P. F.**, Acceleration of cheese ripening, *Food Biotechnol.*, 2(2), 133, 1989.
4. **El Soda, M. and Pandian, S.**, Recent developments in accelerated cheese ripening, *J. Dairy Sci.*, 74, 2317, 1991.
5. **El Soda, M.**, Accelerated maturation of cheese, *Int. Dairy J.*, 3, 531, 1993.
6. **Bangham, A. D., Standish, M. M., and Watkins, J. C.**, Diffusion of univalent ions across the lamellae of swollen phospholipids, *J. Mol. Biol.*, 13, 238, 1965.
7. **Lasic, D. D.**, Preparation of Liposomes, in *Liposomes From Physics To Applications*, Lasic, D. D., Ed., Elsevier Science Publishers, New York, 1993, 63.
8. **New, R. C. C.**, Preparation of liposomes, in *Liposomes: A Practical Approach*, New, R. C. C., Ed., IRL Press, Oxford, 1990, 33.
9. **Szoka, F. and Papahadjopoulos, D.**, Comparative properties and methods of preparation of lipid vesicles (liposomes), *Annu. Rev. Biophys. Bioeng.*, 9, 467, 1980.

10. **Alkhalaf, W., Piard, J. C., El Soda, M., Gripon, J. C., Desmazeaud, M., and Vassal, L.,** Liposomes as proteinase carriers for the accelerated ripening of Saint-Paulin type cheese, *J. Food Sci.*, 53(6), 1674, 1988.
11. **El Soda, M.,** Acceleration of cheese ripening: recent advances, *J. Food Prot.*, 49(5), 395, 1986.
12. **Kirby, C. J., Brooker, B. E., and Law, B. A.,** Accelerated ripening of cheese using liposomes-encapsulated enzyme, *Int. J. Food Sci. Technol.*, 22, 355, 1987.
13. **Alais, C.,** Chimie, biochimie et physique du lait, in *Science du Lait,* part 2, Alais, C., Ed., SEPAIC, 4th edition, Paris, 1984.
14. **Fox, P. F.,** Cheese: an overview, in *Cheese: Chemistry, Physics and Microbiology,* Fox, P.F., Ed., Elsevier Applied Science, New York, 1987, chap. 1.
15. **Adda, J., Gripon, J. C., and Vassal, L.,** The chemistry of flavour and texture generation in cheese, *Food Chem.*, 9, 115, 1982.
16. **Scott, R.,** Coagulation and precipitants, in *Cheesemaking Practice,* 2nd ed., Elsevier Applied Science Publishers, New York, 1986, chap. 12.
17. **Scott, R.,** Cheese ripening, in *Cheesemaking Practice,* 2nd ed., Elsevier Applied Science Publishers, New York, 1986, chap. 16.
18. **Fox, P. F. and Law, J.,** Enzymology of cheese ripening, *Food Biotechnol.*, 5(3), 239, 1991.
19. **Foda, E. A., Hammond, E. G., Reindold, G. W., and Hotchkiss, D. K.,** Role of fat in flavor of Cheddar cheese, *J. Dairy Sci.*, 57, 1137, 1974.
20. **Hawke, J. C.,** The formation and metabolism of methyl ketones and related compounds, *J. Dairy Res.*, 33, 225, 1966.
21. **Steffen, P.,** Teigeigenschaften und Teigfehler in Emmentalen Käse, *Schweiz. Milchztg.*, 101, 72, 1975.
22. **Chen, A. H., Larkin, J. W., Clark, C. J., and Irwin, W. E.,** Textural analysis of cheese, *J. Dairy Sci.*, 62, 901, 1979.
23. **Noomen, A.,** The role of the surface flora in the softening of cheese with a low initial pH, *Neth. Milk Dairy J.*, 37, 229, 1983.
24. **Creamer, L. K. and Olson, N. F.,** Rheological evaluation of maturing Cheddar cheese, *J. Food Sci.*, 47, 631, 1982.
25. **Visser, S.,** Proteolytic enzymes and their relation to cheese ripening and flavor: an overview, *J. Dairy Sci.*, 76, 329, 1993.
26. **El Soda, M., Panell, L., and Olson, N.,** Microencapsulated enzyme systems for the acceleration of cheese ripening, *J. Microencapsulation*, 6, 319, 1989.
27. **Johnson, J. and Etzel, M.,** Spray-drying and freeze-drying for the large-scale production of microbial products, in *American Institute of Chemical Engineers Annual Meeting,* Los Angeles, 1991, 261a.
28. **El Abboudi, M., El Soda, M., Johnson, M., Olson, N., Simard, R., and Pandian, S.,** Peptidases deficient mutants: a new tool for study of cheese ripening, *Milchwissenschaft,* 47, 625, 1992.
29. **Kosikowski, F.,** Fundamentals of cheesemaking and curing, in *Cheese and Fermented Milk Foods,* Kosikowski, F., Ed., Kosikowski and Associates, New York, 1977, chap. 7.
30. **Law, B. A. and King, J. S.,** Use of liposomes for proteinase addition to Cheddar cheese, *J. Dairy Res.*, 52, 183, 1985.
31. **Kirby, C. J. and Law, B. A.,** Recent developments in cheese flavour technology: application of enzyme microencapsulation, *Dairy Ind. Int.*, 52(2), 19, 1987.
32. **El Soda, M., Fathallah, S., and Ezzart, N.,** Acceleration of Cheddar cheese ripening with liposome trapped extracts from *Lactobacillus casei*, *J. Dairy Sci.*, 66 (suppl. 1), 78, 1983.
33. **Rolland, J.,** Les sous produit du lait, in *Science et Technologie du Lait,* La fondation de technologie laitière du Québec, Ed., Les presses de l'Université Laval, Québec, 1984, chap. 13.
34. **Alkhalaf, W., Vassal, L., Demazeaud, M. J., and Gripon, J. C.,** Utilisation de la Rulactine en tant qu'agent d'affinage dans des fromages à pâte pressée, *Le Lait,* 67, 173, 1987.
35. **Lemieux, L. and Simard, R. E.,** Bitter flavour in dairy products. I. A review of the factors likely to influence its development, mainly in cheese manufacture, *Le Lait,* 71, 599, 1991.
36. **Stadhouders, J., Hup, G., Exterkate, F. A., and Visser, S.,** Bitter flavour in cheese 1. Mechanism of the formation of the bitter flavour defect in cheese, *Neth. Milk Dairy J.*, 37, 157, 1983.
37. **Piard, J. C., El Soda, M., Alkhalaf, W., Rousseau, M., Desmazeaud, M., Vassal, L., and Gripon, J. C.,** Acceleration of cheese ripening with liposome entrapped proteinase, *Biotechnol. Lett.*, 8(4), 241, 1986.
38. **Scolari, G., Vescovo, M., Sarra, P. G., and Bottazi, V.,** Proteolysis in cheese made with liposome entrapped proteolytic enzymes, *Le Lait,* 73, 281, 1993.
39. **Alkhalaf, W., El Soda, M., Gripon, J. C., and Vassal, L.,** Acceleration of cheese ripening with liposomes entrapped proteinase: influence of liposomes net charge, *J. Dairy Sci.*, 72, 2233, 1989.
40. **Dufour, P., Laloy, E., Vuillemard, J. C., and Simard, R. E.,** Characterization of chymotrypsin in liposomes, in *Bioencapsulation III. The Reality of a New Tool,* CERIA, Brussells, 1993, 54.
41. **Kirby, C. J.,** Microencapsulation and controlled delivery of food ingredients, *Food Sci. Technol. Today,* 5(2), 74, 1991.
42. **Hwang, K. J.,** Liposomes pharmacokinetics, in *Liposomes from Biophysics to Therapeutics,* Ostro, M. J., Ed., Marcel Dekker, New York, 1987, 109.
43. **Kirby, C. J. and Law, B. A.,** Developments in the microencapsulation of enzymes in food technology, in *Chemical Aspects Of Food Enzymes,* Andrews, A. T., Ed., Royal Society of Chemistry, London, 1987, 106.

44. **Laloy, E., Dufour, P., Vuillemard, J. C., and Simard, R. E.,** Effect of composition parameters and extrinsic factors on the proteinase release from liposomes, in *Bioencapsulation III: the Reality of a New Tool,* CERIA, Brussells, 1993, 59.
45. **El Soda, M., Johnson, M., and Olson, N. F.,** Temperature sensitive liposomes: a controlled release system for the acceleration of cheese ripening, *Milchwissenschaft,* 44(4), 213, 1989.
46. FIL, List of criteria for evaluation of accelerated ripening of cheese, *Bull. IDF,* 261, 35, 1991.
47. **Larivière, B., El Soda, M., Soucy, Y., Trépanier, G., Paquin, P., and Vuillemard, J. C.,** Microfluidized liposomes for the acceleration of cheese ripening, *Int. Dairy J.,* 1, 111, 1991.
48. **Spangler, P. L., El Soda, M., Johnson, M. E., Olson, N. F., Amundson, C. H., and Hill, C. G., Jr.,** Accelerated ripening of Gouda cheese made from ultrafiltered milk using a liposome entrapped enzyme and freeze shocked lactobacilli, *Milchwissenschaft,* 44(4), 1989.
49. **Law, B. A. and Wigmore, A. S.,** Accelerated ripening Cheddar cheese with a commercial proteinase and intracellular enzymes from starter *Streptococci, J. Dairy Res.,* 50, 519, 1983.
50. **Hayashi, K., Revell, D. F., and Law, B. A.,** Accelerated ripening of Cheddar cheese with the aminopeptidase of *Brevibacterium linens* and a commercial neutral proteinase, *J. Dairy Res.,* 57, 571, 1990.
51. **Degnan, A. J., Buyong, N., and Luchansky, J. B.,** Antilisterial activity of pediocin AcH in model food system in the presence of an emulsifier or encapsulated within liposomes, *Int. J. Food Microbiol.,* 18, 127, 1993.
52. **Degnan, A. J. and Luchansky, J. B.,** Influence of beef tallow and muscle on the antilisterial activity of pediocin AcH and liposome-encapsulated pediocin AcH, *J. Food Prot.,* 55(7), 552, 1992.
53. **Kirby, C. J., Whittle, C. J., Rigby, N., Coxon, D. T., and Law, B. A.,** Stabilization of ascorbic acid by microencapsulation in liposomes, *Int. J. Food Sci. Technol.,* 26, 437, 1991.
54. **Hirotsuka, M., Taniguchi, H., Narita, H., and Kito, M.,** Calcium fortification of soy milk with calcium-lecithin liposome system, *J. Food Sci.,* 49, 1111, 1984.
55. **Matsuzaki, M., McCafferty, F., and Karel, M.,** The effect of cholesterol content of phospholipid vesicles on the encapsulation and acid resistance of β-galactosidase from *E. coli, Int. J. Food Sci. Technol.,* 24, 451, 1989.
56. **Vuillemard, J. C.,** Recent advances in the large-scale production of lipid vesicles for use in food products: microfluidization, *J. Microencapsulation,* 8, 547, 1991.
57. **Koide, K. and Karel, M.,** Encapsulation and simulated release of enzymes using lecithin vesicles, *Int. J. Food Sci. Technol.,* 22, 707, 1987.
58. **Perret, S., Golding, M., and Williams, P.,** A simple method for the preparation of liposomes for pharmaceutical applications: characterization of the liposomes, *J. Pharm. Pharmacol.,* 43, 154, 1991.
59. **Langler, J. E., Lisbey, L. M., and Day, E. A.,** Identification and evaluation of selected compounds in Swiss cheese flavor, *J. Agric. Food Chem.,* 15, 386, 1967.
60. **Hemme, D., Bouillanne, C., Metro, F., and Desmazeaud, M. J.,** Microbial catabolism of amino acids during cheese ripening, *Sci. Alim.,* 2, 113, 1982.

Chapter 12

Use of Liposomes for Wool Dyeing

Alfonso de la Maza, Luisa Coderch, Albert M. Manich, Pilar Bosch, and Jose L. Parra

CONTENTS

I. INTRODUCTION

Wool is probably the most reactive of all textile fibers towards dyestuffs of many different types, since it belongs to the class of natural proteins containing a variety of functional groups. For the benefit of those whose experience does not include keratins, it will be desirable to briefly review the macroscopic wool structure and its chemical composition. The wool fiber can be regarded as an assembly of cuticle and cortical cells held together by the "cell membrane complex" (CMC), which forms the only continuous phase in the keratin.[1,2] (Figure 1.)

The CMC plays an important role in the adhesion between cells of both the cuticle and cortex in the keratinised fiber, in the transport of dyestuffs and processing of chemicals into the fiber and in determining its surface properties.[3] This fiber is, in fact, highly heterogeneous, being a composite material made up of numerous markedly different proteins, including the high and low sulphur proteins and the so called nonkeratinous proteins. The individual peptide chains are held together by a number of different crosslinks which contribute to the overall stability of the protein composite: ionic "salt" links exert maximum influence in the so-called isoionic region; hydrophobic interactions and hydrogen bonding are strongest at low temperatures; but maximum stability is imparted by covalent crosslinks such as the disulphide bonds of cystine and the isopeptide bonds of N^{ε}-glutamil lysine.[4]

The wool dyeing process is undoubtedly fiber degradative since it is usually carried out in water at the boil for periods of 1–3 h under a variety of pH conditions, with values ranging from acidic to neutral. These dyeing conditions are needed to achieve levelness, good dye penetration in the individual fibers

0-8493-4013-6/96/$0.00+$.50

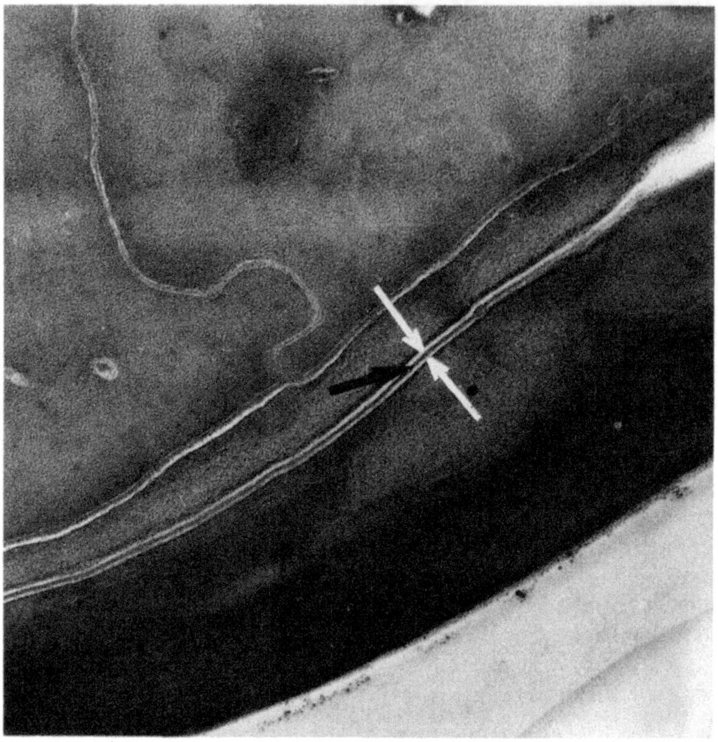

Figure 1 Transmission electron microscope microphotograph of transverse section of merino wool fiber showing several cortical cells separated by cell membrane complex (CMC)(arrowheads) comprising the β-layers (white arrows) and δ-layer (black arrow).

and optimum fastness properties. Over the last decade, a number of investigations have been carried out using different carriers capable of reducing the degradative effect brought about in conventional wool dyeing. Thus, the technology of microencapsulation has given rise to a number of innovations utilizing the basic principles of targeting, slow release and protection of this sensitive fiber.[5] Nevertheless, some technological problems related to the staining of wool with disperse dyes at high temperatures (120–130°C) still exist, especially in the dyeing of wool/polyester blends.[6,7] The selection of the disperse dye, the dyeing conditions and the use of suitable carrier are considered to be very important factors in preventing these problems.

The ability of phospholipids to solubilize small amphiphile molecules such as dyes and drugs plays a major role in pharmaceutical applications[8] and biological staining.[9] The use of liposomes as vehicles for different reagents to wool fibers could be founded on the relevant importance of the hydrophobic interactions in the structural organization of wool, the overall stability of this protein composite and on the lipidic nature of the CMC.

Merino wool fibers contain about 1% by weight of lipids (Table 1).[10] For a typical fiber, this material consists of three major lipid classes: polar lipids, sterols and fatty acids. The sterols consist predominantly of cholesterol and desmosterol in a ratio of about 2:1. The polar lipids consist basically of ceramides, cholesteryl-sulphate and glycosphingolipids. Small amounts of phospholipids have also been detected. The major fatty acids present are stearic, palmitic, 18-methyleicosanoic and oleic acids, with myristic acid being a significant minor component.[1] These lipids build the hydrophobic barrier of CMC and are structured as the two lipid bilayers, similar to those found in membranes of keratinized stratum corneum of the skin which are capable of forming multiple bilayer structures.[11]

Under transmission electron microscopy (TEM), the CMC between cells appears as two parallel nonstaining layers referred to as the β-layers (formed by lipids), separated by a densely stained layer called the δ-layer which is built basically of proteins (Figure 1). Dyeing and diffusion properties of fiber, in particular, are believed to be governed by the lipid structure of the intercellular spaces that might act

Table 1 Internal Lipid Composition of Wool (%)

Polar lipids	31.9
Cholesterol	32.8
Diglycerides	1.5
Free fatty acids	21.4
Triglycerides	1.5
Cholesteryl ester	1.5
Nonidentified	11.7

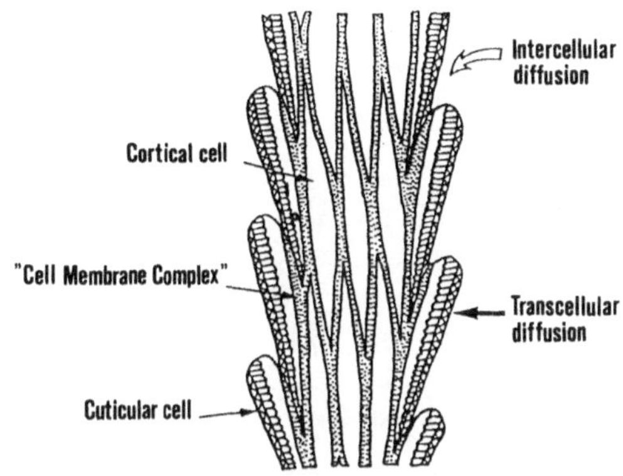

Figure 2 Scheme of the morphological structure of a wool fiber and possible routes of dye diffusion.

as "solvents" for hydrophobic textile chemicals.[2] TEM studies provide evidence that dyes do, in fact, preferentially diffuse along easily swollen regions such as the CMC (intercelullar diffusion), rather than through the cuticle cells (transcellular diffusion).[12] (See Figure 2.)

The fact that the bilayer structuration of CMC is similar to that of the liposomes and the important role played by the CMC in the processing of chemicals into the fibers have led to possible applications of liposomes as carriers in wool finishing. In this sense, liposomes made with pure phosphatidylcholine or containing lipids present in the CMC such as cholesterol[1,13] have been used as vehicles for aqueous chlorine solutions in wool chlorination processes. These applications result in an improvement in both the regularity and the homogeneity of these oxidative treatments,[14,15] minimizing wool degradation and improving the subsequent treatments in wool processing.

Polar dyes are commercial "milling acid dyes" whose main characteristic is that they diffuse much more slowly than typical levelling acid dyes, thereby exhibiting higher wet fastness. However, their migration and coverage properties are inferior, and the addition of levelling agents is normally required.[16] It is therefore possible that liposomes could be used as carriers for such dyes.

In this chapter we shall describe studies involving the use of phosphatidylcholine liposomes including increasing amounts of cholesterol as carriers of two structurally different milling acid dyes to untreated fibers in wool dyeing. The physicochemical stability of some large unilamellar and multilamellar vesicle suspensions (LUV and MLV) was studied during the dyeing process. We also examined the application of these structures to the dyeing of untreated wool, with special emphasis on the kinetic aspects of dye adsorption and dye–fiber bonding forces on wool fibers.

II. METHODOLOGY

A. MATERIALS

Botany wool fabrics knitted from R64/2 tex (count 2/28) yarns were used. Samples were soxhlet extracted for 2 h with methylene chloride and rinsed with water purified by the Milli-Ro system (Millipore) and dried at room temperature.

Two commercially available Ciba-Geigy milling acid dyes were selected (Polar Red B, CI Acid Red 249 and Polar Blue G, CI Acid Blue 90); their molecular structures are given in Figure 3. The selection criterion employed was to choose structurally different dyes belonging to the milling acid groups; the method of application on wool fabrics was similar in both cases.

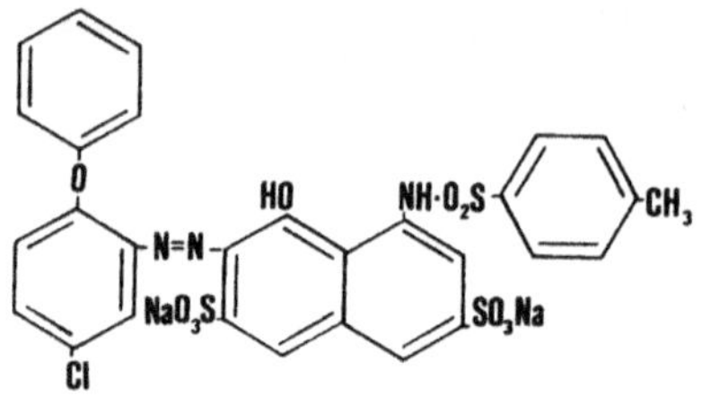

Polar Red B, C.I. Acid Red 249

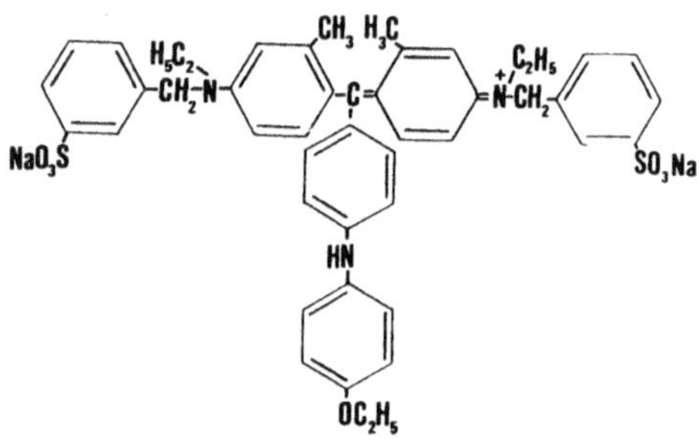

Polar Blue G , C.I. Acid Blue 90

Figure 3 Chemical structure of the milling acid dyes Polar Red B (CI Acid Red 249) and Polar Blue G (CI Acid Blue 90).

Triton X-100 (octylphenol ethoxylated with 10 units of ethylene oxide and an active matter of 100%) was supplied by Tenneco S.A. (Spain).

Phosphatidylcholine (PC) was purified from egg lecithin (Merck) according to the method of Singleton[17] and shown to be pure by thin-layer chromatography (TLC).

Cholesterol (CH) was purchased from Sigma Chemical Co. (St. Louis, MO). Lipids were stored in chloroform under nitrogen at –20°C until use.

Polycarbonate membranes of 400 nm and 800 nm and membrane holders used for liposome extrusion were purchased from Nucleopore (Pleasanton, CA).

B. PREPARATION OF LIPOSOMES CONTAINING DYES

The liposomes used for these studies were large unilamellar vesicles (LUV) and multilamellar vesicles (MLV) of a variety of lipid compositions. The conditions of each preparation are outlined below.

1. LUV

Large unilamellar vesicle suspensions of a defined size (400 nm) at different lipid concentrations (from 0.5 to 4.0 m*M*) containing the milling acid dyes Polar Blue G or Polar Red B were prepared by reverse phase evaporation under nitrogen atmosphere as described by Paternostre and Rigaud and co-workers.[18,19] This procedure was essentially derived from the general procedure of Szoka et al.[20] A film was formed by removing the organic solvent from chloroform solution of PC/CH mixtures (from 10:0 to 8:2 molar ratios), by rotary evaporation in a nitrogen atmosphere and under vacuum (350 mmHg). This film was redissolved in diethyl ether and an aqueous phase containing the dyebath components (dye 1%, sodium sulphate 5%, and acetic acid to pH 5.5) to be introduced into the liposomes was added to the lipid solution. The resulting two-phase system was sonicated (Labsonic 1510, B. Braun) at 70 W for 3 min at 5°C, thereby obtaining an emulsion. The organic solvent was removed at 20°C by rotary evaporation under controlled pressure (400 mmHg). As most of the solvent was removed, the material first formed a viscous gel before turning into an aqueous solution. Liposome suspensions were extruded through 800- and 400-nm polycarbonate membranes to obtain a uniform size distribution. After preparation, the resulting liposome suspensions containing dyestuff were left to equilibrate for 15 min and immediately applied to the wool dyeing processes.[21]

2. MLV

Multilamellar vesicle liposomes of a defined size (400 nm) containing the milling acid dyes Polar Red B or Polar Blue G in the same range of lipid concentrations and also including CH in the same molar ratios were prepared, following a method described by Bangham.[22] An aqueous phase containing the dyebath components (dye 1%, sodium sulphate 5%, and acetic acid to pH 5.5) was added to the lipid film previously formed by removing the organic solvent from chloroform solution of PC or PC/CH lipid mixtures. The solutions were then swirled to transfer the lipid from the walls of the flask and to disperse large lipid/dye aggregates; glass beads were added to facilitate dispersions. The resulting milky suspensions were vortexed for 5 min and sonicated for 15 min at 30°C and 75 W (Labsonic 1510 B. Braun). Likewise, the vesicle suspensions were sequentially extruded through 800- and 400-nm polycarbonate membranes to obtain a uniform size distribution.

C. CHARACTERIZATION AND STABILITY OF LIPOSOME SUSPENSIONS

1. Lipid Composition

The lipid concentration of liposome vesicles as well as the relative proportion of PC:CH in bilayers after preparation were determined using the Iatroscan MK-5 TLC-FID analyser. Coupling thin-layer chromatography (TLC) to an automated detection system, based on flame ionization detection (FID), has considerably enhanced the sensitivity of TLC and allows quantification of separated materials.[23] This method has been used to quantify several kinds of lipids from different sources.[24] Previous experience of lipid analysis from keratinized tissues[25] prompted us to choose this procedure to quantify these particular lipid mixtures even when they formed liposomes in water solutions.

2. Encapsulation Efficiency

The percentages of encapsulated dye in liposomes (expressed in percent volume) were determined using a spectrophotometric method. After preparation, liposome suspensions were cleared of unencapsulated dye by separation through Sephadex G-50 medium resin (Pharmacia Sweden) column chromatography,[26] and then the concentration of entrapped dye was evaluated by spectrophotometry after the solubilization of the supernatant lipid bilayers by addition of Triton X-100.[27,28]

3. Physical Stability

The aggregation state of the vesicles was estimated as a measure of the physical stability of the liposome suspensions. This was done by monitoring the variation of the mean vesicle size distribution of liposome suspensions as a function of time. This parameter, together with the polydispersity indexes of the liposome suspension, was determined using a photon correlator spectrometer (Malvern Autosizer 4700c PS/MV). Vesicle size distributions were made by particle number measurements. Samples were adjusted to the appropriate concentration range with a dyebath solution. Measurements were made at 25°C with a detection angle of 90°.

4. Chemical Stability

The acidic conditions used in wool dyeing may produce hydrolysis at the ester linkage of phospholipidic bilayers, with the production of lysophosphatide and free fatty acids. The level of free fatty acids in liposome suspensions after dyeing was assessed as a measure of the chemical stability of these suspensions. Fatty acids were quantitatively determined as methyl ester derivatives[29] in a Hewlett Packard 5840A gas chromatograph equipped with a flame ionization detector, using an internal standard of heptadecanoic acid (Fluka no. 51610). Methyl esters of fatty acids were prepared using the diazomethane method.[30]

D. DYEING PROCEDURE

In general, the dyeing conditions used in the present work for milling acid dyes (dyebath pH, range of temperatures and time of dyeing) were similar to those recommended by the manufacturer for these dyes in standard wool dyeing. Wool knitted samples (1 g) were treated with LUV or MLV liposome suspensions freshly prepared, containing the milling acid dyes either Polar Blue G or Polar Red B, at different PC:CH molar ratios (from 10:0 to 8:2); the lipid concentration of bilayers ranging from 0.5 to 4.0 m*M*. The dye was applied at 1% on weight fabric (o.w.f.) with a 5% sodium sulphate solution, acetic acid to pH 5.5 and liquor ratio of 60:1.

Dyeing was initiated at 50°C and the temperature was raised by 0.9°C/min to 90°C. Dyeing was continued for 120 min. Next, samples were rinsed with water for 10 min and dried at room temperature. Laboratory dyeing was carried out in a Multi-Mat dyeing machine (Renigal).

Dyebath exhaustion was determined using a Shimadzu UV-265FW spectrophotometer. Liposome aliquots (0.5 ml) were periodically added to quartz cuvettes filled with 2 ml of aqueous solution of Triton X-100 (10 g/l), supplemented with sodium sulphate (5%) and acetic acid at pH 5.5. The interaction between the nonionic surfactant Triton X-100 and liposome structures resulted in a solubilization of lipid vesicles via mixed micelle formation,[27,28] turning the liposome suspensions into clear solutions.

Figure 4 shows the effect of the cleavage of liposome vesicles by Triton X-100 on the absorption spectra of both dyes used at the highest PC concentration (4.0 m*M*). It may be seen that the wavelength of maximum absorption of each dye used in this research does not change in the presence (dashed line) or absence (solid line) of the phospholipid-surfactant mixed micelles. A similar effect was observed when using liposomes containing CH in the bilayer in the range of PC:CH molar ratios form 10:0 to 8:2.

E. DYE EXTRACTIONS OF SAMPLES

After the dyeing process, the superficial dye bonded to the fibers by nonpolar forces (hydrophobic interactions, van der Waals forces and hydrogen bonds) was extracted with pure ethanol at 25°C for 60 min.[21] Subsequent extractions with ammonia solution (0.5% at 60°C for 15 min) stripped the dye diffused inside the fiber and bonded ionically.[31]

F. SMOOTHNESS OF DYED SAMPLES

The surface smoothness of the wool samples treated with dye–liposome systems was determined by measuring the dynamic coefficient of friction using an Instron Textile Tester 1122 with a device designed to meet the requirements of the American Society for Testing and Materials Standards (ASTM).[32]

The sample was attached to the rubber-covered surface of a 220-g sled which was drawn across a metallic table. The sled was connected through a low friction pulley to the Instron load cell, which detected the friction strength. The moving crosshead supplied the motive force to the sled. The measurements were taken at slip rate levels of 1 and 10 mm/min.

III. RESULTS AND DISCUSSION

A. STABILITY OF LIPOSOME SUSPENSIONS

The level of free fatty acids in liposome suspensions, possibly formed during the dyeing of untreated wool samples, was estimated as a measure of the chemical stability of these suspensions. A sequential analysis of the fatty acid composition of phospholipids using gas liquid chromatography was carried out during dyeing in order to determine this parameter. The results showed that a nonsignificant hydrolysis reaction of the ester bond in the phospholipid molecule took place in these liposome suspensions.

The possible aggregation or solubilization of liposomes during dyeing was estimated as a measure of the physical stability of these suspensions and was determined by measuring the variations in mean vesicle size distribution and polydispersity of these suspensions, using a quasi-elastic light-scattering method.[33]

LUV and MLV liposome suspensions containing the Polar Blue G or Polar Red B dyes were treated under the same conditions as in the dyeing process but in the absence of wool and using the same lipid bilayer concentrations at different lipid compositions (PC:CH molar ratios from 10:0 to 8:2). There was a small decrease in the particle size distribution in the initial stage of the dyeing process in all cases. After 30 min, vesicle size increased slightly, reaching maximum growth after approximately 100 min of dyeing (vesicle size about 430–435 mn), the polydispersity indexes remaining below 0.15 after treatment in all cases. Increasing amounts of CH in liposomes enhanced the stability of these systems with respect to the aggregation, reducing both the mean particle size distribution and the polydispersity indexes during dyeing. This behaviour is in agreement with the results reported by Scherphof et al. in studies on liposome stability.[34] The chemical structure of dye and the type of vesicles used do not seem to affect the physical stability of the dye–liposome system. Furthermore, the mean vesicle size distribution and the polydispersity index were maintained at around 400 nm and below 0.24, respectively, for more than 24 h.

As a consequence, the LUV and MLV liposome suspensions including the studied milling acid dyes were physically stable during the dyeing process for the lipid concentration range and lipid compositions studied.

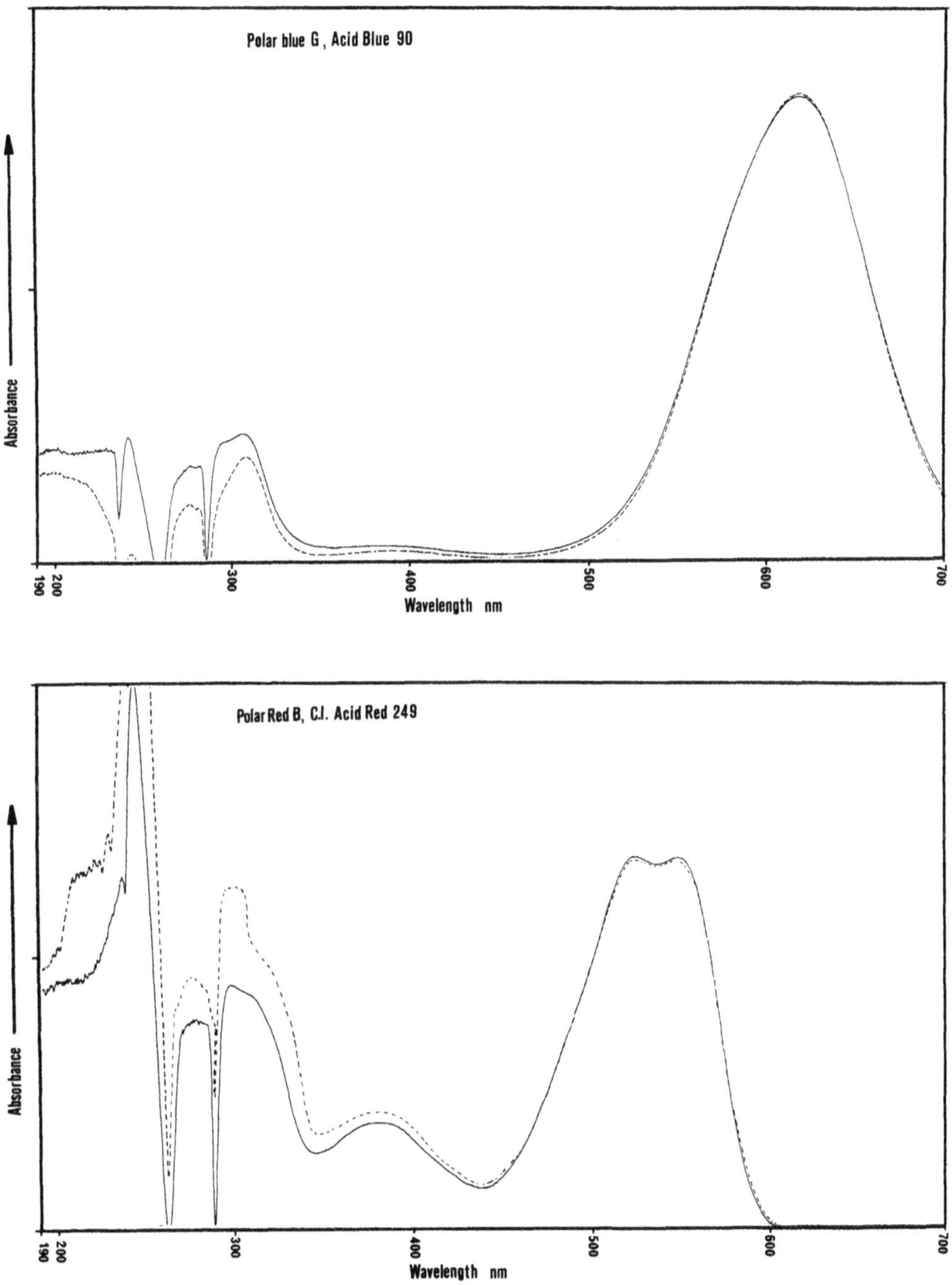

Figure 4 Absorption spectra of the milling acid dyes Polar Blue G (CI Acid Blue 90)(Figure 4A) and Polar Red B (CI Acid Red 249)(Figure 4B) in the presence (dashed line) and absence (solid line) of phospholipid/Triton X-100 mixed micelles.

B. ENCAPSULATION EFFICIENCY

The relative concentrations of encapsulated dye in LUV and MLV liposomes for both the acid dyes investigated at different lipid compositions (PC:CH molar ratios 10:0 and 8:2) were plotted vs. the bilayer lipid concentration. The results obtained are indicated in Figure 5. The encapsulation efficiency increased in direct proportion to the lipid concentration of liposomes for both lipid structures studied. However, we were unable to find any relationship between the dye encapsulation percentages and the chemical structure of the dyes. Increasing amounts of CH in the bilayers slightly decreased the percentages of dye encapsulated in all cases. The higher encapsulation efficiency of the LUV suspensions (maximum value of about 25% for 4.0 m*M* lipid concentration) could be explained by the unicompartmental architecture of these species whose vesicles show a higher internal volume.[35]

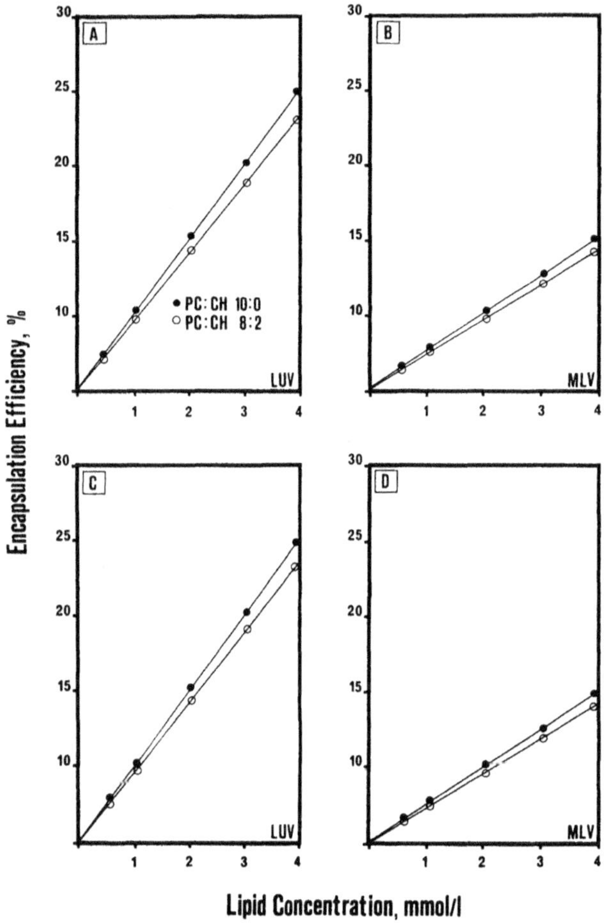

Figure 5 Percentages of encapsulation efficiency of LUV and MLV liposome suspensions containing Polar Blue G (graphs A and B) and Polar Red B (graphs C and D), vs. bilayer lipid concentrations for PC:CH molar ratios 10:0 and 8:2.

C. DYEING KINETICS

We made kinetic studies of dye exhaustion on untreated wool samples for both dyes studied via LUV and MLV liposomes, the PC concentration ranging from 0.5 to 4.0 m*M*. The results obtained are plotted in Figure 6.

As can be seen, the use of liposomes as carriers in wool dyeing resulted in an inhibition of dye exhaustion for both liposome structures studied, this effect being slightly greater for LUV suspensions. This behaviour appeared to be closely connected with the lipid concentration of the bilayers. The maximum inhibition was reached for 4.0 m*M* and remained constant throughout the dyeing process.[19,36]

The maximum percentages of dye exhaustion after dyeing were obtained for Polar Red B (around 80% for LUV liposomes), whereas the Polar Blue G reached lower values in all cases (around 75% also for LUV liposomes). These differences could be attributed to the chemical structure of dyes which probably induces different dye–phospholipid interactions. These interactions may also affect the subsequent action of these dye/liposome systems with the wool fibers, especially in the migration of dye from the solution to the wool surface and in the subsequent adsorption of these systems on the fiber surface.

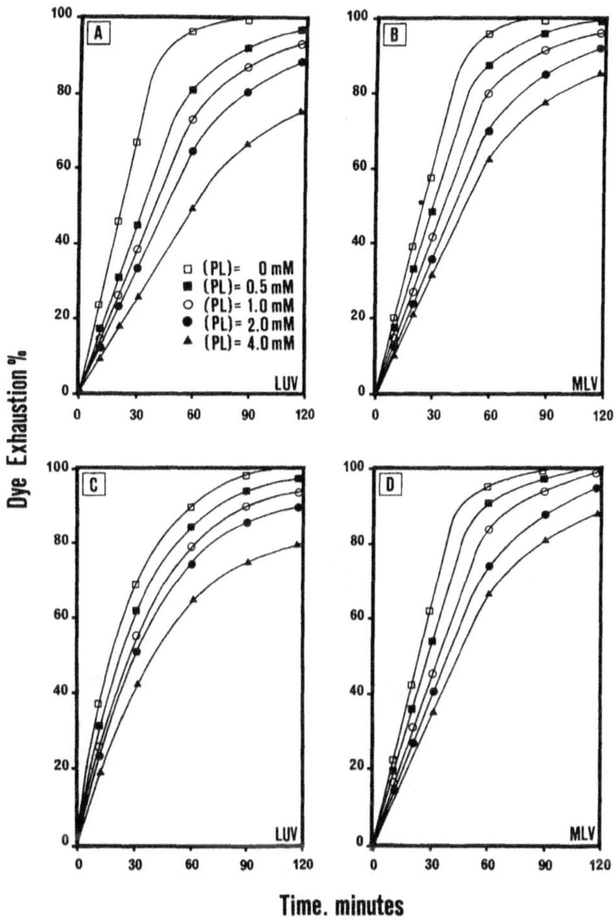

Figure 6 Exhaustion kinetics of Polar Blue G (graphs A and B) and Polar Red B (graphs C and D) dyes on untreated wool samples in dyeing via LUV and MLV liposomes (at different PC concentration) vs. time.

As cholesterol is one of the main components of the internal lipids of wool,[1] we studied the effect caused by including this component in lipid bilayers on wool dyeing. Accordingly, we investigated the dye exhaustion kinetics also for both dyes via LUV or MLV liposomes containing increasing amounts of CH in bilayers, the total lipid concentration remaining constant (4.0 m*M*). The results obtained for Polar Blue G dye via LUV liposomes are given in Figure 7. In general, the marked inhibition in the dyeing exhaustion with respect to dyeing in the absence of liposomes is attributable to the use of the highest lipid bilayer concentration which, as discussed above, led to a maximum inhibitory effect. With regard to the influence of the molecular composition of liposomes, this inhibition was directly connected to the presence of CH in bilayers; i.e., the greater the CH concentration in the bilayers, the higher the inhibition in dye exhaustion, especially for the initial and final stages of the dyeing process.[37] The use of MLV liposomes led to similar dyeing exhaustion curves albeit with a slight diminution in their inhibitory capacity of dye exhaustion. Similar dyeing behaviour was obtained using the Polar Red B, the dye chemical structure playing a role similar to the one when using liposomes without CH.

Bearing in mind the dyeing properties of the milling acid dyes applied in a low acid dyebath, dye–liposome interaction could be attributed to the ionic affinity between the polar groups of both lipids and dyes as well as to the hydrophobic interactions between nonpolar moieties of those structures. These interactions could affect the dye molecules entrapped in vesicles and also those present in the surrounding aqueous medium. Thus, the interaction of dye molecules with wool fibers can be modulated by the presence of these new (dye–liposome) complexes, affecting especially the migration and adsorption of dyes on the wool surface and the dye–fiber bonding mechanisms based on electrostatic bonds between dye anions and basic group charges of the keratin structure.

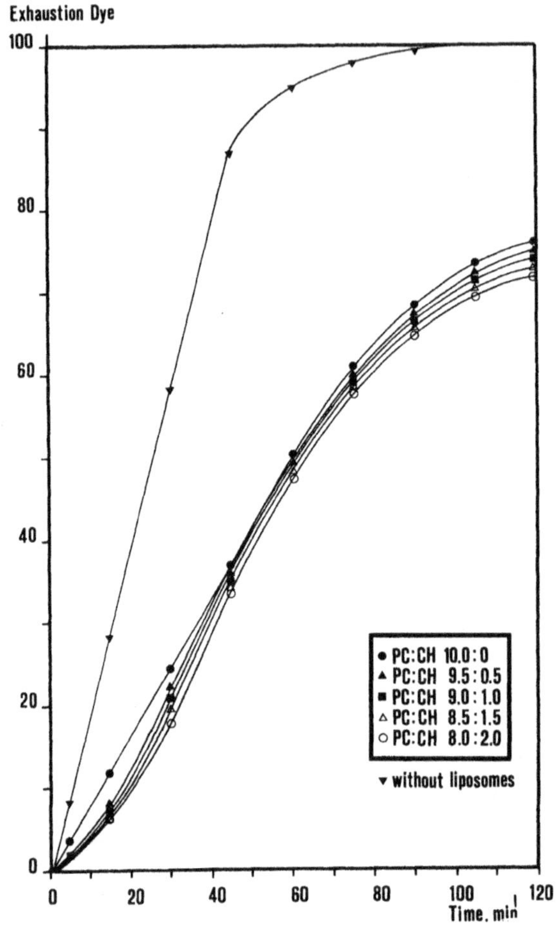

Figure 7 Exhaustion kinetics of Polar Blue G dye on untreated wool samples in dyeing via LUV liposomes, the bilayer lipid concentration remaining constant (4.0 m*M*), and varying the PC:CH molar ratios from 10:0 to 8:2.

D. INFLUENCE OF LIPOSOMES ON BONDING OF DYES ON WOOL

In order to find out whether liposomes as dye carriers brought about changes to dye–fiber bonding forces after dyeing, successive extractions by pure ethanol[21] and ammonia[31] were performed on dyed samples via LUV or MLV liposomes for both dyes and in the range of the lipid concentration studied. The results of these trials are given in Tables 2 (Polar Blue G) and 3 (Polar Red B dye). In general, the amounts of dye extracted via ammonia were clearly higher than those extracted using pure ethanol.

Ethanol extractions of dyed samples via liposomes showed, in general, smaller values than those obtained from samples dyed without liposomes (using the leveling agent recommended by the manufacturer). The amounts of extracted dye were inversely correlated with the lipid concentration of liposomes in all cases. Furthermore, increasing amounts of CH in bilayers led to a progressive diminution in the amounts of extracted dye regardless of the type of liposome used. The use of unilamellar or multilamellar bilayer structures does not seem to affect the total amount of extracted dye.

Similar tendencies were observed in subsequent extractions on the same sample using ammonia. Thus, the amounts of extracted dye were inversely correlated with the lipid concentration in the bilayers. Likewise, increasing concentrations of CH in bilayers drastically reduced the amounts of extracted dye especially that extracted via ammonia. The use of LUV liposomes resulted in a fall in the dye extraction for both dyes used. The chemical structure of the dyes also exerted a certain influence over the dye bonding on the fibers, especially at the highest lipid bilayer concentration where the lowest amounts of extracted dye always corresponded to the Polar Blue G dye.

The overall evidence confirms the role of the CH in liposomes as effective carriers for bonding of milling acid dyes on wool fibers. This effect may be attributed to the fact that the CH affects the

Table 2 Amounts of Extracted Dye (mg dye/g wool) by Pure Ethanol[21] and Ammonia[31] on Dyed Samples via LUV and MLV Liposomes for Polar Blue G dye, the Bilayers Lipid Concentration Ranging from 0.5 m*M* to 4.0 m*M* and for the PC:CH Molar Ratios 10:0, 9:1 and 8:2

PC:CH Molar Ratio	Lipid Conc. (m*M*)	LUV Extracted Dye (mg dye/g wool)		MLV Extracted Dye (mg dye/g wool)	
		A	B	A	B
10:0	4.0	0.012	1.198	0.022	1.798
	3.0	0.018	1.503	0.024	2.016
	2.0	0.022	1.906	0.025	2.200
	1.0	0.034	2.084	0.040	2.400
	0.5	0.049	2.983	0.048	3.020
	0	0.217	3.608	0.217	3.608
9:1	4.0	0.010	0.879	0.016	1.494
	3.0	0.014	1.194	0.020	1.708
	2.0	0.018	1.593	0.021	1.898
	1.0	0.027	1.762	0.033	2.104
	0.5	0.034	2.654	0.040	2.714
8:2	4.0	0.003	0.729	0.014	1.299
	3.0	0.006	1.008	0.017	1.502
	2.0	0.009	1.287	0.018	1.678
	1.0	0.018	1.584	0.027	1.911
	0.5	0.023	2.453	0.036	2.512

Note: A = extractions with pure ethanol;[21] B = extractions with ammonia.[31]

Table 3 Amounts of Extracted Dye (mg dye/g wool) by Pure Ethanol[21] and Ammonia[31] on Dyed Samples via LUV and MLV Liposomes for Polar Red B dye, the Lipid Bilayer Concentration Ranging from 0.5 m*M* to 4.0 m*M* and for the PC:CH Molar Ratios 10:0, 9:1 and 8:2

PC:CH Molar Ratio	Lipid Conc. (m*M*)	LUV Extracted Dye (mg dye/g wool)		MLV Extracted Dye (mg dye/g wool)	
		A	B	A	B
10:0	4.0	0.040	1.480	0.052	2.073
	3.0	0.061	1.780	0.073	2.374
	2.0	0.080	2.080	0.094	2.617
	1.0	0.100	2.180	0.126	2.902
	0.5	0.140	2.800	0.157	3.123
	0	0.200	3.508	0.200	3.508
9:1	4.0	0.038	1.189	0.043	1.764
	3.0	0.056	1.492	0.068	2.083
	2.0	0.073	1.784	0.084	2.301
	1.0	0.089	1.893	0.112	2.599
	0.5	0.123	2.491	0.136	2.831
8:2	4.0	0.028	0.993	0.034	1.571
	3.0	0.046	1.302	0.057	1.881
	2.0	0.063	1.604	0.074	2.216
	1.0	0.082	1.721	0.103	2.402
	0.5	0.117	2.246	0.121	2.601

spontaneous release of encapsulated solutes and decreases the tendency of the vesicles to aggregate once they are formed.[33] Likewise, the presence of CH in the CMC may explain the role of this compound in the diffusion of the dye from the surface towards the center of the fiber and in the improvement of the dye–fiber bonding forces in wool dyeing.

Given that pure ethanol extractions remove ·the dye superficially bonded to the fibers by nonpolar forces, the use of liposomes in dyeing must lead to a decrease in such a dye. Furthermore, the decreasing amounts of dye removed with ammonia with respect to the lipid concentration of liposomes (particularly in the presence of increased amounts of CH) could be attributed to the increasing contribution of the nonpolar forces in the dye–fiber bonds. These interactions, especially those that are hydrophobic in nature, may play an important role in the dye–fiber bonding inside the fiber.

The total percentage of dye bonded to the wool fibers can be expressed by the equation (1):

$$C_b = \frac{C_a - C_e}{C_a} \cdot 100 \tag{1}$$

where C_b is the amount of dye bonded (%), C_a is the amount of dye adsorbed (mg/g wool) and C_e is the amount of dye extracted (mg/g wool). These percentages for both dyes studied using LUV and MLV liposomes (lipid composition PC:CH from 10:0 to 8:2 molar ratios) for different bilayer lipid concentrations are given in Table 4.

Table 4 Percentages of Total Bonded Dye for Polar Blue G and Polar Red B dyes, the Lipid Bilayer Concentration Ranging from 0.5 m*M* to 4.0 m*M* and for the PC:CH Molar Ratios 10:0, 9:1 and 8:2

PC:CH Molar Ratio	Lipid Conc. (m*M*)	Total Bonded Dye (%)			
		Polar Blue G		Polar Red B	
		LUV	MLV	LUV	MLV
10:0	4.0	84.13	77.25	81.00	75.90
	3.0	81.51	75.43	78.36	73.34
	2.0	78.08	73.82	76.00	71.55
	1.0	77.21	72.88	75.95	69.38
	0.5	68.06	67.70	68.81	7.20
	0	62.16	62.16	62.92	62.92
9:1	4.0	87.99	81.91	84.35	79.01
	3.0	84.97	80.08	81.41	76.07
	2.0	81.29	78.68	78.92	74.40
	1.0	80.42	77.29	78.73	72.02
	0.5	71.88	70.14	72.65	69.78
8:2	4.0	89.90	83.84	86.50	80.91
	3.0	87.03	82.05	83.46	77.82
	2.0	84.65	80.77	80.88	74.94
	1.0	82.23	79.10	80.57	73.87
	0.5	73.88	73.51	75.04	71.70

The percentages of total bonded dye were directly correlated with the lipid concentration in bilayers reaching maximum values for 4.0 m*M* (81–84% for LUV and 76–77% for MLV in the absence of CH) and falling to values of about 62% for dyed samples in the absence of liposomes.

The use of LUV liposomes resulted in higher percentages of total bonded dye compared with the levels found with MLV liposomes, regardless of the lipid concentration used in all cases. It is noteworthy that the presence of CH in liposomes increased the percentage of total bonded dye in wool in all cases, i.e., the higher the CH proportion in bilayers, the greater the relative concentration of dye bonded to the fiber.

Figure 8 shows the amounts of bonded dye in wool fibers given as the difference between the amounts of adsorbed dye after dyeing and total extracted dye (Tables 2 and 3) for both dyes studied using LUV or MLV liposomes, vs. lipid concentration (lipid compositions (PC:CH 10:0 and 8:2 molar ratio).

We obtained a maximum amount of bonded dye when lipid concentration was 1.0 m*M* for both dyes investigated. The presence of increasing amounts of CH in the bilayers led to a rise in the total amount of bonded dye. This phenomenon was also observed when using unilamellar vesicle structures in all cases.

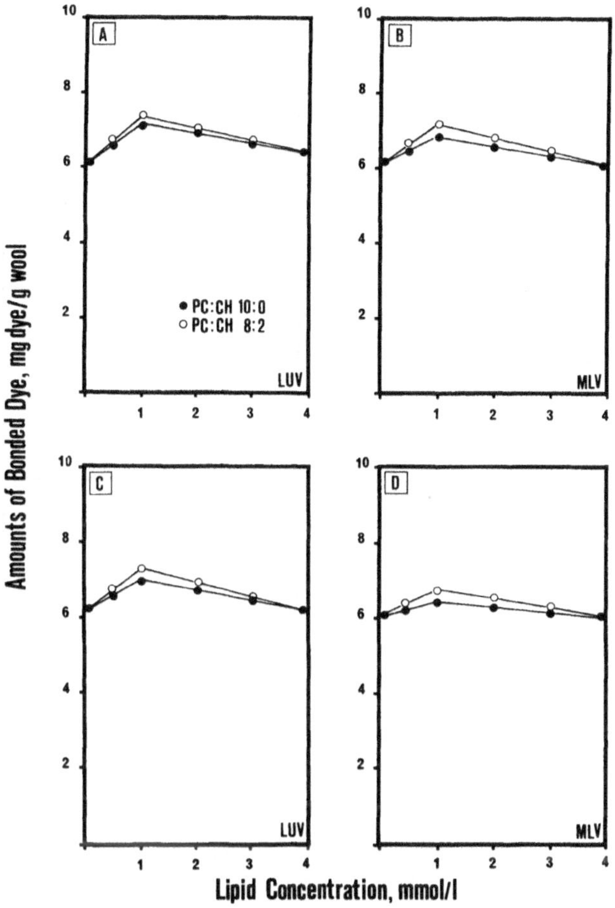

Figure 8 Amounts of bonded dye in wool fibers for Polar Blue G (graphs A and B) and Polar Red B (graphs C and D) via LUV and MLV liposomes vs. lipid concentration (PC:CH 10:0 and 8:2 molar ratio). The results are given as a difference between the amounts of adsorbed dye and total extracted dye.

E. SMOOTHNESS OF DYED SAMPLES

The smoothness of dyed samples expressed in terms of friction strength (cN, sled rate 1 mm/min) between samples and a metallic flat surface vs. the lipid concentration in liposomes are given in Figure 9. The assays were made for both dyes, using LUV or MLV liposomes with or without CH in the bilayers (PC:CH 10:0 and 8:2 molar ratio).

The friction strength slightly diminished as both the lipid concentration in the liposomes and the CH concentration in the bilayers increased. The best results were obtained with 1.0–2.0 mmol/l total lipid concentration for the highest level of CH in the bilayers (PC:CH molar ratio 8:2). This trend was similar for both liposome structures studied, regardless of the chemical structure of the dye used.

It is noteworthy that the lipid concentration for the maximum inprovement in wool smoothness also corresponded to that for the maximum amount of dye bonded to the fiber. (Figure 8). Given that the improvement in smoothness may be correlated with the migration of the lipids present in bilayers to the wool surface, the aforementioned agreement suggests that the maximum lipid adsorption on wool fibers occurred in this interval of lipid concentration. This fact could account for the enhancement in the dye–fiber bonding forces in this range of lipid concentrations.

The marked influence of CH in the smoothness of dyed samples may also be attributed to the fact that this lipid is one of the main components of the internal lipids of wool, especially in the cell membrane complex (CMC). Hence, the protection of the CMC during dyeing by the inclusion of suitable lipids in liposomes may enhance the physical properties of treated samples by especially improving their smoothness.

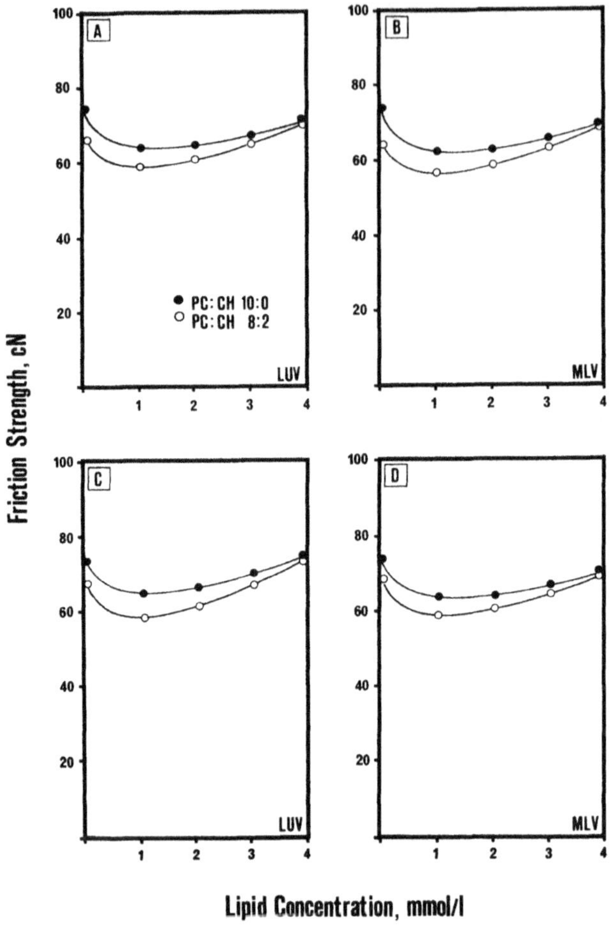

Figure 9 Smoothness of dyed samples via LUV and MLV liposomes containing the dyes Polar Blue G (graphs A and B) and Polar Red B (graphs C and D) vs. lipid concentration (PC:CH 10:0 and 8:2 molar ratio), and expressed in terms of friction strength between samples and a metallic flat surface.

F. DYEING STUDIES USING ONLY THE ENCAPSULATED DYE

Given that the main part of dye molecules included in liposome suspensions are placed in the aqueous medium which surrounds the vesicles (maximum encapsulation efficiency of LUV liposomes of about 25%), we carried out a series of experiments based on the liposome dyebath exhaustion using only the dye encapsulated into the liposomic structures. To this end, wool samples were treated with LUV or MLV liposomes which contained the milling acid dyes after freeing the unencapsulated dye by separation through Sephadex G-50 resin.[26] The PC:CH molar ratios varied from 10:0 to 8:2 and the total lipid concentration remained constant (4.0 m*M*).

Figure 10 shows the exhaustion kinetic of the Polar Blue G dye by means of LUV liposomes, the PC:CH molar ratios ranging from 10:0 to 8:2. Dye exhaustion kinetics showed different tendencies, depending on the CH concentrations present in the bilayers. Thus, the initial dyeing phase showed an increasing inhibition of dye exhaustion, which reached the maximum for the PC:CH molar ratio 8.5:1.5. But increasing amounts of CH in the bilayers (8.0:2.0 molar ratio) reduced this inhibitory effect. However, the final dye exhaustion appeared to be directly correlated with the CH present in bilayers, reaching the highest exhaustion for the PC:CH molar ratio 8.5:1.5. Once more, exceeding CH concentrations resulted in decreased dye exhaustions. This behaviour was observed for both dyes regardless of the type of liposome used. The improved final dye exhaustion obtained using exclusively the encapsulated dye for the increased CH liposomes may confirm the role of the cell membrane complex (CMC) in the transport of dyestuffs, given that the CMC presents a bilayer structuration (including CH Table 2) which is similar to that of the liposomes.

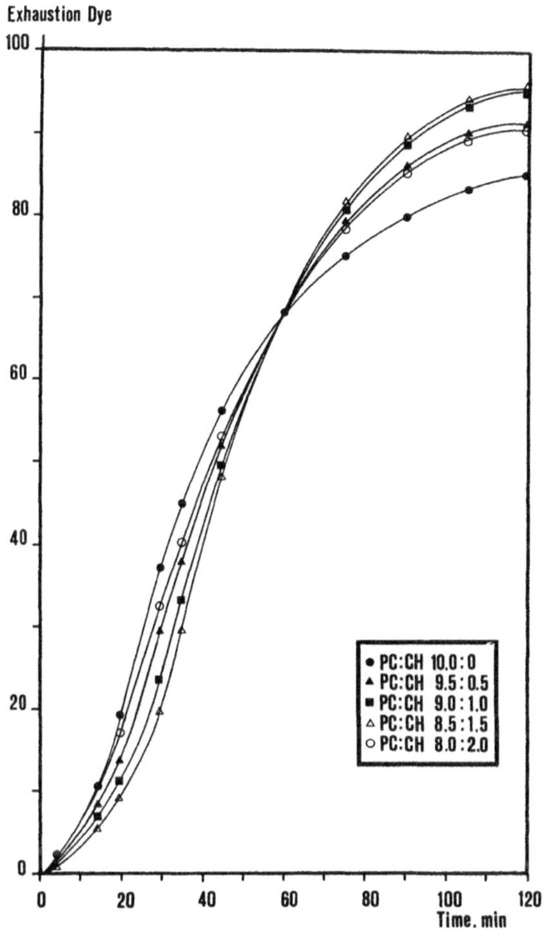

Figure 10 Exhaustion kinetics of Polar Blue G dye in dyeing by means of LUV liposomes, using only the encapsulated dye, the PC:CH molar ratio in bilayers ranging from 10:0 to 8:2.

A comparison of the dye exhaustion percentages obtained with complete liposomes or using only the encapsulated dye (Figures 7 and 10) reveals that, in both cases, initial dye exhaustion follows similar tendencies except for the lipid molar ratio PC:CH 8:2. However, the most striking result is that the final dyeing exhaustions show an opposite tendency, depending on the kind of application used. Thus, in the former case the presence of increasing amounts of CH resulted in the final inhibition of dye exhaustion, whereas in the latter case this presence led to an increase in a final dye exhaustion on wool samples.

These findings emphasize the role of CH in liposomes as effective carriers for transport and bonding of dyes in wool and the importance of the location of the dye molecules in liposome suspensions, i.e., in the lipid bilayers or in the surrounding aqueous medium. The opposite tendency also stresses the role of lipid bilayers as carriers during the subsequent steps of the dyeing process: migration of dyes from the solution to the wool surface, adsorption on the fiber surface, diffusion of the dye from the surface towards the center of the fiber, and electrostatic bonding of the dye molecules in the keratin substrate.

Table 5 shows the percentages of total bonded dye (calculated following Equation 1) for both dyes studied, using LUV or MLV liposomes (4.0 m*M* lipid concentration) the PC:CH molar ratios ranging from 10:0 to 8:2. An overall comparison of Tables 4 and 5 shows that the exclusive use of the encapsulated dye results in increased dye bonded percentages with respect to those obtained using the complete liposomes. These findings emphasize the importance of the lipidic pathway involved in the use of liposomes as suitable carriers in the transport and bonding of dyestuff to hydrophobic domains. Increasing amounts of CH in the bilayers also raised the total percentage of the bonded dye. As observed in Table 5, the use of Polar Blue G dye resulted in higher percentages of bonded dye regardless of both the type of liposome used and the CH concentration present in bilayers.

Table 5 Percentages of Total Bonded Dye for the Polar Blue G and the Polar Red B Dyes Using Only the Encapsulated Dye, the Lipid Bilayer Concentration Remaining Constant (4.0 m*M*) and for the PC:CH Molar Ratios 10:0, 9:1 and 8:2

PC:CH Molar Ratio	Total Bonded Dye (%)			
	Polar Blue G		Polar Red B	
	LUV	MLV	LUV	MLV
10:0	94.82	92.93	91.80	89.89
9.5:0.5	96.13	94.90	93.36	91.23
9.0:1.0	97.51	95.68	94.48	92.58
8.5:1.5	97.96	95.77	95.07	92.78
8.0:2.0	97.30	95.03	95.55	92.64

In the light of our results we shall consider two major aspects related to the dyeing of wool fabrics via liposomes with milling acid dyes: kinetics and bonding of dyes into the fiber.

Depending on both PC concentration and relative concentration PC:CH, a variable inhibition of dye exhaustion during dyeing can be obtained. In fact, the function of leveling agents in a dyeing process is based on their regulatory action leading to an equilibrium between vehiculization of the dye to the fiber and the appropriate bonding in the keratinic structure. The special characteristics of liposome bilayers could encourage this leveling function; moreover, dye delivery by liposomes could be a versatile strategy for modulating the dyeing process in order to obtain a progressive dye exhaustion on the fiber with a suitable and durable dye bonding on the active sites of the keratinic structure. We postulate that the dye transport by liposomes is as important as the lipidic pathways existing in the CMC realms.

The behaviour of liposomes as a delivery system of dye molecules present only in the intraliposomic domain or in the whole space (liposome inside and outside) is clearly different. In the first case, the inhibition of the dye exhaustion observed during the initial steps of dyeing could be attributed to a certain delay on the part of the dye molecules, since they must first cross through the lipid bilayer before interacting with the fiber. Nevertheless, the final exhaustion and bonding onto the fiber could be encouraged because of the physicochemical similarity of the lipid structure of CMC and the phospholipid bilayer of liposomes. The improved final bonding of dye molecules to wool fibers could be explained by this mechanism.

It is also important to note the correlation between the maximum amount of bonded dye to the fiber and the maximum smoothness of the dyed samples for a 1–2 m*M* lipid concentration. It seems that a suitable lipid concentration exists for enabling not only the passage of the dye molecule through CMC but also the adsorption of phospholipids onto the fiber in order to improve the additional smoothness characteristics of the dyed wool samples.

We think that two key concepts emerge from this review: dyeing modulation (in a kinetic and dye bonding sense) and biomimetic strategies (based on the similarity of lipid structuration of liposomes and wool CMC). Bearing both these concepts in mind, it may be possible to optimize dyeing experimental conditions by also improving fastness and smoothness of wool dyed samples.

IV. CONCLUSIONS

From our findings, we conclude that a new method of wool dyeing by means of LUV or MLV liposomes may be considered suitable for modulating the dye exhaustion of two commercial milling acid dyes, Polar Red B (CI Acid Red 249) and Polar Blue G (CI Acid Blue 90), in a low acid dyebath, thereby increasing the dye bonding in untreated wool fibers.

LUV and MLV liposome suspensions at pH 5.5 are physicochemically stable during the dyeing process for both dyes studied, the bilayer lipid concentration ranging from 0.5 to 4.0 m*M*. The presence of increasing amounts of CH in liposomes improves the stability of these structures for both particle size distribution and polydispersity index.

The inhibition of dye exhaustion on untreated wool fabrics is directly dependent on both the liposome lipid concentration and the structure of liposomes. The use of LUV liposome structures results in a greater decrease in dye exhaustion compared with the use of MLV liposomes. This inhibition is also directly dependent on the CH concentration in bilayers, especially for the initial and final stages of dyeing. The exclusive use of the encapsulated dye results in changes in the dyeing kinetics. Thus, although initially the dye exhaustion is inhibited, the final dye exhaustion rises with the CH concentration in bilayers, the maximum effect being achieved for PC:CH molar ratio 8.5:1.5. The control of dye exhaustion allows modulation of dye adsorption on wool fibers, thereby improving both regularity and homogeneity of the dye adsorption and bonding on untreated wool fibers. Given that this control is especially important in the initial dyeing stages (where excessive dye adsorption leads to staining irregularities), the bilayer lipid concentration and particularly the PC:CH ratio may act as a specific leveling mechanism for each dye.

As for the bonding of dyes in wool fibers, despite the aforementioned inhibition of dye exhaustion, the increasing concentrations of lipids in the bilayers enhance the percentage of bonded dye into the fibers. This tendency is also observed when the CH concentration in bilayers is increased. The use of LUV liposomes results in a greater increase in dye bonding compared with the use of MLV liposomes, regardless of the chemical dye structure. The exclusive application of encapsulated dye enhances the percent of bonded dye in all cases.

The lipid concentration resulting in a maximum amount of bonded dye is 1.0 m*M* in all cases. Given that the presence of CH in the bilayers always improves the dye–fiber bonding, the maximum bonding efficacy is achieved at this lipid concentration for PC:CH molar ratio 8:2.

As regards the smoothness of dyed samples, the friction strength decreases slightly as both the lipid concentration and the CH concentration in bilayers increase. The best results are obtained for the 1–2 m*M* lipid concentration. This trend is similar for both the liposome structures studied and may be correlated with the migration of the lipids present in bilayers to the wool surface. The fact that the lipid concentration for maximum smoothness corresponds to that for the maximum amount of bonded dye emphasizes the importance of this concentration, which may be regarded as the optimum conditions in wool dyeing applications via liposomes on untreated wool fibers.

In this chapter, the importance of the role played by liposomes in wool dyeing using two different milling acid dyes has been demonstrated, taking into account both the dyeing modulation and the strategic biomimetic aspects. It is not unreasonable to predict that an optimization of experimental conditions of dyeing using liposomes could be obtained, thereby improving additionally fastness and smoothness of wool dyed samples. Some other liposome strategies could be applied to wool processing taking into consideration the special lipo-keratinic structure of wool fibers, in particular the existence of lipids in the cell membrane complex. Investigations related to both the solubilization of hydrophobic dyes (disperse dyes) and their transport by liposomes to wool and wool/polyester blends, using milder experimental dyeing conditions are in course.

ACKNOWLEDGMENTS

The original research performed in authors' lab has been supported in part by D.G.I.C.Y.T. (Dirección General de Investigación Cietífica y Técnica)(Proyecto No. PB91-0065), Spain. The authors acknowledge the expert technical assistance of Mr. G. von Knorring.

REFERENCES

1. **Rivett, D.E.,** Structural lipids of the wool fibre, *Wool Sci. Rev.* 67, 1, 1991.
2. **Leeder, J.D.,** The cell membrane complex and its influence on the properties of the wool fibre, *Wool Sci. Rev.* 63, 3, 1986.
3. **Brady, P.R.,** Diffusion of dyes in natural fibres, *Rev. Prog. Coloration,* 22, 58, 1992.
4. **Lewis, D.M.,** Damage in wool dyeing, *Rev. Prog. Coloration,* 19, 49, 1989.
5. **Nelson, G.,** Microencapsulates in textile coloration and finishing, *Rev. Prog. Coloration,* 21, 72, 1991.
6. **Wang, J. and Åsnes, H.,** One-bath dyeing of wool/polyester blends with acid and disperse dyes. Part 1. Wool Damage and Dyeing Conditions, *J. Soc. Dyers Colour,* 107, 274, 1991.
7. **Wang, J. and Åsnes, H.,** One-bath dyeing of wool/polyester blends with acid and disperse dyes. Part 2. Disperse Dye Distribution on Polyester and Wool, *J. Soc. Dyers Colour,* 107, 314, 1991.
8. **Eppstein, D.A. and Felgner, P.,** in *Liposomes as Drug Carriers,* Gregoriadis G., Ed., Wiley, 1988, 311.
9. **Almog, R. and Saulsbery R.A.,** The solubility of a dye-detergent complex in phospholipid vesicles, *J. Colloid Interface Sci.,* 159, 328, 1993.
10. **Herrling, J. and Zahn, H.,** Investigation on the composition of the cell membrane complex and its modification during industrial processing of wool, in *Proc. 7th Int. Wool Text. Res. Conf.,* Volume 1, Tokyo, 1985, 181.

11. **Downing, D.T.,** Lipid and protein structures in the permeability barrier of mammalian epidermis, *J. Lipid Res.*, 33, 301, 1992.
12. **Leeder, J.D., Rippon, J.A., Rothery, F.E., and Stapleton, I.W.,** Use of the transmission electron microscope to study dyeing and diffusion processes, in *Proc. 7th Int. Wool Text. Res. Conf.*, Vol. 5, Tokyo 1985, 99.
13. **Coderch, L., Soriano, C., de la Maza, A., Erra, P., and Parra, J.L.,** Chromatographic characterization of internal wool polar lipids, *J. Ann. Oil Chem. Soc.*, 72, 715, 1995.
14. **de la Maza, A., Parra, J.L., and Bosch, P.,** Using liposomes in wool chlorination: stability of chlorine liposomes and their application on wool fibers, *Textile Res. J.*, 61, 357, 1991.
15. **de la Maza, A. and Parra, J.L.,** Unilamellar lipid bilayers including cholesterol in wool chlorination: stability of chlorine liposomes and their application on wool, *Textile Res. J.*, 63, 44, 1993.
16. **Bone, J.A., Shore, J., and Park, J.,** Selecting dyes for wool: technical and economic criteria, *J. Soc. Dyers Colour,* 104, 12, 1988.
17. **Singleton, W.S., Gray, M.S., Brown, M.L., and White, J.L.,** Chromatographically homogeneous lecithin from egg phospholipids, *J. Am. Oil Chem. Soc.*, 42, 53, 1965.
18. **Paternostre, M.T., Roux, M., and Rigaud, J.L.,** Mechanisms of membrane protein insertion into liposomes during reconstitution procedures involving the use of detergents, 1. solubilization of large unilamellar vesicles (Prepared by Reverse Phase Evaporation) by Triton X-100, Octyl Glucoside and Sodium Cholate, *Biochemistry,* 27, 2668, 1988.
19. **Rigaud, J.L., Bluzat, A., and Buschlen, S.,** Incorporation of bacteriorhodopsin into large unilamellar liposomes by reverse phase evaporation, *Biochem. Biophys. Res. Commun.*, 111, 373, 1983.
20. **Szoka, F., Olson, F., Heath, T., Vail, W., Mayhew E., and Papahadjopoulos, D.,** Preparation of unilamellar liposomes of intermediate size by a combination of reverse phase evaporation and extrusion through polycarbonate membranes, *Biochim. Biophys. Acta,* 601, 559, 1980.
21. **de la Maza, A., Parra, J.L., Manich, A., and Coderch, L.,** Liposomes in wool dyeing — the stability of dye-liposome systems and their application to untreated wool fibres, *J. Soc. Dyers Colour.*, 108, 540, 1992.
22. **Bangham, A.D., Standish, M.M., and Watking, J.C.,** Diffusion of univalent ions across lamellae of swollen phospholipids, *J. Mol. Biol.*, 13, 238, 1965.
23. **Ackman, R.G.,** Flame ionization detection applied to thin-layer chromatography on coated quartz rods, in *Methods in Enzymology,* Vol. 72, Lowenstein, J.L., Ed., Academic Press, New York, 1981, 205.
24. **Ackman, R.G., McLead, C.A., and Banerjee, A.K.,** *J. Planar Chromatog.*, 3, 450, 1990.
25. **Coderch, L., Soriano, C., Pinazo, A., and Parra, J.L.,** Degradative wool shrinkproofing processes. Part 2. lipid modification, *Textile Res. J.*, 62, 704, 1992.
26. **Weinstein, J.N., Ralston, E., Leserman, L.D., Klausner, R.D., Dragsten, P., Henkart, P., and Blumenthal, R.,** Self-quenching of carboxifluorescein fluorescence: uses in studying liposomes stability and liposome-cell interaction, Vol. 3, in *Liposome Technology,* Gregoriadis, G., Ed., CRC Press, Boca Raton, FL, 1986, chap 13.
27. **Helenius, A. and Simons, K.,** Solubilization of membranes by detergents, *Biochim. Biophys. Acta,* 415, 29, 1975.
28. **Lichtenberg, D., Robson J., and Dennis, E.A.,** Solubilization of phospholipids by detergents, *Biochim. Biophys. Acta,* 737, 285, 1983.
29. **Christie, W.W.,** *Gas Chromatography and Lipids*, Ayr Press, Ayr, Scotland, 1989.
30. **De Boer, T. and Backer, H. J.,** Diazomethane, *Org. Synth. Coll.*, 4, 250, 1963.
31. **Trotman, E.R.,** *Dyeing and Chemical Technology of Textile Fibres,* Charles Griffin and Company, Bocks, England, 1984, chap. 12.
32. *Coefficients of Friction of Surfaces,* Part 48, ASTM D 1894–78, ASTM, Philadelphia, 1980.
33. **Frøkjaer, S., Hjorth, E.L., and Wørts, O.,** Stability testing of liposomes during storage, in *Liposome Technology,* vol. 1, Gregoriadis, G., Ed., CRC Press, Boca Raton, FL, 1986, chap. 17.
34. **Scherphof, G.L., Damen, J., and Wilschut, J.,** Interaction of liposomes with plasma proteins, in *Liposome Technology,* vol. 3, Gregoriadis, G., Ed., CRC Press, Boca Raton, FL, 1986, chap. 14.
35. **Deamer, D.W. and Uster, P.S.,** Liposome preparation: methods and mechanisms, in *Liposomes*, Ostro, M.J., Ed., Marcel Dekker, New York, 1983, chap. 1.
36. **de la Maza, A., Parra, J.L., Bosch, P., and Coderch, L.,** Large unilamellar vesicle liposomes for wool dyeing: stability of dye-liposome systems and their application on untreated wool, *Textile Res. J.*, 62, 406, 1992.
37. **de la Maza, A., Parra, J.L., and Manich, A.,** Lipid bilayers including cholesterol as vehicles for acid dyes in wool dyeing, *Textile Res. J.*, 63, 643, 1993.

Chapter 13

Liposomes as a Model System to Study Shark Repellency

Eliahu Kalmanzon, Eliahu Zlotkin, and Yechezkel Barenholz

CONTENTS

I. INTRODUCTION

Certain fish secrete amphipathic substances into their surroundings that repel their predators.[1,2] These substances can be classified as (a) low molecular weight compounds resembling synthetic detergents such as pahutoxin, secreted by boxfishes,[3,4] or pavoninin, secreted by flat fishes;[1,2] or (b) polypeptide substances such as grammistins and pardaxins, derived from the skin secretion of soap fishes and flat fishes, respectively.[5,6]

The common denominator of all these compounds is their ability to disrupt and change the barrier properties of cell membranes of the alien organism. Experiments have shown that the skin secretion of the flat fish, *Pardachirus marmoratus*, can repel sharks.[7,8] Based on the assumption that the biological effects of the skin secretion and its derived toxin pardaxin are a result of their amphipathic–surfactant properties, it has been suggested that commercial surfactants may also possess shark repellent abilities.[9] This hypothesis did not withstand experimental examination. Out of 15 different synthetic surfactants only the sodium and lithium salts of dodecyl sulfate (SDS, LDS) had a clear repellent effect.[10] Two effects of SDS on the behavior of sharks have been observed: (1) an ability to induce a fast turning and escape response in aggressive lemon sharks (*Negaprion brevirostris*) and blue sharks (*Prionace glauca*) while in the process of attacking ("feeding assay")[10,11] and (2) an ability to rapidly terminate, in an excitatory manner, the tonic immobility of sharks which had been induced by restraining them in an inverted position.[12] In both assays, the SDS solutions were introduced into the buccal cavity of the shark.

The aim of this work[13,14] was to determine whether liposomes could be used as a model for: (1) the screening of shark repellents, (2) understanding the mechanism of shark repellency by detergents, and particularly for the study of the possible relationship between the shark-repelling capacity of SDS and its physicochemical mode of interaction with lipid bilayers in the natural shark habitat, sea water. For comparative purposes, another well-characterized detergent, Triton X-100, has been included. Triton X-100 was previously shown to possess 20–30 times lower repellent activity toward sharks.[10]

0-8493-4013-6/96/$0.00+$.50

II. SHARK REPELLENCY OF DETERGENTS

Shark repellency was tested using nonionic and ionic detergents. The latter include anionic and cationic detergents. Their structures are represented in Figure 1.

Table 1 demonstrates that of all the detergents tested, SDS is by far the superior shark repeller.

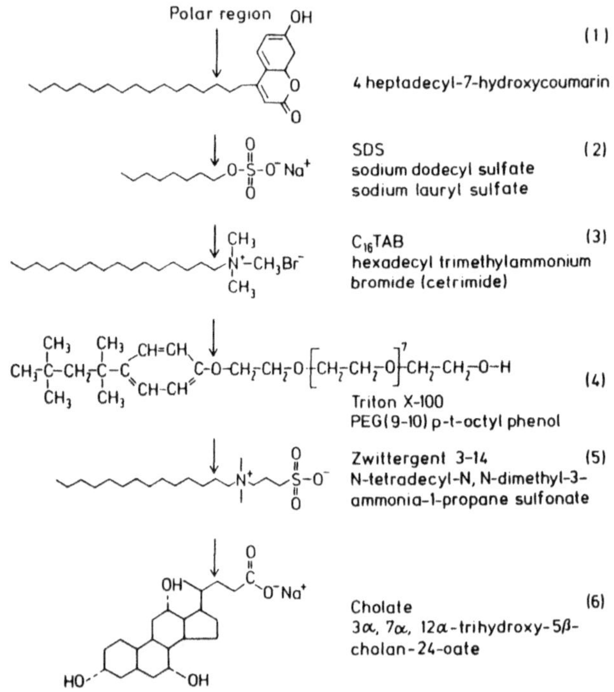

Figure 1 Structure and nomenclature of detergents used. (From Kalmanzon, E., et al., *Tenside Surf. Det.*, 26, 338, 1989. With permission.)

Table 1 Shark Repellency and Fish Lethality of Different Surfactants and *Pardachirus* Secretion[a]

Substance	Killifish Lethality LD_{50} (μg ml^{-1})	Shark Feeding Assay Range (mg ml^{-1})	Shark Tonic Immobility Assay ED_{50} (mg ml^{-1})
Sodium dodecyl (SDS)	3.0	0.2–2.0	0.45
Lithium dodecyl (LDS)	6.0	0.2–2.0	0.62
Lyophilized *Pardachirus* secretion[b]	16.0	0.8–3.0	0.66
Polyethoxylated octylphenol (Triton X-100)	36.0	6.0–8.0	10.0
Dodecyl trimethyl ammonium bromide (DTAB)	60.0	3.0–8.0	8.0
Sodium cholate	100.0	8.0–10.0	8.1
Ethoxylated (20) sorbitan monolaurate (Tween-20)	100.0	10.0–20.0	10.0

[a] The substances are listed in order of their fish lethality.

[b] Dissolved immediately prior to the experiment.

From Zlotkin, E. and Gruber, S. H., *Arch. Toxicol.*, 56, 55, 1984. With permission.

A. DETERMINATION OF MICELLE ELECTRICAL SURFACE POTENTIAL

The determination of electrical surface potential of the detergent micelles is based on the use of 4-heptadecyl-7-hydroxycoumarin (HC) (Molecular Probes, Junction City, OR). This molecule serves as a surface pH indicator at the interface between the hydrophobic and polar phases, such as at the surface of a detergent micelle, lipid vesicles, and biological membranes.[15–17] HC, due to its hydrophobic tail, is incorporated into the hydrophobic region of the lipid assembly with its fluorophore head present at the lipid–water interface. The location of the head and tail in relation to the detergent molecule is shown in Figure 1. A molar ratio of 1:250 (HC: detergent) was used. The micelles were suspended in solutions of either sea water or distilled water at various pH values. Under these conditions, and based on the critical micelle concentration (CMC) and the aggregation number of the micelles of all detergents used here, it is expected to get less than one molecule of HC available per micelle.[15] Such suspensions contain a mixture of the ionized and un-ionized forms of HC in proportions which are determined by the surface pH. These two species can be distinguished by their different excitation (un-ionized 325 nm, ionized 375 nm) and emission (un-ionized 415 nm, ionized 450 nm) spectra peaks. Titration of the fluorescent probe incorporated into the micelles was carried out in the pH range 3–12, using HCL and NaOH. To determine the degree of dissociation (*a*) as a function of the pH (see Figure 2 for SDS), the area integrated under each peak (representing a distinct pH value) was used. The value of the pK_a of hydroxycoumarin in water (pK_w) was taken as 7.75.[15] The apparent pK values ($pK_{a'}$) were obtained from the curve describing *a* as a function of the pH (Figure 2). The emission at pH 12 served as a 100% HC ionization value. The pK_a is equivalent to pH where the degree of HC ionization (α) is 0.5.[15,16] The surface potential ($\Delta\psi$) was calculated from the shift in $pK_{a'}$ introduced by changes in surface charge (pK_{ch}) relative to pK_0 for the uncharged surface of Triton X-100;[15] giving rise to the following equation:

$$pK_{ch} = pK_0 - (\Delta\psi F/2.3RT)$$

where R, F, and T are the gas constant, Faraday constant, and absolute temperature, respectively. This equation can be rewritten in the form:

$$\Delta\psi = -(pK_{ch} - pK_0)2.3RT/F$$

where $\Delta\psi$ is the outer potential at the charged surface of the micelle.[15] It is worth noting that the value for pK_0, 8.65,[17] obtained by using small unilamellar vesicles (SUV) of egg phosphatidylcholine which are uncharged in the above pH range, is very similar to the pK_a, 8.85, of the nonionic detergent Triton X-100 micelles obtained by us (Table 2) and by Fernandez and Fromherz.[15]

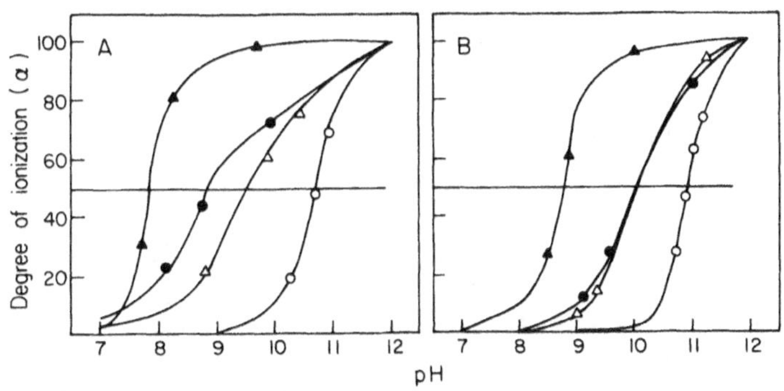

Figure 2 The emission spectrum of HC incorporated into SDS micelles in sea water at various pH values, when excited at 340 nm. The degree of ionization, α, for each pH value was obtained as a ratio between the area under the curve corresponding to a distinct pH and the area under the curve for pH 12 (which represents 100% of HC ionization). (From Kalmanzon, E., et al., *Tenside Surf. Det.*, 26, 338, 1989. With permission.)

Figure 2 shows the emission spectrum of HC incorporated into SDS micelles over a pH range of 8 to 12. Similar experiments were performed with other detergents (Table 2). The emission spectra were recorded by fixing the excitation wavelength at 340 nm. The plots of *a* as a function of pH are shown in Figure 3 for four detergents in sea water (A) and distilled water (B). The $pK_{a'}$ values were taken from these curves. The contribution of fixed charges to the outer electrical potential at the micelle surface after correction for changes in the dielectric component at the surface was obtained using the nonionic detergent Triton X-100.[15] Surface potentials ($\Delta\psi$) for the various micelles calculated from $pK_{a'}$ values are shown in Table 2.

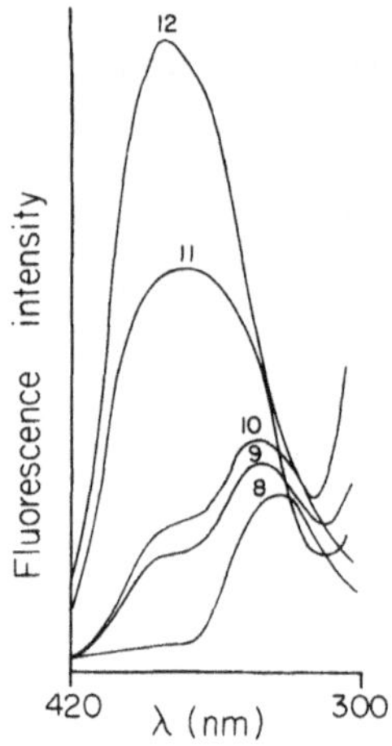

Figure 3 The change in the degree of ionization as a function of pH of the detergents SDS (–o–), Triton-X-100 (–•–), sodium cholate (–Δ–), and Zwittergent 3—14 (–▲–) in distilled water (A) and sea water (B). The value of a = 0.5, indicated by the horizontal line, was used in order to obtain the apparent pK_a values, which were further employed for the estimation of the surface potential. (From Kalmanzon, E., et al., *Tenside Surf. Det.*, 26, 338, 1989. With permission.)

Table 2 Apparent pK Values[a] ($pK_{a'}$) and Surface Potential Values[b] of the Micelles of the Various Detergents

Detergent	Distilled Water $pK_{a'}$	Distilled Water $\Delta\psi$ (mv)	Sea Water $pK_{a'}$	Sea Water $\Delta\psi$ (mv)
SDS	10.75	−111.1	10.9	−46.8
Triton X-100	8.85	0	10.1	0
Sodium cholate	9.0	−8.8	10.1	0
Zwittergent, 3-14	7.85	+58.45	8.8	+76.0

[a] $pK_{a'}$ were estimated from the ionization curves (see Figure 2).
[b] For definition of $\Delta\psi$ see text.
[c] pK_w = 7.75.[15]

The fluorophore of the fluorescent indicator (the hydroxycoumarin group) is located at the interface between the hydrophobic and hydrophilic portion of the micelle[15] (see Figure 1). The exact location of the hydroxycoumarin moiety will determine the region from which the surface potential is sensed. In the case of Triton X-100 and SDS, this region is well defined because the micelles are large, the aggregation numbers for both are above 50, and the interfaces between hydrophobic and hydrophilic regions are easy to identify. For sodium cholate this is not the case, due to the small aggregation number of the micelles and their shape.[18] Therefore the fluorophore may be further from the carboxyl groups than it is from the SDS sulfate group. This, together with the fact that the carboxyl group is a weaker acidic group than the sulfate group over broad ranges of pH, makes the SDS much more negatively charged than the sodium cholate (Table 2). For zwitterionic detergents such as Zwittergent 3-14, it seems that the hydroxycoumarin group is located near the quaternary amino group of the detergent molecule and therefore it senses mainly the positive charge (Figure 1). The sulfate group extends three carbons away from the micelle surface towards the aqueous phase and is therefore not sensed by the hydroxycoumarin (Figure 3). Sea water, due to its high ionic strength (see Table 3), as expected, reduces the negative charge of the SDS micelles, although the surface potential remains significantly negatively charged.

B. DETERMINATION OF CRITICAL MICELLE CONCENTRATION

The CMC was determined as previously reported,[19] using the fluorophore, 1,6-diphenylhexa-1,3,5-triene (DPH), which emits fluorescence only when present in a hydrophobic medium[20] such as the core of a detergent micelle.[19] The measurable fluorescence emission at 430 nm, upon excitation at 360 nm, is

Table 3 The Main Ionic Composition of the Red Sea Water

Ion	%	m*M*
Cl^-	22.39	632.4
Na^+	12.29	536.9
K^+	0.460	11.79
Mg^{2+}	1.487	61.19
Ca^{2+}	0.526	13.15
SO_4^{2-}	3.156	32.87
HCO_3^-	0.157	2.57

plotted as a function of detergent concentration. The lowest concentration at which the fluorescence intensity starts to increase indicates the formation of micelles and is therefore defined as the CMC.

The determination of CMC of SDS, DSS, ODS, sodium cholate, Triton X-100, Tween-20, DTAB, and Zwittergent 3-14 in various conditions of salinity is shown in Table 4. The following aspects deserve attention:

1. The fact that the CMC values for all detergents are slightly lower than those reported[21] is related to the method of CMC determination, the use of fluorophore which becomes part of the micelle.
2. The occurrence of two CMC values (two breaking points) such as in SDS in distilled water (1100 and 2000 μ*M*) may suggest a stage of pre-micellization.
3. The nonionic detergents such as Tween-20 and Triton X-100 show the lowest CMC values. The relatively small but distinct reduction in CMC of Triton X-100 in transition from distilled water to sea water is not entirely understood; however, it is in accord with the salt-induced shift in the HC pK_a and may reflect changes in the dielectric component at the micelle surface[15] (Table 2).
4. The detergents ODS (C18), SDS (C12), and DSS (C10) belong to the same family of compounds, but differ in the length of their hydrocarbon chain, which results in very different CMC values. ODS is insoluble due to its large hydrophobic mass which cannot be shielded from contact with water. DSS, shorter by two carbon atoms than SDS, shows a 60 times higher CMC value (Table 4) due to its reduced hydrophobicity. For straight chain aliphatic hydrocarbons the free energy (for micellization) was shown to be a strictly linear function of the number of C atoms in the chain.[22]
5. SDS as a charged detergent has demonstrated a drastic reduction (15–30-fold) in its CMC value in sea water when compared with distilled water. This reduction is due to the ionic strength of sea water, which neutralizes the electrostatic antagonism among the negatively charged polar heads of SDS molecules, as can be demonstrated from the effect of sea water on the outer electrical potential at the charged micelle surface (Table 2). Furthermore, the above CMC reduction is not affected by the presence of the bivalent ions (Mg^{2+}, Ca^{2+}, SO_4^{2-}) (Table 3) since NaCl in a concentration equal to that in sea water (4% NaCl) yielded a similar CMC value to sea water (Table 4). Indeed, Na^+ and Cl^- are the main ions in sea water and together contribute more than 90% of the total ions (Table 3).

III. THE EFFECTS OF SDS AND TRITON X-100 ON LIPOSOMES

SUV were employed throughout the study.[14] First, multilamellar large vesicles (MLV) of phospholipids and cholesterol were prepared by thin lipid hydration[23,24] in either sea water, sea water containing 0.1 *M* sodium 5,6-carboxyfluorescein (CF) at pH 8.0, or 0.01 *M* Hepes buffer (pH 8.0) containing 0.1 *M* of either the K or Na salt of CF. The MLV were then exposed to ultrasonic irradiation[23] using a Heat System Ultrasonics Inc. (Plain View, NY) 350-W Sonicator and 3/4-inch probe coated with sapphire (in order to reduce contamination with metal particles). The SUV were fractionated by differential centrifugation.[21,22] In all measurements of CF release or leakage, the pH of the external medium and the intraliposomal pH were identical (pH 8.0).

Three kinds of liposomes were used: (a) neutral liposomes composed of phosphatidylcholine (70 mol%), cholesterol (30 mol%), (PC-Chol liposomes); (b) negatively charged liposomes composed of phosphatidylcholine (63 mol%), cholesterol (30 mol%), and dicetyl phosphate (7 mol%) (PC-Chol-DCP liposomes); and (c) bovine brain sphingomyelin (b.b.SPM) liposomes. The b.b.SPM liposomes are highly resistant to fusion and to spontaneous leakage. Therefore, they were more convenient for the studies of CF release by potassium diffusion potential.[25]

Table 4 CMC Determination of Detergents Under Various Salinity Conditions

Detergent	CMC in D.W.[a] (μM)	CMC in S.W.[b] (μM)
SDS	1,100, 2,000	70
SDS 4% NaCl	—	80
SDS 50% S.W.[b]	—	96
SDS 10% S.W.[b]	—	270
DSS	35,000	4,000
ODS	Insoluble	Insoluble
DTAB	7,600	1,975, 2,260
Sodium cholate	5,000	3,000, 4,250
Tween-20	1–3	1–3
Triton X-100	120	60
Zwittergent 3-14	275	120,15

[a] Distilled water.
[b] Sea water.

From Kalmanzon et al., *Tenside Surf. Det.* 26, 338, 1989. With permission.

A. DETERMINATION OF SDS LIPOSOME/SEA WATER PARTITION COEFFICIENT (K_P)

When a detergent is added to a liposome suspension it divides into three phases: (a) monomers in the aqueous solution, (b) micelles (pure detergent micelles or detergent-lipid mixed micelles), and (c) a bilayer component. The determination of the monomer concentration is achieved through the determination of the CMC. The partitioning of the detergent into liposomes is measured using an equilibrium dialysis assay.

The assay was based on the employment of a radioactive tracer ([^{35}S]SDS) in an equilibrium dialysis system using a modification of the methods of Sikaris et al.[26] Both PC-Chol and PC-Chol-DCP liposomes were used.

Mixtures of radioactive and nonlabeled SDS in a final concentration range of 1–150 μ*M* were introduced into sea water (final volume 4.5 ml) in which a dialysis bag filled with 0.5 ml of a 1 m*M* suspension of liposomes in sea water was immersed. At various time intervals, aliquots of SDS solution were removed and their radioactivity counted. The time-dependent reduction in radioactivity of the medium due to diffusion into the dialysis bag (until equilibrium between the external medium and the dialysis bag solution was achieved at 16 h) was monitored. The concentration of SDS at equilibrium inside ($[SDS]_{in}$) and outside ($[SDS]_{out}$) the dialysis bag was determined. The excess SDS inside the dialysis bag was determined as:

$$[SDS]_L = [SDS]_{in} - [SDS]_{out}$$

where $[SDS]_L$ is the SDS concentration in liposomes. $[SDS]_L$ and $[SDS]_{out}$ values were used to calculate K_p, the liposome/sea water SDS partition coefficient.[26] Since Triton X-100 is not well dialyzable, its K_p could not be determined in the above experimental system. Figure 4 presents the time course of the detergent diffusion into the dialysis bag. The plateau value of [^{35}S] radioactivity inside and outside the dialysis bag was used to calculate $[SDS]_{in}$ and $[SDS]_{out}$, from which the K_p was calculated. The average K_p values were obtained from experiments employing SDS concentrations in the range of 10–150 μ*M* in sea water, where $K_p = 3084 \pm 435$ (mean ± S.D.; n = 4) for the negatively charged PC-Chol-DCP liposomes and $K_p = 3469 \pm 578$ (mean ± S.D.; n = 4) for the neutral PC-Chol liposomes. These data indicate that: (1) the high K_p value represents a strong affinity of SDS to lipid bilayers; (2) the K_p value is independent of the CMC value since detergent concentrations above and below CMC yielded the same partition coefficient; and (3) in sea water the detergent–lipid affinity is only slightly reduced by the presence of negatively charged phospholipid in the bilayer, possibly due to the large reduction in surface potential at such high ionic strength.

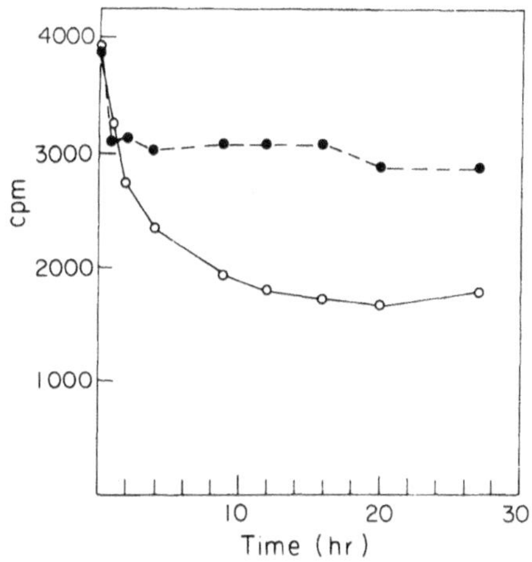

Figure 4 A typical assay of equilibrium dialysis demonstrating the binding of SDS (80 μ*M*) to the liposomal phospholipid. o———o, radioactivity in the external medium with liposomes inside the dialysis bag. •– – – –•, radioactivity in the control assay (no liposomes present in the dialysis bag). (From Kalmanzon, E., et al., *Biochim. Biophys. Acta,* 1103, 148, 1992. With permission.)

B. DETERMINATION OF LIPOSOME FUSION AND/OR SIZE CHANGE

The method for following fusion and size increase is based on resonance energy transfer (RET) between pairs of membranal fluorescent markers. N-NBD-PE served as the donor (excitation 470 nm, emission 545 nm) and N-RH-PE as acceptor (excitation 540 nm, emission 585 nm), with both present in the same membrane. In both molecules the fluorophore is covalently attached to the free amino group of the phosphatidylethanolamine and is therefore a part of the phospholipid head group.[27,28] The advantage of these molecules is that they do not diffuse between membranes.[27] The principle of RET is that its efficiency is dependent on the average distance between the acceptor and the donor fluorophores. The dilution of the two fluorophores by unlabeled molecules increases the average distance between donor and acceptor, thereby causing reduction of RET. Upon excitation of the N-NBD-PE (470 nm),[27,29] RET is expressed as the increase in fluorescence intensity of N-NBD-PE (545 nm) concomitant with the reduction in fluorescence intensity of N-NBD-PE (585 nm).

Generally, there are four causes for dilution-related RET reduction in the system in which fluorescent-labeled liposomes, unlabeled liposomes, and detergent molecules coexist:

1. Fusion of labeled with unlabeled liposomes.
2. Solubilization of labeled liposomes by detergent to form mixed micelles.
3. Intervesicular transfer of the fluorophore molecules to unlabeled liposomes or to detergent micelles.
4. Partition of detergent molecules into the labeled vesicle bilayer without bilayer solubilization.

All four cause surface dilution and therefore reduction in RET. Dequenching due to reduction in RET does not occur upon vesicle aggregation, since aggregation does not affect surface concentration of donors and acceptors.

The data presented in Figure 5 show that SDS caused similar fluorescence dequenching of N-NBD-PE fluorescence and a parallel decrease in the acceptor N-RH-PE fluorescence intensity in the presence (a) as well as absence (b) of unlabeled liposomes. Therefore, the above changes in fluorescence cannot result from vesicle fusion or dilution by other lipid molecules due to intervesicular transfer. Changes in fluorescence should be attributed to an increase in liposome size due to the incorporation of detergent molecules into the lipid bilayer. (See Figure 4.)

C. EFFECTS OF SDS AND TRITON X-100 ON THE TURBIDITY OF LIPOSOME DISPERSION

Increase in turbidity of liposome suspensions occurs upon liposome aggregation and/or size increase. The opposite occurs upon liposome solubilization, which is followed by major turbidity reduction.

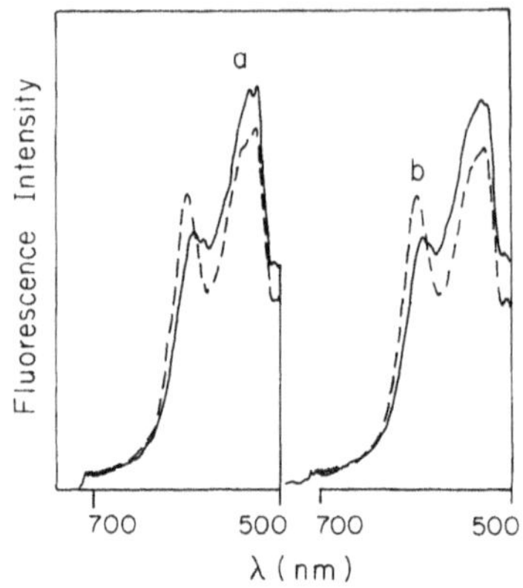

Figure 5 Measurement of change in resonance energy transfer (RET). Indication of SDS-dependent size increase of liposomes labeled by the fluorescent markers N-NBD-PE (donors) and N-RH-PE (acceptors) in sea water. The measurement of reduction in RET is performed in two parallel systems: (a) a system to follow fusion, which includes labeled liposomes, nonlabeled liposomes, and the detergent; and (b) a control system to follow solubilization, which includes labeled liposomes and detergent only. Both fluorescent markers (N-NBD-PE and N-RH-PE) were present in the same PC-Chol SUV at 1 mol% each. SDS concentration was 100 μ*M*. ——, the spectrum obtained before SDS addition; - - - -, the spectrum obtained after SDS addition. (From Kalmanzon, E., et al., *Biochim. Biophys. Acta,* 1103, 148, 1992. With permission.)

Therefore, change in the specific turbidity was used as a sensitive parameter to study detergent-dependent size changes in liposomes. The change in turbidity was measured as follows: a 3-ml suspension of 1 m*M* PC-Chol liposomes in sea water was placed in a spectrophotometer. The absorbance at 400 nm was recorded as a baseline, then the desired amount of either SDS or Triton X-100 in minimal volume was added. The dispersions were mixed, and their turbidity was recorded with time. Measurements were carried out in detergent concentrations below and above CMC.

Turbidity measurements showed that SDS induced a transient decrease in turbidity within 1 min, followed by a gradual increase in turbidity during the subsequent 30 min of observation (data not shown). This may indicate an increase of liposome size. On the other hand, Triton X-100 at the concentration tested caused an obvious decrease in the turbidity (data not shown), probably due to a solubilizing lytic effect leading to mixed micelle formation. These results suggest that the increase in vesicle size which occurs upon SDS addition is not related to fusion or intervesicular lipid transfer but rather to the partitioning of SDS into the vesicles.

D. QUANTIFICATION OF DETERGENT-INDUCED RELEASE OF LIPOSOMAL CONTENTS

SUV were prepared in the presence of 100 m*M* potassium or sodium salt of 5,6-carboxyfluorescein (CF) in 0.01 m*M* Hepes buffer, pH 8.0, and were kept in the dark at 4°C. Prior to the experiment, the unencapsulated CF was removed by gel exclusion chromatography using minicolumns of Sephadex G-50 medium (Pharmacia, Sweden) equilibrated with an isoosmotic solution of 0.01 m*M* Hepes buffer, pH 8.0, containing sodium chloride.[27,28] At 100 m*M* CF, the fluorescence intensity is almost completely quenched.[29] Monitoring of CF release was done, using a Perkin Elmer LS-5 spectrofluorometer, by adding SUV made of 25 nmol phospholipids loaded with CF to 2 ml final volume of thermally equilibrated 0.01 m*M* isoosmotic NaCl-Hepes buffer, pH 8.0. Fluorescence intensity was recorded continuously, first in order to obtain baseline CF release data, and then in order to follow changes related to either detergent or valinomycin addition at the desired concentration. The effect of the added agent was calculated from the change in slope of the time-dependent fluorescence intensity. Major CF dilution occurs upon its release from the aqueous compartment of the liposomes to the external medium, leading to a fluorescence dequenching to the range in which fluorescence (emission 520 nm; excitation 490 nm) is directly proportional to CF concentration. 100% release was obtained by complete solubilization of the vesicle using high Triton X-100 concentration.[29] Fluorescence intensity was corrected for temperature, as described elsewhere.[14]

The data presented in Figure 6 demonstrate the effect of SDS and Triton X-100 on the release of CF from neutral PC-Chol and negatively charged PC-Chol-DCP liposomes in sea water. Triton X-100 in concentrations close to its CMC (80 μ*M*, Figure 6C, D) induced a fast and complete release of CF in both kinds of liposomes. On the other hand, SDS, even in concentrations of about seven times its CMC (Figure 6A, B), induced only a gradual time-dependent release in both kinds of liposomes. Both detergents were more effective on the negatively charged liposomes. This may be related to charge-reduction in packing density at the head group region which increases the rate and the level of detergent incorporation into the vesicle bilayer.[14] Figure 7 demonstrates the concentration dependence of the rate of SDS-induced CF release, using the time required to obtain 50% release (data were taken from Figure 6B). This indicates a strong concentration-dependence only below the CMC of SDS. The slow and incomplete SDS-induced CF release suggests that the process is nonlytic. It was important to demonstrate that nonlytic release can be driven by a specific change such as membrane potential and that it is not necessarily a reflection of nonspecific structural membrane damage. This was demonstrated by the coupling between the CF release and potassium diffusion potential induced by specific transport of this ion by the ionophore valinomycin. The data presented in Figure 8 demonstrate that leakiness of an anionic form of CF is coupled to the valinomycin-stimulated efflux of potassium ions.[14] The specificity of the above process was demonstrated by the obligatory requirement for the potassium diffusion potential. No release occurred when sodium-containing liposomes were used (Figure 8). This suggests that CF may be released from an intact liposome through being coupled to a carrier-mediated cation release.

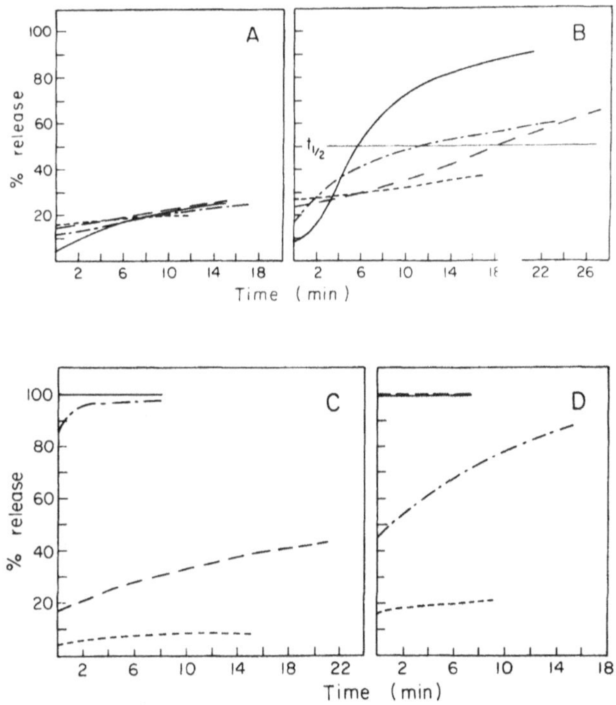

Figure 6 Release of CF from liposomes. The liposomes were prepared in sea water (pH 8.0) containing 0.1 *M* sodium carboxyfluorescein. (A) SDS with neutral liposomes. (B) SDS with negatively charged liposomes. (C) Triton-X-100 with neutral liposomes. (D) Triton-X-100 with negatively charged liposomes. The following concentrations of SDS were used: ----, 10 μ*M*; — — —, 50 μ*M*; •–•–•–•, 90 μ*M*; ——, 500 μ*M*. The following concentrations of Triton-X-100 were used: ----, 10 μ*M*; — — —, 40 μ*M*; •–•–•–•, 80 μ*M*; ——, 400 μ*M*. (From Kalmanzon, E., et al., *Biochim. Biophys. Acta*, 1103, 148, 1992. With permission.)

IV. SDS AND TRITON X-100 EFFECT ON SHARKS AND FISHES

A. DETERMINATION OF THE SHARK-EFFECTIVE CONCENTRATIONS OF SDS

The test system was essentially based on the shark tonic immobility (TI) assay described by Gruber and Zlotkin[12] employing *Mustelus mosis* sharks (Carcharinidae) (70–80 cm long) which were caught in Eilat

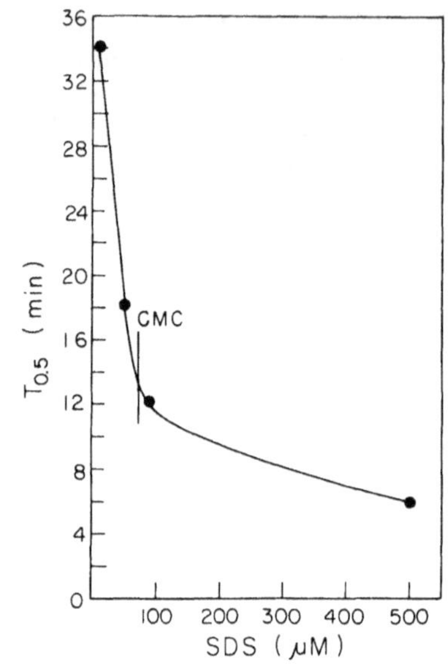

Figure 7 SDS concentration-dependent rate of CF release. The time corresponding to 50% of maximal release of CF ($t_{1/2}$) from the negatively charged liposomes (PC-CHOL-DCP) is presented. Data taken from Figure 6B.

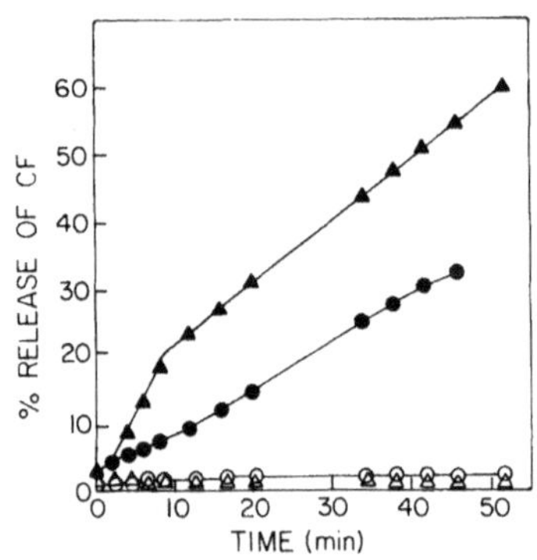

Figure 8 Potassium diffusion potential-dependent CF release. SPM-SUV were used. Filled symbols describe the SPM-SUV loaded with K-CF, pH 8.0 (0.01 *M* K-Hepes buffer) and medium containing NaCl at pH 8.0 (0.01 *M* Na-Hepes) (conditions to create potassium diffusion potential) either at 5°C (●) or at 14°C (▲). Empty symbols describe SPM-SUV loaded with Na-CF, pH 8.0 (0.01 *M* Na-Hepes buffer in both the intraliposomal and the external medium) in medium containing NaCl; conditions of no potassium diffusion potential; at 5°C (○) and 14°C (Δ). The reaction was carried out in the thermally equilibrated cuvette inside the spectrofluorometer. SPM liposomes (25 nmol) loaded with 100 m*M* CF was added. After recording background release rates, 0.5 nmol of valinomycin was added in DMSO (final DMSO concentration was below 0.5%), and increase in fluorescence intensity (520 nm) was followed with time. (From Kalmanzon, E., et al., *Biochim. Biophys. Acta,* 1103, 148, 1992. With permission.)

Bay, the Red Sea, and kept in a large container of circulating sea water. The sharks were restrained in an inverted position for 5–10 s, followed by a quiescent state of immobility of at least 10 min duration (Figures 9, 10). The device for the termination of the threshold effective concentration of SDS is presented and described in Figure 9; for details see legends to Figure 9 and 10.

B. DETERMINATION OF FISH LETHALITY

Gambusia affinis fishes (length 30–40 mm; weight 286 ± 31 mg; mean ± S.D., n = 10) were collected from local ponds and kept in aerated containers of fresh water. The *Gambusia* fishes employed in experiments were acclimated to 10% sea water for 3 days and displayed completely normal behavior.

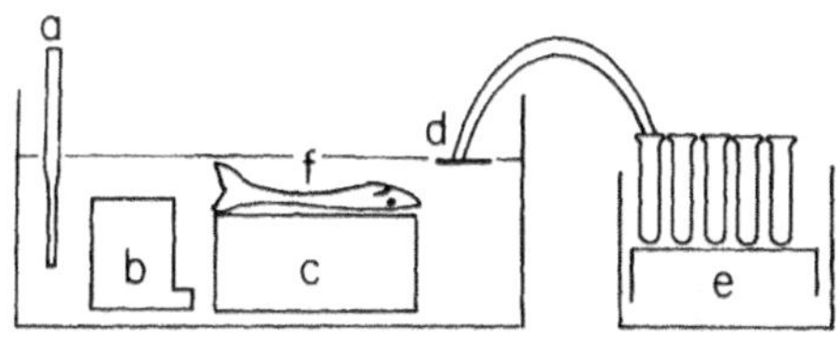

Figure 9 The experimental system for the determination of the threshold effective concentration of SDS to sharks. A rectangular container filled with 70 liters of sea water included a fixed pipette (a) for the addition of the test substance, (b) an immersible pump (output 140 l/min) for the distribution of the test substance, (c) a support unit on which the tonically immobilized shark was placed (Figure 10), (d) a siphon arrangement of tubing which continuously collected water in the vicinity of the shark's head and transferred it at a flow rate of 3–5 ml/s into a fraction collector, (e), which collected samples at 2-s intervals. The sample corresponding to the shark's righting response was processed for the chemical determination of SDS concentration.[36] When selecting the samples, the volume of the siphon tubing was taken into consideration. (From Kalmanzon, E., et al., *Biochim. Biophys. Acta,* 1103, 148, 1992. With permission.)

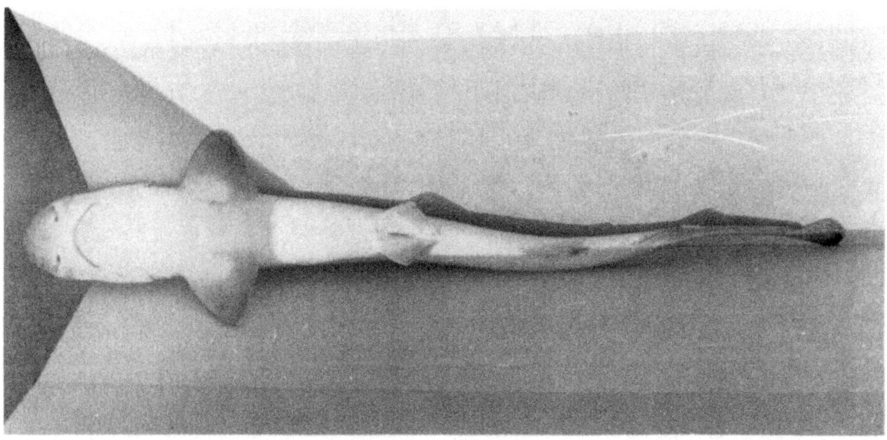

Figure 10 Shark tonic immobility. The shark *Iago ommensis* placed in an inverted position on the support (Figure 9) located in the test container, displaying a typical posture of tonic immobility. (From Kalmanzon, E., et al., *Biochim. Biophys. Acta,* 1103, 148, 1992. With permission.)

Standard LD_{50} determinations were performed. Groups of five fishes were used for each detergent concentration. Each group was placed in a separate glass beaker in a volume of 100 ml of 10% sea water. Death, determined by the arrest of opercular movements, was monitored after 24 h. The kinetics of Triton X-100-induced fish death was determined as described in the legend to Figure 11.

C. SCANNING ELECTRON MICROSCOPY (SEM)

G. affinis fishes were exposed to various concentrations of SDS or Triton X-100 in fresh water for 30 min, followed by dissection of the gills and their processing for SEM examination. This included fixation, post-fixation, dehydration, drying, mounting, and gold coating (Leybold evaporator) according to an established method.[30] The specimens were examined in a Philips 505 scanning electron microscope (accelerating voltage 30 kV).

The vast majority of fish gill surface is composed of a mosaic of respiratory cells accompanied by chloride and mucus cells.[31] The respiratory cells, in addition to their main role in exchange of gases, fulfill osmoregulatory and excretory functions. As shown in Figure 12, both the SDS and Triton X-100 have strongly affected the integrity of respiratory cells as expressed in the disappearance and disarrangement of their surface crenellations.

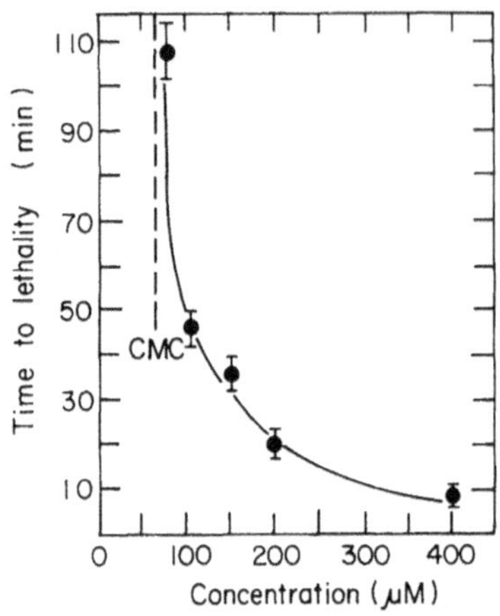

Figure 11 The time course of *Gambusia* lethality as a function of Triton X-100 concentration. Triton X-100 at concentrations ranging from 10 to 400 μ*M* was applied to five *Gambusia* fish, and the time to lethality was recorded. Average time to death is shown on the **y** axis; bars represent variability among the three fishes. No lethality was observed below 60 μ*M* Triton X-100 during 24 h of contact with the detergent. (From Kalmanzon, E., et al., *Biochim. Biophys. Acta,* 1103, 148, 1992. With permission.)

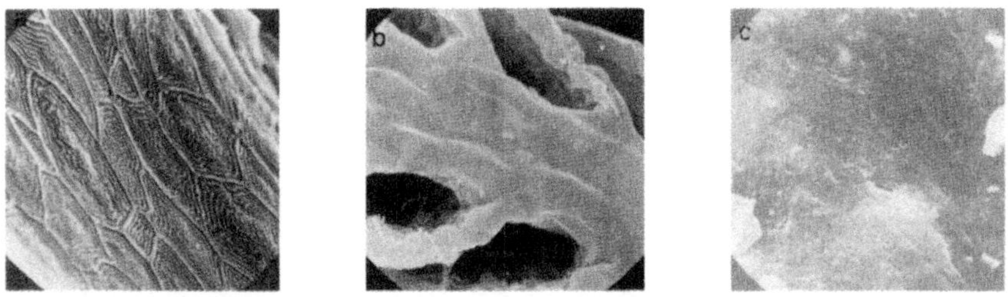

Figure 12 The effect of detergents on the morphology of the fresh water fish gill epithelium. Scanning electron microscopy of the surface of the respiratory cells in gill filaments of *Gambusia affinis*: left, Control untreated — note the regular arrangement of the surface crenelation; center, SDS — note an obvious destruction of the surface membrane of the respiratory cells; right, Triton X-100 — note an obvious destruction of the surface membrane of the respiratory cells. In both detergents, the regular arrangement of the surface crenelations have disappeared (there were no differences in the effects at various detergent concentrations; see also Table 4). Scales correspond to 10 μm per 7 mm on the photographs. (From Kalmanzon, E., et al., *Biochim. Biophys. Acta,* 1103, 148, 1992. With permission.)

V. DISCUSSION

The shark repellency of pardaxin and SDS is, apparently, a result of their effect on the sensitive membranes in sharks. It is suggested that both agents act by means of introducing pores into the lipid bilayers, thereby modifying the optimal membrane potential.[13,32]

Due to the difficulty of routine use of sharks, it is important to find an alternative model system. In this review, we examine the use of liposomes as such a model system. The advantage of liposomes stems from the fact that the exact composition, structure, and size of liposomes are fully controllable and that they can be loaded with a broad spectrum of compounds in order to study the effect of external agents on membrane integrity, such as the ability to induce fusion, size increase, or pore formation. The relevancy of the effect of detergents on liposomes to their potential as shark repellent is discussed.

Among the detergents employed in this study (Table 4) which were previously examined for their shark repellency, SDS was shown to be 20–30, 30–40, and 25–40 times more effective than Triton X-100, sodium cholate, and Tween-20, respectively.[10,11] The question is how far the data presented here can explain this unique capacity of SDS.

SDS was the detergent with the highest decrease in CMC in transition from distilled water to sea water (28.5-fold). This is a result of a decrease in the free energy of micellization ($\Delta F = -RT \ln CMC$) of 3.35-fold[19] and may be the cause for the increase in lipid/water partition coefficient of SDS in sea water. Such a phenomenon was not unexpected since SDS possesses a strong negative charge, and the electrolyte content of sea water neutralizes part of the electrostatic antagonism among the negatively charged polar heads of SDS molecules. It would be difficult to attribute the shark repellent capacity of SDS to the CMC value since Triton X-100 showed identical values and Tween-20 had an even lower CMC value than SDS (Table 4). Such surfactants in sea water should reveal a combination of negative electrical surface potential and high partition coefficient into lipid bilayers as prerequisites to obtain repellency. Detergents form micellar aggregates above their CMC. The CMC varies for different detergents in a given medium condition and varies for the same detergent in different medium conditions.[21] The CMC is affected by medium, pH, ionic strength, and ionic composition mainly for ionic detergents.[21,33] The solubilizing capacity of detergent–membrane interaction is dependent on detergent concentration and on the mole ratio of detergent to membrane lipid.[24,34] In general, the membrane–detergent phase diagram can be divided into three main regions: (1) the lamellar (bilayer) zone at low detergent to membrane-lipid mole ratio; (2) the mixed micelles region, which is a result of membrane solubilization; and (3) the region in which lamellar and micellar phases coexist. In the lamellar zone, the detergent partitions into the bilayer, leading to its expansion and to an increase in membrane permeability by pore formation without grossly affecting membrane integrity.[19,33] The actual meaning of these pores is not yet known and may be related to effects such as increase in membrane "fluidity", reduction in the order of membrane phospholipid acyl chains, increase in the rotational motion of the lipid molecules, and alteration of the lateral distribution of membrane proteins and lipids.[24] These pores are responsible for the increase in membrane leakiness. The incorporation of a detergent into the lipid bilayer below its CMC indicates that the free energy (ΔF) required for the detergent to interact with the lipid bilayer is lower than the free energy of micellization. Our work with the sharks and the model membranes deals with the zone of the phase diagram in which the detergent and membrane lipid coexist in the bilayer.

The similarity between the concentration of SDS required to determine tonic immobility in lemon and in *Mustelus* sharks and its CMC in sea water suggests that its shark repellency may result from general detergent solubilizing capacity. However, this conclusion is strongly opposed by the following considerations and data:

(a) As shown by the data summarized in Table 5, the LD_{50} concentrations of SDS to which the killifish and the present *Gambusia* fish were exposed is definitely far below its CMC value. On the other hand, the lethal effect of Triton X-100 closely corresponds to its CMC values in the above media (Figure 11). This suggests the toxicity induced by the two detergents is related to two different mechanisms: solubilization for Triton X-100 and another mechanism for SDS.

(b) Triton X-100, in contrast to its minimal effect as a shark repellent,[10] caused immediate liposome lysis above its CMC. This effect was demonstrated by (1) fast and almost complete release of CF from liposomes (Figure 6) and (2) a decrease in the turbidity of the liposomal suspension.

(c) SDS, on the other hand, has revealed only a moderate effect on liposome leakiness, demonstrated by a slow, progressive, and time-dependent CF release. This was the case below and above its CMC. The rate of leakage described by the $t_{1/2}$ value was strongly dependent on SDS concentration below the CMC and concentration independent above the CMC (Figure 7). Namely, SDS monomers and not micelles are responsible for this time-dependent leakage. Also, SDS does not induce reduction in turbidity. Therefore, the SDS-induced leakage is different from that obtained by Triton X-100 under our experimental conditions. It is also worth noting that the uptake of SDS by the lipid bilayer of the liposomes occurs above and below SDS CMC and that the Kp value is independent of the SDS CMC. The connection between the data concerning liposomal leakage, liposome solubilization, and fish toxicity (see [a]) is obvious. In both cases, SDS, in contrast to Triton X-100, was effective in concentrations below its CMC value.

(d) One explanation for this difference is that the SDS effect is related to its ability to form pores in lipid bilayers. The possible relation between the SDS-induced CF release and pore formation is supported by the data presented in Figure 8. They indicate that CF release is not necessarily a consequence of membrane destruction but may also result from a specific process such as the coupling to a net cation efflux. This hypothesis is supported by the consideration that SDS in the medium of sea water exhibits two properties, the combination of which may enable it to form cation channels or act as a cation carrier: firstly, its CMC, and secondly, its negative charge on the micellar surface. The leaking of cations may thus induce CF release. It is noteworthy that channel formation by detergents has been previously proposed by Schlieper and Robertis.[35]

(e) The above concept of pore formation is further supported by present data indicating the incorporation of SDS molecules into the vesicle bilayer. This is suggested by (1) the increase in spatial distance between the phospholipid molecules as assessed from the resonance energy transfer release between fluorophores in the bilayer (Figure 5) which is not due to fusion, intravesicular phospholipid transfer, or vesicle aggregation, and (2) the increase of vesicle specific turbidity induced by SDS concentrations below and above CMC, which is a measure of vesicle size growth.

Table 5 The Ichthyotoxicity and CMC of SDS and Triton X-100 Under Various Salinity Conditions

		SDS (μ*M*)		Triton X-100 (μ*M*)	
Test	**Medium**	**CMC**	**Effect**	**CMC**	**Effect**
Lethality of killifish (*Floridicthyes carpio*) (LD_{50})[a]	S.W.	70	10.4	60	56.25
Lethality of *Gambusia affinis* fish (LD_{50})[b]	10% S.W.	270	66.7	65	62.5
Severe histopathological changes in fish gill epithelium[c]	F.W.	1,100	1,400	70	225

Note: S.W., Sea water; F.W., fresh water.

[a] From *Zlotkin* and *Gruber* (1984).[10]

[b] From *Kalmanzon* et al. (1992).[14]

[c] From Figure 12.

From Kalmanzon, et al., *Tenside Surf. Det.*, 26, 338, 1989 and Kalmanzon, et al., *Biochim. Biophys. Acta*, 1103, 148, 1992. With permission.

Thus it may be proposed that the exceptional and unique shark repellent potency of SDS is not simply a consequence of its detergent-solubilizing properties, but rather represents specific interactions with biological membranes at high ionic strength, presumably through a pore-forming process. We suggest that SDS forms negatively charged pores in the lipid bilayer which resemble inverted micelles. These pores can serve as cation "channels" and thus induce disturbances in externally exposed shark sensory-neuronal tissues ("pain production"). This hypothesis may explain the significantly superior effectiveness of SDS as a broad-spectrum shark repellent, as opposed to nonionic detergents such as Triton X-100, positively charged detergents such as dodecyl trimethyl ammonium bromide (DTAB), and negatively charged detergents such as cholic acid salts, which are uncharged in sea water.

This study demonstrates clearly the role of liposomes as a model for the evaluation of the mechanism of the action of membrane-active agents.

ABBREVIATIONS

CF, 5,6-carboxyfluorescein; Chol, cholesterol; CMC, critical micelle concentration; DCP, dicetyl phosphate; DPH, 1,6-diphenylhexa-1,3,5-triene; DSS, decyl sodium sulfate; DTAB, dodecyltrimethyl ammonium bromide; Hepes, *N*-2-hydroxyethylpiperazine-*N*′-2-ethanesulfonic acid; HC, 4-heptadecyl-7-hydroxycoumarin; *N*-*N*-(7-nitrobenz-2-oxa-1,3-diazol-4-yl) phosphatidylethanolamine; N-RH-PE, *N*-(lissamine rhodamine B sulfonyl) phosphatidylethanolamine; ODS, octadecyl sulfate; PC, phosphatidylcholine; RET, resonance energy transfer; SDS, sodium dodecyl sulfate; b.b.SPM, bovine brain sphingomyelin; SUV, small unilamellar vesicles; TI, tonic immobility; Zwittergent 3-14, *N*-tetradecyl-*N,N*-dimethyl-3-ammonia-1-propane sulfonate.

ACKNOWLEDGMENT

This work was supported in part by grant of PHS-NIH, HL-17576.

REFERENCES

1. **Hashimoto, Y.,** *Marine Toxins and Other Bioactive Marine Metabolites,* Japan Scientific Societies Press, Tokyo. 1979, 312.
2. **Tachibana, K., Sakaitanai, M., and Nakanishi, K.,** Pavoninins: shark-repelling ichthyotoxins from the defense secretion of the pacific sole, *Science,* 226, 703, 1984.
3. **Boylan, D. B. and Scheuer, P. J.,** Pahutoxin: a fish toxin, *Science,* 155, 52, 1967.
4. **Fusetani, N. and Hashimoto, K.,** Occurrence of pahutoxin and homopahutoxin in the mucous secretion of the Japanese boxfish *Ostracion immaculatus, Toxicon,* 25, 459, 1987.
5. **Hashimoto, Y. and Oshima, Y.,** Separation of Grammistins A, B and C from a soapfish *Pogonoperca punctata, Toxicon,* 10, 279, 1972.
6. **Thompson, S. A., Tachibana, K., Nakanishi, K., and Kubota, I.,** Melittin-like peptides from the shark-repelling defense secretion of the sole *Pardachirus pavoninus, Science,* 233, 341, 1986.
7. **Clark, E.,** The Red Sea's sharkproof fish, *Natl. Geogr.,* 145, 718, 1974.
8. **Clark, E.,** Shark repellent effect of the Red Sea Moses sole, in *Shark Repellents from the Sea — New Perspectives,* Zahuranec, B. J., Ed., Westview, Boulder, CO, 1983, 135.
9. **Zlotkin, E. and Barenholz, Y.,** On the membranal action of Pardaxin, in *Shark Repellents from the Sea — New Perspectives,* Zahuranec, B. J., Ed., Westview, Boulder, CO, 1983, 157.
10. **Zlotkin, E. and Gruber, S. H.,** Synthetic surfactants: A new approach to the development of the shark repellents, *Arch. Toxicol.,* 56, 55, 1984.
11. **Gruber, S. H., Zlotkin, E., and Nelson, D. R.,** Shark repellent: behavioral bioassays in laboratory and field, in *Toxins, Drugs and Pollutants in Marine Animals,* Bolis, L., Zadunaisky, J., and Gilles, R., Eds., Springer-Verlag, Berlin, 1984, 26.
12. **Gruber, S. H. and Zlotkin, E.,** Bioassay of surfactants as shark repellents, *Naval Res. Rev.,* 34, 18, 1982.
13. **Kalmanzon, E., Zlotkin, E., and Barenholz, Y.,** Detergents in sea water, *Tenside Surf. Det.,* 26, 338, 1989.
14. **Kalmanzon, E., Zlotkin, E., Cohen, R., and Barenholz, Y.,** Liposomes as a model for the study of the mechanism of fish toxicity of sodium dodecyl sulfate in sea water, *Biochim. Biophys. Acta,* 1103, 148, 1992.
15. **Fernandez, M. S. and Fromherz, P. J.,** Lipoid pH indicators as probes of electrical potential and polarity in micelles, *Phys. Chem.,* 81, 1755, 1977.
16. **Pal, R., Petri, A. W., Barenholz, Y., and Wagner, R. R.,** Lipid and protein contributions to the membrane surface potential of vesicular stomatitis virus probed by a fluorescent pH indicator, 4–heptadecyl-7–hydroxycoumarin, *Biochim. Biophys. Acta,* 646, 23, 1983.
17. **Fernandez, M. S.,** Determination of surface potential in liposomes, *Biochim. Biophys. Acta,* 646, 23, 1981.
18. **Small, D. M.,** *The Physical Chemistry of Lipids: Handbook of Lipid Research,* Vol. 4, Plenum, New York, 1986.
19. **Hertz, R. and Barenholz, Y.,** The relations between the composition of liposomes and their interaction with Triton–X–100, *J. Colloid Interface Sci.,* 60, 188, 1977.
20. **Shinitzky, M. and Barenholz, Y.,** Fluidity parameters of lipid regions determined by fluorescence polarization, *Biochim. Biophys. Acta,* 515, 678, 1978.
21. **Mukerjee, P. and Mysels, K. J.,** *Critical Micelle Concentration of Aqueous Surfactant Systems,* NSRDS–NBS36, U.S. Department of Commerce, National Bureau of Standards, U.S. Government Printing Office, Washington, D.C., 1971.
22. **Tanford, C.,** *The Hydrophobic Effect: Formation of Micelles and Biological Membranes,* Wiley, New York, 1980.
23. **Barenholz, Y., Gibbs, D., Litman, B. J., Thompson, T. E., and Carlson, E. D.,** A simple method for the preparation of homogeneous phospholipid vesicles, *Biochemistry,* 16, 2806, 1977.
24. **Lichtenberg, D. and Barenholz, Y.,** in *Methods of Biological Analysis,* Vol. 33., Glick, D., Ed., Wiley, New York, 1988, 337.
25. **Cohen, R.,** Ph.D. thesis, 1984 Hebrew University of Jerusalem, Israel (in Hebrew, English summary.)
26. **Sikaris, K. A., Thulburn, K. R., and Sawyer, W. H.,** Resolution of partition coefficients in the transverse plane of the lipid bilayer, *Chem. Phys. Lipids,* 29, 23, 1981.
27. **Struck, P. K., Hoekstra, D., and Pagano, R. E.,** Use of resonance energy transfer to monitor membrane fusion, *Biochemistry,* 20, 4093, 1981.
28. **Amselem, S., Loyter, A., Lichtenberg, D., and Barenholz, Y.,** The interaction of Sandai virus with negatively charged liposomes: virus-induced lysis of carboxyfluorescein-loaded small unilamellar vesicles, *Biochim. Biophys. Acta,* 820, 1, 1985.
29. **Lichtenberg, D., Freire, E., Schmidt, C. F., Barenholz, Y., Felgner, P. L., and Thompson, T. E.,** Effect of surface curvature on stability, thermodynamic behavior, and osmotic activity of dipalmitoylphosphatidylcholine single lamellar vesicles, *Biochemistry,* 20, 3462, 1981.
30. **Gamliel, H., Gurfel, D., Leizerowits, R., and Polliack, A.,** Air drying of human leucocytes for scanning electron microscopy using the GTGO procedure, *J. Microsc.,* 131, 87, 1983.
31. **Sardet, C., Pisam, M., and Moetz, J.,** The surface epithelium of teleostean fish gills, *J. Cell. Biol.,* 80, 96, 1979.

32. **Shai, Y., Hadari, Y. R., and Finkels, A.,** pH-dependent pore formation properties of pardaxin analogues, *J. Biol. Chem.*, 22346, 266, 1991.
33. **Elworthy, P. H., Florence, A. T., and MacFarlene, C. B.,** *Solubilization by Surface Active Agents*, Chapman and Hall, London, 1968.
34. **Helenius, A. and Simones, K.,** Solubilization of membranes by detergents, *Biochim. Biophys. Acta*, 415, 29, 1975.
35. **Schlieper, P. and Robertis, E.,** Triton X–100 as a channel-forming substance in artificial lipid bilayer membranes, *Arch. Biochem. Biophys.*, 184, 204, 1977.
36. **Stumm, W. and Morgan, J. J.,** *Aquatic Chemistry, An Introduction Emphasizing Chemical Equilibrium in Natural Waters*, Wiley, New York, 1981, 567.
37. **Waters, J. and Taylor, C. G.,** The colorimetric estimation of anionic surfactants in *Anionic Surfactant Chemical Analysis*, Cross, J., Ed., Marcel Dekker, New York, 1977, 8, 193.

Chapter 14

Use of Antigen-Coupled Liposomes for Homogeneous Immunoassays of Polyclonal Antibody

Shigeo Katoh, Masaaki Kishimura, and Hideki Fukuda

CONTENTS

I. INTRODUCTION

Various assay methods, which depend on the specific interaction between antigen and antibody, have been developed and used to detect trace amounts of biomaterials. Among them, radio immunoassay (RIA) and enzyme-linked immunosorbent assay (ELISA) are the most popular but require separation of free and antigen-bound antibodies and many repeated incubation and washing steps.

In recent years, the great increase in the number of samples in clinical assay and bioprocesses requires a rapid and automated immunoassay method. For this purpose homogeneous assay methods are desired, and many kinds of homogeneous methods have been developed. Among them liposome immune lysis assay (LILA), which depends on immune lysis of artificial phospholipid vesicles by complement, is completely homogeneous and simple and quick in operation.[1-3] In this method antigens or antibodies are attached to the outer surface of liposomes encapsulating a suitable marker. Addition of complement, which is an immune defense system in the bloodstream and activated by the antigen–antibody complex or other activators, causes formation of the membrane attack complex and following immune lysis of liposomes. Depending on this principle, assay methods of antibodies, antigens and also the complement activity have been developed. The LILAs have, however, yet to be used in the clinical field and bioprocesses, mainly because the lack of quantitative evaluation of the lysis of immune liposomes in LILA.

In this report, we describe the measurement of the concentration of antibodies and also the measurement of the complement activities by use of immune liposomes. In the former case, the effects of the concentration, association constant and its heterogeneity of polyclonal antibodies, as well as the preparation conditions of immune liposomes on the sensitivity and reliability of the assay were quantitatively studied. In the latter case the complement activities of both the total and alternative pathways were measured by the same system.

II. HOMOGENEOUS IMMUNOASSAY OF POLYCLONAL ANTIBODIES

A. PREPARATION OF ANTIGEN-COUPLED LIPOSOMES

For preparation of liposomes used for antigen coupling, *N*-[4-(*p*-maleimidophenyl) butyryl] dipalmitoylphosphatidylethanolamine (MPB-DPPE) was synthesized by the procedures previously reported.[3]

0-8493-4013-6/96/$0.00+$.50

Small unilamellar vesicles (SUV) were prepared by the sonication method.[4] A lipid mixture containing dipalmitoylphosphatidylcholine (DPPC, 75 µmol), cholesterol (Chol, 75 µmol), dicetylphosphate (DCP, 7.5 µmol) and various amounts of MPB-DPPE (0.75–7.5 µmol) in chloroform was dried to a lipid film in a round-bottom flask. After addition of 15 cm^3 of 0.2 *M* carboxyfluorescein (CF) as a marker, the mixture was shaken vigorously for 10 min at 55°C and then sonicated with a probe-type sonicator for 15 min under a nitrogen atmosphere. The liposomes thus prepared were separated from untrapped CF by gel chromatography with Sephadex G-25. The average diameter of these liposomes determined by quasi-elastic light scattering analysis was about 600 Å.

Antigens (cytochrome c, myoglobin, α-chymotrypsinogen and bovine serum albumin) were modified with *N*-succinimidyl 3-(2-pyridyldithio) propionate, and pyridyldithio groups introduced to the antigens were reduced with dithiothreitol, as previously reported.[5]

The liposome suspension prepared (0.5–5.0 µmol-DPPC/cm^3) was mixed with an equal volume of the modified antigen solution (about 1 mg/cm^3) and reacted 18 h at room temperature under nitrogen. Uncoupled antigen was separated by gel chromatography with Sephacryl S-1000. The antigen-coupled liposomes were stored in a phosphate buffer saline (PBS) containing 0.1% gelatin at 4°C and stable for 6 months.

The amount of antigen coupled to liposomes depended on the molar ratio of MPB-DPPE constituting the liposomes as well as on the molar ratio of modified antigen to MPB-DPPE in the reaction mixture, as shown in Figure 1. It depended also on the molecular weight of the antigen.[6]

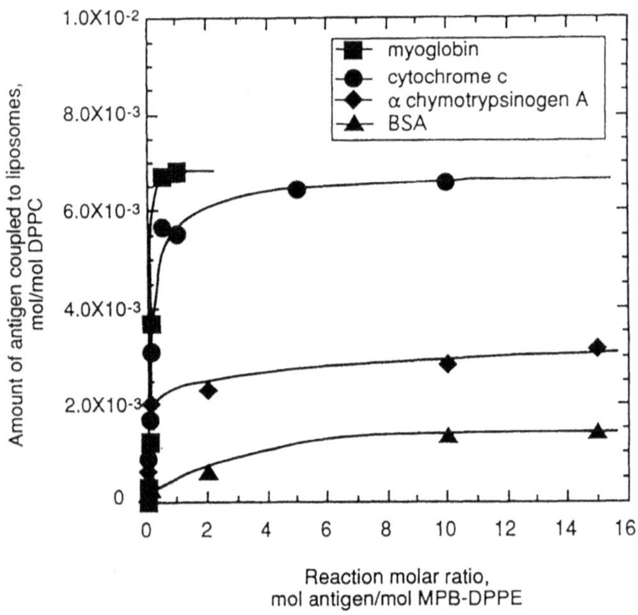

Figure 1 Factors affecting the amount of antigen coupled to liposomes; effects of molar ratio of modified antigen to MPB-DPPE and molecular weight of antigens.

B. PREPARATION OF ANTIBODIES

The antisera against the antigens were raised in rabbits. A mixture of equal volume of an antigen solution and complete Freund's adjuvant (1 cm^3 each) was injected into rabbits. Booster injections were repeated twice. Antisera were pooled, and specific antibodies were purified by affinity chromatography using antigen-coupled Sepharose 4B.

C. MEASUREMENT OF IMMUNE LYSIS OF ANTIGEN-COUPLED LIPOSOMES

The antigen-coupled liposomes were diluted with PBS^{2+} (PBS plus 0.1% gelatin, 0.5 m*M* $MgCl_2$ and 0.15 m*M* $CaCl_2$, pH 7.4). To 0.8 cm^3 of the appropriately diluted antigen-coupled liposomes (0.21 nmol-DPPC/cm^3 unless otherwise noted), a specific amount of the purified antibody was added to obtain a desired final concentration (0.02–20 µg/cm^3) and preincubated at 37°C for 10 min. After addition of guinea pig whole complement (GPWC, 1–3 CH50/cm^3), the mixture, which final volume was 1 cm^3, was incubated at 37°C. The fluorescent intensity of CF released from the liposomes was measured with a spectrofluorophotometer (excitation: 490 nm, emission: 514 nm). The total amount of releasable CF in the liposomes was determined by completely lysing the liposomes with addition of 4 cm^3 of 1-propanol.

The marker release was calculated by the following equation.

$$\text{Marker release (\%)} = \frac{F_1 - F_0}{F_T - F_0} \times 100 \quad (1)$$

where F_0 is the fluorescent intensity of the liposome suspension before addition of GPWC. F_1 and F_T are the intensities of released CF at t min after addition of GPWC and of total releasable CF, respectively.

The complement system is activated by both the classical pathway of complement (CPC) and the alternative pathway of complement (APC). The activation of CPC mainly depends on the formation of the antigen–antibody complex on cell surfaces, and shows much stronger lysis of cell membranes by the membrane attack complex (MAC) than APC. On the other hand, APC is activated without the formation of the antigen–antibody complex. Except the initial stages of the activation, both CPC and APC follow the same pathway to form the membrane attack complex. Therefore, preincubation of antigen-coupled liposomes with the antibody and the addition of GPWC cause the activation of both CPC and APC. The schematic sequence of this immunoliposome assay is shown in Figure 2: (1) Equilibrium binding of antibody in sample with antigen on liposome; (2) cascade activation of complement by antigen–antibody complex; (3) marker release by immune lysis of liposome caused by MAC.

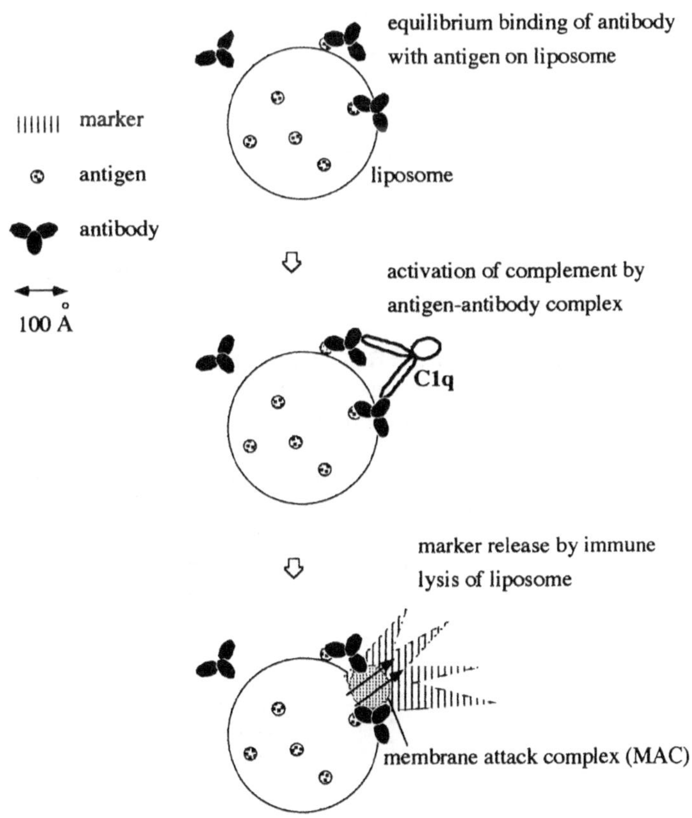

Figure 2 Schematic sequence of immunoliposome assay utilizing complement system.

Figure 3 shows time courses of the marker release from myoglobin (Mb)-coupled liposomes with addition of 2 μg/cm³ of anti-Mb antibody and 1–3 CH50/cm³ of GPWC.[7] In the same figure, the time course of the marker release from rabbit γ-globulin (RGG)-coupled liposomes without addition of antibodies is also shown, in which APC is mainly activated. The marker release in all cases increased remarkably with time up to 30 min and reached plateau at about 60 min. Therefore, an incubation time of 60 min was adopted as a standard condition in this work. In spite of a much higher concentration of complement (11 CH50/cm³), the marker release from RGG-coupled liposomes without addition of the antibody was similar to that from the Mb-liposomes at 3 CH50/cm³, which shows that lysis of liposomes by both CPC and APC similarly leveled off after 60 min by consumption or inactivation of some components of complement, though CPC had much higher ability to lyse liposomes by the formation of the MAC.

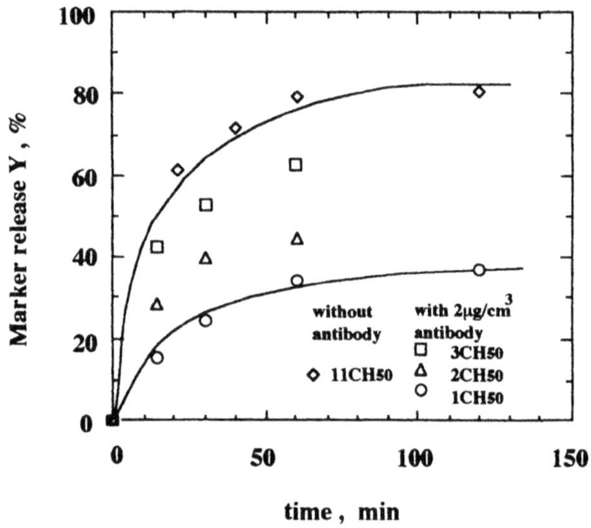

Figure 3 Time course of marker release from Mb-liposomes (2.8 × 10⁻³ mol/mol-DPPC) with addition of anti-Mb and rabbit γ-globulin-coupled liposomes (1.6 × 10⁻³ mol/mol-DPPC) without addition of antibody.

In the range of 1–3 CH50/cm³ the marker release without addition of the antibody was less than 20% after 60 min. Addition of 4% of fresh rabbit serum to the liposomes also caused 20% marker release, probably because of complement attributed to rabbit serum.[7]

D. EFFECTS OF ANTIBODY CONCENTRATION ON MARKER RELEASE

Figure 4 shows the effects of the antibody concentration of anti-cytochrome c (Cy) antibody on the marker release from cytochrome c-coupled liposomes (Cy-liposomes) under incubation with 1, 2 and 3 CH50/cm³ of GPWC for 60 min.[7] The concentrations of cytochrome c coupled to liposomes and of liposomes added were 3.0 × 10⁻³ mol-cytochrome c/mol-DPPC and 0.21 nmol-DPPC/cm³, respectively. The marker release increased in the range from 0.1–10 μg-antibody/cm³, depending on the concentration of GPWC, and was reproducible to about ±15%. Addition of nonspecific IgG (6 μg-antibody/cm³, 2 CH50/cm³) did not increase the marker release, as shown in the figure.

According to Pauling et al.,[8] the effective association constant K_0 for polyclonal antibody can be estimated by assuming that the free energy of antigen–antibody combination can be described by the normal distribution function.

$$y = 1 - \frac{1}{\sqrt{\pi}} \int_{-\infty}^{\infty} \frac{e^{-\alpha^2}}{1 + K_0 C e^{\alpha\sigma}} d\alpha \tag{2}$$

$$\alpha = \ln(K/K_0)/\sigma \tag{3}$$

where C is the liquid phase concentration of free antigen; K is the association constant; σ is the heterogeneity of K; and y is the fractional saturation of antibody. Since the adsorption equilibrium

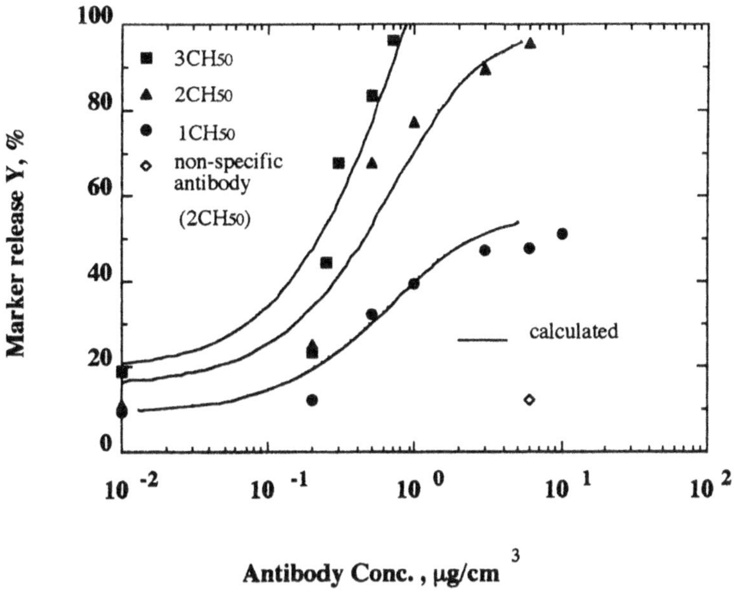

Figure 4 Marker release from Cy-liposomes with addition of anti-Cy antibody (3.0×10^{-3} mol/mol-DPPC; 0.21 nmol-DPPC/cm^3).

between antigen and polyclonal antibody is expressed by Equations (2) and (3), the fraction of antigen bound with antibody can be calculated by use of these equations with the assumption that one molecule of antigen on liposomes can bind with one binding-site of antibody because of relatively low molecular weight, i.e., small size, of the antigen.

As shown in Figure 5, the calculation was repeated until the value of the fractional saturation of the antigen y', which was calculated by use of Equations (2), (3), the values of K_0, σ and the mass balance for the antigen, agreed with the assumed fractional saturation of antigen y'_{ass}.

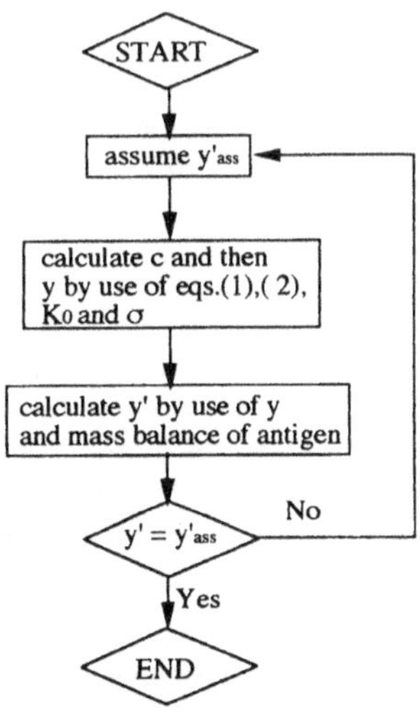

Figure 5 Flow chart for calculation of fractional saturation of antigen y' on liposomes.

In Figure 6, the marker release was plotted against the concentration of the antigen–antibody complex for different concentrations of cytochrome c coupled to liposomes.[7] The marker release increased linearly with the concentration of the antigen–antibody complex on liposomes, and the slopes of three lines shown in Figure 6 were proportional to the GPWC concentrations added. Similar results were obtained for the marker release from myoglobin-coupled liposomes with addition of anti-myoglobin polyclonal antibody.

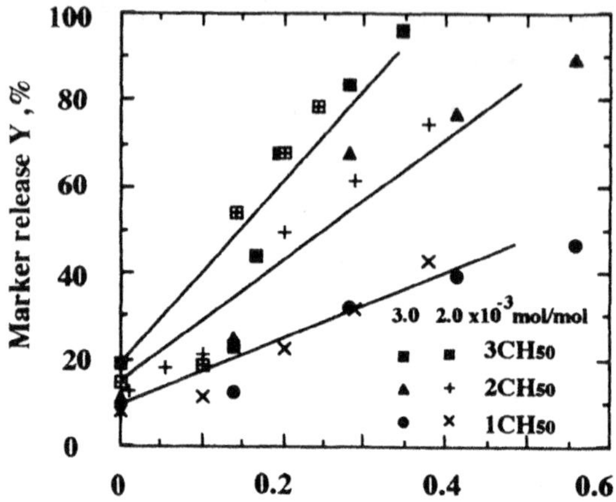

Figure 6 Effect of concentration of coupled antigen on marker release (Cy-liposomes, 2.0 and 3.0 × 10^{-3} mol/mol-DPPC).

These results show that the marker release is proportional to the concentration of the antigen–antibody complex formed on liposomes. Naturally, the complex formation depends on the concentrations of antigen and antibody and the value of K_0. In polyclonal antibodies, it depends also on the value of σ, because a larger value of σ means a broader distribution of the association constant and results in increase in the antibody fraction, which has much larger affinity than K_0 .

By use of the correlation lines for marker release Y and the concentration of the antigen–antibody complex in Figure 6, the relationship between the marker release and the antibody concentration was calculated for the values of K_0 and σ shown in Table 1. In the calculation the value of y′ was first assumed, and then the fractional saturation of antibody y was obtained by Equations (2) and (3). The total concentration of the antibody added was determined by a mass balance equation. As shown by the solid curves in Figure 4, the calculated curves agreed well with the experimental data.

Table 1 Average Association Constant and Heterogeneity of Polyclonal Antibodies

Antibody	Average Association Const. K_0, M^{-1}	Heterogeneity σ
Anti-cytochrome c	3.6 × 10^7	3
Anti-myoglobin	2.6 × 10^6	4

To clarify the effects of σ on the marker release, the relationship between the marker release and the antibody concentration was calculated also for K_0 = 3.6 × 10^7 M^{-1} and σ = 0, 3, 6 (2CH50/cm^3). As shown in Figure 7,[7] the calculated marker release increases with the values of σ. Naturally the marker release increases with K_0. Thus the marker release depended on both the association constant and its heterogeneity, and monoclonal antibodies (σ = 0), for example, need much higher affinity than polyclonal antibodies to attain the same level of sensitivity.

III. MEASUREMENT OF COMPLEMENT ACTIVITIES

The complement system is activated by both the classical pathway of complement (CPC) and the alternative pathway of complement (APC). The activation of CPC mainly depends on the formation of the antigen–antibody complex on cell surfaces, and shows much stronger lysis of cell membranes than APC. On the other hand, APC is activated without the formation of the complex. Except the initial stages of the activation, both CPC and APC follow the same pathway to form the membrane attack complex (MAC). Therefore, preincubation of antigen-coupled liposomes with a specific antibody and the addition

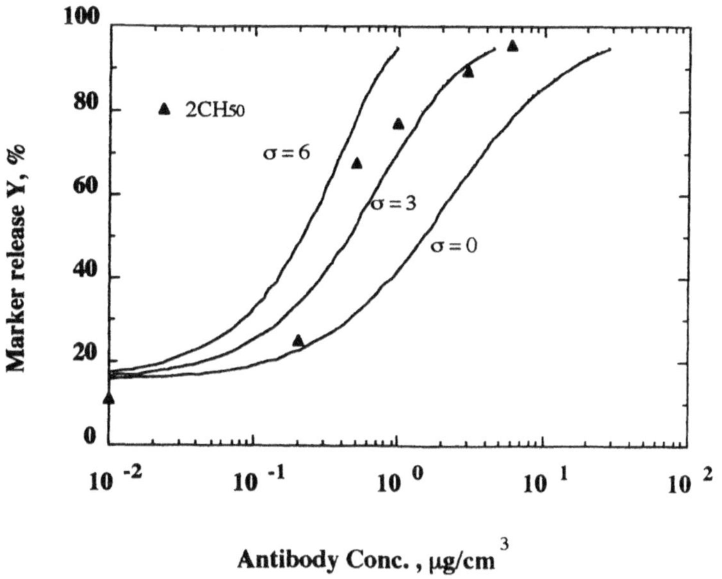

Figure 7 Effect of σ on marker release (K_0: 3.6 × 10^7 M^{-1}, σ, 0, 3, 6)

of GPWC cause the activation of both CPC and APC. From the results obtained above, the rate of liposome lysis was proportional to the activity of complement and also to the concentration of the antigen–antibody complex formed on the surface of liposomes. Since lysis of the liposomes was observed without formation of the antigen–antibody complex, as shown in Figure 3, it may be possible to measure the complement activities of both CPC and APC by use of liposomes coupled with antigen.

A. PREPARATION OF RABBIT γ-GLOBULIN (RGG)-COUPLED LIPOSOMES AND ASSAY OF COMPLEMENT ACTIVITIES

RGG was coupled to MPB-DPPE in the membrane of CF-containing liposomes prepared by the same method described above. The concentrations of coupled RGG were 0.8 – 2.3 × 10^{-3} mol/mol-DPPC.

The RGG-coupled liposomes were diluted with PBS^+ (PBS + 0.1% gelatin + 0.5 m*M* $MgCl_2$ + 0.15 m*M* $CaCl_2$, pH 7.4) to 0.72 – 2.5 nmol DPPC/cm^3. To 1 cm^3 of the diluted liposome suspension, 0.04 cm^3 of a complement sample (guinea pig whole complement, GPWC) was added, and the mixture was incubated at 37°C with or without addition of anti-RGG antibody. The marker release was measured with a spectrofluorophotometer and calculated by Equation (1).

B. MARKER RELEASE FROM LIPOSOMES BY COMPLEMENT WITH AND WITHOUT ANTI-RGG ANTIBODY

Figure 8 shows the relationship between the marker release and the complement concentration without addition of anti-RGG antibody.[3] The marker release increased with the complement concentration. With heat-inactivated GPWC, no release of CF was observed, as shown by solid circles. A comparison of the marker release with and without anti-RGG antibody is shown in Figure 9.[3] The marker release with addition of anti-RGG antibody was significantly enhanced, depending on amount of the antibody added. With nonspecific sheep γ-globulin, the marker release was almost equivalent to that without addition. Since CPC shows a high activity in lysing bilayer membranes in comparison with APC, this enhancement of the marker release was caused through the CPC by the existence of the antigen–antibody complex.

From these results, the lysis with addition of anti-RGG antibody corresponds to the total complement activity, while the lysis without the antibody seems to be caused by APC. To confirm this point, the effect of factor I (C3b-C4b inactivator) on the lysis of RGG-coupled liposomes was studied. Factor I affects the activities of both CPC and APC through inhibition of C3b and C4b activities. APC, however, is much more strongly inhibited, because its activation is initiated by the formation of stable C3b. Figure 10 shows the degree of inhibition of CPC and APC by factor I of various concentrations.[9] The degree of inhibition was expressed by the decrease in the marker release shown by the following equation:

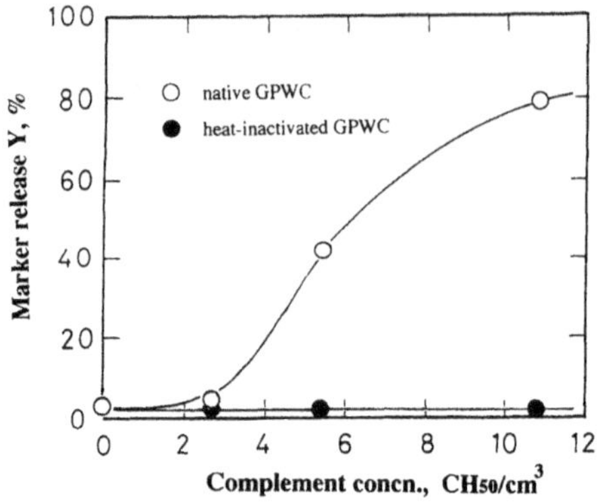

Figure 8 Effect of complement concentration on marker release RGG-coupled liposomes by native and heat-inactivated (56°C, 30 min) GPWC without addition of specific antibody (1.6×10^{-3} mol/mol-DPPC).

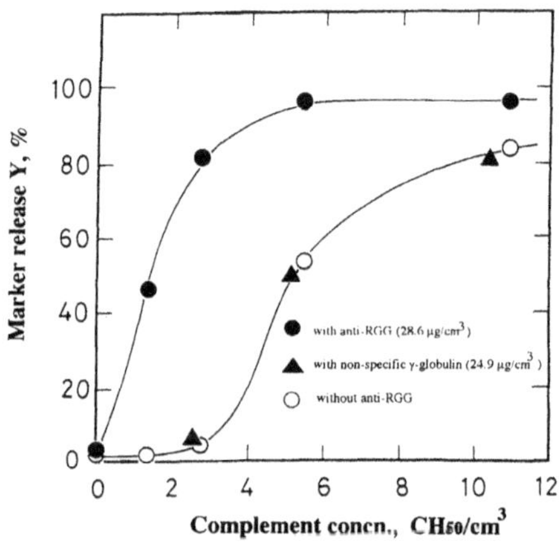

Figure 9 Effect of anti-rabbit IgG antibody added to RGG-coupled liposomes (1.6×10^{-3} mol/mol-DPPC).

$$\text{Decrease in marker release (\%)} = \frac{Y_0 - Y}{Y_0} \times 100 \tag{4}$$

where Y and Y_0 are the marker release with and without addition of factor I, respectively. The decrease in the marker release increased with the concentration of factor I added. The degree of the decrease without addition of anti-RGG antibody, however, was much larger than that with anti-RGG antibody, indicating that APC was mainly responsible for the lysis without the antibody.

The RGG-coupled liposomes were lysed by the complement system with and without addition of anti-RGG antibody. The lysis in the former case is well correlated to the total complement activity determined by the conventional method. Without addition of anti-RGG antibody, APC is mainly responsible for lysis of the liposomes. Therefore, both the total and APC activities can be measured within 2 h by use of RGG-coupled liposomes.

IV. CONCLUSIONS

By use of antigen-coupled liposomes, specific antibodies are detected, and the total and APC activities of complement can also determined within 2 h. These homogeneous liposome assays are suitable for automated procedures and the treatment of large numbers of samples.

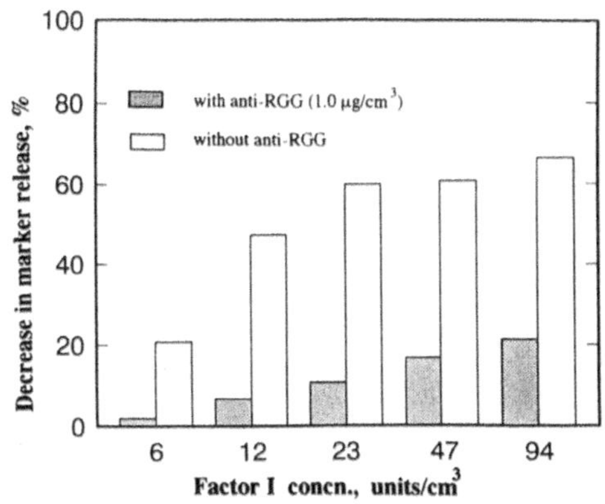

Figure 10 Effect of factor I on liposome lysis by human complement (1.6×10^{-3} mol/mol-DPPC).

REFERENCES

1. **Kinsky, S. C.,** Preparation of liposomes and a spectrophotometric assay for release of trapped glucose marker, *Methods Enzymol.*, 32B, 501, 1974.
2. **Umeda, M., Ishimori, Y., Yoshikawa, K., Takada, M., and Yasuda, T.,** Homogeneous determination of C-reactive protein in serum using liposome immune lysis assay (LILA), *Jpn. J. Exp. Med.*, 56, 35, 1986.
3. **Kishimura, M., Yamaji, H., Fukuda, H., Terashima, M., Katoh, S., and Sada, E.,** A simple method for measuring the complement activities of both classical and alternative pathway by using rabbit γ-globulin-coupled liposomes, *J. Ferment. Bioeng.*, 68, 395, 1989.
4. **Huang, C.,** Studies on phosphatidylcholine vesicles: formation and physical characteristics, *Biochemistry*, 8, 344, 1969.
5. **Barbet, J., Patrick, M., and Leserman, L. D.,** Monoclonal antibody covalently coupled to liposomes: specific targeting to cells, *J. Supramol. Struct. Cell Biochem.*, 16, 248, 1981.
6. **Tomioka, K., Fukuda, H., and Katoh, S.,** Characterization of antigen-coupled liposomes for homogenous immunoassay of polyclonal antibodies, *Chem. Eng. J.*, 54, B33, 1994.
7. **Katoh, S., Sohma, Y., Mori, Y., Fujita, R., Sada, E., Kishimura, M., and Fukuda, H.,** Homogeneous immunoassay of polyclonal antibodies by use of antigen-coupled liposomes., *Biotechnol. Bioeng.*, 41, 862, 1993.
8. **Pauling, L., Pressman, D., and Grossberg, A. L.,** The serological properties of simple substrates. VII. A quantitative theory of the inhibition by haptens of the precipitation of heterogeneous antisera with antigens, and comparison with experimental results for polyhaptenic simple substances and for azoproteins, *J. Am. Chem. Soc.*, 66, 784, 1944.
9. **Kishimura, M., Fukuda, H., Katoh, S., Sada, E., and Taniguchi, H.,** Lysis of rabbit γ-globulin-coupled liposomes by the complement system, *J. Ferment. Bioeng.*, 74, 81, 1992.

Chapter 15

Liposomes in Immunodiagnostics

Anup K. Singh and Ruben G. Carbonell

CONTENTS

I. INTRODUCTION

Liposomes have been applied to numerous fields of research owing to their unique structure. Their similarity to biological membranes makes them ideal for studying cell membrane properties and their capacity to encapsulate large number of molecules in the core or in the bilayer makes them ideal vehicles for drug delivery. These features, combined with the availability of a large number of functional groups on the surface to which various receptor molecules can be attached, render them suitable for targeting for drug and genetic material delivery and immunodiagnostic applications. Liposomes can be made out of thousands of natural or synthetic lipids that are commercially available. Depending on the lipids chosen, they can be cationic, anionic, or neutral and can have headgroups containing reactive moieties such as amines, carboxyls, sugars, or sulfhydryls. An overwhelming number of research articles exist on various applications of liposomes. This article reviews applications of liposomes for immunodiagnostic assays. Readers are also referred to some excellent review articles on this topic published earlier by Ho and Huang,[1] Alving and Richards,[2] Monroe,[3] and Monji et al.[4] There has been lot of work done in recent years on application of liposomes in immunodiagnostics, and this paper emphasizes recent work while trying to present a general summary of approaches taken in the development of immunoliposome assays.

Immunoassays can be used to detect an analyte such as an antibody, an antigen, a hormone or a drug in a clinical sample. They are also routinely being used in food analysis and for the detection of pollutants and toxins. These assays rely on the binding of an analyte to a molecule for which it has specific affinity, followed by detection of the resulting complex. Examples of substances exhibiting such specific molecular recognition include enzymes, antibodies, lectins, transport proteins, and cell surface receptors. For example, to detect an antigen in a sample, an antibody can be added which is specific to that antigen. The antibody can be tagged with a reporter or marker molecule to produce a desired signal (alternatively, a tagged secondary antibody which binds to the primary antibody can be introduced). The choice of

0-8493-4013-6/96/$0.00+$.50

signal-generating molecules is virtually unlimited — enzymes, fluorophores, spin labels, and radiolabels are most commonly used.

Liposomes have exhibited a potential for signal enhancement and higher sensitivities when used in immunodiagnostic applications. Liposomes provide a large interior volume for entrapment of thousands of small molecules. Hydrophilic groups on the outer surface of liposomes can offer sites for attachment of ligands (antibodies or antigens) as well as reporter molecules. Consequently, for a small percentage of ligands on the surface, there could be thousands of marker molecules associated with a liposome (entrapped inside or immobilized on the surface) as opposed to the conventional reporter–ligand conjugate carrying at the most a few reporter molecules per ligand molecule. In principle, this should allow the liposome immunoassays to perform much better in terms of detection limit than their nonliposomal counterparts. An example of this principle is illustrated in the Figure 1.

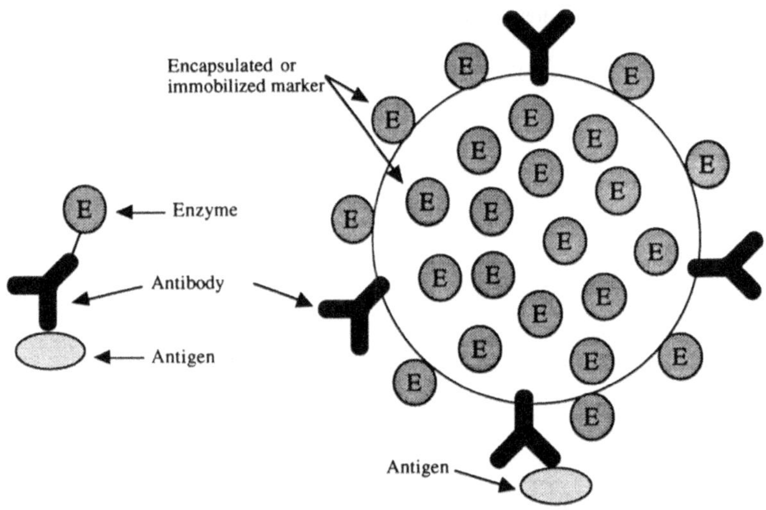

Figure 1 Signal enhancement achieved by using liposomes in an immunoassay.

II. PREPARATION OF LIPOSOMES

Numerous methods exist to prepare liposomes,[5] and virtually all possible liposome compositions and preparation techniques have been used to form liposomes for potential application in immunoassays. The important criteria to decide on the type of liposomes to be used are stability, encapsulation volume (in case of encapsulated marker), and surface area per unit volume (in case of surface immobilized markers). Unilamellar liposomes have been the most commonly used liposomes as they are the most stable liposomes and can be stored for months without aggregation. Multilamellar liposomes (or, multilamellar vesicles, MLVs) have been used in some instances, owing to the simplicity of their preparation, but these liposomes tend to aggregate and precipitate upon storage. Other advantages of unilamellar liposomes over MLVs are reproducibility in preparation and economy in lipid utilization. MLVs are also very inefficient in entrapping water-soluble markers since a very small fraction (<0.1) of the total internal volume is aqueous phase. Large unilamellar liposomes (diameter > 1000 Å) are preferred when marker molecules have to be encapsulated as they provide large interior volumes. When proteins are to be immobilized on the outer surface of the liposomes, small unilamellar liposomes have been used as they provide a high surface area per unit volume. The common methods to form unilamellar liposomes are extrusion of multilamellar liposomes through polycarbonate membranes; bath or probe sonication of multilamellar liposomes; and reverse phase evaporation of lipids. The choice of lipids to form liposomes is virtually infinite. Researchers have used both unsaturated and saturated, and synthetic and natural lipids. Unsaturated lipids are prone to oxidative degradation, and hence liposomes containing these should be stored in the absence of oxygen or should contain an antioxidant lipid such as tocopherol. Use of synthetic saturated lipids minimizes the possibility of oxidation and should be used in preference to unsaturated lipid whenever possible. Very small unilamellar liposomes (diameter < 400 Å) are prone to fusion as a means of relieving stress arising from high curvature of the membrane. Even large liposomes can fuse if they have packing defects in the bilayer. Since fusion is more prominent at or near the phase transition temperature, liposomes should be stored away from T_c or should contain enough cholesterol to reduce or completely eliminate transition. If liposomes are to be stored for very long durations, aggregation brought about by Van der Waals atrraction could be a problem. The rate of aggregation of liposomes is proportional to their size (MLV > LUV > SUV) and can be reduced significantly by incorporating a small amount of anionic lipid (5–10% PG or PA) in the liposomes.

For liposomes to be used in an immunoassay, they generally need to carry either an antibody or an antigen on the outer surface. Small antigens or haptens are usually coupled covalently to a lipid in an organic phase, phosphatidylethanolamine (PE) being the leading example. The resulting hapten-PE conjugate is mixed with other lipids to prepare liposomes. For larger proteins such as antibodies, there have been two approaches for immobilization on a liposome surface. The antibody can be conjugated covalently, through its amines, carbohydrate chains, or thiolated amines, to preformed liposomes which contain a lipid or a modified lipid bearing a reactive head group such as an amine, sulfhydryl, maleimide or carbohydrate.[6] The second approach is to insert the antibody noncovalently into the bilayer. Antibody molecules by themselves are normally not hydrophobic enough to get incorporated in the bilayer, so that they are commonly conjugated to a lipid such as PE or cholesterol or to a hydrophobic molecule such as palmitic acid. The resulting amphiphile is then added during the hydration step of liposome formation or incubated with liposomes after they are formed. The conjugated antibody is anchored on the bilayer through the lipid portion.

III. TYPES OF IMMUNOLIPOSOME ASSAYS

Immunoassays with liposomes have been performed in numerous formats which can be categorized as shown in Figure 2.

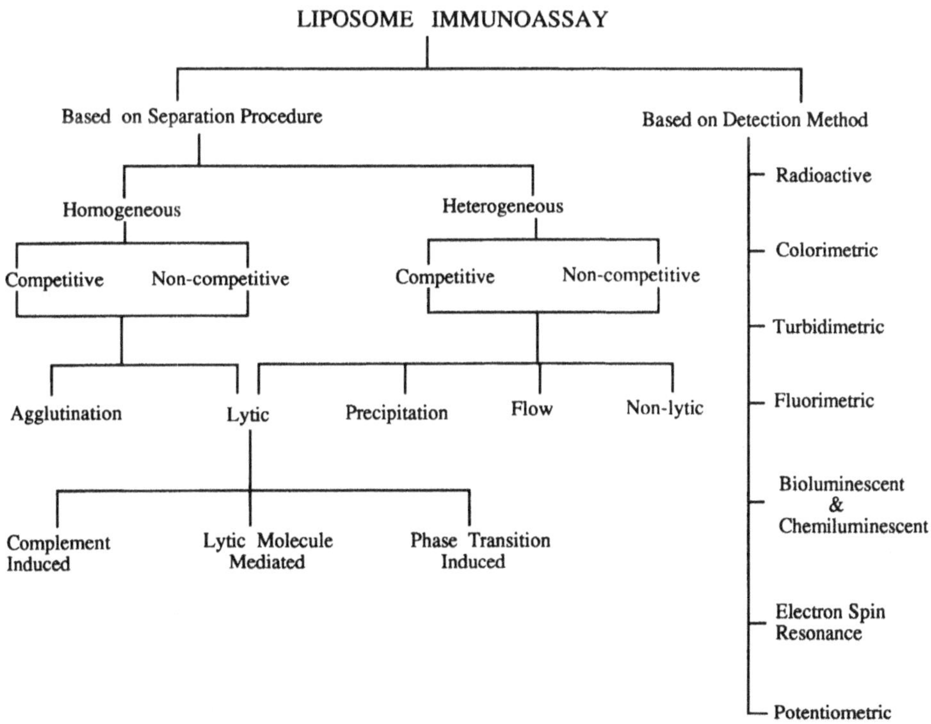

Figure 2 Types of liposome immunoassays.

The first broad division is between heterogeneous and homogeneous assays.[7] In *heterogeneous* assays, a procedure is required to separate or wash the immunologically bound label from the excess unbound label prior to measurement. In *homogeneous* assays, measurement of the extent of the reaction is performed in solution without separation of free and antibody-bound components. An important point to note is that this distinction is based solely on the requirement of a washing/separation step, and the presence or absence of a solid phase as a support is immaterial. Heterogeneous assays generally take longer to perform than homogeneous assays but their sensitivity and accuracy is much better and they have a wider range of application.

Each of these assays can also be divided into competitive (inhibition) and noncompetitive (direct) immunoassays.[7] In the competitive method for determining antigens, the amount of antibody present in the assay solution (in homogeneous format) or adsorbed on the solid phase (in heterogeneous format) is insufficient to bind all the antigen. A known concentration of liposome sensitized with antigen is added to the sample containing free antigen, whose concentration is to be determined. This mixture is incubated with the antibody (in solution or immobilized on the solid phase), thus allowing the liposome-bound and the free antigen to compete for the limited quantity of antibody. The liposomes also contain encapsulated or immobilized marker molecules and the extent of binding is determined by detecting the marker molecules. In a competitive or inhibition assay, the signal is inversely proportional to the concentration of the antigen in the sample. In a non-competitive or direct assay the antibody molecules are in excesss, and the signal is directly proportional to the concentration of analyte in the sample. In a sandwich assay, a type of direct assay, the antigen carrying at least two epitopes is sandwiched between two different but specific antibodies, one of which is labeled. Both types are discussed in detail in later sections.

A. HOMOGENEOUS IMMUNOLIPOSOME ASSAYS

These can be divided into two broad categories based on the mechanism of carrying out the assay: lytic and agglutination.

1. Lytic Assays

In this assay, liposomes carry a ligand on the surface and marker molecules inside. The liposome lyses when exposed to sample-containing analyte molecules which bind specifically to the ligand. Lysis of the liposomes results in the release of the marker molecules that can then be detected analytically. The magnitude of the signal is directly proportional to the concentration of the analyte in the solution. This method can also be used to develop an inhibition assay where liposomes derivatized with analyte molecules compete with free analyte to bind to a limited amount of ligand in the reaction mixture. This type of assays is ideal for automation as the measurement of released reporter molecules from lysed vesicles can be performed in a single assay mixture. The assay also has a high sensitivity since relatively few antigen–antibody events on the surface of liposomes can lead to the release of a large number of reporter molecules. This assay can be further subdivided into different classes depending on the method of lysis and the type of marker molecules used. The lysis can be achieved in the following ways:

a. Complement-Mediated Lysis

These assays are based on the ability of complement to lyse liposomes carrying antigen–antibody complexes on the surface. The complement system is a chain of proteins that recognize and bind to antibody-antigen complexes at membrane surfaces and whose activation eventually results in the destruction of cells or invading microorganisms.[8] There are at least 19 of these proteins and together they make up about 10% of the globular fraction of serum. If the complement sequence proceeds to completion, membrane attack complexes (MAC) are formed. These are ring-shaped structures made up of proteins with a central electron-dense area surrounded by a lighter ring of poly C9 proteins. MACs are incorporated into the lipid bilayer of cells (or liposomes) resulting in channels through which the cell is lysed.

There are two pathways for activation of complement system — the classical pathway and the alternative pathway. An antigen–antibody complex formation is necessary to initiate the classical pathway. The first component of the classical pathway consists of three proteins: C1q, C1r, and C1s, bound together by calcium dependent bonds. C1q binds to an antibody that is attached to antigen but not to free immunoglobulin molecules. In free immunoglobulin, the binding site on the CH2 chain is masked by the Fab regions, but when Fabs are bound to an antigen, the site is free to bind to C1q. To activate the C1 complex, single antigen-bound molecules of IgM or paired antigen-bound molecules of IgG are needed. Hence, two IgG molecules must be located very close together to activate the C1 complex. As a result, IgG is much less active than IgM in activating complement. In the alternative pathway of complement activation, the formation of antigen–antibody complex is not needed. One of the intermediate proteins, C3, breaks down slowly but spontaneously in plasma, generating small amounts of C3b which is essential to propagate the complement cascade. The C3b, in the presence of "activating" surfaces, surfaces deficient in sialic acids, e.g., bacterial cell walls, some tumor cells, and some virus-infected cells, bind to factor B leading to production of a complex C3bB. This complex can convert C3 to C3b, thus propagating the complement sequence. The alternative pathway is much slower compared to the classical pathway and is not a dominant factor in cell lysis.

Complement-dependent assays have been the most widely used liposome assays since Kinsky et al.[9] and Haxby et al.[10] first used them for the detection of Forssman agent. Most of these assays are based on the principle that antigen-specific lysis can be inhibited by the addition of a soluble antigen to compete with the antigen on liposomes to bind to the antibody. In the first step of the assay, antibody is added to the serum sample containing the antigen. It is imperative to heat-inactivate the native complement proteins in the serum sample before using it for the assay. After an incubation of 5–10 min, marker-loaded liposomes sensitized with the antigen being assayed are added. During the 10–20 min incubation period, both the free antigen and liposomal surface antigen compete for binding to the antibody molecules present in the solution. The higher the concentration of antigen in the sample, the fewer the number of liposomes that bind to antibodies. In the final step, guinea pig complement is added. The complement readily binds to any immune complex formed on the liposome surface, leading ultimately to lysis of liposomes. The release of marker molecules can be monitored using appropriate equipment. Since it is an inhibition or competitive assay, the signal generated is inversely proportional to the concentration of antigen in the sample. In the following section, a few examples of these assays are discussed. For literature prior to 1986, readers are referred to the review article by Ho and Huang.[1]

Canova-Davis et al.[11] used complement-mediated release of enzyme molecules from liposomes to develop an inhibition-type assay for theophylline. Unilamellar liposomes were prepared by the reverse phase evaporation method[12] with PC: PG: Cholesterol: theophylline-DPPE: α-tocopherol (45.2: 12.9: 40.4: 1.4: 0.1 molar ratios). The encapsulated marker was a mixture of the substrate glucose-6-phosphate (G6P) and the enzyme glucose-6-phosphate dehydrogenase (G6PD). When the theophylline-sensitized liposomes were incubated with antitheophylline antibody and complement, the antibody bound to the liposomes, activating the complement system. The liposomes lysed subsequently, releasing the G6P and G6PD, which reacted with the co-factor NAD^+ present in the assay solution to give a colored product. When free theophylline was present in the assay system, a decrease in the enzymatic activity was observed as fewer liposomes were lysed due to binding of some of the antibody molecules to the free analyte.

A few other recent examples of complement-based inhibition assays have been presented. Bowden et al.[13] measured total complement activity in human serum in the presence of anti-DNP antibody using sonicated liposomes carrying dinitrophenol (DNP) on the surface and containing encapsulated AP (alkaline phosphatase). Ligler et al.[14] and Gaber et al.[15] detected T-2 mycotoxin using unilamellar liposomes containing carboxyfluorescein (CF) in the core prepared by extrusion and sensitized with T-2. Fiechtner et al.[16] assayed for digoxin using MLVs derivatized with digoxigenin and containing a hydrophilic derivative of fluorescein, 5(6)-carboxyfluorescein trismethyloamide, in the core. Kubotsu et al.[17] measured anticonvulsant drugs such as phenytoin, phenobarbital, and carbamazepine using unilamellar liposomes prepared by extrusion carrying the drug on the surface and G6PD entrapped inside. Pashkov et al.[18] developed a complement-based immunoassay for latrotoxin using biotinylated liposomes prepared by reverse phase evaporation and containing entrapped calcein. Paul et al.[19] developed an assay for plant toxin gelonin using reverse phase evaporation liposomes carrying gelonin on the surface and containing entrapped calcein.

Direct or noncompetitive lytic assays, in which the signal is directly proportional to the analyte concentration, have also been developed. Umeda et al.[20] reported a sandwich assay for measuring C-reactive protein (CRP) in human sera. Multilamellar liposomes were prepared from DPPC, cholesterol and DTP-DPPE containing 0.2 *M* CF as entrapped marker. Goat anti-CRP IgG was coupled to liposomes using a heterobifunctional cross-linker, SPDP (*N*-succinimidyl pyridyl dithiopropionate). Human serum containing CRP and liposomes was added to wells of a microtiter plate, followed by addition of rabbit anti-CRP antibody. CRP in the serum binds to both the primary antibody immobilized on the liposome and the secondary rabbit anti-CRP IgG, forming a liposome–CRP-secondary antibody complex. The reaction mixture was incubated for 1 h at 37°C after adding guinea pig complement. The resulting specific lysis of liposomes, which is proportional to the concentration of CRP in serum sample, was monitored by measuring fluorescence. A few other examples of direct assays are mentioned. Hosoda and Yasuda[21] assayed for α_2 plasmin inhibitor (α_2PI) using MLVs conjugated with anti-α_2PI monoclonal antibodies and containing CF as entrapped marker. Tatsu et al.[22] developed an assay for anti-DNP antibody and complement using MLV carrying DNP on the surface and containing entrapped CF. Katoh et al.[23] detected anticytochrome c and antimyoglobin antibodies, using sonicated liposomes conjugated with cytochrome c (or myoglobin) by SPDP chemistry and containing entrapped CF. Yu et al.[24] and Kim and Lim[25] assayed for immune lysis of liposomes in the presence of complement and anti-BSA serum, using liposomes derivatized with BSA and carrying entrapped AP.

Many researchers are using monoclonal antibodies routinely in immunoassays because they provide homogeneity in specificity and affinity and can be produced more easily in large quantities. However, most of these monoclonals can not activate the complement system efficiently and hence a secondary polyclonal immunoglobulin needs to be added to the system, which activates the complement system upon binding to the primary antibody.[14,21,26]

Complement-dependent lytic assays have been shown to have a comparable or even higher sensitivity than conventional ELISA (enzyme-linked immunosorbent assay) or RIA (radioimmunoassay) in a number of cases, yet their commercial application has been rare owing to the instability of the complement proteins upon storage. Inactivation of any one of the complement proteins can inhibit the cascade reaction leading to reduction or total loss of lytic activity. Most of these assays require rather large concentration of complement to overcome the slow kinetics of liposome lysis and high amount of guinea pig complement is relatively expensive. In some cases the nonspecific lysis, lysis of liposomes without addition of antibody to the mixture of sensitized liposomes and complement, was as much as 20–30% of total lysis.[23] The authors have postulated that the alternative pathway of complement, which does not require the formation of antigen–antibody complex, may be responsible for the nonspecific lysis of liposomes.

Complement-dependent lytic assays require attachment of antigen to the liposomes, and consequently a unique formulation of sensitized liposomes has to be used for each antigen being assayed. The use of an avidin–biotin or streptavidin–biotin system can alleviate this problem. Pashkov et al.[18] used a streptavidin bridge to immobilize biotinylated protein antigens on biotinylated liposomes for a subsequent application in an assay to detect latrotoxin. Biotinylated liposomes were prepared by the reverse phase evaporation method of Szoka and Papahadjopoulos[12] from a lipid mixture of PC, cholesterol, DCP, and biotin-PE. To sensitize the liposomes with the analyte, latrotoxin (LT), biotinylated liposomes were added to streptavidin. After removing the excess streptavidin, biotinylated LT was added, forming a complex consisiting of biotinylated liposome–streptavidin–biotinylated antigen.

b. Cytolysin Mediated Lysis

Liposomes can also be lysed by potent cytolytic agents, such as mellitin from bee venom,[27,28] central Asia cobra venom,[29] and marine worm protein, *Cerebratulus lacetus* toxin A-III.[30] Cytolytic agents can be considered as biological ionic detergents. Sessa et al.[31] reported that mellitin, a cationic peptide, is extremely surface active and penetrates artificial and natural lipid layers very avidly. Hydrophobic attraction between the acyl chains of lipids are the major cohesive forces holding a liposomal bilayer together. Owing to its strong surface activity and capacity of its peptide chain to form hydrogen bonds with the acyl chains of lipids, mellitin overcomes these forces thus disrupting the bilayer.

Litchfield et al.,[27] Freytag and Litchfield,[28] and more recently Nakamura et al.,[32] and Haga et al.[33] developed assays for digoxin based on the lysis of liposomes with hapten–cytolysin conjugates. The principle behind such an assay is that the cytolysin–hapten conjugate can lyse a liposome, releasing the contents, but if the conjugate binds to an antibody, then the lytic activity is blocked because of conformational changes. Litchfield et al.[27] prepared unilamellar liposomes containing entrapped AP by dialysis of octyl glucoside. In the homogeneous assay for digoxin, affinity purified digoxin antibodies and ouabin (an analog of digoxin)–mellitin conjugate were added to the assay solution containing digoxin and *p*-nitrophenyl phosphate (substrate). The conjugated ouabin competed with free digoxin to bind to a limited amount of the antibody. After incubating for 5 min, liposomes with entrapped AP were added and the enzymatic reaction was monitored by measuring absorbance at 410 nm. Only the ouabin–mellitin conjugate not bound to the antibody lysed the liposomes and hence the absorbance signal was directly proportional to the concentration of digoxin in the sample. Nakamura et al.[32] and Haga et al.[33] improved the sensitivity of the assay by using chemiluminescence-based detection. Nakamura et al.[32] entrapped the enzyme GO (glucose oxidase) in unilamellar liposomes formed by extrusion. Lysis by mellitin–ouabin conjugate released the enzyme that reacted with glucose to produce hydrogen peroxide, which in turn reacted with isoluminol in the presence of peroxidase to produce chemiluminescence. The detection limit was 3 nM of digoxin, three times lower than that reported by Litchfield et al.[27] Haga et al.[33] further improved the sensitivity by a factor of 300 by encapsulating a low molecular weight co-factor FAD (flavine adenine dinucleotide) instead of the high molecular weight enzyme GO.

Another approach is to use Sendai virus, which possesses hemolytic activity. When an acylated antibody against the analyte was incorporated into the viral membrane, the resulting target virus can bind and lyse liposomes bearing the analyte. Since the free soluble antibody or soluble antigen can inhibit lysis, this principle allows one to assay for either the analyte or cognate antibody.[34]

The advantage of a lytic molecule-based system is that a unique liposome preparation for each assay system is not required, as neither the analyte nor the antibody needs to be conjugated to liposomes. The cytolytic agents are fast and effective and a very small amount is needed to lyse the liposomes. In general, cytolysin-based assays require fewer steps and are faster and considerably more versatile than the complement-dependent approaches. A major disadvantage of this technique, however, is that extreme caution has to be taken while using these immensely potent venoms. The cytolysin needs to be linked covalently to either an antibody or antigen without any significant loss in immunological and cytolytic activity. Mellitin is a 26-amino acid polypeptide containing three lysine residues, two at the C terminal and one near the N terminal, surrounding a hydrophobic segment.[31] The lysine amino moieties can be covalently linked to the amino or sugar groups of proteins without disturbing the structure of the internal hydrophobic portion which penetrates and disrupts the lipid bilayer.

c. Liposome Lysis by Detergents

Ullman et al.[35] devised a homogeneous enzyme immunoassay in which the reactive components are combined in a single liquid suspension but are prevented from reacting because one of the reagents is encapsulated in liposomes. The antigen is covalently coupled to an enzyme and the conjugate is encapsulated in a liposome. These liposomes are added to a solution containing antibody against that antigen. The principle behind this assay is that the enzymatic activity of conjugate is inhibited by binding of antibody. Unilamellar liposomes were prepared with POPG, POPC, and cholesterol by extrusion through a 0.2 μm polycarbonate filter. The enzyme–antigen conjugate, G6PD–theophylline, was encapsulated in the liposomes. The reagent containing liposomes and antitheophylline antibody was added to the sample, followed by addition of a detergent and the substrate (G6P). The detergent disrupted the lipid bilayer of the liposomes, releasing the antigen–enzyme conjugate which competed with antigen in the sample to bind to the antibody present. Only the antigen–enzyme conjugate not bound to the antibody reacted with the substrate to give a product, and hence measured enzyme activity was proportional to the concentration of theophylline in the sample.

The detergents used for vesicle lysis are mostly nonionic. The ones most effective in complete and rapid lysis are Triton™ X-100, sodium deoxycholate, and octyl glucoside. Ionic detergents, owing to their charge, penetrate rather slowly into the inner bilayer of the liposome, leading to incomplete or slow lysis.

d. Lysis Due to Phase Transition

Unsaturated phosphatidylethanolamine does not make stable vesicles on its own, but if it is mixed with another lipid with a complimentary geometry, for example, phosphatidylcholine (PC), or a transmembrane protein like glycophorin, it makes a stable bilayer. Babbitt et al. [36] provide a list of amphiphiles, termed "liposome stabilizers", used for making a stable bilayer of unsaturated phosphatidylethanolamine that includes fatty acids, fatty acyl amino acid, gangliosides, acidic phospholipids, acidic/basic double-chain amphiphiles, and amphipathic proteins. If the stabilizer amphiphile is chosen such that it is an immunoreactant, i.e., an antigen or antibody, then the liposomes could be used in an immunoassay. Ho and Huang[37] used a hapten, DNP, conjugated to a lipid to stabilize liposomes formed by DOPE, an unsaturated PE. Using light scattering data they determined that stable liposomes could be formed at concentrations of DNP-PE above 12%. The DNP-liposomes were applied to a glass slide coated with anti-DNP IgG. Lysis of liposomes occurred with the release of contents as judged by the fluorescent enhancement of an entrapped self-quenching dye, calcein. The proposed mechanism, as shown in Figure 3, is that the haptenated lipids diffuse laterally to the contact area between the liposome and the solid surface leading to multiple binding events between the haptens on the liposome and the IgG molecules on the solid surface. As the stabilizer is localized near the binding site, the remainder of the bilayer experiences depletion of the stabilizer molecules, leading to a transition to the hexagonal phase. The bilayer gets disrupted, resulting in leakage of the entrapped dye. Addition of free DNP inhibits binding of liposomes to immobilized antibody, and this observation was used to develop an inhibition assay for the hapten. The detection limit was 10 pmol of free DNP in 40 μl of sample. This type of assay can also be designed in a reciprocal manner where antibody is used as a stabilizer of the PE bilayer. Ho et al.[38,39] employed anti-HSV-gD, a mouse antibody against herpes simplex virus (HSV) glycoprotein antigen gD, to make stable PE liposomes. They acylated the antibody with palmitic acid and then used it in conjunction with other lipids to make sonicated unilamellar liposomes. These liposomes lysed specifically when incubated with HSV, releasing the entrapped calcein or AP. Ho et al.[40] could detect HSV at a concentration of 3.2×10^3 pfu/ml in a 5 μl sample, improving the sensitivity by 10-fold over the standard sandwich ELISA.

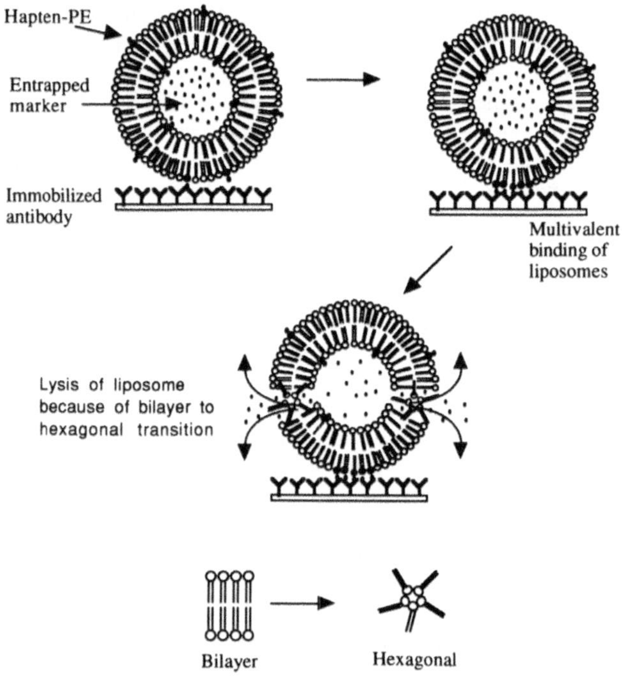

Figure 3 Lysis of liposomes induced by phase-transition when haptenated liposomes bind multivalently to antibodies. (Adapted from Ho, R.D.J. and Huang, L., *J. Immunol.*, 134(6), 4035, 1985.)

Wright and Huang[41] also reported that unsaturated PE-liposomes undergo contact-dependent lysis based upon the ability of base lipid, e.g., DOPE, to adopt nonlamellar phases. Under conditions where PE-based liposomes are brought into close proximity, they undergo a transition from a bilayer phase to a reversed hexagonal phase which results in loss of liposome integrity and lysis. The authors used N-biotinyl-PE to stabilize the DOPE-liposomes containing calcein as the fluorescent marker. The liposomes carried palmitylated murine monoclonal antibodies as receptors on the surface. Addition of magnetic chromium dioxide particles coated with antimouse antibody led to binding of liposomes to the magnetic particles, forming an aggregate and releasing the entrapped calcein. In a practical application, both liposomes and magnetic particles would be coated with antibodies against a common antigen and the antigen would become the bridging agent inducing lysis. Babbitt et al.[36] observed that DOPE liposomes, stabilized by an amphiphile and in addition having a haptenated lipid, aggregate and lyse in the presence of antihapten antibody. The aggregation and lysis of the liposomes induced by the antibody against the hapten could be inhibited by free hapten and this formed the basis of an inhibition assay for theophylline.

Another approach to achieve rapid destabilization of PE liposomes is enzymatic degradation of the hydrophilic portion of the stabilizer. For example, removal, by action of papain, of the Fab fragments of the antibody stabilizing the PE liposomes causes lysis of otherwise stable liposomes.[38] The enzymatic cleavage by trypsin of the stabilizer glycophorin destabilizes the bilayers of a DOPE–glycophorin liposome.[42] Pinnaduage and Huang[43] used this principle to develop an immunoassay for digoxin. Liposomes were made from PE and GM_1 containing the enzyme G6DPH in the aqueous core. The liposomes can be lysed by β-galactosidase as GM_1 is a substrate for this enzyme. The enzyme β-gal is divided into two inactive components, E_d and E_a, which spontaneously reconstitute to give the active enzyme. E_d is covalently linked to an antigen so that the antibody-bound E_d-antigen conjugate can no longer participate in the enzyme reconstitution. In the presence of free antigen in the sample, E_d–antigen complex is not bound by antibody and active enzyme molecules are reconstituted. The active β-gal degalactosylates the GM_1 on the PE-liposomes, causing lysis and subsequent release of reporter enzyme molecules.

It has been shown[44] that PE- and cardiolipin-containing lipid bilayers can be induced to form an inverted micelle structure known as hexagonal phase by binding of divalent cations such as Ca^{2+} or Mg^{2+}. Janoff et al.[45] designed an assay based on this principle to detect autoantibody in the serum of systemic lupus erythematosis patients. A divalent cation-sensitive dye, Arsenazo III, was encapsulated inside the liposomes which contained 30 mol% cardiolipin. Introduction of Mg^{2+} in the liposome solution caused cardiolipin to transform to the hexagonal phase, resulting in the lysis of liposome. As a result, Arzenazo III complexed with Mg^{2+} and the resultant color change was quantitated on a spectrophotometer. If autoantibody was present in the sample, it bound to the phosphate group of cardiolipin, and hence binding of Mg^{2+} to cardilipin was inhibited, resulting in a smaller color change.

2. Agglutination Assays

These assays are based on the turbidimetric detection of liposomal aggregates resulting from cross-linking of ligand-bearing liposomes by molecules which bind specifically to that ligand. Fry et. al.[46] used galactocerebroside (GC)-sensitized liposomes to detect antigalactocerebroside antibody in serum. The GC-liposomes bind to several anti-GC antibodies, and agglutination starts when liposomes cross-link upon addition of a secondary antihuman antibody. Kung et. al.[47] used liposomes to enhance the 'agglutination' signal of a typical latex assay for the detection of human rheumatoid factor. Latex agglutination assays are one of the fastest and easiest *in vitro* diagnostic assays. The latex spheres are coated, covalently or by physical adsorption, with a ligand (antigen or antibody) and incubated with a sample containing the analyte. Two essential conditions for agglutination are the polyvalency of the analyte and the presence of a relatively large number of ligands on the latex particles. One analyte molecule binds to more than one latex sphere, and each sphere in turn binds to multiple analyte molecules leading to formation of a network which eventually grows large enough to precipitate. These assays are fast but are rather insensitive. One way to amplify these specific agglutination is to use liposomes bearing a second ligand which also recognizes the analyte.[48] The liposomes lead to a second binding event in which analyte molecules, bound to the ligands on the latex spheres, serve as receptors for the second ligand attached to the liposomes. This "sandwiching effect" leads to agglutination of those latex particles which have bound the analyte but not yet formed into aggregates large enough to be visible. This is illustrated schematically in Figure 4. Kung and co-workers[47] reported that, with addition of liposomes, the sensitivity of the agglutination test improved by a factor of 2- to 4-fold.

B. HETEROGENEOUS IMMUNOLIPOSOME ASSAYS

As discussed earlier, heterogeneous assays require the separation or washing of unbound from bound label prior to detection of the appropriate signal. The heterogeneous assays can be performed in two formats: competitive and noncompetitive. In the competitive method for determining antigens, the amount of antibody adsorbed on the solid phase is insufficient to bind all the antigen. A known concentration of reporter-labeled antigen is added to the sample containing unlabeled antigen, whose concentration is to be determined. This mixture is then incubated with the antibody immobilized on the solid phase, thus allowing the labeled and the unlabeled antigen to compete for the limited quantity of antibody. After washing, the signal is detected using appropriate equipment. The larger the signal, the lower the concentration of the antigen in the sample. There are two ways that liposomes work towards increasing the sensitivity of a competitive assay: (1) the ability of liposomes to entrap very large number of marker molecules leads to a tremendous increase in signal corresponding to each antigen–antibody binding event, and (2) the sensitivity of a competitive assay is limited by the association constant of antibody for the analyte. In general, a higher association constant results in a more sensitive assay. It has been shown[49,50] that the apparent binding constant of liposomes for an antibody is several orders of magnitudes larger than free analyte, owing to multivalency of the liposomes.

In a noncompetitive assay, the antigen in the sample is allowed to react with an excess of the solid phase-coupled antibody. Then an excess of reporter-labeled antibody binds to another site on the antigen. The signal is measured after washing the unbound labeled antibody and is directly proportional to the concentration of analyte in the sample. The noncompetitive assays are in general more sensitive than a

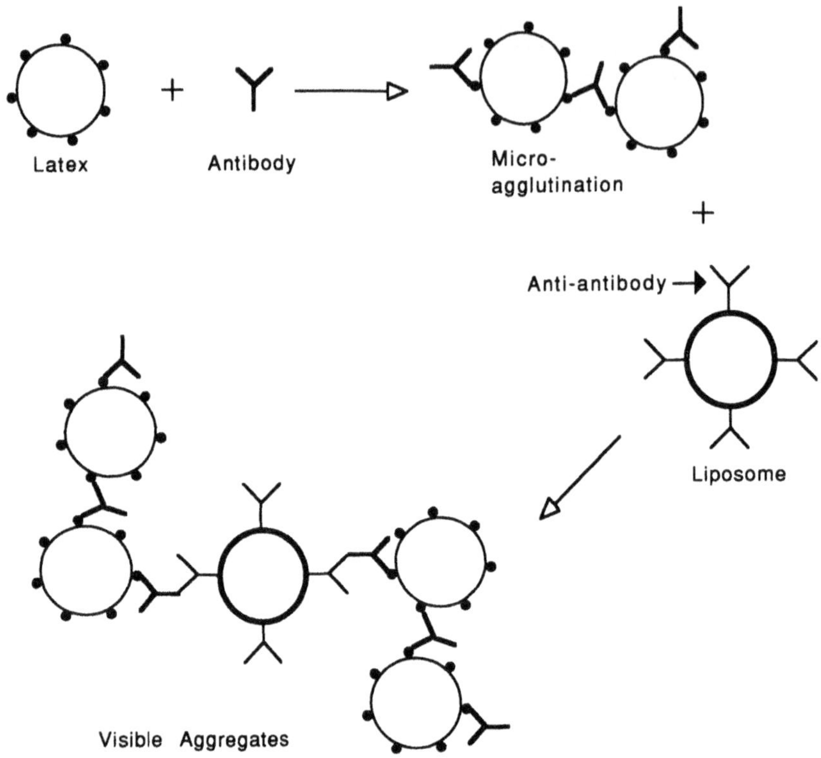

Figure 4 Enhancement of latex agglutination by addition of liposomes. (Adapted from Kung, V. T., Maxim, P. E., Veltri, R. W., and Martin, F. J., *Biochim. Biophys. Acta,* 839, 105, 1985.)

competitive assay as theoretically it is possible to detect a single analyte molecule as long as there is enough signal produced.

Each of the two types, competitive and noncompetitive assays, can be further subdivided into four groups: lytic, precipitation, flow, and nonlytic assays.

1. Lytic Assays

These assays are essentially the same as homogeneous lytic assays, except that one of the immunoreactants is bound to a solid surface and the liposomes are lysed after washing off the excess reagents. O'Connell et al.[51] developed a competitive assay for digoxin using this approach. Liposomes with digoxigenin on the surface were made with sulforhodamine-B encapsulated inside. Polypropylene tubes were coated with rabbit antidigoxin serum. After washing with glycine buffer, they were incubated with a mixture of analyte (digoxin) and digoxigenin liposomes which compete to bind to the antibody on the tube surface. The tube was washed with buffer to get rid of nonspecifically bound liposomes, and Triton™ X-100 was added to lyse the bound liposomes. The released dye was then measured in a spectrophotometer. Wagner and Baffi[52] developed a similar competitive assay for digoxin, employing a fluorescent rare earth metal as marker. Europium was encapsulated in liposomes containing DSPE-digoxin in the bilayer. A mixture of liposomes and digoxin serum standard was added to antidigoxin coated tubes. After washing, the liposomes were lysed with detergent, and time-resolved fluorescence was used to detect the europium concentration. Time-resolved fluorescence assays take advantage of the large difference in fluorescence lifetimes between the labeling fluor and the nonspecific background fluorescence. Plant et al.[53] developed an assay for theophylline using biotinylated liposomes. The liposomes containing 100 m*M* CF were prepared by the injection method[54] from a lipid mixture of DMPC, DCP, cholesterol, and biotin-DPPE. The biotinylated antitheophylline antibody was linked to the biotinylated liposomes, using avidin as a cross-linker. BSA-theophylline conjugate was nonspecifically adsorbed on the inside of a polystyrene cuvette. A mixture of theophylline and antibody liposomes was added to the cuvette. After discarding the contents, the cuvette was washed with buffer. Fluorescence was measured after solubilizing the specifically bound liposomes with β-D-glucopyranoside. Free theophylline blocked binding of liposomes to the solid support and hence the signal decreased with increasing concentrations of theophylline in solution. The assay with liposomes was compared with an analogous ELISA, and the use of liposomes increased the sensitivity by approximately two orders of magnitude, the lowest detectable concentration being $10^{-7} M$ and $10^{-5} M$ for liposome assay and conventional ELISA, respectively.

Noncompetitive or sandwich assays based on liposome lysis have also been developed.[55,56] Rongen et al.[55] designed a microtiter plate assay for immunoregulatory cytokine interferon-γ (IFN-γ) using liposomes. Biotinylated liposomes containing CF were prepared by the extrusion method [12] from a lipid mixture of DPPC, DPPG, cholesterol, and biotin-x-DPPE. Microtiter wells were coated with the capture anti-IFNγ antibody. The following reagents were added in succession, with the wells being incubated and washed after every step: sample containing IFN-γ, a second anti-IFN-γ antibody conjugated with biotin, avidin, and biotinylated liposomes. The specifically bound liposomes were lysed with Triton™ X-100, and released CF was detected in a fluorescence detector, the signal being directly proportional to concentration of analyte. The detection limit was $6 \times 10^{-12} M$ of IFN-γ, which was comparable to the colorimetric ELISA for IFN-γ. Vonk and Wagner[56] amplified the signal of a time-resolved fluorescence assay by using liposomes containing large quantities of europium. Sandwich assays using microtiter plates were developed for choriogonadotropin and thyrotropin. The detection limit of liposome assay compared favorably with those obtained by using direct labeling of the antibody (Delfia™: LKB).

2. Precipitation Assays

These are very similar to the homogeneous agglutination assays. After agglutination of liposomes carrying ligands and marker molecules on the surface, the solution is centrifuged or filtered to isolate the precipitate. The precipitate is resuspended and the signal is measured. Axelsson et al.[57] used this approach to detect antirat transplantation antigen (RT-1) antibody in serum. Liposomes were prepared from a mixture of PC, RT-1 antigen, and either ^{125}I-labeled PE or FITC-labeled PS. RT-1 gets incorporated into the bilayer as it is a membrane glycoprotein. In the assay, the liposomes were incubated with serum from rat immunized against RT-1. A secondary rabbit anti-rat IgG serum was added and the formed immune complexes were precipitated by centrifugation. The precipitate was recovered and radioactivity (or fluorescence) was measured. The specific antibody in the serum could be detected at a dilution titer of 1/3000, which is comparable to that obtained in a ^{51}Cr-release assay.

3. Liposome Flow Assays

a. Flow Injection Assays

In the field of immunodiagnostics, a recent trend has been to develop analytical techniques which can be readily automated. Flow injection analysis is one technique that can be readily automated for clinical measurements. Locasio-Brown et al.,[50,58] Yap et al.,[59] and Choquette et al.[60] combined the potential of a flow injection analysis for easy automation with high sensitivities provided by use of liposomes to develop automated immunoassays for clinical analytes. Locasio-Brown et al.[50] developed a competitive assay system using theophylline, a therapeutic drug, as the model analyte. The immunoreactor column consisted of a glass tube packed with antitheophylline antibody-derivatized nonporous silica particles. Liposomes containing carboxyfluorescein in their core were prepared from a mixture of DMPC, cholesterol, DCP, and theophylline-PE. The sample containing free theophylline plus liposomes was injected into the column. Free theophylline and theophylline–liposomes competed to bind to the limited amount of antibodies immobilized on the column, with higher levels of sample analyte causing fewer liposomes to bind to the column. The column was washed with buffer to get rid of any nonspecifically bound liposomes. A detergent, octyl glucopyranoside, was circulated in the column to disrupt the bound liposomes releasing the entrapped CF which was detected downstream. The column was regenerated by washing with carrier buffer. The minimum detectable concentration of theophylline was 5.4 ng/ml. The reactor system and the competitive binding of liposomes to the antibody-derivatized surface are illustrated in Figure 5.

Wu and Durst[61] used the same principles to develop a double-amplification immunoassay for theophylline. The double amplification was achieved by means of liposome-encapsulated peroxidase molecules which upon release react with an organofluorine substrate to produce fluoride ions which can be detected potentiometrically. The competition between the free theophylline and theophylline-derivatized liposomes for immobilized antibody molecules in a flow-through immunoreactor column resulted in unbound liposomes being carried downstream to the detector. These HRP-containing liposomes were

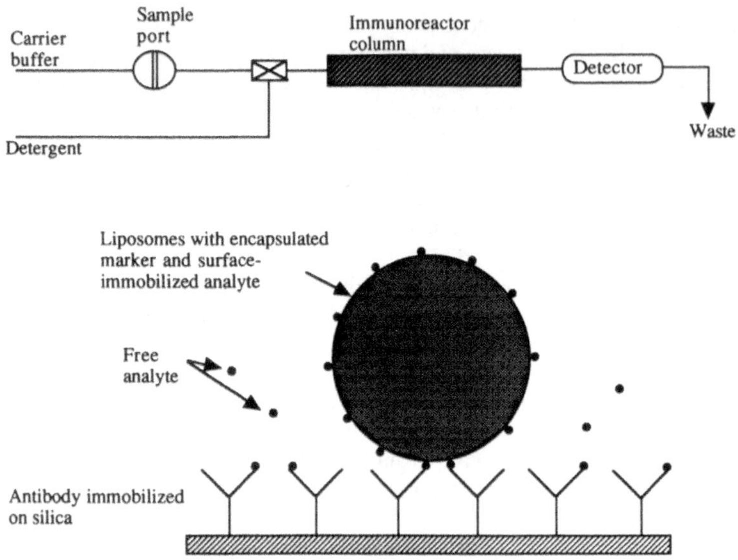

Figure 5 (Top) Schematics of flow injection immunoassay (FIIA) system. (Bottom) Competitive binding of liposomes and analyte in the immunoreactor column. (Figure 5A adapted from Locascio-Brown, L., Plant, A. L., Horváth, V., and Durst, R. A., *Anal. Chem.* 62, 2587, 1990.)

ruptured in the presence of *p*-fluorophenol and H_2O_2 to release HRP. This enzyme converts *p*-fluorophenol to produce fluoride ions which can be measured with an ion-selective electrode. The range of concentration over which theophylline could be measured was 0.2–4000 ng/ml. The column was regenerated by an acidic buffer, 0.1 *M* glycine-HCl (pH 2.4).

Choquette et al.[60] used liposomes in combination with regenerable planar waveguide immunosensor to detect theophylline. The use of theophylline-labeled liposomes in the competitive assay led to an order of magnitude enhancement in signal over theophylline derivatized with fluorescein. Liposomes were made with encapsulated CF and were sensitized with theophylline. A mixture of liposomes and the sample containing the analyte was passed over a waveguide sensor containing immobilized antitheophylline antibody. The liposomes bound competitively to the sensor surface and the fluorescence emission intensity was measured following excitation of the waveguide with a 488-nm argon-ion laser. To regenerate the sensor, liposomes were lysed with a detergent and antibody-antigen complex was dissociated by flowing buffer. A schematic of this technique is shown in Figure 6.

b. Immunomigration Assays

These are similar to flow-injection assays, except that solution flow is controlled by capillary action. Durst et al.[62] and Siebert et al.[63] developed an assay for the herbicide alachlor using immunomigration techniques. A mixture of alachlor-sensitized liposomes (containing sulforhodamine-B) and free alachlor was allowed to migrate up a strip of nitrocellulose membrane on which anti-alachlor antibody and egg white avidin zones have been immobilized. In the antibody zone, liposomes and alachlor compete for binding sites, while the second avidin zone binds all the liposomes that do not bind to the antibody zone. The binding of liposomes in either zone is quantified by the amount of color of the entrapped dye. In the antibody zone, the presence of free alachlor inhibits the binding of liposomes and hence the color is inversely proportional to the amount of analyte in the sample. In the avidin zone, the intensity of color is directly proportional to the concentration of alachlor.

4. Nonlytic Assays

In most of the examples discussed earlier, liposomes contained marker molecules which were released after lysis of liposome following antigen–antibody complexation. One inherent problem associated with these assays is leakage of marker due to nonspecific lysis of liposomes and leakage upon storage. Scherier et al.[64] reported that nonspecific lysis can occur in a rather unpredictable fashion, ranging from a few percentage to peak values of up to 70% of total releasable marker, presumably depending on such variables as serum source, age and lipid composition. Jones et al.[49,65,66] and Singh et al.[67,78] proposed that

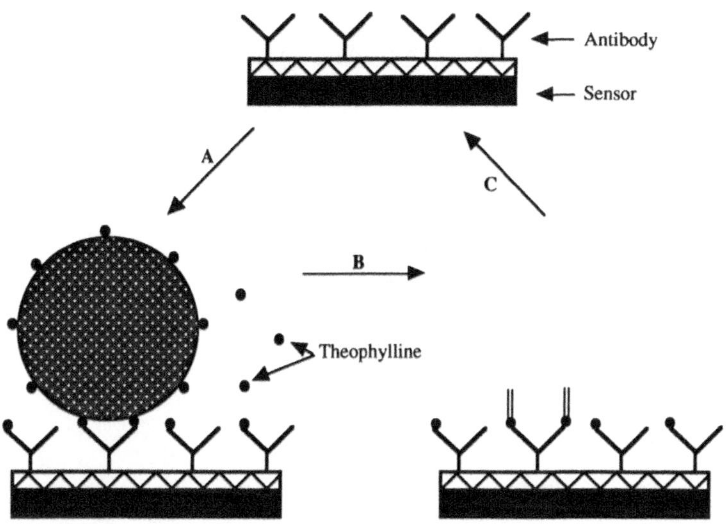

Figure 6 Planar waveguide assay format A) competitive binding of liposome and theophylline to the sensor B) lysis and removal of specifically bound liposomes with detergent C) regeneration with flowing buffer. (Adapted from Choquette, S. J. Locascio-Brown, L., and Durst, R. A., *Anal. Chem.*, 64, 55, 1992.)

if marker molecules were covalently attached to the outside of the liposome, then leakage would be reduced to minimal levels. Also, for large marker molecules like enzymes (HRP, AP, etc.), surface immobilization leads to a much higher number of marker per liposome (~200) compared to encapsulation (~10 molecules per liposome).

Jones et al.[49,66] developed a competitive assay for biotin in a microtiter plate format, using liposomes with covalently attached biotin and horseradish peroxidase on the surface. The wells of the plates were coated with antibiotin antibody (ABA). After blocking with casein, the wells were incubated with sample containing the analyte (biotin) and the liposomes. The biotinylated liposomes competed with the free biotin to bind to the limited number of antibody molecules on the solid surface. After washing to remove nonspecifically bound liposomes, substrate was added and the measured signal was inversely proportional to the amount of biotin in the sample. The lowest detectable antigen concentration for liposomes (~10^{-9} *M*) was an order of magnitude lower than the value found for a conventional biotin-HRP conjugate (10^{-8} *M*). The assay with liposomes, unlike that with biotin-HRP conjugate, depended very strongly on the surface density of immobilized antibody on the plate, presumably owing to the multiple-point attachment of liposomes to the surface. Singh et al.[67] extended the idea to a sandwich-type ELISA for detection of d-dimer, a fibrin dimer formed at early stages of thrombosis. The noncompetitive or sandwich ELISA is one of the most sensitive immunoassays as it is not limited by the association constant of the antigen for the antibody and the detection limit is equal to the minimum concentration of marker molecule detectable. Figure 7 shows a schematic of a heterogeneous sandwich assay using liposomes. The anti-d-dimer antibody was adsorbed on the microtiter well surface and the remaining sites on the solid surface were blocked by bovine serum albumin. The wells were incubated with serum containing the analyte. After washing to get rid of nonspecifically bound antigen, liposomes derivatized with anti-d-dimer antibody and HRP were added. The wells were washed again to remove any nonspecifically bound liposomes, and signal was measured in an absorbance plate-reader after adding the substrate, TMB. A control assay was done in exactly same manner, except antibody-HRP conjugate was used instead of liposomes. The least detectable doses obtained with liposomes and HRP-antibody conjugate were 2.4 and 21.5 ng/ml, respectively. Hence, the liposome assay led to detection limit an order of magnitude lower than its nonliposomal counterpart.

Based on the concept of surface immobilized markers and antibodies, a fluoroimmunoassay was developed by Singh et al.[78] using liposomes with fluorescein and anti-d-dimer antibodies attached to the bilayer. Unilamellar liposomes were prepared by sonication or extrusion from a mixture of lipids, one of the lipids being labeled with fluorescein. Monoclonal anti-d-dimer IgGs were covalently conjugated to the outside of liposomes. The liposomes contained approximately 10^4 fluors and 10–20 antibodies per liposome. These liposomes were used in a sandwich, microtiter-plate based immunoassay for

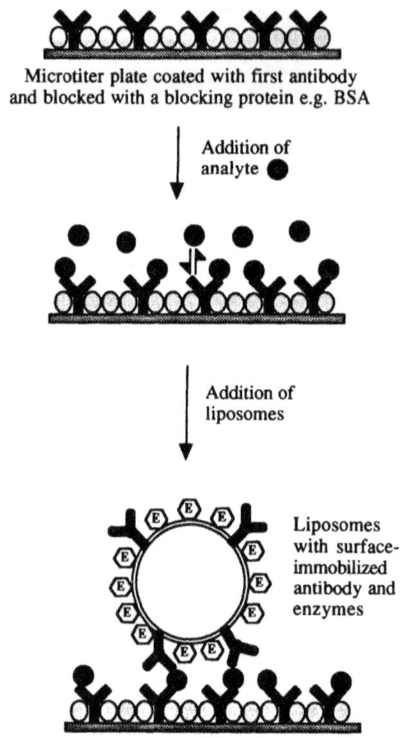

Figure 7 Enzyme-linked immunosorbent assay (ELISA) using liposomes. (Adapted from Singh, A. K., Kilpatrick, P. K., and Carbonell, R. G., *Biotechnol. Progr.,* in press.)

detection of d-dimer and led to two orders of magnitude lowering of detection limit compared to conventional fluorantibody conjugate.

Rosenberg et al.[68] used liposomes in an electrochemical assay where the lipid bilayer acts as a diffusion-limiting membrane. The conventional enzyme electrodes for measuring glucose in blood samples are limited by the saturation kinetics of the GO not allowing clinically relevant glucose concentrations to be measured (0–25 m*M*). If the GO is encapsulated in a liposome prior to exposure to the blood sample, then glucose has to diffuse through the bilayer to react with the enzyme, thereby maintaining a smaller concentration of glucose inside the liposome than in the bulk solution. Liposomes were prepared by reverse phase evaporation [12] with DMPC, DPPC, or DSPC and contained glucose oxidase in the aqueous core. Liposomes were immobilized on a nitrocellulose membrane and interfaced with an oxygen electrode polarized for H_2O_2 detection. When the surface of the membrane was in contact with the sample containing glucose, glucose diffused across the liposomal bilayer reacting with the entrapped GO to produce H_2O_2, which diffused out of the liposome to be measured by the electrode. The signal was directly proportional to the concentration of glucose in the sample and a linear response was obtained up to 40 m*M* of glucose.

IV. MARKERS USED AND DETECTION METHODS

A variety of marker molecules and detection techniques have been used in liposome immunoassays as listed in Table 1. The most common have been fluorophores like CF, Fl, calcein, and sulforhodamine B. These are highly water soluble and can be encapsulated in liposomes in very high concentration. While inside the liposomes they are concentration quenched and do not exhibit high fluorescence, but when they are released from liposomes, they get diluted and become highly fluorescent. Although fluorophores have been widely used as marker molecules, sometimes they tend to leak out of liposomes with time, and the leakage rate is dependent on various factors — pH and ionic strength of the buffer, types of lipids, temperature, and choice of fluorophore. Fietchner et al.[16] reported that this problem could be

Table 1 Markers and Detection Method

Marker	Detection Method	Example
Enzyme	Spectrophotometric	HRP[66,67]
		AP[24,25,39]
		G6PD[11,17]
	Chemiluminescence (luminometer)	GO[32]
	Potentiometric	HRP[61,69]
		GO [68]
Fluorophore	Fluorescence spectrophotometry	Sulforhodamine B[51]
		Calcein[36,37]
		CF[14,15,20,22,23]
		Eu[52,56]
Cofactor, substrate	Chemiluminescence (luminometer)	FAD[33]
	Potentiometric	Glucose[77]
Ion	Potentiometric	TPA^{+}[69,71]
		Ferrocyanide[73]
Spin-label	Electron spin resonance spectrometry	Tempocholine chloride[75]

circumvented by using hydrophilic derivatives of fluorophores — e.g., 5(6)-carboxyfluorescein tris-methyolamide instead of carboxyfluorescein.

Other very widely used markers are enzymes. Enzyme molecules, being relatively large, are less susceptible to leakage and also improve the sensitivity by turning over a large number of substrate molecules. HRP, AP, and GO are the most commonly used enzymes. The detection method used depends on the choice of substrate. The released enzyme (or, immobilized enzyme) reacts with substrate to produce a colored, fluorescent, or chemiluminescent product which can be detected by appropriate spectrometry. In the case of potentiometric substrates, the enzyme catalyzes a redox reaction; for example, released HRP[69] will catalyze the reaction:

$$2\ NADH + O_2 + 2H^+ \rightarrow 2NAD^+ + 2H_2O$$

and oxygen in the solution is depleted. The consumption of oxygen can be detected by a Clark-type oxygen electrode with an oxygen-permeable Teflon® membrane.

Another example of electrochemical detection is the detection of released ion by an ion-selective electrode. Shiba and his co-workers[70,71] developed a thin-layer potentiometric liposome immunoassay. They entrapped the ion tetrapentyl-ammonium (TPA+) in liposomes; it was later released by complement-induced lysis and its concentration was measured by a selective microelectrode directly onto the plate surface. Katsu[72] compared a series of tetraalkylammonium ions, R_4N^+, where R = methyl, ethyl, propyl, butyl, and pentyl, respectively, to find a less leaky ionic marker to be used in liposomal membrane permeability assays. Tetraalkylammonium ions with the alkyl chain shorter than that of a propyl group did not permeate significantly across PC bilayers and hence should be used as entrapped ion markers. Kannuck[73] developed a homogeneous immunoassay by using liposomes with entrapped potassium ferrocyanide. The assay was based on an inhibition format. The liposomes were lysed by addition of complement, and the released ferrocyanide was monitored by differential pulse voltametry. Monroe[74] provides a good treatise on liposomal electrochemical immunoassays in his review article.

Another choice for an encapsulated marker could be a spin label. At high concentration of spin label, the electron spin resonance (ESR) spectrum is broadened due to spin exchange interaction between neighboring spin molecules. Upon release, the spin labels are diluted, and ESR intensity is greatly amplified. Hsia and Tan,[75] and Humphries and McConnell[76] encapsulated a spin label, tempocholine chloride, in liposomes and used it in complement-based immune lysis homogeneous assays. The spin markers lead to highly sensitive assays but the expensive instrumentation needed for detection renders them prohibitive for general use.

One recent development in liposome assays has been use of fiber optic-based systems. Optical fibers have attracted a lot of attention as choice for an immunosensor owing to their flexibility, noise immunity, and smaller size in comparison with nonfiber optics and electric conductors. Tatsu et al.[22] synthesized a fiber optic sensing tip containing liposomes immobilized in agar gel. The liposomes contained CF in the aqueous core and DNP on the surface. The dye within the liposome is not fluorescent because of concentration quenching. In the assay for anti-DNP antiserum, the tip of the fiber was immersed in anti-DNP antiserum and fluorescence was measured after adding complement to the sample cell. A schematic of this system is shown in Figure 8.

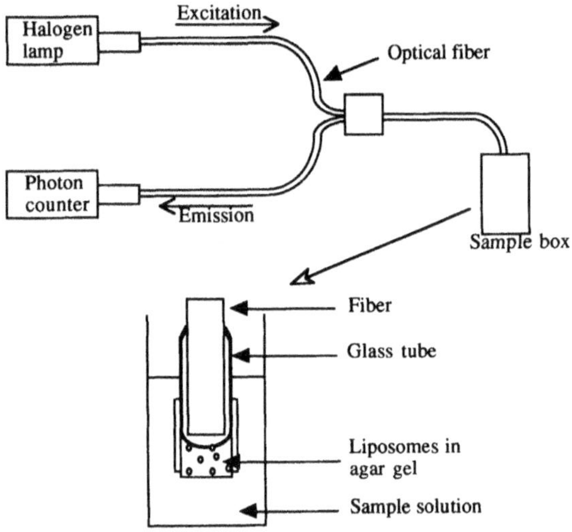

Figure 8 Schematic diagram of the fiber-optic liposome immunosensing system. (From Tatsu, Y., Yamamura, S., and Yoshikawa, S., *Biosensors Bioelectronics,* 7, 741, 1992. With permission.)

V. CONCLUSIONS

The underlying concept of applying liposomes to immunodiagnostics is based on the hypothesis that liposomes provide a large signal enhancement or lower detection limit. In fact, the sensitivity of liposome immunoassays, heterogeneous and homogeneous both, has been shown to be better than their nonliposomal counterparts in most instances. Table 2 compares detection limits of a few immunoliposome assays with other established assays. The use of liposomes results in signal amplification, too. Jones et al.,[49] in a competitive assay to detect biotin, reported that at low antibody surface densities the signal generated by vesicle-biotin-HRP conjugate was 100 times higher than that generated by the conventional biotin-HRP conjugate. Choquette et al.[60] demonstrated that the use of theophylline-labeled liposomes in a competitive assay provides an order of magnitude signal enhancement over theophylline derivatized with fluorescein. In the competitive-binding assay for digoxin,[51] signals were amplified 10^3- to 10^4-fold by using liposomes as compared to labeled hapten.

Liposome immunoassays have come a long way since their first use by Kinsky et al.[9] The first type of assays to be developed were based on complement-mediated lysis of liposomes, and this format

Table 2 Comparison of Immunoliposome Assays with Their Nonliposomal Counterparts

	Liposome Assay		Standard Assay		
Analyte	**Type**	**Detection Limit**	**Type**	**Detection Limit**	**Ref.**
D-dimer	Heterogeneous FIA	5.6 ng/ml	Heterogeneous FIA	674 ng/ml	Singh et al.[78]
D-dimer	Sandwich ELISA	2.4 ng/ml	Sandwich ELISA	21.5 ng/ml	Singh et al.[67]
Biotin	Competitive ELISA	10^{-9} *M*	Competitive ELISA	10^{-8} *M*	Jones et al.[66]
Theophylline	Competitive ELISA	10^{-7} *M*	Competitive ELISA	10^{-5} *M*	Plant et al.[53]
HSV-1	Homogeneous lytic	3.2×10^3 pfu/ml	Sandwich ELISA	$\sim 2 \times 10^4$ pfu/ml	Ho et al.[40]
Rheumatoid factor	Agglutination	0.7 IU/ml	Agglutination	6.25 IU/ml	Kung et al.[47]

attracted the most attention. This assay is very efficient but there are some inherent drawbacks associated with it, such as labile nature of complement and nonspecific lysis. Later, researchers developed lytic assays that do not need complement and they accomplished the lysis by either using cytolytic agents or inducing phase transition of bilayer subsequent to multivalent binding of liposomes to an antibody or antigen. Most of the forementioned assays were developed in a homogeneous format. Recently, heterogeneous assays have attracted lot of attention as they are inherently more sensitive than homogeneous assays. Two main types are flow-injection assays which can be readily automated and require a minimal amount of time and sample and assays developed in microtiter plates where a large number of samples can be assayed simultaneously and with very high sensitivity. Liposome immunoassays have not been applied commercially on a large scale yet, but their potential is evident from the fact that there are hundreds of patents held by companies as well as individual researchers on the use of liposomes in immunoassays.

NOMENCLATURE

ABA	Antibiotin antibody
AP	Alkaline phosphatase
β-gal	β-Galactosidase
BSA	Bovine serum albumin
CF	Carboxyfluorescein
DCP	Dicetyl phosphate
DMPE	Dimyristoyl phosphatidylethanolamine
DNP	Dinitrophenol
DOPE	Dioleoyl phosphatidylethanolamine
DPPC	Dipalmitoyl phosphatidylcholine
DPPE	Dipalmitoyl phosphatidylethanolamine
DSPE	Distearoyl phosphatidylethanolamine
DTP-DPPE	Dithiopyridyl-DPPE
ELISA	Enzyme-linked immunosorbent assay
Eu	Europium
FAD	Flavine adenine dinucleotide
G6DPH	Glucose-6-phosphate dehydrogenase
G6P	Glucose-6-phosphate
GC	Galactocerebroside
GO	Glucose oxidase
HRP	Horseradish peroxidase
HSV	Herpes simplex virus
LUV	Large unilamellar vesicle
MAC	Membrane attack complex
MLV	Multilamellar vesicle
NAD	Nicotinamide adenine dinucleotide
PA	Phosphatidic acid
PC	Phosphatidylcholine
PE	Phosphatidylethanolamine
PG	Phosphatidylglycerol
pNP	*p*-Nitrophenyl phosphate
POPC	Palmitoyl oleoyl phosphatidylcholine
POPG	Palmitoyl oleoyl phosphatidylglycerol
RIA	Radioimmunoassay
SPDP	*N*-Succinimidyl pyridyl dithiopropionate
SUV	Small unilamellar vesicle
TMB	3,3′,5,5′ Tetramethyl benzidine dihydrochloride
TPA+	Tetrapentyl ammonium

REFERENCES

1. **Ho, R. J. and Huang, L.,** Immunoliposome assays: perspectives, progress and potential, in *Liposomes as Drug Carriers,* G. Gregoriadis (Ed.), John Wiley & Sons, New York, 1988, 527.
2. **Alving, C. R. and Richards, R. L.,** Immunologic aspects of liposomes, in *Liposomes,* M. J. Ostro (Ed.), Marcel Dekker, New York, 1983, 209.
3. **Monroe, D.,** Novel liposome immunoassays for detecting antigens, antibodies, and haptens, *J. Lip. Res.*, 1(3), 339, 1989.
4. **Monji, N., Cole, C. A., and Castro, A.,** Application of liposomes in non-isotopic immunoassays: a review, *Rev. Immunoassay Technol.*, 2, 73, 1988.
5. **New, R. R. C.,** Preparation of liposomes, in *Liposomes: a Practical Approach,* R. R. C. New (Ed.), IRL Press, Oxford, 1989, 33.
6. **Martin, F. J., Heath, T. D., and New, R. R. C.,** Covalent attachment of proteins to liposomes, in *Liposomes: a Practical Approach,* R. R. C. New (Ed.), IRL Press, Oxford, 1989, 163.
7. **Miyai, K.,** Classification of immunoassays, in *Principles and Practice of Immunoassay,* C. P. Price and D. J. Newman (Eds.), Stockton Press, New York, 1991, 246.
8. **Tizard, I. R.,** *Immunology: An Introduction* (3rd ed.). Saunders College Publishing, Orlando, 1992.
9. **Kinsky, S. C., Haxby, J. A., Zopf, D. A., and Kinsky, C. B.,** Complement-dependent damage to liposome prepared from pure lipids and Forssman hapten, *Biochemistry,* 8, 4149, 1969.
10. **Haxby, J. A., Gotze, O., Muller-Eberhard, H. J., and Kinsky, S. C.,** Release of trapped marker from liposomes by the action of purified complement component, *Proc. Natl. Acad. Sci. U.S.A.*, 64, 290, 1969.
11. **Canova-Davis, E., Redemann, C. T., Vollmer, Y. P., and Kung, V. T.,** Use of reversed-phase evaporation vesicle formulation for a homogeneous liposome immunoassay, *Clin. Chem.*, 32(9), 1687, 1986.
12. **Szoka, F. and Papahadjopoulos, D.,** Procedure for preparation of liposomes with large internal aqueous space and high capture by reverse-phase evaporation, *Proc. Natl. Acad. Sci. U.S.A.*, 75(9), 4194, 1978.
13. **Bowden, D. W., Rising, M., Akots, G., Myles, A., and Broeze, R. J.,** Homogeneous, liposome-based assay for total complement activity in serum, *Clin. Chem.*, 32(2), 275, 1986.
14. **Ligler, F. S., Bredehorst, R., Talebian, A., Shriver, L. C., Hammer, C. F., Sheridan, J. P., Vogel, C., and Gaber, B. P.,** A homogeneous immunoassay for the mycotoxin T-2 utilizing liposomes, monoclonal antibodies, and complement, *Anal. Biochem.*, 163, 369, 1987.
15. **Gaber, B. P., Ligler, F. S., and Bredehorst, R.,** Liposome based immunoassays for detection of small and large molecules, in *Biotechnological Applications of Lipid Microstructures,* B.P. Gaber, J.M. Schnur, and D. Chapman (Eds.), Plenum Press, New York, 1988, 209.
16. **Fiechtner, M., Wong, M., Bieniarz, C., and Shipchandler, M. T.,** Hydrophilic fluorescein derivatives: useful reagents for liposomes immunolytic assays, *Anal. Biochem.*, 180, 140, 1989.
17. **Kubotsu, K., Goto, S., Fujita, M., Tuchiya, H., Kida, M., Takano, S., Matsuura, S., and Sakurabayashi, I.,** Automated homogeneous liposome immunoassay systems for anticonvulsant drugs, *Clin. Chem.*, 38(6), 808, 1992.
18. **Pashkov, V. N., Tsurupa, G. P., Griko, N. B., Skopinskaya, S. N., and Yarkov, S. P.,** The use of streptavidin-biotin interaction for preparation of reagents for complement-dependent liposome immunoassay of proteins: detection of latrotoxin, *Anal. Biochem.*, 207, 341, 1992.
19. **Paul, A., Madan, S., Vasandani, V. M., Ghosh, P. C., and Bachhawat, B. K.,** Liposome immune lysis assay (LILA) for gelonin, *J. Immunol. Methods,* 148, 151, 1992.
20. **Umeda, M., Ishimori, Y., Yoshikawa, K., Takada, M., and Yasuda, T.,** Liposome immune lysis assay, *J. Immunol. Methods,* 95, 15, 1986.
21. **Hosoda, K. and Yasuda, T.,** Homogeneous immunoassay for α_2 plasmin inhibitor and α_2 PI-plasmin complex, *J. Immunol. Methods,* 121, 121, 1989.
22. **Tatsu, Y., Yamamura, S., and Yoshikawa, S.,** Fluorescent fibre-optic immunosensing system based on complement lysis of liposome containing carboxyfluorescein, *Biosens. Bioelectron.*, 7, 741, 1992.
23. **Katoh, S., Sohma, Y., Mori, Y., Fujita, R., Sada, E., Kishimura, M., and Fukuda, H.,** Homogeneous immunoassay of polyclonal antibodies by use of antigen-coupled liposomes, *Biotech. Bioeng.*, 41, 862, 1993.
24. **Yu, B. S., Choi, Y. C., and Chung, H.,** Development of immunoassay methods by use of liposomes, *Biotechnol. Appl. Biochem.*, 9, 209, 1987.
25. **Kim, C. and Lim, S.,** Liposome immunoassay with antigen-coupled liposomes containing alkaline phosphatase, *J. Immunol. Methods,* 159, 101, 1993.
26. **Six, H. R., Uemura, K., and Kinsky, S. C.,** Effect of immunoglobulin class and affinity on the initiation of complement-dependent damage to liposomal model membranes sensitized with dinitrophenylated phospholipid, *Biochemistry,* 12(20), 4003, 1973.
27. **Litchfield, W. J., Freytag, J. W., and Adamich, M.,** Highly sensitive immunoassays based on the use of liposomes without complement, *Clin. Chem.*, 30, 1441, 1984.
28. **Freytag, J. W. and Litchfield, W. J.,** Liposome-mediated immunoassays for small haptens (digoxin) independent of complement, *J. Immunol. Methods,* 70, 133, 1984.

29. **Aripov, T. F., Salakhutdinov, B. A., Salikhova, Z. T., Sadykov, A. S., and Tashmukhamedov, B. A.,** Structural changes of liposomes phospholipid packing induced by cytotoxin of the central Asia cobra venom, *Gen. Physiol. Biophys.*, 3, 489, 1984.
30. **Blumenthal, K. M.,** Release of liposomal markers by cerebratulus toxin A-III, *Biochem. Biophys. Res. Commun.*, 121(1), 14, 1984.
31. **Sessa, G., Freer, J. H., Colacicco, G., and Weissman, G.,** Interaction of a lytic polypeptide, mellitin, with lipid membrane systems, *J. Biol. Chem.*, 244(13), 3575, 1969.
32. **Nakamura, T., Hoshino, S., Hazemoto, N., Haga, M., Kato, Y., and Suzuki, Y.,** A liposome immunoassay based on a chemiluminescence reaction, *Chem. Pharm. Bull.*, 37(6), 1629, 1989.
33. **Haga, M., Hoshino, S., Okada, H., Hazemoto, N., Kato, Y., and Suzuki, Y.,** An improved chemiluminesence-based liposome immunoassay involving apoenzyme, *Chem. Pharm. Bull.*, 38(1), 252, 1990.
34. **Heath, B., Martin, F., and Huang, A.,** The interaction of sendai virus with target cells: basis of novel homogeneous immunoassays, *Biophys. J.*, 49, 119, 1986.
35. **Ullman, E. F., Tarnowski, T., Felgner, P., and Gibbons, I.,** Use of liposome encapsulation in a combined single-liquid reagent for homogeneous enzyme immunoassay, *Clin. Chem.*, 33(9), 1579, 1987.
36. **Babbitt, B., Burtis, L., Dentinger, P., Constantinides, P., Hillis, L., McGirl, B., and Huang, L.,** Contact-dependent, immunocomplex-mediated lysis of hapten-sensitized liposomes, *Bioconjugate Chem.*, 4, 199, 1993.
37. **Ho, R. D. J. and Huang, L.,** Interactions of antigen-sensitized liposomes with immobilized antibody: a homogeneous solid-phase immunoliposome assay, *J. Immunol.*, 134(6), 4035, 1985.
38. **Ho, R. J. Y., Rouse, B., and Huang, L.,** Destabilization of target-sensitive immunoliposomes by antigen binding-arapid assay for virus, *Biochem. Biophys. Res. Commun.*, 138(2), 931, 1986.
39. **Ho, R. J. Y., Rouse, B. T., and Huang, L.,** Target-sensitive immunoliposomes: preparation and characterization, *Biochemistry*, 25, 5500, 1986.
40. **Ho, R. J. Y., Rouse, B. T., and Huang, L.,** Interactions of target-sensitive immunoliposomes with herpes simplex virus, *J. Biol. Chem.*, 262(29), 13979, 1987.
41. **Wright, S. E. and Huang, L.,** Immunomagnetic particle induced lysis of antibody-conjugated liposomes, *J. Lip. Res.*, 2(2), 257, 1992.
42. **Hu, L., Ho, R. J. Y., and Huang, L.,** Trypsin induced destabilization of liposomes composed of dioleoylphosphatidylethanolamine and glycophorin, *Biochem. Biophys. Res. Comm.*, 141(3), 973, 1986.
43. **Pinnaduage, P. and Huang, L.,** A homogeneous, liposome-based signal amplification for assays involving enzymes, *Clin. Chem.*, 34(2), 268, 1988.
44. **Rand, R. P. and Sengupta, S.,** Cardiolipin forms hexagonal structures with divalent cations, *Biochim. Biophys. Acta*, 255, 484, 1972.
45. **Janoff, A. S., Carpenter-Green, S., Weiner, A. L., Seibold, J., Weissman, G., and Ostro, M. J.,** Novel liposome composition for a rapid colorimetric test for systematic lupus erythematosus, *Clin. Chem.*, 29, 1587, 1983.
46. **Fry, J. M., Lisak, R. P., Brudy, R. O., and Alving, C. R.,** Serological techniques for detection of antibody to galactocerebroside, *J. Immunol. Methods*, 11, 185, 1976.
47. **Kung, V. T., Maxim, P. E., Veltri, R. W., and Martin, F. J.,** Antibody-bearing liposomes improve agglutination of latex particles used in clinical diagnostic assays, *Biochim. Biophys. Acta*, 839, 105, 1985.
48. **Martin, F. J. and Kung, V. T.,** Use of liposomes as agglutination-enhancement agents in diagnostic tests, in *Methods in Enzymology*, Vol. 149 Academic Press, New York, 1987, 200.
49. **Jones, M. A., Kilpatrick, P. K., and Carbonell, R. G.,** Preparation and characterization of bifunctional unilamellar vesicles for immunosorbent assays, *Biotechnol. Prog*, 9, 242, 1993.
50. **Locascio-Brown, L., Plant, A. L., Horváth, V., and Durst, R. A.,** Liposome flow injection immunoassay: implications for sensitivity, dynamic range, and antibody regeneration, *Anal. Chem.*, 62, 2587, 1990.
51. **O'Connell, J. P., Campbell, R. L., Fleming, B. M., Mercolino, T. J., D., J. M., and McLaurin, D. L.,** A highly-sensitive immunoassay system involving antibody-coated tubes and liposome-entrapped dye, *Clin. Chem.*, 3119, 1424, 1985.
52. **Wagner, D. B. and Baffi, R. A.,** Vesicle including a metal marker for use in an assay, *U.S. Pat. 4,707,453*, 1987.
53. **Plant, A. L., Brizgys, M. V., Lacsio-Brown, L., and Durst, R. A.,** Generic liposome reagent for immunoassays, *Anal. Biochem.*, 176, 420, 1989.
54. **Batzri, S. and Korn, E. D.,** *Biochim. Biophys. Acta*, 298, 1015, 1973.
55. **Rongen, H. A. H., van der Horst, H. M., Hugenholtz, G. W. K., Bult, A., and van Bennekon, W. P.,** Development of a liposome immunosorbent assay for human interferon-γ, *Anal. Chim. Acta*, 287, 191, 1994.
56. **Vonk, G. P. and Wagner, D. B.,** Encapsulation of europium in liposomes for use in an amplified immunoassay detection system, *Clin. Chem.*, 37(9), 1519, 1991.
57. **Axelsson, B., Eriksson, H., Borrenbeck, C., Mattiason, B., and Sjorgen, H. O.,** Liposome immunoassay: use of membrane antigen inserted into labeled lipid vesicles as target in immunoassay, *J. Immunol. Methods*, 41, 351, 1981.
58. **Locasio-Brown, L., Plant, A. L., Chesler, R., Kroll, M., Ruddel, M., and Durst, R. A.,** Liposome-based flow-injection immunoassay for determining theophylline in serum, *Clin. Chem.*, 39(3), 386, 1993.

59. **Yap, W. T., Lacasio-Brown, L., Plant, A. L., Choquette, S. J., Horvath, V., and Durst, R. A.**, Liposome flow injection immunoassay: model calculations of competitive immunoreactants involving univalent and multivalent ligands, *Anal. Chem.*, 63, 2007, 1991.
60. **Choquette, S. J., Locascio-Brown, L., and Durst, R. A.**, Planar waveguide immunosensor with fluorescent liposome amplification, *Anal. Chem.*, 64, 55, 1992.
61. **Wu, T. and Durst, R. A.**, Liposome-based flow injection enzyme immunoassay for theophylline, *Mikrochim. Acta*, 1, 187, 1990.
62. **Durst, R. A., Seibert, S. T. A., and Reeves, S. G.**, Immunosensor for extra-lab measurements based on liposome amplification and capillary migration, *Biosens. Bioelectron.*, 8(6), xiii-xv, 1993.
63. **Siebert, S. T. A., Reeves, S. G., and Durst, R. A.**, Liposome immunomigration field assay device for Alachlor determination, *Anal. Chim. Acta*, 282, 297, 1993.
64. **Schreier, H., Valentino, C., Heath, B. P., and Kung, V. T.**, Prevention of nonspecific lysis in liposomal and erythrocyte immunoassay systems by small lipid vesicles and erythrocyte ghosts, *Life Sci.*, 45(20), 1919, 1989.
65. **Jones, M. A., Singh, A., Kilpatrick, P. K., and Carbonell, R. G.**, Preparation and characterization of ligand-modified labelled liposomes for solid phase immunoassays, *J. Lip. Res.*, 3(3), 793, 1993.
66. **Jones, M. A., Kilpatrick, P. K., and Carbonell, R. G.**, Competitive immunosorbent assays for biotin using bifunctional unilamellar vesicles, *Biotechnol. Prog.*, 10, 174, 1994.
67. **Singh, A. K., Kilpatrick, P. K., and Carbonell, R. G.**, Non-competitive immunoassays using bifunctional unilamellar vesicles (or liposomes), *Biotechnol. Prog.*, 11, 333, 1995.
68. **Rosenberg, M. F., Jones, M. N., and Vadgama, P. M.**, A liposomal enzyme electrode for measuring glucose, *Biochim. Biophys. Acta*, 1115, 157, 1991.
69. **Haga, M., Sugawara, S., and Itagaki, H.**, Drug sensor: liposome immunosensor for theophylline, *Anal. Biochem.*, 118, 286, 1981.
70. **Shiba, K., Umezawa, Y., Watanabe, T., Ogawa, S., and Fujiwara, S.**, Thin-layer potentiometric analysis of lipid antigen–antibody reaction by TPA^+ ion loaded liposomes and TPA^+ ion selective electrode, *Anal. Chem.*, 52, 1610, 1980.
71. **Shiba, K., Watanabe, T., Fujiwara, Y., and Momoi, H.**, Liposome immunoelectrode, *Chem. Letts.*, 2, 155, 1980.
72. **Katsu, T.**, The use of tetraalkylammonium ion-selective electrodes for the liposome marker release assay, *Anal. Chem.*, 65, 176, 1993.
73. **Kannuck, R. M., Bellama, J. M., and Durst, R. A.**, Measurement of liposome-released ferrocyanide by a dual-function polymer modified electrode, *Anal. Chem.*, 60, 142, 1988.
74. **Monroe, D.**, Liposomal electrochemical immunoassay, *Am. Clin. Products Rev.*, 6(12), 8, 1987.
75. **Hsia, J. C. and Tan, C. T.**, Principle and application of spin membrane immunoassay, *Ann. N. Y. Acad. Sci*, 308, 139, 1978.
76. **Humphries, G. K. and McConnell, H. M.**, Immune lysis of liposomes and erythrocyte ghosts loaded with spin label, *Proc. Natl. Acad. Sci. U.S.A.*, 71, 1691, 1974.
77. **Umezawa, Y., Sofue, S., and Takamoto, Y.**, Amperometric detection of glucose released from immune lysis of glucose loaded liposomes, *Anal. Letts.*, 15, 135, 1982.
78. **Singh, A. K., Kilpatrick, P. K., and Carbonell, R. G.**, Application of antibody and fluorophore-derivatized liposomes to heterogeneous immunoassays for d-dimer, *Biotechnol. Prog.*, in press.

Chapter 16

Steric Stabilization of Liposomes Improves Their Use in Diagnostics

Noam Emanuel, Eli Kedar, Ofer Toker, Elijah Bolotin, and Yechezkel Barenholz

CONTENTS

I. INTRODUCTION

A. ADVANTAGES OF LIPOSOME-BASED DIAGNOSTIC ASSAYS

The use of liposomes in improving diagnostic assays for a broad spectrum of medical, environmental, and civil defense purposes, is demonstrated in many scientific papers as described in detail elsewhere.[1–6]

The main advantage of the liposome-based diagnostic assay (LDA) is that it combines the lack of need to use radioactivity with increased sensitivity due to spontaneous amplification. The latter means that, unlike in the case of enzyme immunoassay, the response for LDA is immediate and does not depend on biochemical considerations related to enzyme kinetics. The spontaneous amplification in the LDA is achieved through the fact that in each liposome the ratio of label to ligand is high. This approach takes advantage of one of the two distinct environments of the liposomes: the intraliposomal (aqueous phase) and the liposomal membrane. Both can be prepared in such a way that they will contain a tagged marker molecule which is either a chromophore or a fluorophore and, therefore, can be easily detected (for review see References 1–5). Water-soluble markers can be encapsulated in the intraliposomal aqueous phase either passively[7] or by remote loading using a pH gradient,[7,8] ammonium sulfate gradient,[9,10] or calcium acetate gradient. The latter seems to be very useful for remote loading of the amphipathic weak acids, such as the popular fluorescent marker carboxyfluorescein.[11] (For the theoretical aspects of remote

0-8493-4013-6/96/$0.00+$.50

loading see Reference 12). A second possibility is to introduce nontransferable lipid-like molecules into the liposome bilayer either during or after the liposome preparation. A good example of these lipid analogues is one of many head-group-labeled fluorescent phospholipid analogues which are nontransferable markers,[13] such as FITC-PE described here. Most of these fluorescent lipids are available from various commercial suppliers.

A second approach to label the lipid bilayers can be exemplified by the use of an agent with a high liposome aqueous phase partition coefficient and a low rate of desorption from the liposomes.[14] A good example of such a marker is the dye Oil Red O which is extensively used to detect and quantify lipoproteins in various chromatographic and electrophoretic analyses.[15]

Alternatively, a fluorescently labeled ligand (peptide, antibody, etc.) can be attached to the liposome (or labeled liposome) surface. When labeled liposomes are used the liposome–ligand is doubly labeled. The ligand labeling is obtained by established methodologies.[16] Use of liposomes has many other advantages, including:

1. The fact that liposomes are chemically and physically well-characterized systems and can be prepared from a broad spectrum of amphiphiles including phospholipids (natural, modified, semisynthetic, and synthetic) and other synthetic amphiphiles. All amphiphiles that can serve as a liposome bilayer matrix share similar properties of forming a stable lamellar phase and have the ability to vesiculate. Only formulations that meet these two conditions can be used for liposome preparation.[7] Such amphiphiles are now commercially available at reproducible high quality and reasonable prices from many suppliers. Therefore, based on exact lipid composition, liposomes being in the solid-ordered phase (also referred to as gel state), liquid- ordered phase (for bilayers containing cholesterol) or liquid-disordered phase[17] can be prepared. It should be noted that even for the liquid-ordered phase the permeability of the lipid bilayer is highly dependent on the matrix phospholipid. For symmetric disaturated phospholipids the higher the gel-to-liquid-crystalline phase transition temperature (T_m), the lower the vesicle permeability.[9,18,19]
2. Liposomes can be prepared at the desired size, number of lamellae and method of preparation to fit many needs, such as entrapped volume, kinetics of marker release, etc.[6,7,20,21]
3. Liposomes can be prepared under GMP conditions, even on a large scale, as a sterile preparation well defined regarding structural and chemical features.[7]
4. Quality control protocols for evaluation of liposomes are now available.[21–24]
5. The methodologies to attach ligands (such as antigens or antibodies) to the liposome surface are well established,[25] and new methods can be adopted from the well-developed methodologies of protein conjugation.[26] The versatility of the methodologies available for liposome preparation when combined with the ability to select one of many approaches to the attachment of the ligand to the liposome, enables optimizing the preparation of the ligandoliposomes to be applied for the LDA.

B. WHY STERICALLY STABILIZED LIPOSOMES?

One aspect which may offset the increase in sensitivity due to use of liposomes is a reduction in the signal-to-noise ratio. This is related to the association of the ligand to the liposomes, which may destabilize them[27] or reduce the specificity of the assay.[28,29]

This study is aimed at demonstrating the feasibility of improving the relatively low signal-to-noise ratio by replacing the conventional liposomes used in most LDA described so far with sterically stabilized liposomes (SSL). These are liposomes containing grafted lipid molecules having a head group which, due to its special nature, acts as a steric barrier, reducing the interaction of molecules from the external medium with the liposome immediate surface, thereby improving liposome stability.[30–32] The concept of steric stabilization was well proven *in vivo* using SSL in animal model systems[18,31–33] and in humans.[34] This concept was demonstrated here, using SSL that were conjugated with an antitumor MAb as a recognition ligand.

II. MATERIALS AND METHODS

A. REAGENTS FOR LIPOSOME PREPARATION

Lipids: Hydrogenated phosphatidylcholine (HPC) and hydrogenated phosphatidylethanolamine (HPE), both having iodine value 3.0, were obtained from Lipoid KG, Ludwigshafen, Germany. This lipid acyl chain composition is mainly stearoyl. The gel-to-liquid-crystalline phase transition temperature (T_m) of the HPC as determined by differential scanning calorimetry is 52.5°C.[20]

Polyethylene glycol–distearoyl phosphoethanolamine (^{2000}PEG–DSPE) was a gift of Liposome Technology, Menlo Park, CA. Cholesterol, α-tocopherol (vitamin E), and α-tocopherol succinate were purchased from Sigma (St. Louis, MO). Lipid purity assessed by TLC[23] was greater than 97%. Fluorescein isothiocyanate–phosphatidylethanolamine (FITC–PE) was purchased from Avanti Polar Lipids (Pelham, AL).

Other reagents: bovine serum albumin, trinitrobenzene sulfonate, succinimidyl maleimidophenylbutyrate (SMPB), succinimidyl pyridylthiopropionate (SPDP), HEPES buffer, dithiothreitol (DTT), Dowex 50WX-4, and Protein-A Sephadex G-50 and Sepharose CL-6B and 4B were purchased from Sigma. All the reagents and solvents used were of analytical grade or better.

B. PREPARATION OF ANTIBODIES

A monoclonal antibody NI32/2 MAb (32/2) directed against a polyoma virus tumor-associated antigen expressed on murine A9 fibrosarcoma cells (see below) was isolated from the ascitic fluid of BALB/c mice by precipitation with 50% ammonium sulfate, followed by purification using Protein-A Sepharose Cl-4B (Zymed, San Francisco, CA) affinity columns.[35] An IgG_3-enriched fraction from the serum of normal BALB/c mice, prepared as described above, was used as a control of nonrelevant IgG antibodies. Both purified IgG preparations were diluted to a protein concentration of 2 mg/ml. The binding of the two purified antibodies to A9 and 3T3 (the corresponding nontransformed cells) cells was quantified by flow cytometry,[42] using a Becton Dickenson FAST scan flow cytometry analyzer.

C. PREPARATION OF FLUORESCENT LIPOSOMES AND IMMUNOLIPOSOMES

Fluorescent sterically stabilized immunoliposomes (F–SSIL) that are evaluated in this study were prepared as described in Figure 1A,B. In most cases used in this study the fluorescent labeling of the liposomes was obtained by FITC, either through a lipid label FITC–PE (see Introduction above) or covalently attached to the IgG. In specified cases the fluorescent labeling was introduced through doxorubicin (DOX) which was remote loaded into the intraliposomal aqueous phase through ammonium sulfate gradients.[9] In general, immunoliposomes were prepared by two steps: first F–SSL preparation as described by Haran et al.,[9] followed by either (a) direct covalent attachment of the antibodies to the liposome surface (F–SSIL–32/2), in a modification of a method described by Martin et al.;[25] or (b) binding of antibodies through Protein A attached to the liposome surface (F–SSIL–Protein A–32/2). In both cases the same antibody 32/2 MAb was used. As a control, F–SSL attached to nonspecific IgG_3-enriched immunoglobulins (F–SSIL–IgG), F–SSL lacking antibody, as well as liposomes lacking the stabilizer (F–Lip) were also prepared and used in this study. The preparation procedures are described below:

1. Direct Covalent Attachment of Antibodies to Liposomes

1. F–Lip (nonstabilized liposomes): F–Lip were prepared from HPC:HPE:FITC–PE: cholesterol 55:5:0.6:40 (mole ratio), 0.1 mol% α-tocopherol, and 0.1 mol% α-tocopherol succinate.
2. F–SSIL: Preparation of F–SSIL (Figure 1A) included the following steps:
   a. Preparation of F–SSL. F–SSL were prepared from (HPC:HPE):^{2000}PEG–DSPE: FITC–PE:cholesterol at a mole ratio of 50:5:5:0.6:40. Vesicles also contained 0.1 mol% α-tocopherol and 0.1 mol% α-tocopherol succinate, added to prolong stability.[23] At this mole ratio the fluorescence of FITC–PE is unquenched and is directly proportional to FITC concentration.

      Liposomes were prepared as described by Haran et al.[9] In short, lipids were dissolved in tertiary-butanol at 30–35°C, and the solution was lyophilized. The dried lipid "cake" was hydrated at 60°C in 0.15 *M* sterile NaCl, pH 5.6, to form 10% total lipid of large multilamellar vesicles (MLV). The MLV were downsized at 60–70°C using the Rannie Minilab 8.30H High Pressure Homogenizer (APV Rannie, Albertslund, Denmark) at a pressure of 10,000 psi. The vesicles were characterized for FITC concentration, phospholipid concentration, drug and lipid degradation and size distribution.[23] When stored for 6 months at 4°C, phospholipid hydrolysis and cholesterol oxidation were below 2 and 0.5%, respectively.[23]
   b. Preparation of PDP–IgG. Thiol groups were attached to NI32/2 MAb or control IgG (usually 30 mg protein in 10 ml of phosphate-buffered saline) as PDP by 30 min incubation with SPDP (1:15 molar ratio protein to SPDP) as described previously,[25] except that pH 7.4 was used. The unreacted reagent was separated from the PDP–IgG using Sephadex G-50 columns as described below for the preparation of F–SSIL–MPB. The antibodies were stored under N_2 at 4°C.

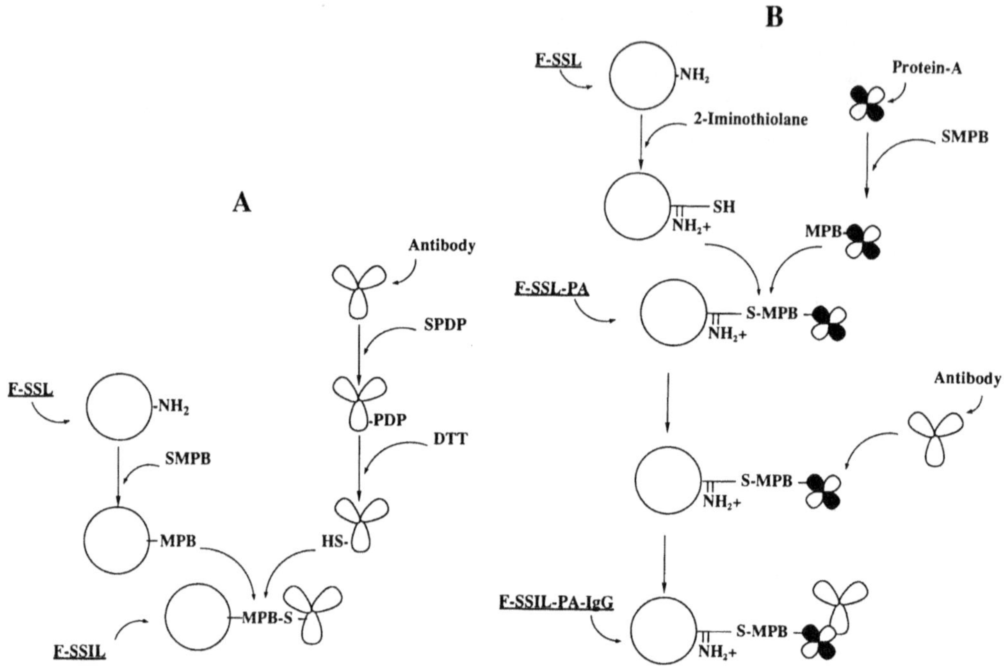

Figure 1 Schematic flow chart describing the preparation of immunoliposomes: F—SSIL—32/2 (Figure 1A) and F—SSIL—Protein A—32/2 (Figure 1B).

c. Preparation of F–SSL–MPB. The binding of SMPB to the HPE of the external surface of the vesicles was carried out immediately before the attachment of the thiolated antibodies to the vesicles. The level of HPE in the external leaflet of the liposomes was determined by its exposure to trinitrobenzene sulfonate.[36,37] Of the HPE, 64% was present in the outer surface of these vesicles, in agreement with previous data on vesicles of similar size containing monosialoganglioside.[38] Forty-three μmoles of SMPB in 150 μl of tetrahydrofuran was added to 3 ml of liposomes (87 μmol total lipid) in 0.1 *M* HEPES buffer, pH 7.8 (mole ratio of SMPB:external HPE was 20:1). The mixture was incubated with stirring at room temperature for 2 h. Unbound SMPB was removed from the liposomes by Sephadex G-50 gel exclusion chromatography using 0.2 *M* sodium phosphate buffer, pH 6.5, for column equilibration and elution. The elution solution was degassed and then purged with nitrogen. No lipid degradation or measured change in vesicle size distribution occurred during the SMPB binding. The F–SSL–MPB were used immediately for the conjugation with activated PDP–IgG, as described below.

d. Binding of PDP–IgG to F–SSL–MPB. Prior to the conjugation of the PDP–IgG to the MPB liposomes, and in parallel to the preparation of F–SSIL–MPB, the pH of the PDP–IgG was lowered to 5.5 with HCl; then DTT was added in 0.2 *M* sodium acetate buffer, pH 5.5, to a final concentration of 25 m*M*. After 30 min at room temperature, the antibodies were isolated using gel exclusion chromatography on Sephadex G-50 eluted with 0.1 *M* sodium citrate buffer pH 6.5. The "activated" PDP–antibodies were incubated together with the F–SSL–MPB in 0.1 *M* citrate buffer pH 6.5 not later than 3 h after the reaction of SMPB and F–SSL started. Routinely, we used the ratio of ~250 μg protein per 1 μmol of vesicle phospholipids. Incubation was carried out for 16–20 h with stirring, under N_2, at room temperature. Unreacted IgG and other low molecular mass molecules were removed by Sepharose CL-6B gel exclusion chromatography using 10% sucrose for column equilibration and elution. The void volume containing the liposomes was collected. To maximally reduce the level of free IgG in the F–SSIL, the Sepharose CL-6B gel exclusion chromatography step was repeated. The ratio protein:PL was identical after the first and second separations, indicating that the level of free IgG in the liposome fraction was negligible. The FITC–PE:PL ratio remained almost constant and identical to that in the F–SSL throughout all the IgG binding steps, suggesting that no FITC–PE was released during the process.

2. Binding of Antibodies through Protein A Attached to the Liposome Surface (F–SSIL–Protein A–32/2)

In general, the preparation of F–SSIL–Protein A–32/2 (Figure 1B) consists of two stages. The first one is the covalent attachment of Protein A to the HPE in the outer leaflet of the F–SSL, and the second stage is the noncovalent attachment of the 32/2 MAb to the liposomal Protein A. The first stage is similar to that described in the direct binding of MAb to the liposome surface (see above) except that in this system the SMPB was bound to Protein A to form Protein A–MPB, while the thiol group was introduced to the liposomal HPE using iminothiolane.[39] For the preparation of Protein A–MPB, we used 1.8 mg of Protein A and 10 to 20 mole excess of SMPB in tetrahydrofuron. Reaction time was 45 min. Liposomal-SH was prepared using 0.4 ml F–SSL (10% lipid) which was mixed for 3 h with 12 mg iminothiolane in HEPES buffer pH 7.5 (0.1 *M* final buffer concentration) at room temperature. Unreacted SMPB, iminothiolane or other low molecular mass components were removed from the Protein A–MPB and F-SSL–SH by gel exclusion chromatography using a Sephadex G-50 5-ml column eluted with citrate buffer pH 6.5. For the covalent attachment of Protein A–MPB to F–SSL–SH by a thioether bond, we mixed the void volume fractions of these two columns containing Protein A–MPB and F–SSL–SH in 0.1 *M* citrate buffer pH 6.5. After 12 h shaking of the mixture at room temperature, the reaction was terminated by addition of 0.02 ml of 1 *M* dithionitrobenzoic acid (Sigma) in potassium phosphate buffer pH 7.8 (final pH) to block all remaining SH on F–SSL.

The F–SSL–Protein A adduct was separated from unreacted Protein A by gel exclusion chromatography using Sepharose CL-6B.[23] A solution of 5% sterile glucose was used as elution medium. All F–SSL–Protein A–32/2 was eluted in the void volume; free Protein A was not present in the void volume.

The level of binding of Protein A to the F–SSL was determined using the protein determination of Minamide and Bamburg,[40] modified by us as described in Section D, below. In some of the liposome preparations the FITC labeling was introduced through the 32/2 MAb (FITC-MAb) and not by FITC–PE. The FITC was attached to the 32/2 MAb at pH 9.0 using 0.8 ml of 0.2 *M* borate buffer containing 2 mg MAb protein to which 0.04 ml of FITC 5% in acetone was added. After shaking at room temperature for 1–2 h, the unreacted FITC was removed from the protein using Sephadex G-50 in 5-ml gel exclusion chromatography columns. The labeled protein was eluted by 0.1 *M* HEPES buffer pH 6.0. The specific labeling of the MAb (FITC/protein mole/mg) was determined from the ratio of absorbance at 495 nm (FITC) and corrected absorbance at 280 nm.[35b]

3. Binding of 32/2 IgG_3 MAb or FITC 32/2 IgG_3 MAb to F–SSL–Protein A

The MAb were mixed with the liposome–Protein A conjugate in a vesicle/MAb ratio of 20 to 1 (calculated as described elsewhere[44]). After 1 h shaking at room temperature, the unreacted MAb was removed from the liposomes–Protein A–32/2 MAb adduct by Sepharose CL-6B gel exclusion chromatography.[23] HEPES buffer (0.1 *M*) pH 7.6 was used for elution.

These liposomes were loaded in their intraliposomal aqueous phase by remote loading of DOX.[9,10] The level of DOX was determined spectrophotometrically or fluorometrically.[9]

D. CHARACTERIZATION AND QUALITY CONTROL

The various liposomal preparations were characterized for:

1. Chemical composition which includes phospholipid composition and concentration,[23] protein concentration[40,44] and total FITC — either per phospholipid or per liposomal DOX. Phospholipid composition and concentration were determined as described elsewhere.[23]

   Protein concentration in the final F–SSIL dispersion was determined by a modification to the method of Minamide and Bamburg.[40] This assay is more sensitive than most other assays currently used for protein determination in the presence of lipids, amino lipids, and antioxidants. To overcome the background of the drug absorbance, the protein was first precipitated with methanol using an Eppendorf centrifuge and the supernatant discarded. The process was repeated twice. Before protein determination, the pellet was suspended in 1% sodium dodecyl sulfate and processed as described in the original method.[40] Under these conditions no lipids or DOX remained in the precipitate, as was tested by using the control of protein-free DOX-loaded SSL.

   DOX and FITC labeling of liposomes were determined spectrofluorometrically. DOX was determined after vesicle solubilization by acidic isopropanol[9,23] using excitation at 472 nm and emission at 586 nm. FITC was determined after vesicle solubilization in alkaline isopropanol (containing borate buffer pH 9.0) using excitation at 495 nm, and at emission at 525 nm.[35b] Under these conditions, the spectral overlap in the emission between the DOX and FITC is minimal and enables accurate quantification of one in the presence of the other.

All absorbance measurements were carried out using a Kontron Uvikon 860 double beam spectrophotometer. All fluorescence measurements were performed using a Perkin-Elmer LS 5B or LS 50B spectrofluorometer.

2. Vesicle size distribution was determined using photon correlation spectroscopy[23] by a N4SD Submicron-Particle Analyzer (Coulter Electronics) using Size Distribution Process Analysis software.

The average number of MAb per vesicle was determined from the ratio of FITC-MAb/PL or FITC-MAb/DOX (see below). FITC and DOX were determined spectrofluorimetrically as described above.

E. CELL LINES

A9 ctc 102 tumor cells (referred to as A9 cells) were used as the specific target cells for binding the IgG_3 mouse MAb NI32/2/4 (32/2) *in vitro*.[41] The A9 murine fibrosarcoma cell line is derived from BALB/c 3T3 cells (H-2^d) transformed *in vitro* by polyoma virus. The A9 cells expressed on their surface a high density of polyoma virus tumor-associated antigen[41] (PTAA). Nontransformed 3T3 cells lacking PTAA were used as a negative control. Cell lines were maintained in DMEM supplemented with 10% fetal calf serum (FCS) that was inactivated for 30 min at 56°C, 2 m*M* glutamine, 10 m*M* HEPES buffer, and antibiotics (Biological Industries, Beit-HaEmek, Israel).

F. FLUORESCENCE IMMUNOASSAY (FIA)

The principle of the assay is to use the F–SSIL having the specific MAb attached to their surface as a means to detect the presence of antigen recognized by the specific MAb. This approach allows one to assess the feasibility of developing a quantitative FIA. For this we used specific F–SSIL having the NI32/2 MAb on their surface. F–SSIL having irrelevant IgG on their surface and F–SSL were used as a control for nonspecific binding.

The F–SSIL were prepared as described above. Cultured cells (A9, which carry the specific antigen, and 3T3, which lack this antigen) were suspended in trypsin–EDTA, then washed with DMEM containing 5% FCS and 0.05% sodium azide. Specific or nonspecific SSIL (0.25 µmol lipid) were added to 10^6 cells in a final volume of 0.1 ml. After a 40 min incubation on ice with gentle shaking, cells were washed twice in the same medium and analyzed using a Becton Dickinson FAST scan flow cytometry analyzer.[42]

Identical conditions were used for competition experiments between the F–SSIL and the free NI32/2 MAb or control IgG anti H-2^d, except that the cells were first incubated with varying amounts of the free IgG for 20 min on ice in DMEM containing 5% FCS and 0.05% sodium azide before adding the immunoliposomes.

III. RESULTS AND DISCUSSION

A. F–SSIL DESIGN

1. Coupling Titration of SPDP or SMPB to HPE in the External Face of F–SSL

The strategy was to prepare the F–SSL containing HPE as anchor for the ligand. These liposomes can be stored and used to prepare the F–SSIL from F–SSL by attachment of the desired ligand such as MAb. We selected two of the many available procedures[25,26] to attach the MAb to the F–SSL surface: (1) through a direct thioether bond which involves covalent binding of SPDP or SMPB to the external HPE to form the anchor for the antibody (Figure 1); (2) by covalent attachment of Protein A by a thioether bond. The attached Protein A then serves as an anchor for the noncovalent Fcγ domain-mediated binding of antibodies. The first approach is more general and applies to most proteins and peptides, while the second approach is limited to antibody only; therefore, the first approach will be described here in more detail.

The first step was to quantify the efficiency of this binding. SPDP and SMPB bind to HPE and have similar molecular masses and physical features; therefore it can be assumed that their binding will be affected to the same extent by factors such as bilayer rigidity, surface potential, and surface modification, e.g., by the presence of PEG in the head group.

The F–SSL, composed of FITC–PE, HPC, HPE, ^{2000}PEG–DSPE and cholesterol, were prepared as described in "Materials and Methods" (Section II). The only variable in the lipid composition was the

HPC:HPE mole ratio. The level of SPDP binding was followed by reacting the liposomal HPE–PDP with DTT and quantifying the product by its absorbance at 343 nm.[25] The binding of SPDP or SMPB was performed under mild conditions: 2 h at pH 7.8. Triethylamine, which is routinely used for binding of PE in organic solvents,[25] was replaced with the buffer having an alkaline pH 7.8.

Figure 2 demonstrates the relationship between the SPDP binding to liposomal HPE and the mol% of HPE in the liposomes. From the level of HPE in the external leaflet (64% of the 5 mol% HPE) it was determined that 44% and 29% of the external HPE reacted with SPDP to form HPE–PDP in liposomes containing 1 and 5 mol% HPE, respectively. Although the fraction of HPE to which SPDP binds decreases with increase in mol% HPE, the binding expressed as mmole HPE–PDP per mole total lipid increases constantly with increased mol% of HPE in the liposomes. Based on these results, all the F–SSL preparations were prepared with 5 mol% of HPE. The validity of using SPDP binding as a model for SMPB binding was demonstrated using quantitative TLC;[23] the levels of HPE–MPB and HPE–PDP formed when SMPB or SPDS were used were found to be equal. An alternative approach for *in vitro* diagnostics is to use thioesters by attaching the PDP moiety to both the ligand and the liposomes. However, thioesters may be less stable, especially in the presence of free thiol groups.[43]

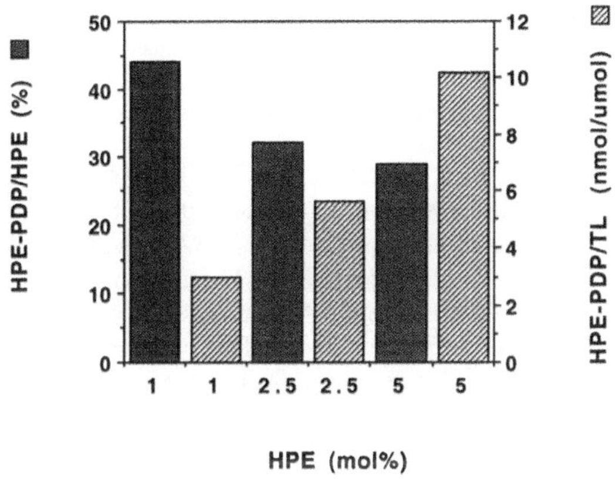

Figure 2 Effect of HPE concentration (mol%) in the D—SSL on PDP binding to HPE. Left-hand scale of y-axis refers to HPE in the external leaflet, and right-hand scale to total lipid (TL).

2. Effects of Mol% ^{2000}PEG–DSPE on Binding of Protein to the Liposome Surface

The long and flexible PEG moieties on the surface of the vesicles sterically stabilize the liposomes by reducing their interaction with plasma components.[30,32] Therefore, it is expected that these PEG groups may also interfere with the covalent attachment of protein molecules to the HPE–MPB on the vesicle surface. This was tested by measuring the degree of bovine serum albumin binding to F–SSL containing 3, 4, or 5 mol% ^{2000}PEG–DSPE. All three F–SSL preparations contained 5 mol% HPE and were of identical size. The binding of SMPB to the surface HPE, the binding of albumin (which has free thiol groups, and therefore needs no further thiolation) to the HPE–MPB, and the removal of the unreacted albumin were performed as described in "Materials and Methods".

The vesicles were characterized after attachment of the albumin. It was found that:

1. The amount of albumin bound was inversely related to the level of ^{2000}PEG–DSPE in the liposome. This relationship can be described as:

$$y = 54.8 - 7.85x$$

   where y is the bound albumin µg/µmole PL and x is the mol% ^{2000}PEG–DSPE.
2. The size distributions of the vesicles containing 4 and 5 mol% PEG–DSPE were unimodal and similar, with a mean size of 90 nm, while the vesicles having 3 mol% PEG–DSPE had a bimodal size distribution, with a mean size of 100 nm for the first population and >300 nm for the second population. Since no leakage of vesicle content occurred[10,44] (Table 1), these data can be explained by the albumin acting as a linker between vesicles, which occurs only at the low level of ^{2000}PEG–DSPE.

Table 1 Physical and Chemical Characterization of F–SSL and F–SSIL

	Liposomal formulation			
	F–SSL	F–SSIL–IgG	F–SSIL–32/2	F–SSIL Protein A–32/2
Mean diameter[a] (nm) ± SD	89 ± 33	103 ± 41	99 ± 69	100 ± 40
Shape of size distribution	Unimodal	Unimodal	Unimodal	Unimodal
Protein/PL (μg/μmol)	—	24.1 ± 1.3	22.3 ± 1.4	12.6 total (4.1 Protein A + 8.5 32/2)
DOX/PL (μg/μmol)	115	92	114	110
IgG/Liposome[b]		17	15	6.0 (8.2)[c]
IgG Fcγ exposed	(–)	(+)	(+)	(–)

[a] Size distribution of the vesicles was determined using the N4SD Submicron Particle Analyzer (Coulter Electronics).[23]

[b] Average number of antibody molecules per liposome. Calculations are based on the following assumptions: liposomes are unilamellar, average area per lipid molecule is 0.5 nm^2, and in the direct IgG binding all liposome-associated protein is IgG, while in the indirect binding through Protein A, the level of Protein A attached was determined first by determination of protein and then of the IgG by using FITG IgG (see Section II, "Materials and Methods").

[c] The number in parenthesis is the average number of Protein A molecules per vesicle.

3. Chemical and Physical Characterization of F–SSL and F–SSIL

F–SSL were prepared as described in "Materials and Methods". One part of the F–SSL dispersion was used as reference. To two other parts, antibodies (either NI32/2 MAb or IgG) were covalently attached either directly or through Protein A, as described. These liposomes are referred to as F–SSL, F–SSIL–32/2, F–SSIL–IgG, and F–SSIL–Protein A–32/2, respectively.

The procedure of attaching IgG or 32/2 causes a slight increase in vesicle size but no immediate vesicle aggregation or fusion. Table 1 also demonstrates that the attachment of Protein A and/or the IgGs had only minimal effect on vesicle size distribution (~100 nm) and no effect on the integrity of the vesicle bilayer, as the DOX/PL ratio remained unaffected by the protein attachment. Direct coupling efficiency of both antibodies to liposomes was almost identical (12.1% and 9.7% for IgG and NI32/2, respectively), as also expressed in similar levels of protein per μmole phospholipid (24 μg and 22 μg, respectively) and average number of IgG molecules per liposome (17 and 15, respectively).

Coupling efficiency of Protein A to the F–SSL was somewhat lower (8.2 Protein A molecules per vesicle), and 6.0 of these Protein A molecules became occupied by 32/2 MAb (see Table 1).

B. EXPOSURE OF Fc DOMAIN IN SSIL

To study the exposure of the Fc domain of the IgG_3 32/2 MAb on the immunoliposome surface, aliquots of F–SSIL–32/2 and F–SSL were passed through protein A Sepharose CL-4B columns.[35a] The liposomes were fluorescently labeled by DOX remote loading into their intraliposomal aqueous phase by ammonium sulfate gradient.[9,10] This labeling enables accurate and very sensitive follow-up of liposome elution. The columns were eluted using 1.5 *M* glycine buffer, pH 8.5, containing 3 *M* NaCl. Elution was followed by quantification of DOX fluorescence. Recovery of F–SSL was 50%, and all the F–SSL were eluted in the void volume (0.8–1.2 ml); a similar profile was obtained for F–SSL–Protein A–32/2 MAb. However, a much lower (15%) total recovery was obtained for F–SSIL–32/2 or IgG_3, which was eluted at volumes larger than the void volume (1.8–2.4 ml). This suggests that the F–SSIL 32/2 interact with the column Protein A and, therefore, Fcγ domains of at least part of the 32/2 MAb or IgG_3 molecules exposed on the SSIL surface are available for binding. This experiment indicates that the binding of the IgG to the F–SSL–Protein A is fully oriented, and their Fab domains remain unreacted, while the direct binding of IgG to the liposome surface is random and leaves free Fcγ domains (Table 1). Immunoliposome targeting, which is mediated through the Fab domain of the antibody molecule, may therefore be more efficient using the Protein A technique for antibody binding to the liposome.

C. STABILITY OF MAb–VESICLE ASSOCIATION

In many cases the FIA is intended to be used for determination of analytes in a biological milieu; therefore, it was important to study the effect of such medium on F–SSIL integrity. We tested the stability of MAb–vesicle association for the two modes of attachment used in this study: (1) F–SSIL 32/2 (MAb covalently attached via thioether bond to the vesicle surface) and (2) F–SSIL–Protein A–32/2 (MAb attached to the vesicle via Protein A). Instability may be relevant especially to the noncovalent (Protein A-mediated) binding.[47] The assessment is based on the ability of the Protein A column (see above) to detach the FITC-MAb from the vesicles, as determined from the ratio of FITC/DOX (see Section II.D above).

In both cases, fresh F–SSIL loaded with DOX at 4°C were compared with F–SSIL stored for 1 month at 4°C. The recovery of vesicles and the ratio of 32/2 MAb molecules per vesicle (FITC/DOX ratio) was unchanged for at least 1 month storage at 4°C, indicating the high stability of the MAb–vesicle association for both types of attachment upon storage. We also studied the stability of the association of 32/2 MAb after incubation for 3 h at 25 °C in 80% human serum or 5% human serum albumin, conditions which are relevant to and imitate real-life immunodiagnostics. It was found that the ratio of 32/2 MAb to vesicle was unaffected by the incubation, indicating the high stability of MAb vesicle association for both types of attachment in a biological milieu.

D. EFFECTS OF ANTIBODY ATTACHMENT ON SSIL INTERACTION WITH SERUM COMPONENTS

One of the major factors which may reduce the sensitivity of liposome-based FIA is aggregation of the liposomes which may be induced by the medium (such as serum) components other than the analytes. It is well established that one mechanism to increase particle size is vesicle aggregation induced by serum components. This was tested using heat-inactivated FCS and fresh guinea pig serum. F–SSIL, F–SSL, and F–Lip (0.4 μmol PL) were incubated with 50% heat-inactivated FCS at room temperature. The kinetics of change of turbidity, measured as absorbance at 440 nm, was used to indicate change in vesicle size.[23]

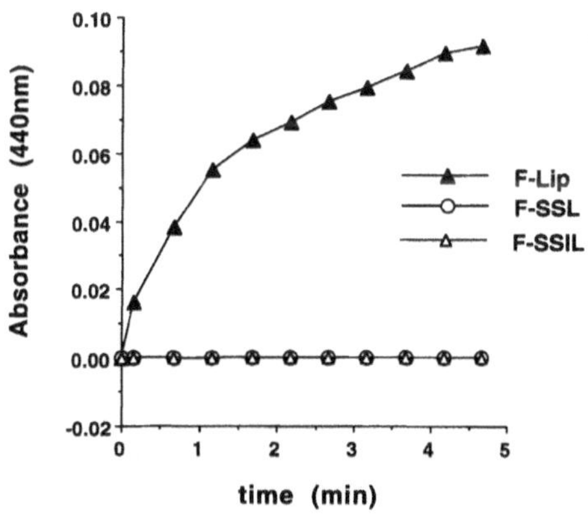

Figure 3 Effect of ^{2000}PEG—DSPE on serum-induced aggregation of liposomes and immunoliposomes. F–Lip (▲—▲), F–SSL (o——o), and F–SSIL (Δ——Δ) in 10% sucrose were mixed with equal volumes of heat-inactivated FCS and incubated at room temperature. The time-dependent increased turbidity (absorbance at 440 nm) was determined with time.

Figure 3 demonstrates that while F–Lip preparation shows a marked and fast increase in turbidity, no such changes were observed for F–SSL and F–SSIL–IgG. Similar results were obtained in fresh guinea pig serum containing active complement system incubated for 15 min at 25°C (conditions under which cells sensitized with the specific antibodies undergo lysis).

This, together with our previous observations that the incubation with either heat-inactivated or fresh serum does not induce leakage of SSIL content,[44] supports previous suggestions that the steric stabilization prevents interaction of serum components with the liposomal membrane (steric barrier effect), thereby improving the stability of SSIL in a biological milieu.[30,32,33]

E. CHARACTERIZATION OF THE BINDING OF F–SSIL–32/2 TO THEIR TARGET CELLS

The main motivation to develop liposome-based FIA is to improve the sensitivity of the immunoassays without reducing the immunospecificity which directly determines the signal-to-noise ratio. The use of

SSL as a replacement for conventional liposomes in such an assay is aimed, in addition, to improve the stability of the liposomes in the presence of biological fluids containing molecules which are "hostile" to liposome stability. These variables may be related to each other, as the loss of integrity is one of the main causes of reduction in signal/noise ratio. In the studies described above we demonstrate that, indeed, the steric stabilization improves liposome physical stability. Furthermore, steric stabilization, if done with suitable geometric consideration, improves specificity to a very large extent, and at the same time it increases signal to noise ratio (see Figures 4–6 below). Introducing the PEG moiety as a liposome surface modification proved to be a very efficient steric barrier which reduces undesired steric interactions of liposomes with serum macromolecules and cells *in vitro* and *in vivo*.[30–33] Preliminary work with SSL to which antibodies have been attached indicates that a larger PEG moiety (^{5000}PEG) reduces the immunological recognition, probably by steric hindrance.[45,46] When using a much shorter PEG moiety such as ≤750 Da PEG, there is either no barrier or the barrier is not sufficient to prevent insertion-induced destabilization. The following experiments were designed to test to what extent the presence of the ^{2000}PEG moiety on the vesicle surface affects immunospecificity.

Specific binding, false positives, and false negatives were studied by using the specific target cells, A9, having the PTAA on their surface to which the 32/2 MAb should bind specifically. These were compared with 3T3 cells which lack the PTAA (Figure 4). All liposomes used in these experiments contained 5 mol% of ^{2000}PEG–DSPE in their bilayer. We compared the binding of F–SSL, F–SSIL–IgG, F–SSIL–32/2, and F–SSIL–Protein A–32/2 to the two cell lines. The binding of the fluorescent liposomes was determined by flow cytometry, and by fluorescence microscopy.

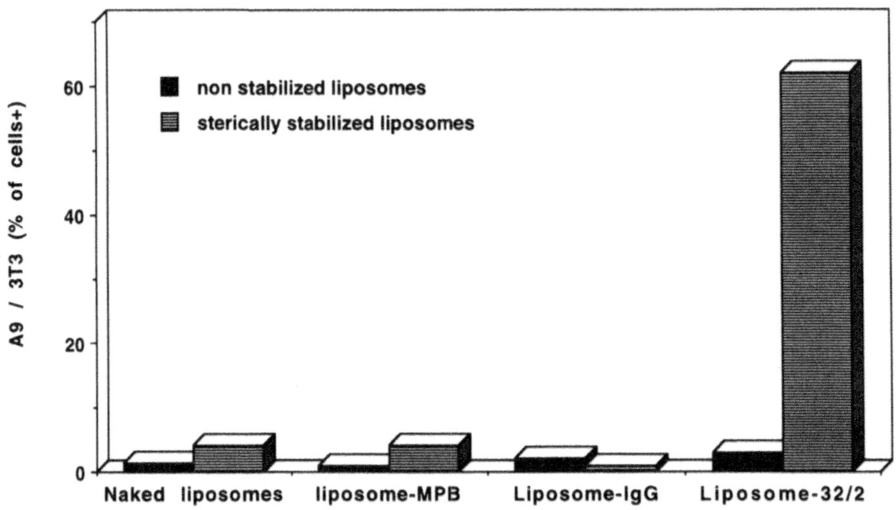

Figure 4 Effect of ^{2000}PEG—DSPE on the specific binding of fluorescent immunoliposomes and their intermediates expressed as ratio of stained A9 cells to stained 3T3 cells: liposomes before and after the binding of SMPB to HPE on their surface (naked liposomes, liposome—MPB) and immunoliposomes with the control IgG or the NI32/2 MAb, all labeled with FITC—PE, were incubated for 45 min in DMEM containing 5% FCS and 0.05% sodium azide with the nonrelevant 3T3 cells or the specific target A9 tumor cells. The unbound liposomes were washed and the fluorescence intensity of the cells was measured using a flow cytometer. The positive cells (cells +) are those which have a fluorescence intensity greater than the autofluorescence of the untreated cells.

Figure 4 demonstrates that of all the fluorescent liposomes tested, immunospecific binding occurred only when A9 tumor cells were reacted with the F–SSIL–32/2 immunoliposomes. It shows the percent of fluorescent-stained cells after interaction with fluorescent liposomes, comparing the binding of various liposome preparations as specified below to A9 (specific target) cells and 3T3 (control) cells, shows that the average binding of specific F–SSIL–32/2 to the A9 tumor cells (having PTAA on their surface) is at least 25 times greater than that of the nonspecific F–SSIL–IgG, and that the binding of F–SSIL–32/2 to 3T3 cells (which lack the PTAA antigen) is minimal. Also, the liposome-related small increase in fluorescence of 3T3 cells was identical for F–SSL, F–SSIL–IgG, and F–SSIL–32/2. Thus, only the F–SSIL–32/2, but not the F–SSIL–IgG or F–SSL, bind to the A9 cells, while none of the three liposome preparations binds appreciably to the nonrelevant 3T3 cells, which suggests that the binding of the SSIL–32/2 is immunospecific.

To obtain more detailed information on the effect of the steric barrier of ^{2000}PEG present on the vesicle surface, we also compared the binding to target cells of fluorescent liposomes at different stages of immunoliposome preparation containing, or lacking, ^{2000}PEG–DSPE (Figure 4).

The following liposome (F–Lip or F–SSL) preparations were compared: (i) F–Lip and F–SSL; (ii) F–Lip and F–SSL after the attachment of MPB moiety; (iii) ii, after binding of control IgG; (iv) ii, after binding of 32/2 MAb. Figure 4 demonstrates that the presence of ^{2000}PEG on the liposome surface improves the signal-to-noise ratio since much more nonspecific binding of liposomes to cells occurred for all liposomes lacking the PEG moiety. Of special interest is the very high binding of liposomes containing HPE–MPB to both cell types. Although there is a 2.5-fold higher binding of F–Lip–32/2 to A9 cells than to 3T3 cells, the specificity is much lower than that of the F–SSIL–32/2 preparation. All of the high background of nonspecific binding obtained with Lip (non-SSL) was eliminated when ^{2000}PEG–DSPE (5 mole %) was included in the liposomes. Only the F–SSIL–32/2 bound efficiently to the specific A9 cells (>60% stained cells); F–SSL and F–SSIL–IgG, which lack the specific MAb, have very low binding (<5% stained cells). This suggests that the steric barrier of ^{2000}PEG on the liposome surface improves the immunospecific binding of the immunoliposomes containing the specific MAb on their surface to the relevant target cells, possibly by preventing nonspecific binding or adsorption of the liposomes to the cell surface. Similar results were obtained for the F–SSIL–Protein A–32/2 (data not shown).

These findings were supported by competition experiments with soluble (nonliposome-attached) antibodies. Figure 5 demonstrates that the soluble specific NI32/2 MAb, but not the nonrelevant anti-H-2^d antibodies, which also bind to A9 cells (H-2^d) with high affinity, inhibits the binding of the F–SSIL–32/2 to A9 cells. The inhibition was dose-dependent and reached saturation (90% inhibition) at 20 ng 32/2 MAb. This confirms that the binding of F–SSIL–32/2 to the A9 cells is indeed immunospecific.

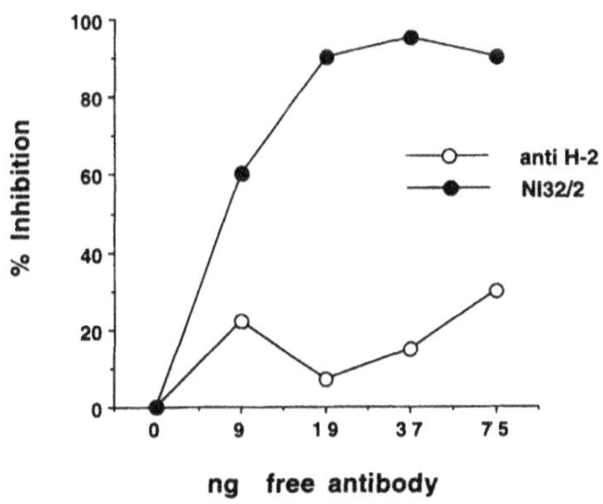

Figure 5 Specific inhibition of SSIL—32/2 binding to A9 tumor cells by free NI32/2 MAb. A9 cells (1 × 10^6) were incubated at 4°C either alone (control) or with different amounts of free NI32/2 MAb (●——●), or anti-H-2^d Ab (○——○) in a final volume of 0.1 ml. After 20 min, the fluorescent FITC-labeled SSIL—32/2 (0.3 μmol PL in 0.1 ml) were added to every sample for an additional 45 min under the same conditions. The mean fluorescence intensity of the cells was analyzed using flow cytometry and percent inhibition was calculated. For more details see Sections II and III.

IV. CONCLUSIONS

The results indicate that use of SSIL for *in vitro* and *ex vivo* diagnostics is superior to that of conventional liposomes, as long as the recognition between the ligand on the liposome surface and the analyte is undisturbed by the PEG moiety. The geometry of the vesicle surface is very important, as previous data showing that the ^{5000}PEG moiety attached to the immunoliposome surface prevents the binding of the SSIL to their target cells,[45,46] while, as we demonstrate here, reducing the PEG size to 2000 Da prevents the interference of the PEG with the immunospecific binding to target antigens on the surface of cells. At the same time, this PEG moiety gave full protection (steric stabilization) to liposomes, and also to a

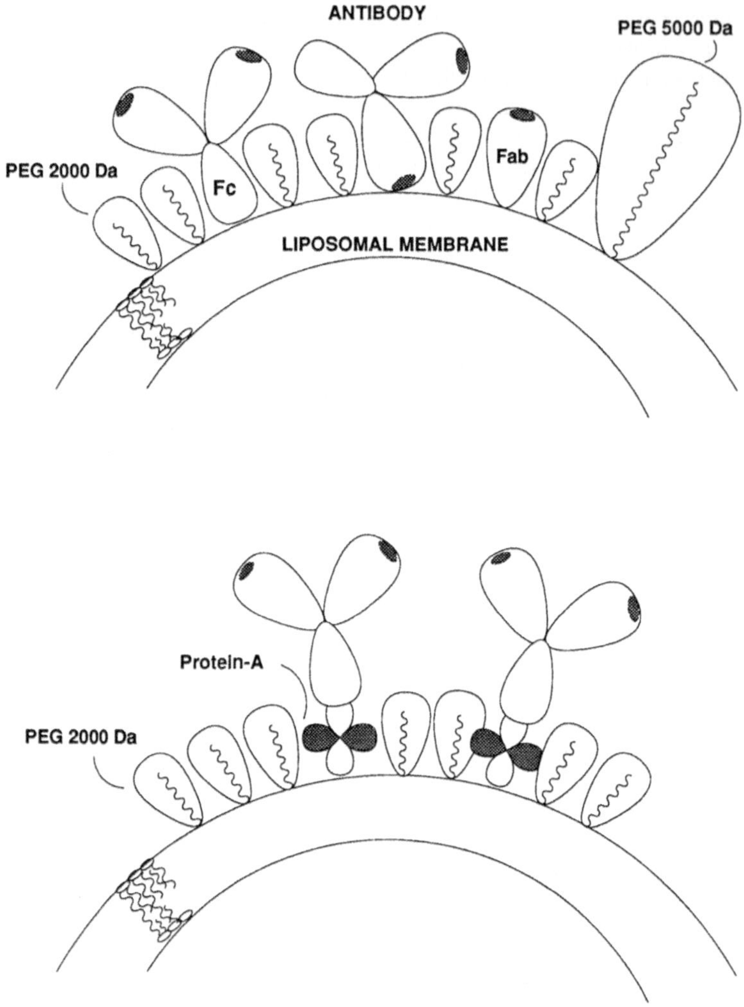

Figure 6 Schematic diagram: spatial relationships of SSIL—32/2 (top) and SSIL—Protein A—32/2 (bottom) of the following ligands: PEG—5000, PEG—2000, intact IgG, and a Fab fragment. The size ratio between the various molecules described in the cartoon is based on values obtained elsewhere.[30,31,33,39,46,51]

large extent it eliminated false positives and false negatives in the binding. This geometric consideration is described in Figure 6.

Neither exposure of the antibody Fc domain, as demonstrated here and in Emanuel et al.,[44] nor the binding of the antibody to the far end of the ^{2000}PEG (attached to the liposomes),[48,49] nor of the Fab domain[44,47] has a major effect on the stability of the SSIL *in vitro*. This, together with the observation that a short PEG moiety (≤750 Da) does not stabilize the vesicles,[32] indicates that the barrier effect is mainly related to the prevention of interaction between medium (serum) components and the vesicle membrane. Therefore, a combination of a sufficient steric barrier together with exposure of the relevant domain of the recognition ligand is the formula to achieve optimal use of liposomes in diagnostics (Figure 6). This study demonstrates that detection and diagnostic assays based on the use of SSIL are highly feasible and ready to be developed for immunodiagnostics for a broad spectrum of analytes. Since the binding is highly specific, the use of fluorescent-labeled liposomes will enable improving assay sensitivity. Here we used liposomes containing a nontransferable fluorescent phospholipid analogue (FITC–PE).[13] Many other similar lipid analogues with a broad range of spectral properties are available.

An alternative approach will be to use liposomes which contain a fluorescent marker in their internal aqueous phase. The latter approach was used extensively in the past.[1,2] Although it may give even higher sensitivity, it may be less favorable due to a lower signal-to-noise ratio, which is due to the fact that *full* retention of liposome content is more difficult to achieve than stability of liposomal bilayer. From the exposure of the Fcγ domain, determined by binding to Protein A columns, it is clear that for the direct covalent attachment the binding of the MAb to the liposome surface is random. This may reduce affinity and avidity of binding. This problem does not exist for the attachment of antibodies through Protein A, as was shown in this work. However, this use of Protein A is limited to the attachment of antibodies having an intact Fcγ domain. Another approach to overcome random binding, which is much more general, is to use ligand molecules in which a single-SH group can be introduced in a specific site which does not interfere with the relevant epitope.

Many other methods of attachment of the ligand, either directly to the liposome surface[25,26] or to the far end of the PEG moiety[48,49] should be compared in order to optimize the ligand binding procedure.

The feasibility demonstrated by our results enables development of SSL-based LDA using homogeneous or heterogeneous methodologies for *in vitro* and *ex vivo* diagnostics based on either the direct or the competition approaches (both are also demonstrated here). Either ELISA in which the reaction involves solid support, or flow-injection liposome-enhanced immunoassay can be applied. Both require instrumentation and can also be automated.[1–5]

An additional attractive option is the use of the stick/strip approach, which can be developed in a way which eliminates the need for instrumentation for qualitative detection and therefore can give fast "yes/no" evaluation under field conditions. The stick/strips can even be prepared in such a way that the final product contains freeze-dried liposomes in which integrity and stability are retained due to the presence of cryoprotectants in the lyophilization medium and the selection of suitable lipid composition.[50] Such stick/strips may be most suitable for field tests (including medical, veterinary, food and environmental uses), especially in Third World countries and in sites lacking laboratory facilities.

ABBREVIATIONS

DMEM	Dulbecco's modified Eagle medium
DOX	Doxorubicin
DTT	Dithiothreitol
FCS	Fetal calf serum
FIA	Fluorescence immunoassay
FITC	Fluorescein isothiocyanate
FITC–PE	Fluorescein isothiocyanate–phosphatidylethanolamine
F–Lip	FITC–PE-labeled Lip
F–SSL	FITC–PE-labeled SSL
F–SSIL	FITC–PE-labeled SSIL
HPC	Hydrogenated soy phosphatidylcholine
HPE	Hydrogenated soy phosphatidylethanolamine
LDA	Liposome-based diagnostic assay
Lip	Nonstabilized liposomes (liposomes lacking PEG–DSPE)
MAb	Monoclonal antibodies
MPB-	4-(*p*-Maleimidophenyl) butyryl
PDP-	3-(2-Pyridyldithio) propionyl
PE	Phosphatidylethanolamine
PEG	Polyethylene glycol
PEG–DSPE	*N*-Carbamyl-poly-(ethylene glycol methyl ether)-1,2-distearoyl-*sn*-glycero-3-phosphoethanolamine triethyl ammonium salt
PL	Phospholipid
PTAA	Polyoma virus tumor-associated antigen
RES	Reticuloendothelial system
SMPB	*N*-Succinimidyl 4-(*p*-Maleimidophenyl)butyrate
SPDP	*N*-Succinimidyl 3-(2-pyridyldithio)propionate
SSIL	Sterically stabilized immunoliposomes
SSL	Sterically stabilized liposomes
32/2	NI32/2 MAb

REFERENCES

1. **Monroe, D.,** Novel liposome immunoassays for detecting antigens, antibodies and haptens. *J. Liposome Res.*, 1, 339, 1989.
2. **Rongen, H. A. H., Van der Horst, H. M., Hugenholtz, G. W. K., Buet, A., and van Bennekom, W. P.,** Development of a liposome immunosorbent assay for human interferon-γ. *Anal. Chim. Acta*, 287,191, 1994.
3. **Katoh, S., Kishimara, M., and Fukuda, H.,** Use of antigen-coupled liposomes for homogeneous immunoassays of polyclonal antibody, in *Handbook of Nonmedical Applications of Liposomes*, Vol. 4, Lasic, D. D. and Barenholz, Y., Eds., CRC Press, Boca Raton, FL, 1995.
4. **Singh, A. K. and Carbonell, R. G.,** Liposomes in immunodiagnostics, in *Handbook of Nonmedical Applications of Liposomes*, Vol. 4, Lasic, D. D. and Barenholz, Y., Eds., CRC Press, Boca Raton, FL, 1995.
5. **Reeves, S. G., Siebert, S. T. A., and Durst, R. A.,** Liposome-amplified immunodetermination of environmental contaminants, in *Handbook of Nonmedical Applications of Liposomes*, Vol. 4, Lasic, D. D. and Barenholz, Y., Eds., CRC Press, Boca Raton, FL, 1995.
6. **Lasic, D. D.,** *Liposomes: From Physics to Applications*, Elsevier, Amsterdam, 1994.
7. **Barenholz, Y. and Crommelin, D. J. A.,** Liposomes as pharmaceutical dosage forms, in *Encyclopedia of Pharmaceutical Technology*, Swarbrick, J. and Boylan, J. C., Eds., Vol. 9, Marcel Dekker, New York, 1994, 1.
8. **Parr, M. J. and Cullis, P. R.,** Transbilayer transport induced by transmembrane pH gradients in liposomes: implications for biological systems, in *Handbook of Nonmedical Applications of Liposomes*, Vol. 2, Barenholz, Y. and Lasic, D. D., Eds., CRC Press, Boca Raton, FL, 1995.
9. **Haran, G., Cohen, R., Bar, L. K., and Barenholz, Y.,** Transmembrane ammonium sulfate gradients in liposomes produce efficient and stable entrapment of amphipathic weak bases. *Biochim. Biophys. Acta*, 1151, 201, 1993.
10. **Bolotin, E. M., Cohen, R., Bar, L. K., Emanuel, N., Ninio, S., Lasic, D. D., and Barenholz, Y.,** Ammonium sulfate gradients for efficient and stable remote loading of amphipathic weak bases into liposomes and ligandoliposomes. *J. Liposome Res.*, 4, 455, 1994.
11. **Weinstein, J. W., Ralston, E., Legerman, L. D., Klausner, R. D., Dragsten, P., Henkart, P., and Blumenthal, R.,** Self-quenching of carboxyfluorescein fluorescence: uses in study of liposome stability and liposome-cell interactions, in *Liposome Technology*, 1st ed., Vol. 3, Gregoriadis, G., Ed., CRC Press, Boca Raton, FL, 1989, 205.
12. **Ceh, B.,** Theory of loading of agents into liposomes, in *Handbook of Nonmedical Applications of Liposomes*, Vol. 3, Barenholz, Y. and Lasic, D. D., Eds., CRC Press, Boca Raton, FL, 1995.
13. **Struck, D. K., Hoekstra, D., and Pagano, R. E.,** Use of resonance energy transfer to monitor membrane fusion. *Biochemistry*, 20, 4094, 1981.
14. **Barenholz, Y.,** Optimization of amphiphile-based colloidal carriers, in *Handbook of Nonmedical Applications of Liposomes*, Vol. 3, Barenholz, Y. and Lasic, D. D., Eds., CRC Press, Boca Raton, FL., 1995.
15. **Mackness, M. I. and Durrington, P. N.,** Lipoprotein separation and analysis for clinical studies, in *Lipoprotein Analysis — A Practical Approach*, Converse, C. A. and Skinner, E. R., Eds., IRL Press, Oxford, 1992, chapter 1.
16. **Brinkley, M.,** A brief survey of methods for preparing protein conjugates with dyes, haptens and cross-linking reagents. *Bioconjugates*, 3, 2, 1992.
17. **Dammen, B., Fogedby, H. C., Ipsen, J. H., Jeppsen, C., Jorgensen, K., Mouristen, O. G., Risbo, J., Sabra, M. C., Sperotto, M. M., and Zuckermann, M. J.,** Computer simulation of the thermodynamic and conformational properties of liposomes, in *Handbook of Nonmedical Applications of Liposomes*, Vol. 4, Lasic, D. D. and Barenholz, Y., Eds., CRC Press, Boca Raton, FL, 1995.
18. **Gabizon, A., Barenholz, Y., and Bialer, M.,** Prolongation of the circulation time of doxorubicin encapsulated in liposomes containing a polyethylene glycol-derivatized phospholipid: pharmacokinetic studies in rodents and dogs, *Pharmaceutical Res.*, 10, 703, 1993.
19. **de Gier, J.,** Osmotic behavior and permeability properties of liposomes, *Chem. Phys. Lipids*, 64, 187, 1993.
20. **Lichtenberg D. and Barenholz Y.,** Liposomes: preparation, characterization and preservation, in *Methods in Biochemical Analysis*, Glick, D., Ed., Vol. 33, Wiley, New York, 1988, 337.
21. **New, R. C., Ed.,** *Liposomes — A Practical Approach*. IRL Press, Oxford, 1990.
22. **Barenholz, Y., Ed.,** *Chem. Phys. Lipids*, Special Issue: *Quality Control of Liposomes*, Vol. 64, 1994.
23. **Barenholz, Y. and Amselem, S.,** Quality control assays in the development and clinical use of liposome-based formulations, in *Liposome Technology*, Vol. 1, 2nd ed., Gregoriadis, G., Ed., CRC Press, Boca Raton, FL, 1993, 527.
24. **Burgess, S. W., Moore, J. D., and Shaw, W. A.,** Lipids as raw materials for liposomal products: properties and characterization, in *Handbook of Nonmedical Applications of Liposomes*, Vol. 3, Barenholz, Y. and Lasic, D. D., Eds., CRC Press, Boca Raton, FL, 1995.
25. **Martin, F. J., Heath, T. D., and New, R. R. C.,** Covalent attachment of proteins to liposomes, in *Liposomes — A Practical Approach*, New, R. R. C., Ed., IRL Press, Oxford, 1990, 163.
26. **Wong, S. S.,** *Chemistry of Protein Conjugation and Cross-Linking*, CRC Press, Boca Raton, FL, 1991.
27. **Bredehorst, R., Ligler, F. S., Kusterbeck, A. W., Chang, E. L., Gaber, B. P., and Vogel, C.-W.,** Effect of covalent attachment of immunoglobulin fragments on liposomal integrity, *Biochemistry*, 25, 5693, 1986.

28. **Peeters, P. A. M., Oussoren, C., Eling, W. M. C., and Crommelin, D. J. A.**, Unwanted interactions of maleimidophenylbutyrate-phosphatidylethanolamine containing (immuno) liposomes with cells *in vitro*, *J. Liposome Res.*, 1, 261, 1990.
29. **Senior, J. H.**, Fate and behavior of liposomes *in vivo*: A review of controlling factors, *Crit. Rev. Ther. Drug Carrier Syst.*, 3, 123, 1987.
30. **Torchilin, V. P.**, Effect of polymers attached to lipid head groups on properties of liposomes, in *Handbook of Nonmedical Applications of Liposomes*, Vol. 4, Lasic, D. D. and Barenholz, Y., Eds., CRC Press, Boca Raton, FL, 1995.
31. **Lasic, D. D. and Martin, F. J.**, *Stealth Liposomes*, CRC Press, Boca Raton, FL., 1995.
32. **Woodle, M. C. and Lasic, D. D.**, Sterically stabilized liposomes. *Biochim. Biophys. Acta*, 1113, 171, 1992.
33. **Huang, L.**, Covalently attached polymers and glycans to alter the biodistribution of liposomes, *J. Liposome Res.*, 2, 289, 1992.
34. **Gabizon, A., Catane, R., Uziely, B., Kaufman, B., Safra, T., Cohen, R., Martin, F., Huang, A., and Barenholz, Y.**, Prolonged circulation time and enhanced accumulation in malignant exudates of doxorubicin encapsulated in polyethylene-glycol coated liposomes, *Cancer Res.*, 54, 987, 1994.

35a. **Hudson, L. and Hay, F. C.**, *Practical Immunology*, 3rd ed., Blackwell, Oxford, 1989, 316.

35b. **Hudson, L., and Hay, F. C.**, *Practical Immunology*, 3rd ed., Blackwell, Oxford, 1989, 34.

36. **Barenholz, Y., Gibbes, D., Litman, B. J., Goll, J., Thompson, T. E., and Carlson, F. D.**, A simple method for the preparation of homogeneous phospholipid vesicles. *Biochemistry*, 16, 2806, 1977.
37. **Amselem, S., Cohen, R., and Barenholz, Y.**, *In vitro* tests to predict *in vivo* performance of liposomal dosage forms, *Chem. Phys. Lipids*, 64, 219, 1993.
38. **Thomas, P. D. and Poznansky, M. J.**, Curvature and composition-dependent lipid asymmetry in phosphatidylcholine vesicles containing phosphatidylethanolamine and gangliosides. *Biochim. Biophys. Acta*, 978, 85, 1989.
39. **Lasch, J., Niedermann, G., Bogdanov, A. A., and Torchilin, V. P.**, Thiolation of preformed liposomes with iminothiolane, *FEBS Lett.*, 214, 13, 1987.
40. **Minamide, L. S. and Bamburg, J. R.**, A filter paper dye-binding assay for quantitative determination of protein without interference from reducing agents or detergents. *Anal. Biochem.*, 190, 66, 1990.
41. **Langer, A. B., Emanuel, N., Even, J., Fridman, W. H., Gohar, O., Gonen, B., Katz, B. Z., Ran, M., Smorodinksy, N. I., and Witz, I. F.**, Phenotypic properties of 3T3 cells transformed *in vitro* with polyoma virus and passaged once in syngeneic animals, *Immunobiology*, 185, 281, 1992.
42. **Carter, N. P. and Meyer, E. W.**, Introduction to the principles of flow cytometry, in *Flow Cytometry — A Practical Approach*, 2nd ed., Ormerod, M. G., Ed., IRL Press, Oxford, 1994, chapter 1.
43. **Matzku, S., Krempel, H., Weckenmann, H.-P., Schirrmacher, V., Sinn, H., and Stricker, H.**, Tumour targeting with antibody-coupled liposomes: failure to achieve accumulation in xenografts and spontaneous liver metastases, *Cancer Immunol. Immunother.*, 31, 285, 1990.
44. **Emanuel, N., Kedar, E., Bolotin, E. M., Smorodinsky, N. I., and Barenholz, Y.**, Targeted delivery of doxorubicin-loaded sterically stabilized immunoliposomes. I. Preparation and characterization, submitted, 1995.
45. **Mori, A., Klibanov, A. L., Torchilin, V. P., and Huang, L.**, Influence of the steric barrier activity of amphipathic poly(ethyleneglycol) and ganglioside GM_1 on the circulation time of liposomes and on the target binding of immunoliposomes *in vivo*, *FEBS Lett.*, 284, 263, 1991.
46. **Klibanov, A. L., Maruyama, K., Beckerleg, A. M., Torchilin, V. P., and Huang, L.**, Activity of amphipathic poly(ethylene glycol) 5000 to prolong the circulation time of liposomes depends on the liposome size and is unfavorable for immunoliposome binding to target, *Biochim. Biophys. Acta*, 1062, 142, 1991.
47. **Kessler, S. W.**, Rapid isolation of antigens from cells with staphylococcal Protein A antibody adsorbent: parameters of the antibody-antigen complexes with Protein A, *J. Immunol.*, 115, 6, 1617, 1975.
48. **Allen, T. M., Agrawal, A. K., Ahmad, I., Hansen, C. B., and Zalipsky, S.**, Antibody-mediated targeting of long-circulating ($Stealth^R$) liposomes, *J. Liposome Res.*, 4, 1, 1994.
49. **Blume, G., Cevc, G., Crommelin, D. J. A., Bakker-Woudenberg, I. A., Kluft, C., and Storm, G.**, Specific targeting with poly(ethylene glycol)-modified liposomes: coupling of homing devices to the ends of the polymeric chains combines effective target binding with long circulation times, *Biochim. Biophys. Acta*, 1149, 180, 1993.
50. **Crowe, J. H. and Crowe, L. M.**, Preservation of liposomes by freeze-drying, in *Liposome Technology*, 2nd ed., Vol. 1, Gregoriadis, G., Ed., CRC Press, Boca Raton, FL, 1993, 229.
51. **Roitt, I.**, *Essential Immunology*, 8th ed., Blackwell, London, 1994, 43.

Chapter 17

Liposome-Amplified Immunodetermination of Environmental Contaminants

Stuart G. Reeves, Sui Ti A. Siebert, and Richard A. Durst

CONTENTS

I. INTRODUCTION

Immunoassays[1] are based on the highly specific recognition of an analyte by an antibody raised against it and the subsequent binding that occurs. They have been used in the clinical field for many years, where advantage is taken of their specificity and speed. Most immunoassays currently in use are ELISAs (enzyme-linked immunosorbent assays), where the antibodies to the analyte of interest are usually bound to a solid phase. Then the amount of analyte that competitively binds to them is measured in a variety of ways, e.g., using an enzyme to produce a detectable product from an externally supplied substrate. Such enzyme-linked assays have the benefits of speed and ease of use, and involve neither the regulatory requirements associated with the use of radioactivity in radioimmunoassays[2] nor the environmental and personal hazards. ELISAs have been used extensively for clinical applications, both as laboratory and extra-laboratory tests. More recently they have gained increasing acceptance in the area of environmental monitoring.[3-5]

The use of liposomes in a variety of immunoassay formats has been demonstrated in the last few years. Liposomes are the spherical vesicles spontaneously formed when certain lipids are dispersed in water.[6] During their formation, they encapsulate in their lumen a portion of the aqueous solution in which they were dispersed, along with any dissolved chemicals. Such chemicals are termed encapsulants, and these can include markers of various types. It is also possible to label the outer surface of the liposome with analyte or antibody molecules, and thus incorporate the liposome as a reagent in an immunoassay. Some of the techniques used in liposome immunoassays will be detailed in this chapter, along with possible future developments in the field.

II. CURRENT ELISA FORMATS IN ENVIRONMENTAL MONITORING

The assays used in environmental monitoring can be roughly divided into field (screening) assays and laboratory assays. The former are designed to be very rapid and easy to use and to give an approximate measurement or a present/absent answer for a large number of samples. They are designed to give the answer on site, so that those samples that are suspect can be sent to a laboratory for confirmatory testing,

0-8493-4013-6/96/$0.00+$.50

either using more sophisticated immunoassays or by classical analytical techniques. A brief description of the types of immunoassays that have been developed is given below. More comprehensive details can be found in the literature.

A. HOMOGENEOUS AND HETEROGENEOUS ASSAYS

A homogeneous immunoassay does not require separation of the antibody/analyte complex from the free analyte before measurement, whereas a heterogeneous assay does. The former format requires that the amount of antibody/antigen interaction in the assay can be measured directly and that the presence of unbound antibody and antigen in the assay mixture does not affect the result. An example of a format for such a liposome-mediated assay is shown in Figure 1. In this assay, the binding of an antibody to a tagged liposome causes lysis of the liposome. Thus the presence of a high level of analyte will lead to a low signal from the released contents of the liposomes and vice versa. While there are methods available that will fit these criteria, most assays, especially in the environmental monitoring field, require that unbound antigen or antibody is washed away from bound analyte, and these assays are termed heterogeneous. Such assays are usually more time consuming, and less easy to automate, but are often more accurate because of the removal of potentially interfering background material.

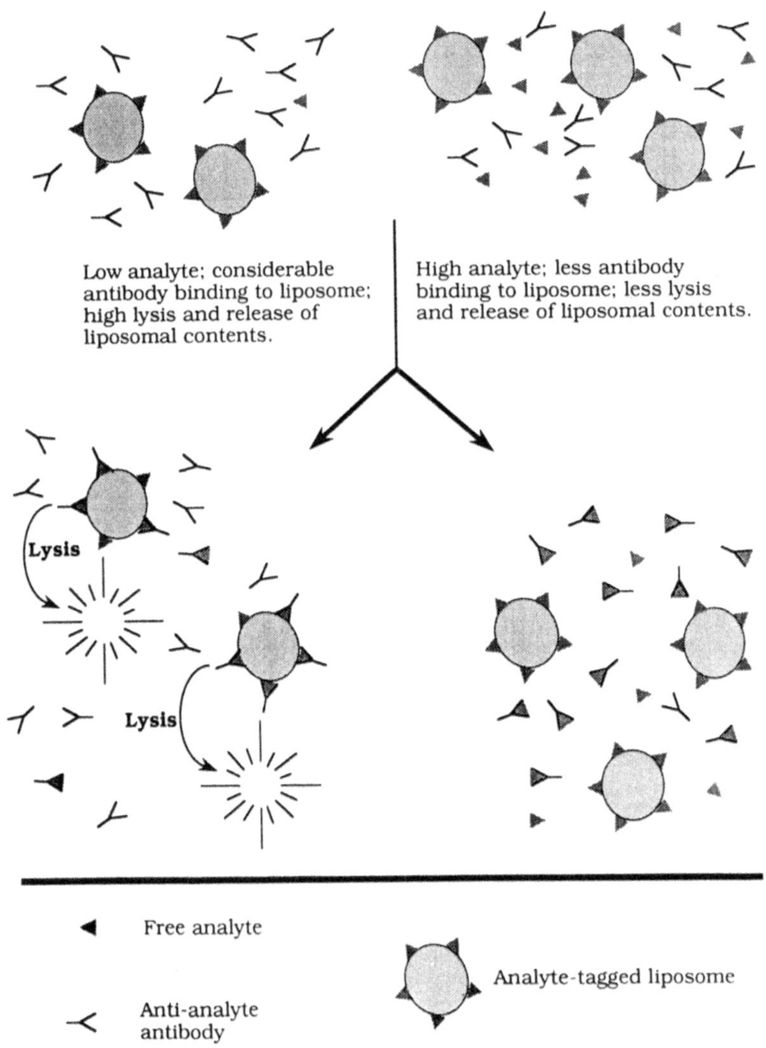

Figure 1 Diagrammatic representation of a liposome-enhanced homogeneous immunoassay.

B. COMPETITIVE AND NONCOMPETITIVE ASSAYS

In a noncompetitive (sandwich) assay, the amount of analyte bound to the excess immobilized antibody is measured directly by a secondary antibody. This requires an analyte that has a minimum of two distinct binding sites (epitopes) and can consequently bind to two different antibodies at once. The amount of sample analyte present and bound is directly proportional to the signal obtained in this form of the assay. Unfortunately, most environmental contaminants are small molecules called haptens and have only one available binding site. As a result, most are measured in a competitive assay format, where sample analyte and labeled analyte compete for a limited number of antibody sites on the solid phase. The greater the proportion of unlabeled sample analyte, the less labeled analyte can bind. Thus the amount of labeled analyte bound and measured is inversely proportional to the amount of free analyte in the test solution. Both of these types of assay are shown diagrammatically in Figure 2.

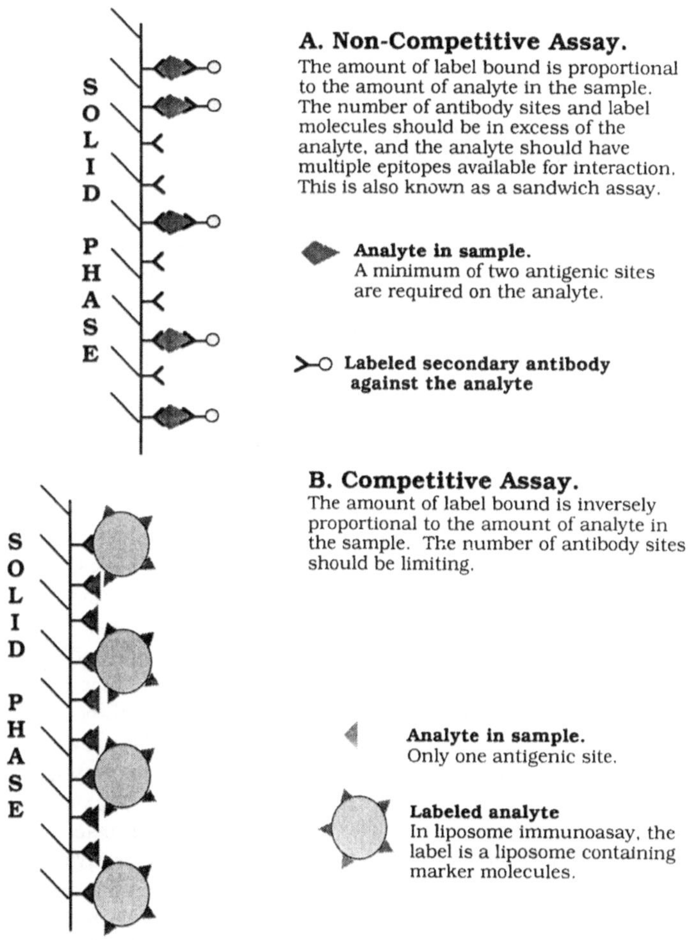

Figure 2 Diagrammatic representation of competitive and noncompetitive immunoassays.

C. LABELING AND AMPLIFICATION

Both competitive and noncompetitive immunoassays require that the binding event is detected, and this is usually done by labeling one of the components of the binding reaction. The first immunoassays that were developed used radioactive labels[2] and, because of high measurement sensitivity, needed no amplification. The desire to develop nonisotopic assays led to the concept of amplification. The original amplification method was by means of an enzyme linked to the analyte or to an antibody. The action of the enzyme on a substrate to produce a measurable product (usually colored) was then allowed to proceed for a defined time. Many assays were developed with alkaline phosphatase and horseradish peroxidase as the amplifying enzymes, both of which can be used in reactions producing colored products that can be measured spectrophotometrically, and these labels are still in common use. Most commercial ELISA kits and methods use this technique, in combination with a microtiter plate or strip and an ELISA plate reader. Other enzymes and other methods of product detection have also been used, with and without amplification, and in this chapter the use of liposomes for instantaneous amplification is described.

III. LIPOSOMES IN IMMUNOASSAY

Liposomes are the spherical, bilayer vesicles formed spontaneously when certain lipids are dispersed in water. During their formation they entrap some of the aqueous solution in which they are formed, and if this solution contains a water-soluble marker, this is also encapsulated. Details of the composition, preparation and processing of liposomes are given elsewhere in this book and in the literature.[6]

A. LIPOSOME MARKERS

The markers used in liposome immunoassay are frequently simple chemicals that are encapsulated in the aqueous core of the vesicle and thus need to be water soluble. This hydrophilicity also reduces the tendency of the marker to cross the lipid bilayer, which would result in marker leakage and decreased stability of the liposome reagent. The markers may be ones that can be measured *in situ* or may be compounds that must be released from the liposome by lysis prior to measurement. Alternatively, enzymes that are bound to the outer lipid layer or that are released upon lysis can be used, and the product of their reaction on an externally supplied substrate can be measured to determine the number of liposomes present. Thus the range of markers and detection protocols that can be used is extremely large.

Perhaps the simplest type of marker in terms of quantitation is a water-soluble dye. This can be measured either visually or spectrophotometrically, with or without lysis. The use of a diode-array or scanning spectrophotometer gives the potential for measurement of multiple dyes at the same time. A greater degree of sensitivity can be obtained by using a fluorescent compound, as fluorescence is orders of magnitude more sensitive than absorbance. Furthermore, use of a colored fluorescent compound provides the option of using either technique.

Certain markers that have been used in immunoassays change their properties upon release from the liposome, and this factor can be used to advantage in the assay. For example, many fluorescent compounds have their fluorescence quenched at the high concentrations at which they are encapsulated, and this quenching is abated upon the dilution caused by lysis.[7] There are also dyes that undergo a wavelength shift in the presence of certain ions, and if these ions are present outside the liposome but not inside, this shift can be induced by liposome lysis.[8]

Another group of water-soluble compounds that are being increasingly used are electroactive species. These can be monitored electrochemically, either potentiometrically or amperometrically after lysis.[9-13]

The markers described so far can be measured with or without lysis of the liposomes, depending on the requirements of the particular detection system being used. Unlike enzymic markers, there is no chemical reaction that has to occur, so they can be measured directly and instantaneously.

In some protocols, enzymes are involved in the detection method. For example, they can be encapsulated in the liposome and then released to act upon their substrate in the external solution. The product of the reaction can then be measured as above. The early techniques involved the use of horseradish peroxidase[12] and alkaline phosphatase,[14] the two enzymes most commonly used in ELISAs. More recently other enzymes have been used with a variety of substrates and modes of detection of products. However, in all these cases, the problems of temporal stability of the enzyme and the substrate and the careful timing needed to obtain precise and reproducible substrate turnover for consistent results still remain. Thus the use of enzymes removes one of the advantages of liposome enhancement, namely, the instantaneous measurement of the marker molecules. On the other hand, this approach does provide a double amplification effect, i.e., the release of large numbers of enzyme molecules from each liposome, followed by the enzymic production of measurable product.

B. LIPOSOME LYSIS

Although in some liposome immunoassays the markers encapsulated in the vesicle are measured directly, in most cases they are measured after release from the liposome by lysis, i.e., perforation or rupture of the lipid bilayer. The methods commonly used to effect lysis are the complement system, cytolytic agents and surfactants (detergents).

Complement lysis[15] utilizes a naturally occurring enzymic protein system present in serum. When an antibody binds to a cell surface antigen in serum, it undergoes a conformational change in the F_c region that induces a cascade of sequential multimolecular events which finally produces a membrane attack complex (MAC) that will lyse lipid membranes.[15] This system can be used in a liposome assay either by adding the complement factors to the assay or by utilizing those already present in the case of serum samples. Complement lysis has the advantage that only the analyte-tagged liposomes that have an antibody bound to them will be lysed, thus fulfilling one of the criteria of a homogeneous immunoassay. The disadvantages of this system lie with its complexity, the slowness of the reaction and the instability of the reagents.

Some of these disadvantages can be overcome by the use of cytolysis.[16] In this form of the assay, hapten, antigen or antibody is covalently linked to a toxin molecule that will lyse added liposomes, and this reagent is incorporated into an immunoassay. Toxin conjugates that have not participated in the antibody/antigen reaction remain free in solution and are capable of lysing added liposomes.

Because of its nonspecificity, the detergent lysis method is usually used in a heterogeneous format. Lysis by this technique is usually extremely rapid and complete, is simpler than complement lysis and the reagents are stable. A variety of different detergents can be used, and different liposome preparations have varying susceptibilities to detergents.[17,18]

C. LIPOSOME IMMUNOASSAY FORMATS

In order to be used in an immunoassay, liposomes must be labeled with one of the participant molecules in the immunorecognition reaction being studied. For example, if a pesticide is to be determined, then the liposome must be labeled with either the pesticide molecule so that it can compete with free pesticide or the antipesticide antibody to permit complexation with the pesticide molecule itself. A diagrammatic representation of an analyte-tagged liposome reacting with an immunoreactor (antibody bound to a solid surface) is shown in Figure 3. The liposome (ca. 0.4 μm diameter) and the antibody are shown approximately to scale, while the analyte molecule is magnified to permit visualization. One way to produce such a liposome is to conjugate the molecule of interest with a lipid and to incorporate this conjugate into the lipid mixture that is used to form the liposomes.[19] A very useful lipid for this purpose is DPPE (dipalmitoyl phosphatidyl ethanolamine), as it has a free amino moiety attached to the hydrophilic phosphate head group. This can be conjugated to a variety of analytes or proteins by using standard chemical coupling techniques.[20] When this approach is taken, the tagged molecules are present on both sides of the lipid bilayer, although only those on the outside of the liposome participate in the immunoreactions.

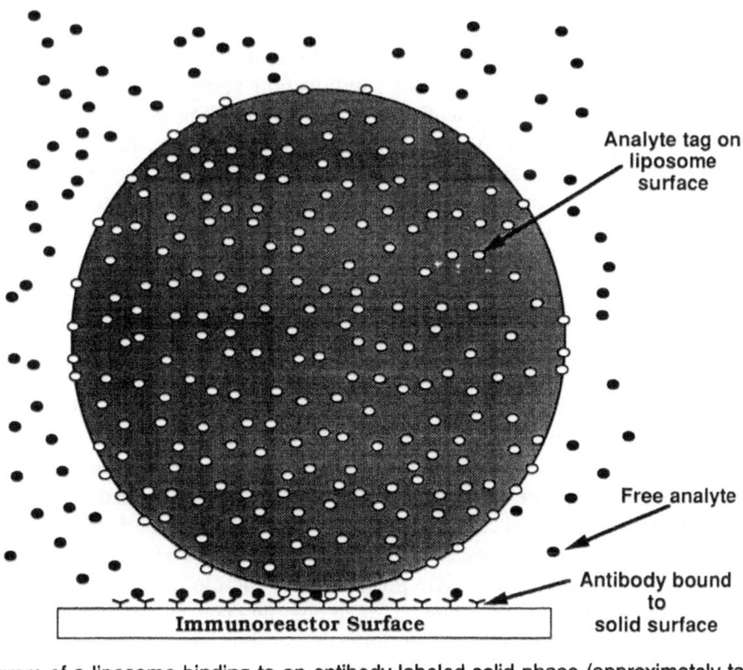

Figure 3 Diagram of a liposome binding to an antibody-labeled solid phase (approximately to scale).

It is also possible to modify the surface of the liposome after it has been prepared, and in this case only the outer membrane of the liposome is modified.[21] However, the modification chemistry has to occur under mild enough conditions so as not to lyse the liposomes, and assumptions about the degree of conjugation that has occurred have to be made in order to calculate the amount of analyte tag present on the surface of the liposome.

The majority of the homogeneous liposome-enhanced immunoassays that have been reported in the literature use some form of complement lysis.[22-25] These have mainly been for clinical analytes, and are beyond the scope of this article. Heterogeneous liposome immunoassays for clinical analytes[26,27] have also been reported. Current developments in the area of liposome immunoassays for environmental contaminants are described below.

D. ADVANTAGES OF LIPOSOMES

One of the major advantages of the use of liposomes and nonenzymic markers is that it is possible to obtain instantaneous amplification. This of course requires the right combination of liposome composition, detergent and temperature. The fact that there is no time required for a chemical reaction to proceed, as there is in enzyme-linked assays, make it easier to use nonequilibrium assays such as those described below. When this is allied to the variety of different markers that are water soluble and thus can be encapsulated, there is potential for a wide range of assays to be created.

A further advantage lies in the degree of control of the liposome properties, and thus of the assay, that is possible in this system. Varying lipid compositions can be used to control the physical characteristics and behavior of the liposomes. The amount of analyte tag bound to the surface can be controlled and defined. The size of the liposome can be controlled, as can be the exact amount of material encapsulated. These parameters are much less controlled and measurable in enzyme-linked assays.

IV. FLOW-INJECTION LIPOSOME IMMUNOASSAY (FILIA)

As the use of immunoassay has become more accepted for routine analysis, the need has arisen for automated systems. It is possible to automate the ELISA system, and many machines are now available that can add reagents, wash plates and collect and analyze data. However, they are still batch assays, and there are often significant well-to-well and plate-to-plate variations. It would be preferable to develop an automated system where the immunoaffinity sites are regenerable and reusable immunoreactor sites. This would reduce much of the variation, and allow calibration standards and samples to be reacted with the same assemblage of antibodies and thus give more accurate data. For this approach, some form of flow system linked to an autosampler and data collection system is required.

An enzyme-linked FILIA system using electrochemical detection to measure theophylline levels in serum has been described.[12] In this double-amplification system, liposome-encapsulated horseradish peroxidase, which is released subsequent to a competitive binding reaction in a regenerable immunoreactor column, cleaves fluoride ions from an organofluorine substrate in the presence of hydrogen peroxide (which also serves as the lysing agent). The fluoride ions are measured potentiometrically with a fluoride ion-selective electrode in a flow-through cell. There has also been an automated flow-injection system described to measure other anticonvulsant drugs in blood samples[28] by homogeneous liposome immunoassay, and a flow injection homogeneous immunoassay for pesticides has been also reported.[29]

A FILIA system has also been developed for the analysis of theophylline in serum samples using fluorescent detection,[30] and a similar system has recently been improved and extended to the analysis of environmental contaminants.[31] The system consists of an immunoaffinity column containing silica glass beads to which antibodies to the analyte of interest (the herbicide alachlor) are covalently bound. Liposomes are then prepared with a lipid mixture containing DPPE that has been coupled to the analyte, and this conjugate is incorporated into the lipid bilayer. A fluorescent dye (sulforhodamine B) is encapsulated into the liposomes. The liposomes are passed through the column concurrently with sample analyte, and competition between the free analyte and the liposome-tagged analyte occurs. The resulting amount of liposome bound is then quantified by passing detergent through the column to lyse the liposomes and measuring the fluorescence of the released dye. The fluorescence intensity is inversely proportional to the amount of free analyte present in the sample.

In this method of analysis, the antibody/antigen binding does not reach equilibrium. This means that control of timing of contact of the antigen with the antibody has to be precise to give reproducible results. It also means that the total amount of liposome bound is smaller than would be the case in a microtiter plate assay, where much longer antibody/analyte contact times allow equilibrium to occur. This is partially offset by the higher concentration of antibody sites in the immunoreactor column and the higher efficiency of interaction resulting from the convective mass transport occurring in the column. This lower binding requires greater sensitivity in the detection method to compensate, but the benefits of speed and ease of automation are gained. Such a system has the potential to assay samples fully automatically every few minutes, with calibration standards or controls included as often as needed to achieve the requisite analytical accuracy.

The current FILIA system developed in this laboratory is shown in Figure 4A. The sampling, the flow of carrier solution, the addition of detergent and the collection of data are fully automated and computer controlled. Commercially available tubing, columns and fingertight fittings are used throughout, which allows easy changing of columns and other modifications. An electronically controlled slider valve directs the flow of solutions. As configured, this system is capable of analyzing a sample every 5 min, with a detection limit of 5 ppb alachlor. Experiments are presently underway to further optimize the system, with a view to lowering the detection limit and increasing the sample throughput rate.

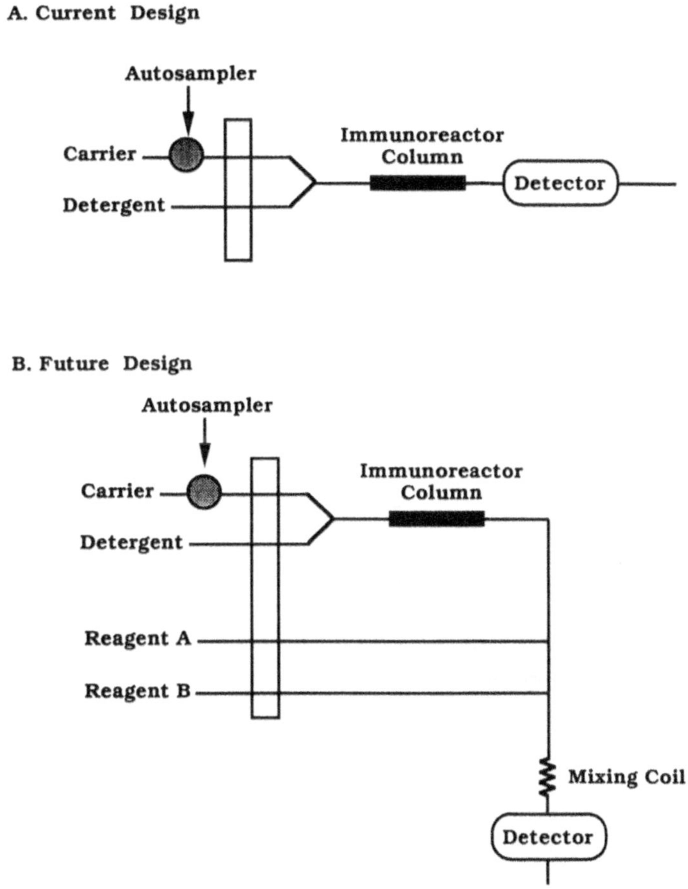

Figure 4 Diagram of flow-injection liposome-enhanced immunoassay systems.

This system can also be modified to allow automated multianalyte analysis, using multiantibody columns (separate columns or a single mixed-bed column), and a series of appropriately tagged liposomes, each with a unique marker encapsulated. These markers could be dyes or fluorophores with different spectral characteristics, chemiluminescent reagents, electroactive species or any combination of these compounds. For example, studies are already underway using liposomes with potassium ferrocyanide encapsulated and electrochemical detection after detergent lysis. Some of the other methodologies suggested above would require post-column addition of reagents in one form or another, but the system is designed to be easily modified as shown, for example, in Figure 4B.

V. LIPOSOME IMMUNOMIGRATION STRIPS

Immunoassays for field detection of environmental contaminants have been developed commercially and are gaining wide acceptance.[32] Liposome immunoassays have been developed[19] to carry out this function, although none is yet available commercially.

This device[19] consists of an immunomigration strip containing an antibody competition zone and a liposome capture zone on plastic-backed nitrocellulose. The strips as currently used are 80 × 5 mm, with the antibody and capture zones being about 25 mm^2. A mixture of analyte-tagged, dye-containing liposomes and free analyte to be measured migrates along the strip, and competitive binding occurs in the antibody zone. The liposomes that do not bind in this zone then continue to migrate and are bound in the liposome capture zone. This zone consists of egg white avidin bound to the nitrocellulose, and this reagent binds all the liposomes that do not bind in the antibody zone. The reaction in the antibody zone is competitive, so the more sample analyte there is, the less liposome and its entrapped dye will be bound in this zone. Thus, the color in this zone will be inversely proportional to the amount of free analyte. The opposite is true in the liposome capture zone, where the actual measurement is made, and the color intensity is directly proportional to the amount of analyte in the sample. This color can be determined by visual matching with a color intensity chart, by reflectance measurement or by scanning densitometry. Such strips can be run in a variety of formats, such as a double strip, shown in Figure 5 with a tolerance-level control to aid in quantitation and provide verification of performance. Alternatively, a triple strip run with low and high level controls will allow interpolation for improved quantitation.

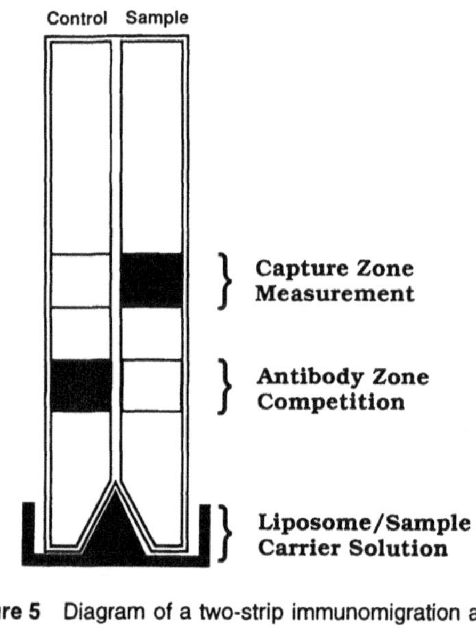

Figure 5 Diagram of a two-strip immunomigration assay.

This type of assay, as with the FILIA described in the previous section, can also be modified for multianalyte assays. For example, several different antibody zones could be arranged in sequence on the strip, and liposomes labeled with the corresponding antigens would flow along the strip, binding to the relevant band. Work is also underway to develop an electrochemical detector for this form of strip[33,34] which would have the added feature of automatically providing a reading when the immunomigration assay was complete.

VI. CONCLUSIONS

The use of liposomes in immunoassays has considerable utility but, as yet, this has not been realized in the marketplace. The instantaneous amplification produced by liposomes provides the potential for performing rapid immunoassays in the field and automated flow-through analyses in the laboratory. The continuous-flow systems with a reusable column allow calibration and avoid the delays associated with enzyme-linked reactions. The potential for using liposome immunoassays for the detection of microorganisms is also being explored. It is anticipated that commercially available liposome-based systems will be developed, and they are likely to find use for clinical, environmental and food safety analyses.

REFERENCES

1. **Price, C. P. and Newman, D. J.,** *Principles and Practice of Immunoassay*, Stockton Press/Macmillan Publishers, London, 1991.
2. **Yalow, R. S. and Berson, S. A.,** Immunoassay of endogenous plasma insulin in man. *J. Clin. Invest.*, 39, 1157-1175, 1960.
3. **Kaufman, B. M. and Clower, M.,** Immunoassay of pesticides. *J. Assoc. Off. Anal. Chem. 74*, 239-247, 1991.
4. *Immunochemical Methods for Environmental Analysis*. Van Emon, J. M. and Mumma, R. O., Eds, American Chemical Society, Washington, D.C., 1990.
5. **Feng, P. C., Wratten, S. J., Horton, S. R., Sharp, C. R., and Logusch, E. W.,** Development of an enzyme-linked immunosorbent assay for alachlor and its application to the analysis of environmental water samples. *J. Agric. Food Chem.*, 38, 159-163, 1990.
6. **New, R. R. C.,** *Liposomes. A Practical Approach*. Oxford University Press, Oxford, 1990.
7. **Chen, R. F. and Knutson, R. J.,** Mechanism of fluorescence concentration quenching of carboxyfluorescein in liposomes: energy transfer to nonfluorescent dimers. *Anal. Biochem.*, 172, 61-77, 1988.
8. **Janoff, A. S., Carpenter-Green, S., Weiner, A. L., Siebold, J., Weissmann, G., and Ostro, M. J.,** Novel liposome composition for a rapid colorimetric test for systemic lupus erythematosus. Clin. Chem., 29, 1587-1592, 1983.
9. **Tie, F., Pan, A. H., Ru, B. G., Wang, W. Q., and Hu, Y. H.,** An improved ELISA with linear sweep voltammetry detection. *J. Immunol. Methods*, 149, 115-120, 1992.
10. **Vlasov, G. S., Torchilin, V. P., Gremyakova, T. A., Likhoded, V. G., Koresteleva, M. D., and Ivanov, N. N.,** Detection of antibodies neutralizing the endotoxins of gram-negative bacteria by the liposomal potentiometric method. *Zh. Mikrobiol. Epidemiol. Immunobiol.*, 10, 87-91, 1985.
11. **Kannuck, R. M., Bellama, J. M., and Durst, R. A.,** Measurement of liposome-released ferrocyanide by a dual function polymer modified electrode. *Anal. Chem.*, 60, 142-147, 1988.
12. **Wu, T-G. and Durst, R. A.,** Liposome-based flow injection enzyme immunoassay for theophylline. *Mikrochim. Acta*, 1, 187-195, 1990.
13. **Durst, R. A. and Kannuck, R. M.,** Chemically modified electrode for liposome-mediated homogenous immunoassay. *Ion-Selective Electrodes*, Vol. 5, E. Pungor, Ed., Pergamon Press, New York, 1989, 65-79.
14. **Kim, C. K. and Lim., S. J.,** Liposome immunoassay (LIA) with antigen-coupled liposomes containing alkaline phosphatase. *J. Immunol. Methods*, 159, 101-106, 1993.
15. **Stroud, R. M., Volanakis, J. E., and Lint, T. F.,** The biochemistry and biological reactions of complement proteins. *Immunochemistry of Proteins*, Vol. 3, M. Z. Atazzi, Ed., Plenum Press, New York, 1979.
16. **Freytag, J. W. and Litchfield, W. J.,** Liposome-mediated immunoassays for small haptens (digoxin) independent of complement. *J. Immunol. Methods*, 70, 133-140, 1984.
17. **Ruiz, J., Goñi, F., and Alonso, A.,** Surfactant induced release of liposomal contents. A survey of results and methods. *Biochim. Biophys. Acta*, 937, 127-134, 1988.
18. **Sila, M., Au, S., and Weiner, N.,** Effects of Triton X-100 concentration and incubation temperature on carboxyfluorescein release from multilamellar liposomes. *Biochim. Biophys. Acta*, 859, 165-170, 1986.
19. **Siebert, S. T. A., Reeves, S. G., and Durst, R. A.,** Liposome immunomigration field assay device for alachlor determination. *Anal. Chim. Acta*, 282. 297-305, 1993.
20. **Wong, S. S.,** *Chemistry of Protein Conjugation and Cross-Linking*, CRC Press, Boca Raton, FL, 1991.
21. **Bredehorst, R., Ligler, F. S., Kusterbeck, A. W., Chang, E. L., Gaber, B. P., and Vogel, C. W.,** Effect of covalent attachment of immunoglobulin fragments on liposomal integrity. *Biochemistry*, 25, 5693-5698, 1986.
22. **Yu, B. S., Choi, Y. K., and Chung, H.,** Development of immunoassay methods by use of liposomes. *Biotech. Appl. Biochem.*, 9, 209-216, 1987.
23. **Kim, C. K. and Lim, S. J.,** Liposome immunoassay (LIA) with antigen-coupled liposomes containing alkaline phosphatase. *J. Immunol. Methods*, 159, 101-106, 1993.
24. **Glagasigij, U., Sato, Y., and Suzuki, Y.,** Highly sensitive immunoliposome assay of theophylline. *Chem. Pharm. Bull.* (Tokyo), 36, 1086-1094, 1988.
25. **Pashov, V. N., Tsurpa, G. P., Griko, N. B., Skopinskaya, S. N., and Yarkov, S. P.,** The use of streptavidin biotin interaction for preparation of reagents for complement-dependent liposome immunoassay of proteins. Detection of Latrotoxin. *Anal. Biochem.*, 207, 341-347, 1992.
26. **O'Connell, J. P., Campbell, R. L., Fleming, B. M., Mercolino, T. J., Johnson, M. D., and McLaurin, D. A.,** A highly sensitive immunoassay system involving antibody-coated tubes and liposome-entrapped dye. *Clin. Chem.*, 31, 1424-1426, 1985.
27. **Gerber, M. A., Randolph, M. F., and DeMeo, K. J.,** Liposome immunoassay for rapid identification of streptococci directly from throat swabs. *J. Clin. Microbiol.*, 28, 1463-1464, 1990.
28. **Kubotsu, K., Goto, S., Fujita, M., Tuchiya, H., Kida, M., Takano, S., Matsuura, S., and Sakurabayashi, I.,** Automated homogeneous liposome immunoassay systems for anticonvulsant drugs. *Clin. Chem.*, 36, 808-812, 1992.
29. **Krämer, P. and Schmid, R.,** Flow injection immunoanalysis (FIIA) — a new immunoassay format for the determination of pesticides in water. *Biosensors Bioelectronics*, 6, 239-243, 1991.

30. **Locascio-Brown, L., Plant, A. L., Chesler, R., Kroll, M., Ruddel, M., and Durst, R. A.,** Liposome-based flow-injection immunoassay for determining theophylline in serum. *Clin. Chem.*, 39, 386-391, 1993.
31. **Reeves, S. G., Rule, G. S., Roberts, M. A., Edwards, A. J., and Durst, R. A.,** Flow-injection liposome immunoanalysis (FILIA) for alachlor. *Talanta,* 41, 1747-1753, 1994.
32. **Van Emon, J. M. and Lopez-Avila, V.,** Immunochemical methods for environmental analysis. *Anal. Chem.*, 64, 79A-88A, 1992.
33. **Durst, R. A., Siebert, S. T. A., Roberts, M. A., Larsson-Kovach, I.-M., and Reeves, S. G.,** Development of liposome-enhanced immunobiosensing devices for field measurements of toxic chemicals. *Bioelectroanalysis*, Vol. 2, E. Pungor, Ed., Academiai Kiado, Budapest, 1993, 15-32.
34. **Roberts, M. A. and Durst, R. A.,** Investigation of liposome-based immunomigration sensors for the detection of polychlorinated biphenyls. *Anal. Chem.,* 67, 482-491, 1995.

Chapter 18

The Use of Liposomes in Biodegradability Testing

Raina M. Miller, Ilia Y. Jimenez, and Richard Bartha

CONTENTS

I. INTRODUCTION

The Toxic Substances Control Act of the U.S. (USEPA, 1979) requires a premanufacture review of newly introduced chemical substances to assess their environmental fate and effects. Similar requirements have been or are being adopted in other developed countries. An important consideration concerning novel organic compounds is their "inherent biodegradability," i.e., their propensity to undergo ultimate biodegradation resulting in CO_2, H_2O, minerals and microbial biomass. Inherent biodegradability is typically tested using ^{14}C-labeled compounds under conditions favorable for microbial activity. In soil tests, a compound is considered inherently biodegradable if within a 2-month period 50% or more of its radiocarbon is released as $^{14}CO_2$ (Code of Federal Regulations, 1992).

When a substance fails such an inherent biodegradability test, the presumption is that it is recalcitrant because microorganisms lack the necessary enzymes for ultimate biodegradation or "mineralization." Undoubtedly, this conclusion is frequently correct. However, it needs to be recognized that an inordinately slow mass transfer from the environment to the enzyme system of the degrading microorganism may also make a compound appear recalcitrant, even though enzymes for its degradation exist. This is most likely to occur in the case of substances that are solids at physiological temperatures and have negligible solubility in water. There are biogenic examples of such substances, including the polymers cellulose, chitin and lignin which are degraded at appreciable rates in the environment. These molecules are dealt with by exoenzymes that break down the parent molecule to water-soluble monomeric or oligomeric molecules that are transported into the microbial cells and metabolized further. No such exoenzymes are known for large hydrophobic hydrocarbon molecules and xenobiotic polymers such as polyethylene, polypropylene, polyvinyl chloride, polystyrene, nylon etc. At molecular weights above 500–600, such substances tend to be very recalcitrant. There is no clear structural reason for this recalcitrance and the oligomers of these substances tend to be biodegradable (Aminabhavi et al., 1990).

Based on these considerations, it is not unreasonable to postulate that mass transfer and transport problems rather than a lack of appropriate enzymes may be responsible for at least some recalcitrance phenomena. The classic biochemical approach for bypassing transport phenomena is the preparation of cell extracts or homogenates, but the preparation processes frequently inactivate membrane-bound enzyme systems crucial for the metabolism of hydrocarbons and their xenobiotic analogs. This chapter will show that liposome packaging of test compounds offers a way to differentiate true recalcitrance from mass transfer and permeability problems.

0-8493-4013-6/96/$0.00+$.50

II. MATERIALS AND METHODS

A. PACKAGING METHODS

1. Sonication

Phosphatidylcholine (L-α-lecithin from egg yolk; in chloroform) and phosphatidylserine (in chloroform) were obtained from Avanti Polar Lipids, Inc., Birmingham, AL. Cholesterol was obtained from Sigma Chemical Co., St. Louis, MO. [7,10-^{14}*C*]Benzo(*a*)pyrene (specific activity 60.7 mCi/mmol) was obtained from Amersham Corp. Arlington Heights, IL [9-^{14}C]Phenanthrene (specific activity 15 mCi mmol^{-1}) was obtained from Sigma, St. Louis, MO.

As described by Miller and Bartha (1989), vesicles were prepared from chloroform solutions of ^{14}C-labeled aromatic substrates and phospholipid or phospholipid–cholesterol mixtures using sonication. Briefly, solvent was evaporated under N_2 with rotation and then placed under vacuum for 1 h to remove residual solvent. The remaining substrate–phospholipid film was suspended in 0.02 *M* Tris buffer (pH 7; 10–15 mg/ml), vortexed, and then sonicated at room temperature until optically clear in a bath sonicator (Laboratory Supplies Co., Inc., Hicksville, New York). After sonication was complete, each sample was centrifuged for 10 min and then allowed to stand at 4°C for at least 12 h. Radioactivity in the supernatant was then assayed to determine packaging efficiency.

2. Extrusion

Alternately, liposomes may be prepared by extrusion through a small pore size filter. This method of preparation was introduced by Olson et al. (1979) and developed further by Hope et al. (1985). The extrusion procedure was recently made much more simple and convenient by MacDonald et al. (1991). In this latest version of the procedure, the phospholipid, with or without the material to be packaged, is dissolved in chloroform or another suitable solvent, and the solvent is evaporated. The lipid mixture is hydrated in a suitable buffer and subjected to 10 successive freeze–thaw cycles. This treatment yields a suspension of multilamellar vesicles with even distribution of phospholipid and packaged material.

This suspension is subsequently passed back and forth through a Nucleopore polycarbonate membrane mounted in a holder between two syringes attached to the opposing sides of the holder (MM Developments, Ottawa, Canada). After 19 passes through the filter, the multilamellar vesicles are converted essentially quantitatively to large unilamellar vesicles. Using a 100 nm pore diameter filter, the average diameter of the resulting liposomes is 70–80 nm. Slightly smaller (47–52 nm) diameter liposomes can be prepared by the use of a 30 nm pore diameter filter.

We used the apparatus and method described by MacDonald et al. (1991) to prepare phosphatidylcholine (Sigma, St. Louis, MO) liposomes for packaging of [4,5,9,10-^{14}C]pyrene, specific activity 40–60 μCi/mmol, (Chemsyn Scientific Laboratories, Lenexa, KA) and [1-*C*^{14}]*n*-octadecane, specific activity 12.6 mCi/mmol (Sigma). For packaging of pyrene, phosphatidylcholine (15.6 mg) and 0.75 mg pyrene were dissolved in 2.4 ml methanol containing 7% chloroform. The solvent was evaporated in a rotary evaporator under vacuum, lyophilized and subsequently hydrated in 2 ml of 10 m*M* Tris buffer, pH 7.0, containing 150 m*M* NaCl. This mixture was emulsified by blending and was then extruded through a 100-nm membrane at room temperature as described above. Octadecane was packaged in a similar manner using 11.7 mg phosphatidylcholine and 0.125 mg octadecane, hydrated in 1 ml buffer.

B. BIODEGRADATION TESTS

Rates of metabolism of free and liposome-packaged substrates were compared by measuring their rates of conversion to $^{14}CO_2$ and in some cases by measuring substrate disappearance by high performance liquid chromatography (HPLC). Both isolated cultures and microbial consortia were used as microbial agents of conversion. The isolates were capable of utilizing either naphthalene or phenanthrene as a sole carbon and energy source, while the consortium used was capable of growing on pyrene or *n*-octadecane as a sole source of carbon and energy (Jimenez and Bartha, 1993, 1994).

In experiments using isolates, a 1- to 2-day culture grown on naphthalene was inoculated into fresh mineral medium containing free or liposome-packaged substrate. The mineral medium used contained 0.025 *M* $CaCl_2$ to promote fusion of liposomes with the degrading cell membrane. The medium was made up in two parts to prevent precipitation: (I) 0.2 g l^{-1} Na_2HPO_4, 0.2 g^{-1} KH_2PO_4, 0.4 g l^{-1} $MgSO_4 \cdot 7H_2O$, 2.0 g l^{-1} NH_4Cl, 0.2 g l^{-1} yeast extract, and (II) 5.54 g l^{-1} $CaCl_2$ and 4.84 g l^{-1} Tris. Each part was adjusted to pH 6.6, autoclaved, cooled and then mixed together.

The bacterial consortium was pre-grown on pyrene, the cells were harvested and washed by centrifugation and suspended in fresh Stanier's mineral medium (Atlas, 1993), without pyridoxin and yeast extract, to yield 0.4 to 0.6 mg protein per ml. Liposome-packaged or free hydrocarbon was added to the medium containing either an isolate or the consortium, to yield equal hydrocarbon concentrations and radioactivity in both samples.

The described cell–substrate suspensions were enclosed in modified 125-ml microfernbach flasks that were incubated at the indicated temperature with rotary shaking (200 rpm). In appropriate time intervals, the flasks were flushed through the $^{14}CO_2$ trapping system described by Marinucci and Bartha (1979) and the trapped radioactivity was counted by liquid scintillation using a Beta-Trac Model 6895 instrument (TM Analytic, Elk Grove Village, IL).

III. RESULTS AND DISCUSSION

A. PACKAGING BY SONICATION

As reported by Miller and Bartha (1989), octadecane and hexatriacontane were packaged at substrate:phosphatidylcholine ratios of up to 1:1 and 1:2 (w/w), respectively. Packaging of aromatic substrates by sonication was found to be more complex than packaging of alkanes. Benzo(*a*)pyrene could not be emulsified by blending in the presence of phosphatidylcholine, a cationic phospholipid. However, in the presence of phosphatidylserine which has a polar head group with both a positive and negative charge, the benzo(*a*)pyrene was readily emulsified into solution. To optimize packaging of benzo(*a*)pyrene, a variety of phospholipid and phospholipid–cholesterol mixtures were tested for efficiency of packaging (Table 1). Of the mixtures tested, a 7:1:1 mix of phosphatidylcholine, phosphatidylserine, and cholesterol was most effective in packaging benzo(*a*)pyrene. It was also found that packaging efficiency was dependent upon the concentration of substrate. As shown in the last two columns of Table 1, when the amount of benzo(*a*)pyrene was increased from 0.5 mg to 1.0 mg/9 mg phospholipid mix, the packaging efficiency decreased from an optimized ratio of 1:9 to 1:26 mg benzo(*a*)pyrene/per milligram of phospholipid mix.

Table 1 Packaging of Benzo(*a*)pyrene[a]

Packaging Material	Amount Packaging Material Used[b] (mg ml^{-1})					
Phosphatidylcholine	0	0	0	1	7	3.5
Phosphatidylserine	10	3	3.5	1	1	.5
Cholesterol	0	1.5	1	1	1	.5
	Packaging Efficiency (mg B(a)P:mg Packaging Material)					
	1:125	1:82	1:45	1:35	1:9	1:26

[a] Amount of benzo(*a*)pyrene used was 0.5 mg/ml.
[b] Concentration units given refer to the concentration of packaging material in the buffer used for sonication.

The packaging efficiency can also be expressed as a molar solubilization ratio (MSR), a relationship often used in surfactant literature to assess relative effectiveness of surfactants. The MSR is defined as the number of moles of organic compound solubilized per mole of surfactant. The MSR can be normalized with respect to aqueous solubility of the substrate to give a relative indication of how much packaging enhances substrate solubility. Table 2 summarizes the normalized MSR values for the substrates and packaging materials tested in this study. Two points are made clear by this Table: (1) packaging increases the apparent solubility of alkanes more than the apparent solubility of the aromatics used in this study, and (2) the lower the initial substrate aqueous solubility, the more packaging enhances solubility.

B. PACKAGING BY EXTRUSION

The extrusion process through a 100-nm pore size filter yielded phosphatidylcholine liposomes containing either 5% (w/w) pyrene or 1.07% (w/w) *n*-octadecane content. The MSR value for phosphatidylcholine solubilization by extrusion was much lower than that for sonication; however, conditions for

Table 2 Molar Solubilization Ratios for Substrate–Phospholipid Systems

Substrate	Solubility[a] (mol l^{-1})	Packaging Material and Method	MSR[b] (molar solubilization ratio)	Mormalized MSR × 10^{-3} (MSR/solubility)
Octadecane	2.4×10^{-8}	Phosphatidylcholine (sonication)	3.1[c]	130,000
Octadecane[d]	2.4×10^{-8}	Phosphatidylcholine (extrusion)	0.033	1,400
Hexatriacontane	$<1.2 \times 10^{-8}$	Phosphatidylcholine (sonication)	0.8[c]	33,000
Phenanthrene	7.9×10^{-6}	Phosphatidylcholine/ phosphatidylserine/ cholesterol (sonication)	0.022	2.4
Pyrene[d]	7.9×10^{-7}	Phosphatidylcholine (extrusion)	0.19	235
Benzo(*a*)pyrene	1.2×10^{-8}	Phosphatidylcholine/ phosphatidylserine/ cholesterol (sonication)	0.016	1,344

[a] Values for water solubility taken from Verschueren (1983) with the exception of hexatriacontane. For hexatriacontane, the solubility of octadecane was used for purposes of calculation.
[b] MSR is defined as the number of moles of organic compound solubilized per mole of surfactant.
[c] From Miller and Bartha, 1989.
[d] Packaging conditions were not optimized.

extrusion packaging were not optimized so that it is not known whether sonication or extrusion is the more efficient method for packaging (Table 2).

C. BIODEGRADATION — ALKANES

As demonstrated by Miller and Bartha (1989), the rate of degradation of alkanes can be significantly enhanced by liposome-packaging of the alkanes. For example, the apparent K_s (2453 ± 148 mg l^{-1}) for growth on liposome-packaged *n*-octadecane was over 40-fold less than the K_s (60 ± 12 mg l^{-1}) for growth on *n*-octadecane alone. Similarly, the apparent K_s (2698 ± 831 mg l^{-1}) for growth on liposome-packaged *n*-hexatriacontane was over 65-fold less than the K_s (41 ± 7 mg l^{-1}) for growth on *n*-hexatriacontane alone. These large increases in the rate of alkane degradation show that the inherent biodegradability of alkanes can be assessed much more rapidly using liposome-packaged instead of free substrate.

In order to demonstrate that extrusion-prepared liposomes were also effective in delivering hydrocarbon to degrading cells, *n*-octadecane was extrusion-packaged and the rate of $^{14}CO_2$ evolution from liposome-packaged and free *n*-octadecane was compared (Figure 1). By 120 h, the free substrate was 38% mineralized, while the packaged substrate was 76% mineralized, a two-fold increase. The inoculum used in this set of experiments was the pyrene-degrading consortium. In this and several similar experiments (data not shown), liposome packaging clearly and consistently increased the rate of *n*-octadecane mineralization. Therefore, we conclude that the liposomes prepared by the extrusion technique are effective vehicles of alkane delivery to the membrane-bound enzyme systems of microorganisms.

When the increase in octadecane mineralization rates from extrusion-prepared liposomes (2-fold) and sonication-prepared liposomes (16-fold) are compared, it seems as if sonication had a larger effect on increasing mineralization. However, as shown in Table 3, the packaging efficiency or molar solubilization ratio (MSR) also seems to affect degradation rates. In the sonicated system which was optimized for packaging efficiency, the MSR was 3.1, a much higher value than the MSR of 0.03 for the extrusion system which was not optimized for packaging efficiency. Similar results were shown in a system that used a rhamnolipid biosurfactant as the packaging material (Zhang and Miller, 1992, 1994). Shaking produced a much lower MSR than sonication for the same materials. The higher packaging efficiency for sonication was reflected in the larger increase in mineralization rates (18-fold vs. 3.5-fold). These data illustrate the importance of optimizing packaging efficiency to determine degradation potential of packaged substrates.

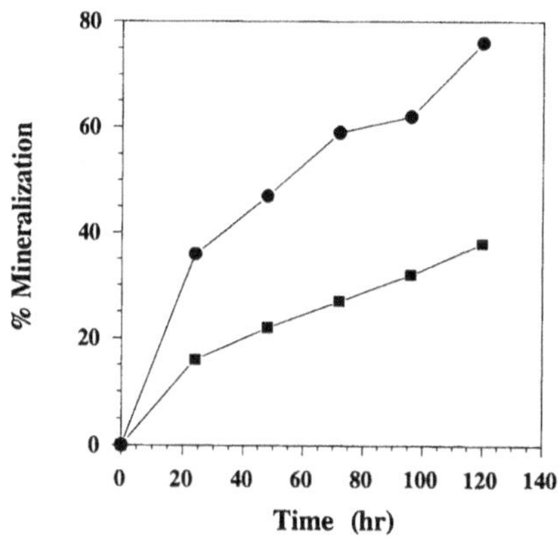

Figure 1 Mineralization of radiolabeled *n*-octadecane (1.5 ug ml^{-1}), with (●) or without (■) liposome-packaging by a pyrene-grown microbial consortium incubated at 28°C.

Table 3 Comparison of Octadecane Packaging Efficiency and Mineralization Rates

Packaging Material and Method	MSR	Initial Substrate Concentration (mg l^{-1})	Degrader	Ratio of Packaged:Free Substrate Mineralization Rate
Phosphatidylcholine (sonication)	3.1	400	Isolate	16
Phosphatidylcholine[a] (extrusion)	0.03	1.5	Consortium	2
Rhamnolipid (sonication)	2.0	780	ATCC 9027	18
Rhamnolipid (shaking)	0.16	400	ATCC 9027	3.5

[a] Not optimized for packaging.

D. BIODEGRADATION — AROMATICS

Work with alkanes suggested that liposome packaging might be a useful approach for determining whether recalcitrance of aromatics is due to enzymatic or structural constraints. Therefore, this system was tested with several aromatic compounds, phenanthrene, pyrene and benzo(*a*)pyrene. Since the octadecane-degrading isolate originally used for degradation of alkanes did not degrade aromatic compounds, several aromatic-degrading isolates were tested. Of the organisms tested, a Gram-positive organism isolated from a soil enrichment on naphthalene was most promising. Using this organism, the effect of liposome packaging of phenanthrene on phenanthrene biodegradation was determined. Packaging was found to inhibit phenanthrene mineralization in comparison to mineralization of free substrate. For example, after a 24 h incubation, 17.0% of the phenanthrene was mineralized in control samples compared to 8.1% mineralization of the phenanthrene which was liposome packaged. This was a pattern repeated with two other aromatic degrading organisms tested. These data suggest that the positive effect of liposome packaging on solubilization of phenanthrene did not overcome the limitations to biodegradation imposed by requiring uptake of packaged phenanthrene. One reason for this may be that the effective increase in apparent phenanthrene solubility was minimal, 1.6 mg l^{-1} to 5 mg l^{-1}, an increase of only 3.1-fold.

Similar to results for phenanthrene biodegradation by isolates, the pyrene-degrading consortium also metabolized packaged pyrene at a slower rate than free pyrene (Figure 2). In order to explain the inhibition caused by packaging, several additional liposome preparations were made and tested, and the ratio of liposomes to cells was varied over a wide range (results not shown). Nevertheless, liposome packaging consistently inhibited rather than stimulated the pyrene mineralization rate. Considering the possibility that the liposomes may have damaged the cell membranes, the consortium was incubated for 24 h with empty liposomes. Subsequently, the microorganisms were separated from the liposomes by differential centrifugation. Unexposed cells and cells preexposed to liposomes were then both dosed with equal amounts of free pyrene, and the rate of $^{14}CO_2$ production was measured (Figure 3). There was no difference in the activity of the unexposed and liposome-exposed cells, thus ruling out damage to the cell membranes by the liposomes. In this set of experiments the effective increase in pyrene solubility was 0.16 mg l^{-1} to 1.5 mg l^{-1}, a 9.4-fold increase.

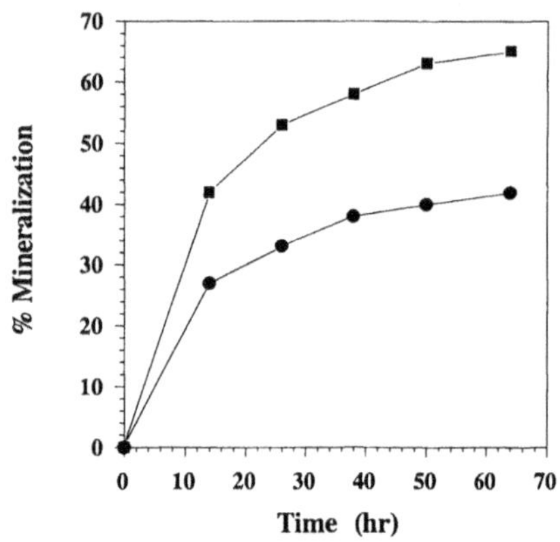

Figure 2 Mineralization of radiolabeled pyrene (1.5 ug ml^{-1}), with (●) or without (■) liposome-packaging by a pyrene-grown microbial consortium incubated at 28°C.

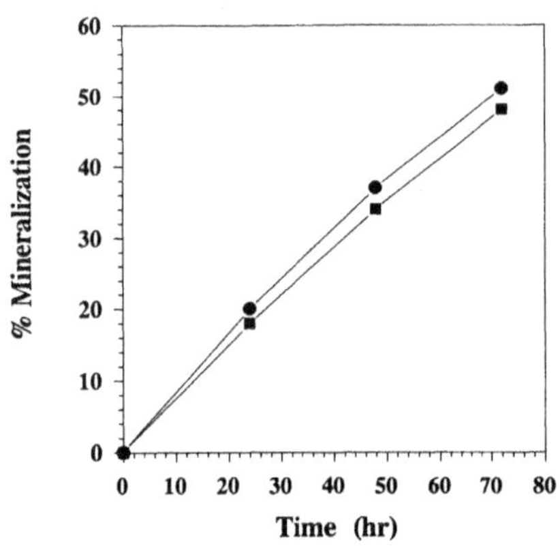

Figure 3 Mineralization of free radiolabeled pyrene (0.3 ug ml^{-1}) by a pyrene-grown microbial consortium incubated at 28°C, with (■) and without (●) a 24 h pre-exposure to empty (pyrene-free) liposomes.

As shown in Table 4, biodegradation of benzo(*a*)pyrene was also tested in this system. Two separate experiments are reported in Table 4, both of which show that mineralization of benzo(*a*)pyrene did not occur under either control or liposome-packaged conditions. Of interest, however, is that both radioactivity (as measured by ^{14}C) and benzo(*a*)pyrene (as measured by HPLC) were quantitatively recovered for both control samples (packaged benzo(*a*)pyrene time 0 recovery and free benzo(*a*)pyrene incubated for 4 days). In contrast, for samples incubated with packaged benzo(*a*)pyrene either with or without phenanthrene as a cosubstrate, the recovery of radioactivity was greatly decreased, ranging from 67% to 81%, and the recovery of benzo(*a*)pyrene as measured by HPLC ranged from 1% to 10%. These data suggest that although mineralization did not occur, the majority of packaged benzo(*a*)pyrene in the incubated samples was transformed, perhaps cometabolically. For benzo(*a*)pyrene, the inhibition of biodegradation (transformation) caused by packaging that was observed for phenanthrene was not apparent. This is probably because of the extremely low initial bioavailability of the benzo(*a*)pyrene. The effective increase in bioavailability due to packaging in this experiment was 0.003 mg l^{-1} to 5 mg l^{-1}, an increase of over 1600-fold. This value is several orders of magnitude larger than the increase for pyrene (9.4-fold) or phenanthrene (3.1-fold). Thus, for benzo(*a*)pyrene, in contrast to the alkanes (up to

C_{36}) and smaller aromatic molecules tested, structural and enzymatic constraints rather than mass transfer and transport constraints seem to be the limiting factor in biodegradation for the isolates tested in this study.

Table 4 Biodegradation of Benzo(*a*)pyrene by a Gram-Positive Isolate

Experiment[a]	$^{14}CO_2$	Volatiles	Organic Extraction	Aqueous Extraction	Total ^{14}C Recovered	Total BaP Recovered[b]
BaP-liposomes, 0 time recovery	—	—	90.2	2.5	92.7	83.6
	—	—	87.9	0.2	88.1	100.6
BaP and inoculum	0.3	0.01	90.8	8.1	99.2	102.8
	2.0	0.3	86.6	5.2	94.1	74.1
BaP-liposomes and inoculum	0.1	0.04	43.1	23.9	67.1	10.4
	1.0	0.2	64.1	8.7	74.0	0.8
BaP-liposomes, phenanthrene,	0.6	0.06	49.3	31.1	81.1	8.4
and inoculum	1.0	0.1	66.7	5.6	73.4	1.1

Note: Values expressed as % of added radioactivity

[a] Two separate experiments were performed under similar conditions. Each flask contained 5 ml minimal medium, and received a 5 ml inoculum from a 2-day preculture grown on 0.1% naphthalene. Substrate was added in a 1 ml volume. Each experiment was run for 4 days. Values for each experiment are shown in each column; the top value is from experiment 1 and the bottom value is from experiment 2.

[b] Recoveries were calculated from HPLC peak areas compared with BaP standard curves.

E. MECHANISMS OF BIODEGRADATION ENHANCEMENT BY LIPOSOMES

Hydrophilic solutes enclosed into the inner lumen of the lipsome can be delivered to eukaryotic cells by fusion or endocytosis (Huang, 1983). Because of the nature of the bacterial cell envelope, this type of delivery appears to be irrelevant for most biodegradation studies. The situation is more complex for delivery of hydrophobic molecules incorporated into the lipid bilayer itself (Brown, 1992; Huang, 1983). In the latter case, liposomes may simply dissolve and disperse the otherwise crystalline solid and thereby facilitate the diffusion of the packaged material throughout the aqueous medium to the cell membrane. This is obviously a relatively inefficient mode of delivery that can be achieved, alternately, by creating emulsions through sonication alone, or by detergents as described by Miller and Bartha (1989). Somewhat more efficient transfer of the packaged hydrophobic test material to the cell membrane can be visualized through the transient collision model (Brown, 1992). In this case only the approximately 1.5 nm hydration barrier needs to be crossed for delivery of the test substance. This occurs, presumably, simply by concentration gradient, since the test substance is in higher concentration in the liposomes than in the cell membrane. The most efficient delivery would occur by fusion of the liposome with the bacterial cell membrane. This process can be visualized most readily in the case of Gram negative bacteria that possess an outer membrane (Prescott et al., 1993). The great majority of bacteria that degrade hydrocarbons and other xenobiotic substances happen to be Gram negative (Bartha, 1990). In the case of Gram-positive bacteria, delivery by fusion would require the liposomes to pass through the thick mucopeptide cell wall before fusion with the cell membrane can occur. Since even the small unilamellar liposomes have diameters above 20 nm (Szoka and Papahadjopoulos, 1980), it is not easy to imagine how this occurs. However, the previously discussed monomer diffusion and transient collision delivery models could function in the case of both Gram-positive and Gram-negative microorganisms.

Studies have shown that fusion of liposomes to each other and to some bacterial species can be enhanced in several ways: (1) by passage up and down through the critical temperature (T_c) of the bilayer; (2) by the addition of fusogenic agents, such as lysophosphatidylcholine and other single-chain polar lipids; (3) by the addition of millimolar concentrations of Ca^{+2} or Mg^{+2}; and (4) by freezing and thawing the vesicle preparation (Gibson and Strauss, 1984). Results indicate that some of the isolated bacteria required such enhancements while others did not (Huang, 1983). For example, the *Pseudomonas* sp. isolated by Miller and Bartha (1989) required no enhancement for degradation of *n*-octadecane and *n*-hexatriacontane. However, the Gram-positive isolate degraded both free and packaged phenanthrene and benzo(*a*)pyrene more rapidly in the presence of 0.025 *M* $CaCl_2$. For some isolates tested, a freeze-thaw cycle was more effective in promoting uptake of liposomes than $CaCl_2$ (data not shown).

IV. CONCLUSIONS

Lipsome-packaging provides the possibility of a biodegradability assay for some solid substrates with extremely low water solubility. Liposome-packaging facilitates the delivery of such substrates to the membrane of the degrading cells and is able to assess the true inherent biodegradability of a material, bypassing mass transfer limitations. In addition to liposome-packaging, some but not all degradation systems may need to be optimized for liposome fusion to cell membranes by the addition of $CaCl_2$ or by freeze–thaw cycles. The results reported in this study suggest that liposome packaging as a delivery measure is effective for test substrates with water solubilities similar to or less than those of *n*-octadecane and benzo(*a*)pyrene (0.006 and 0.003 mg l^{-1}, respectively). In the case of substrates with substantially higher water solubilities such as phenanthene and pyrene, liposome-packaging tends to lower rather than increase the degradation rate as compared to the free substrate, although it does not prevent biodegradation completely. Factors other than water solubility may play a role in determining whether liposome delivery of a substrate in biodegradation tests is effective, but too few studies have been performed to date to allow additional generalization at this time.

As to the practical benefit of liposome-enhanced biodegradation, this should be looked upon as a testing techique rather than a bioremediation measure. If liposome-packaging of a hydrophobic solid pollutant results in enhanced biodegradation rates, this leads to the conclusion that the bioremediation of such a pollutant is feasible if mass transfer problems can be alleviated by means of detergents, dispersants or solvents compatible with biological systems.

REFERENCES

Aminabhavi, T.M., R.H. Balundgi, and P.E. Cassidy. 1990. A review on biodegradable plastics. *Polym. — Plast. Technol. Eng.* 29:253-262.

Atlas, R.M. 1993. *Handbook of Microbiological Media.* CRC Press, Boca Raton, FL, p. 841.

Bartha, R. 1990. Isolation of microorganisms that metabolize xenobiotic compounds. *In*: *Isolation of Biotechnological Organisms from Nature* (D.P. Laberda, Ed.) McGraw-Hill New York, pp. 283-307.

Brown, R.E. 1992. Spontaneous lipid transfer between organized lipid assemblies. *Biochim. Biophys. Acta* 1113:375-389.

Code of Federal Regulations. 1992. Inherent biodegradability in soil. CFR 40 §796.3400, p. 224-229. Government Printing Office, Washington, D.C.

Gibson, S.M. and G. Strauss. 1984. Reaction characteristics and mechanisms of lipid bilayer vesicle fusion. *Biochim. Biophys. Acta* 769:531-542.

Hope, M.J., M.B. Bally, G. Webb, and P.R. Cullis. 1985. Production of large unilamellar vesicles by a rapid extrusion procedure. Characterization of size distribution, trapped volume and ability to maintain a membrane potential. *Biochem. Biophys. Acta* 812:55-65.

Huang, L. 1983. Liposome-cell interactions in vitro. *In*: *Liposomes* (M.J. Ostro, Ed.). Marcell Dekker, New York, pp. 87-125.

Jimenez, I.Y. and R. Bartha. 1993. Rapid pyrene metabolism by a microbial consortium. 93rd General Meeting, American Society Microbiol., May 16-20, Atlanta, GA, Abstr. Q340.

Jimenez, I.Y. and R. Bartha. 1994. Characterization of a microbial consortium capable of rapid pyrene utilization. 94th General Meeting, American Society Microbiol. May 23-27, Las Vegas, NV. Abstr. No. Q-180.

MacDonald, R.C., R.I. MacDonald, B. Ph. M. Menco, K. Takeshita, N.K. Subbarao, and L. Hu. 1991. Small-volume extrusion apparatus for preparation of large, unilammellar vesicles. *Biochim. Biophys. Acta* 1061:297-303.

Marinucci, A.C. and R. Bartha. 1979. Apparatus for monitoring the mineralization of volatile ^{14}C-labeled compounds. *Appl. Environ. Microbiol.* 38:1020-1022.

Miller, R.M. and R. Bartha. 1989. Evidence from liposome encapsulation for transport-limited microbial metabolism of solid alkanes. *Appl. Environ. Microbiol.* 55:269-274.

Olson, F., C.A. Hunt, F.C. Szoka, W.J. Vail, and D. Papahadjopoulos. 1979. Preparation of liposomes of defined size distribution by extrusion through polycarbonate membranes. *Biochim. Biophys. Acta* 557:9-23.

Prescott, L.M., J.P. Harley, and D.A. Klein. 1993. *Microbiology* (second ed.) William C. Brown Publishers, Dubuque, IA. pp. 40-68.

Szoka, F.Jr. and D. Papahadjopoulos. 1980. Comparative properties and methods of preparation of lipid vesicles (liposomes). *Annu. Rev. Biophys. Bioeng.* 9:467-508.

U.S. Environmental Protection Agency. 1979. Toxic Substances Control Act Premanufacture Testing of New Chemical Substances. *Fed. Regist.* 44:16240-16292.

Zhang, Y. and R.M. Miller. 1992. Enhanced octadecane dispersion and biodegradation by a *Pseudomonas* rhamnolipid surfactant (biosurfactant). *Appl. Environ. Microbiol.* 58:3276-3282.

Zhang, Y. and R.M. Miller. 1994. Effect of a *Pseudomonas* rhamnolipid biosurfactant on cell hydrophobicity and biodegradation of octadecane. *Appl. Environ. Microbiol.* 60:2101-2106.

Chapter 19

Microbe Entrapment in Giant Liposomes

Gregory Gregoriadis, Ihsan Gursel, and Sophia G. Antimisiaris

CONTENTS

I. INTRODUCTION

Liposomes were shown in 1974 to act as immunological adjuvants for diphtheria toxoid[1] and in subsequent numerous studies[2,3] for a wide spectrum of protein and peptide antigens from bacteria, viruses, protozoa and other sources. In several studies,[2] animals immunized with liposomal antigens were protected from infection as a result of liposome-induced humoural or cellular immunity. The structural versatility of liposomes also provided opportunities for the tailoring of the system towards optimal adjuvanticity. Thus, appropriate choice of structural characteristics,[4-8] ligand-mediated targeting of liposomes to immunocompetent cells[9] and the use of co-entrapped[10] or free[11] cytokines such as interleukin-2 have all contributed to further amplification of adjuvanticity.

The vast majority of such studies[2,3] were carried out with liposomes prepared by techniques which generate vesicles of submicron average diameter.[12] These are known to accommodate peptides and proteins but not larger vaccines such as attenuated or killed viruses and bacteria. Most of these vaccines (e.g., measles, polio virus, *Bordetella pertussis,* Bacillus Calmette-Guérin and *Salmonella typhi*) are highly immunogenic,[13] although there are situations where their administration in liposomes that are sufficiently large to contain them may be a preferable alternative. For instance, in the case of vaccines composed of a mixture of soluble and particulate (e.g., microbial) antigens or formulations also containing cytokines, simultaneous delivery of all materials incorporated in the same liposomes to antigen-presenting cells may be advantageous in terms of improving immunogenicity. In this respect, we have developed a procedure for the entrapment of particles (latex microspheres) and microbes (*Bacillus subtilis* and Bacillus Calmette-Guérin) into giant liposomes.[14] The procedure is mild and is based on the freeze drying of "empty" preformed vesicles in the presence of materials (soluble or particulate) destined for entrapment. Upon rehydration with water, generated vesicles (up to 8 microns in diameter) contain up to 30% or more of the materials, depending on the type of vesicles used in the initial dehydration step. Further, entrapped microorganisms retain their viability during the procedure. Data obtained suggest that the present technique could be used for the production of stable, liposome-based live or attenuated microbial vaccines as such or in mixture with (co-entrapped) soluble antigens.

0-8493-4013-6/96/$0.00+$.50

II. MATERIALS AND METHODS

The sources and grades of egg phosphatidylcholine (PC), distearoyl phosphatidylcholine (DSPC), phosphatidyl glycerol (PG), cholesterol (Chol), triolein (TO), immunopurified tetanus toxoid, oil red O and trehalose have been described elsewhere.[4,8,14] Killed and live *B. subtilis* spores and killed Bacillus Calmette-Guérin (BCG) were gifts from Dr. Bruce Jones (Public Health Laboratories Service, Porton Down, Salisbury, Wilts) and Dr. J. L. Stanford (Dept. of Medical Microbiology, UCL Medical School, London), respectively. Tetanus toxoid, *B. subtilis* and BCG were radiolabeled with ^{125}I as described.[14] Labeling of *B. subtilis* with fluorescein isothiocyanate (FITC) (Sigma) was carried out by the method of Mann and Fish.[15] All other reagents were of analytical grade.

A. ENTRAPMENT OF *B. SUBTILIS* SPORES AND OTHER MATERIALS INTO LIPOSOMES

Method A: The solvent spherule evaporation method with modifications[14] was applied. In brief, 1 ml of a 0.15 *M* sucrose solution containing ^{125}I-labeled material (e.g., 100–150 μg tetanus toxoid and 1×10^7 live *B. subtilis* spores) was mixed by vortexing with 1 ml of chloroform into which PC or DSPC, Chol, PG, TO (4:4:2:1 molar ratio; 9 μmol total lipid) were dissolved. The water-in-chloroform emulsion formed was mixed by vortexing with a diethyl ether-in-water emulsion made from 0.5 ml of a solution of the lipids as above in diethyl ether and 2.5 ml of a 0.2 *M* sucrose solution. The double emulsion (water-in-oil-in-water) thus prepared was placed in a conical flask and the organic solvents were evaporated at 37°C by flushing nitrogen. Generated liposomes were either centrifuged twice at 600 × g over a 5% glucose solution with the pellet resuspended in 0.1 *M* sodium phosphate buffer supplemented with 0.9% NaCl, pH 7.4 (PBS) to separate entrapped from non-entrapped toxoid or subjected to sucrose gradient fractionation (see below) for the separation of non-entrapped *B. subtilis*. In some experiments, FITC-labeled polystyrene particles (0.5 and 1.0 μm diameter, Polysciences) or FITC-labeled *B. subtilis* were entrapped as above in liposomes which were used in fluorescence microscopy studies. To improve visualization, liposomes were made to contain oil red O by dissolving the dye (1 mg per 9 μmol total lipid) in the chloroform solution of lipids prior to the generation of giant vesicles.

Method B: To entrap potentially labile materials into giant liposomes in the absence of the organic solvents used in method A, giant liposomes containing sucrose only were preformed as in method A and centrifuged over a 5% glucose solution at 600 × g for 5 min. The liposomal pellet, resuspended in 1 μλ PBS, was mixed with 1 μλ of the ^{125}I-labeled materials destined for entrapment (live *B. subtilis*, 1×10^7 spores; tetanus toxoid, 100–150 μg and killed BCG; 1×10^6 bacteria) and freeze-dried overnight under vacuum. Rehydration of the freeze-dried material was initially carried out by the addition of 0.1 ml distilled water at the appropriate temperature[14] and then, after 30 min, by the successive addition of 0.1 ml PBS and of 0.8 ml PBS 30 min later. Separation of entrapped from non-entrapped material was carried out as above (method A).

B. SUCROSE GRADIENT FRACTIONATION

A discontinuous gradient was prepared from two sucrose solutions (made up by dissolving 59.7 and 117 g of sucrose, respectively, in 100 ml water) in swing-out bucket centrifuge tubes. Preparations of entrapped and non-entrapped *B. subtilis* or BCG (methods A and B) were layered on top of the gradient and centrifuged for 1.5 h at 90,000 × g at 4°C. Subsequently, 1 ml fractions were pipeted out from the top of the gradient and assayed for ^{125}I radioactivity and phospholipid content.[16]

C. CHARACTERIZATION OF GIANT VESICLES

The mean diameter (volume distribution) of liposomes was measured in a Malvern Mastersizer equipped with a laser beam (613 nm). Light microscopy studies with oil red O-stained liposomes containing FITC-labeled *B. subtilis* or FITC-labeled polystyrene particles were carried out using a Nikon microscope or a Leica confocal microscope, both equipped with a fluorescence light source. For tests on spore viability following the entrapment procedure, giant vesicles containing spores presumably live, or the same vesicles after treatment with Triton X-100 (to free the spores) were serially diluted in nutrient broth and then spread on nutrient agar plates to estimate the total count, viable count and the average number of spores per individual vesicle (for details see Table 2). Spore viability as well as other properties were also studied in spore-containing liposomes subjected to several cycles of freeze-drying in the presence of a cryoprotectant (trehalose).

III. RESULTS AND DISCUSSION

A. ENTRAPMENT OF SOLUBLE AND PARTICULATE MATTER

Figure 1 shows that complete separation of entrapped from non-entrapped *B. subtilis* spores was achieved by sucrose gradient fractionation. Judging from ^{125}I radioactivity *(B. subtilis)* and phospholipid (liposomes) values determined in the fractions after centrifugation (Figure 1A), spore-containing liposomes were recovered in the upper four fractions of the gradient with free spores sedimenting to the bottom fraction. Moreover, incubation of "empty" giant liposomes with *B. subtilis* spores for 24 h prior to gradient fractionation resulted in the quantitative recovery of the spores in the bottom fraction (Figure 1B), suggesting that spores had not adsorbed to the surface of liposomes during the entrapment procedure.

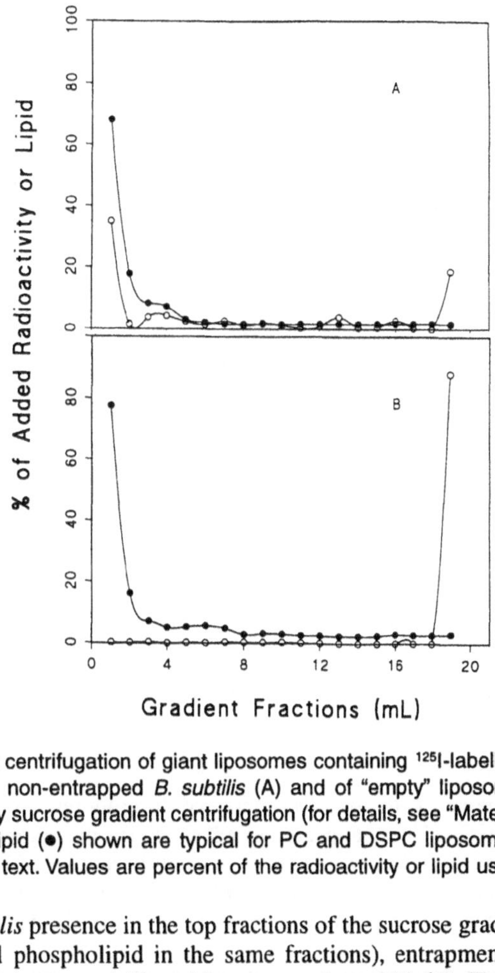

Figure 1 Sucrose gradient centrifugation of giant liposomes containing ^{125}I-labelled live *B. subtilis*. Separation of liposome-entrapped from non-entrapped *B. subtilis* (A) and of "empty" liposomes from added free live *B. subtilis* (B) was carried out by sucrose gradient centrifugation (for details, see "Materials and Methods"). Patterns of ^{125}I-radioactivity (○) and lipid (●) shown are typical for PC and DSPC liposomes prepared by either of the techniques described in the text. Values are percent of the radioactivity or lipid used for fractionation.

On the basis of *B. subtilis* presence in the top fractions of the sucrose gradient (which coincided with the presence of liposomal phospholipid in the same fractions), entrapment of spores in 12 separate experiments using method A was variable, with a mean value of 31.6% (Table 1). However, values for the entrapment of tetanus toxoid by the same method (i.e., in the presence of organic solvents) were nil (Table 1). This finding and the prospect of damage or otherwise inactivation of proteins and live microbes that could be used for entrapment in subsequent work, prompted the modification of the procedure to one that could ensure entrapment of potentially labile materials in the absence of organic solvents. Such modification (as applied in method B), was based on previous work[17] which showed that freeze-drying of buffer-loaded empty small unilamellar vesicles in the presence of solutes destined for entrapment and subsequent controlled rehydration leads to the generation of multilamellar liposomes[18] entrapping up to 80% or more of the solutes.[17] Thus, sucrose-loaded giant liposomes were prepared as above (method A) and freeze-dried in the presence of killed or live *B. subtilis* spores, killed BCG, tetanus toxoid, or a mixture of live spores and tetanus toxoid (as in method B). Rehydration of the powder obtained led to the formation of giant vesicles containing substantial proportions of *B. subtilis* (9–34.6%), BCG (27.8%), and tetanus toxoid (3.5–16.5% of the material used; Table 1 and legend). No significant difference in the entrapment values of *B. subtilis* obtained with killed and live spores using the present new procedure (method B) and the one described earlier (method A) was observed.

Table 1 Entrapment of *B. subtilis* Spores and Tetanus Toxoid in Giant Liposomes

Giant Vesicles	*B. subtilis* Killed	*B. subtilis* Live	Tetanus Toxoid
(PC)A	31.6 ± 24.2(12)		0.0(4)
(PC)B	26.7 ± 12.1 (7)		8.4 ± 2.6(4)
(DSPC)A			0.0(4)
(DSPC)B	21.3 ± 8.9 (6)		11.1 ± 1.9(4)
(PC)A		22.5 ± 9.8(6)	4.1 ± 3.8(4)
(PC)B		34.6 ± 5.2(6)	16.5 ± 11.0(4)
(DSPC)A		20.1 ± 12.4(6)	3.3 ± 3.1(4)
(DSPC)B		26.8 ± 4.4(6)	12.4 ± 9.8(4)
(PC)B[a]		9.0 ± 5.4(3)	3.5 ± 3.1(3)
(DSPC)B[a]		10.2 ± 3.6(3)	16.5 ± 6.0(3)

Note: ^{125}I-labelled *B. subtilis* and ^{125}I-labelled tetanus toxoid were entrapped separately or together (in this case toxoid was unlabelled) in giant vesicles made of PC or DSPC by method A or B. Results based on radioactivity measurements (or the fluorescamine method for the toxoid co-entrapped with spores) are expressed as % ± SD of material used for entrapment. In one experiment, entrapment of ^{125}I-labelled BCG in PC giant vesicles was 27.8% (method B). Numbers in parantheses denote number of preparations.

[a] Denotes preparation with co-entrapped *B. subtilis* and tetanus toxoid.

B. SIZE OF SPORE-CONTAINING GIANT LIPOSOMES

Size distribution measurements of spore-containing liposomes (method A) revealed a mean diameter of 6.5 ± 2.4 μm (with a lower and upper range of 2.1 –16.4 μm; results not shown). Similar size values were obtained for spore-containing vesicles prepared by method B (Table 2). Moreover, freeze-drying of giant vesicles in the presence of trehalose and reconstitution with water had no significant effect on their original size (results not shown).

C. LIGHT MICROSCOPY

Nearly all giant vesicles stained with oil-red-O contained differing numbers of FITC-labeled *B. subtilis* spores or polystyrene particles. These appeared to adhere to (or precipitate towards) the inner wall of the vesicles, probably because of the hydrophobic nature of the surface of both spores and particles and bilayers. Further evidence of particle localization within liposomes was given by confocal microscopy which enabled their visualization in different sections of space within the vesicles.

D. THE EFFECT OF THE ENTRAPMENT PROCEDURE ON THE VIABILITY OF *B. SUBTILIS* SPORES

Results in Table 2 show that liposome-entrapped spores (following treatment with Triton X-100 to liberate spores) were able to form colonies. As expected, the number of colonies formed for each of the detergent-treated preparations was much greater than that observed with intact (control) liposomes: it was anticipated that, in the latter case, spores within a single (intact) liposome would only produce one colony (see legend to Table 2). Table 2 also suggests that up to 20 spores could be found within individual vesicles with no apparent relationship between spore numbers and type of phospholipid used. There was, however, a positive relationship between number of spores entrapped and vesicle size. For example, among the six different giant vesicle preparations used, the lowest spore number (three) per vesicle was obtained with the smallest vesicles (6.2 μm) and the highest (twenty) with vesicles of the highest mean size (8.4 μm diameter).

Table 2 Estimated Average Number of *B. subtilis* Spores Per Vesicle

Giant Vesicles	Number of Colonies: Triton X-100	Number of Colonies: Control	Spores per Vesicle	Vesicle Mean Size (μm)
(PC)A	96	29	3	6.2
(DSPC)A	162	38	4	6.5
(PC)B	203	10	20	8.4
(DSPC)B	250	21	12	7.2
(PC)B[a]	160	12	13	8.0
(DSPC)B[a]	109	15	7	6.4

Note: *B. subtilis* spores were entrapped in giant liposomes composed of PC or DSPC by method A or B. Number of colonies developed on agar plates are shown for preparations previously treated with Triton X-100 to liberate the spores or intact (control) liposomes. Estimation of spores per vesicle was carried out by dividing colony numbers after Triton X-100 with colony numbers without Triton X-100. It should be noted that the number of colonies after Triton equals the number of viable spores liberated from the vesicles and that number of colonies without Triton must be equal to the number of vesicles (entrapped spores in each vesicle, regardless of their spore number, produce a single colony).

[a] Denotes preparation with co-entrapped *B. subtilis* and tetanus toxoid.

E. THE EFFECT OF REPEATED DEHYDRATION ON THE STABILITY OF SPORE-CONTAINING LIPOSOMES

B. subtilis spore-containing liposomes were freeze-dried in the presence of trehalose (0.25 *M*). Following controlled rehydration, up to 93% of the entrapped material was still entrapped in the reconstituted vesicles. Repetition of this process (up to five times) showed that liposomes could still retain most of their spore content (Figure 2). The extent of spore retention by the vesicles was found to correlate with spore viability: at the end of five cycles of dehydration-rehydration, colony counts (initially 44 colonies) remained nearly the same (39 colonies) (Figure 2).

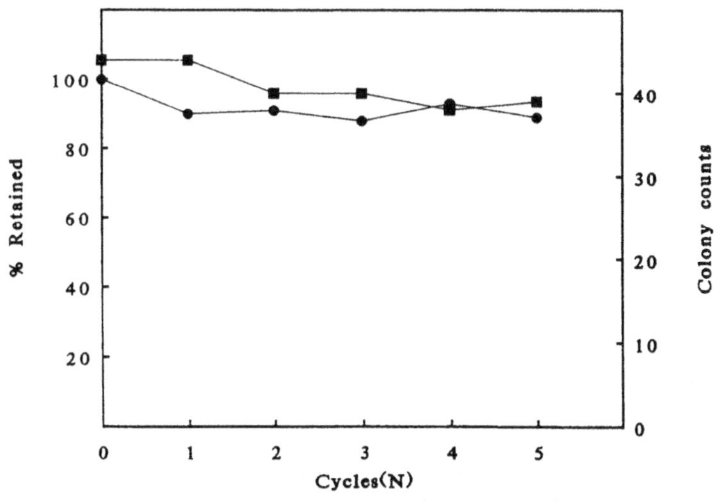

Figure 2 Retention of entrapment (•) and viability (■) values of *B. subtilis* spores following five rehydration-dehydration cycles. Spore retention was calculated on the basis of ^{125}I-radioactivity and represents the percentage of the initial amount used. For further details see "Materials and Methods" and legend of Table 2.

IV. CONCLUSIONS

A procedure has been developed for the incorporation of killed and live microbes, other particulate matter and soluble antigens into giant liposomes of several microns in diameter. In the case of live *B. subtilis,* entrapment under the conditions employed did not affect their viability. Vesicles containing microbes can be freeze-dried (for storage) in the presence of a cryoprotectant with most of their content recovered within the vesicles on reconstitution with buffered saline. The present procedure compares favorably with others[19-21] reported to produce micron-size liposomes, but only in conjunction with organic solvents or detergents. These may be detrimental to labile agents and the presence of detergent in formulations may be toxic for *in vivo* use. In contrast, conditions applied in the present study are mild enough to suggest application of the procedure for the entrapment of live or attenuated microbes, together with soluble antigens or cytokines if required. Previous work with smaller, multilamellar liposomes[22] and our recent unpublished data (M. Gursel, I. Gursel and G. Gregoriadis) with giant liposomes as prepared here, suggests that such liposomes containing soluble or particulate antigens are incapable of interacting with relevant antibodies *in vitro,* or in preimmunized animals.[22] Thus, giant liposomes could serve not only as immunoadjuvants for entrapped microbial and soluble vaccines but also as carriers of vaccines in cases where there is a need to prevent interaction of the latter with maternal antibodies or preformed antibodies to vaccine impurities.

ACKNOWLEDGMENTS

We thank the Ministry of Defence, Chemical and Microbiological Research Establishment, Porton Down, Salisbury for financial support and Mrs. Concha Perring for excellent secretarial assistance.

REFERENCES

1. **Allison, A. C. and Gregoriadis, G.,** Liposomes as immunological adjuvants, *Nature,* 252, 252, 1974.
2. **Gregoriadis, G.,** Immunological adjuvants: a role for liposomes, *Immunol. Today,* 11, 89–97, 1990.
3. **Alving, C. R.,** Liposomes as carriers of antigens and adjuvants, *J. Immunol. Methods,* 140, 1–13. 1991.
4. **Davis, D. and Gregoriadis, G.,** Liposomes as immunological adjuvants with immunopurified tetanus toxoid: influence of liposomal characteristics, *Immunology,* 61, 229–234, 1987.
5. **Therien, H.-M., Shahum, E., and Fortin, A.,** Liposome adjuvanticity: influence of dose and protein-lipid ratio on the humoural response to encapsulated and surface-linked antigen, *Cell Immunol.,* 136, 402–413, 1991.
6. **Bakouche, Q. and Gerlier, D.,** Enhancement of immunogenicity of tumour virus antigens by liposomes: the effect of lipid composition, *Immunology,* 57, 219–223, 1986.
7. **Latiff, N. A. and Bacchawat, B. K.,** The effect of surface-coupled antigen of liposomes in immunopotentiation, *Immunol. Lett.,* 15, 45–51, 1987.
8. **Davis, D., Davies, A., and Gregoriadis, G.,** Liposomes as adjuvants with immunopurified tetanus toxoid: the immune response, *Immunol. Lett.,* 14, 341–348, 1987.
9. **Garçon, N., Gregoriadis, G., Taylor, M., and Summerfield, J.,** Mannose mediated targeted immunoadjuvant action of liposomes, *Immunology,* 64, 743–745, 1988.
10. **Tan, L. and Gregoriadis, G.,** The effect of interleukin-2 on the immunoadjuvant action of liposomes, *Biochem. Soc. Trans.,* 17, 693–694, 1989.
11. **Mbawuike, I. N., Wyde, P. R., and Anderson, P. M.,** Enhancement of the protective efficacy of inactivated influenza A virus vaccine in aged mice by IL-2 liposomes, *Vaccine,* 8, 347–352, 1990.
12. **Gregoriadis, G. (Ed.),** *Liposome Technology,* 2nd ed., Vols. 1–3, Boca Raton, FL, 1993.
13. **Mimms, C. A., Playfair, J. H. L., Roitt, I. M., Wakelin, D., and Williams, R.,** *Medical Microbiology,* Mosby, St. Louis, MO, 1993, Chap. 36.
14. **Antimisiaris, S. G., Jayasekera, P., and Gregoriadis, G.,** Liposomes as vaccines carriers: incorporation of soluble and particulate antigens in giant vesicles, *J. Immunol. Meth.,* 166, 271–280, 1993.
15. **Mann, K. G. and Fish, W. W.,** Protein polypeptide chain molecular weights by gel chromatography in guadinium chloride, *Methods Enzymol.,* 26, 28–42, 1972.
16. **Stewart, J. C. M.,** Colorimetric determination of phospholipids with ammonium ferrothiocyanate, *Anal. Biochem.,* 104, 10–14, 1979.
17. **Kirby, C. and Gregoriadis, G.,** Dehydration-rehydration vesicles: a new method for high yield drug entrapment, *Biotechnology,* 2, 979–984, 1984.
18. **Gregoriadis, G., Garçon, N., da Silva, H., and Sternberg, B.,** Coupling of ligands to liposomes independently of solute entrapment: Observations on the formed vesicles, *Biochim. Biophys. Acta,* 1147, 185–193, 1993.

19. **Ishii, F.,** Production and size control of large unilamellar liposomes by emulsification, in *Liposome Technology,* Vol. 1, Gregoriadis, G. (Ed.), CRC Press, Boca Raton, FL, 1993, pp. 111–121.
20. **Gould-Forgerite, S. and Mannino, R. J.,** Preparation of large unilamellar vesicles with high entrapment yield by rotary dialysis or agarose plug diffusion, in *Liposome Technology,* Vol. 1, Gregoriadis, G. (Ed.), CRC Press, Boca Raton, FL, 1993, pp. 67–80.
21. **Philippot, J. R. and Liautard, J. P.,** A mild method for the preparation of very large unilamellar liposomes, in *Liposome Technology,* Vol. 1, Gregoriadis, G. (Ed.), CRC Press, Boca Raton, FL, 1993, pp. 81–98.
22. **Gregoriadis, G. and Allison, A. C.,** Entrapment of proteins in liposomes prevents allergic reactions in preimmunized mice, *FEBS Lett.,* 45, 71–74, 1974.

Chapter 20

Liposomes as a Model for Membrane Structures and Structural Transformations: A Liposome Album

Brigitte Sternberg

CONTENTS

I. INTRODUCTION

Biological membranes consist of a multitude of components which together carry out the diverse functions that characterize a particular membrane. Although the lipid bilayer, embedding transmembrane proteins and superficial peripheral proteins and enzymes, is the structural element present in most of the membranes, its continuity can be interrupted to varying degrees by a heterogeneous lateral organization of its molecular constituents. This process can lead to the formation of specialized membrane structures localized in certain domains, and each such area can undergo structural fluctuations associated with its function.

0-8493-4013-6/96/$0.00+$.50

Due to their biochemical and structural similarity with biomembranes, liposomes are excellent models for the reconstitution of a particular membrane structure.[1] This allows us to define the minimum number of these components that are absolutely necessary for a particular function to be carried out via the structure formed. It also allows us to take apart a functioning structure and reassemble it from its constituent pieces. Once such a structure has been assembled, we can investigate its properties in detail, free from the influence of any other membrane constituents.

Liposomes (Figure 1) are also an appropriate model for studying intrinsic structures build up by special lipids under certain conditions (temperature, pH, shear pressure, concentration of certain constituents), such as fluid and liquid-crystalline lipid phases, domain formation, formation of horseshoe structures, and cochleate cylinders. However, under certain conditions, such bilayer structures can also be transformed to nonbilayer structures, such as micelles, hexagonal lipid tubules (H_I and H_{II}), or cubic phases. Finally, liposomes are also good models for studying induced membrane structures, such as induced by drugs, lytic peptides or oligonucleotide, but also induced during interaction with cells.[2]

Figure 1 Freeze-fracture electron micrograph of a multilamellar liposome (MLV). The fracture plane is running along the hydrophobic area of an inner bilayer of the MLV showing intramembranous particles of the membrane-spanning protein, bacteriorhodopsin, excluded from lipid domains. An outer shell of cross-fractured bilayers is visible with quite regular water layers between them. (From Sternberg, B., Gale, P., and Watts, A., *Biochim. Biophys. Acta*, 980, 117, 1989. With permission.) All scales represent 100 nm, and the shadow direction runs from bottom to top in all electron micrographs.

All membrane structures described here, whether they are reconstituted, intrinsic or induced, have been studied extensively by a multiplicity of techniques, among them freeze-fracture electron microscopy (FFEM). FFEM has taken its place as a standard technique in membrane structure research because the fracture plane follows the hydrophobic interior of the membrane and therefore allows examination of the hydrophobic membrane region. Membrane structures, built up by lipids and/or intrinsic proteins, are visualized by this technique, but the technique is also used to investigate transformations in superstructures of lipids such as transformations from bilayer to micelles or other nonbilayer structures.[3]

This chapter shows freeze-fracture electron micrographs of reconstituted (Section II), intrinsic (Sections III and IV), and induced membrane structures (Sections V and VI) to underline the potential of liposomes as a model system to mimic membrane structures and to study membrane structure transformations.

II. MODELING AND INTERPRETATION OF STRUCTURES FOUND IN BACTERIA MEMBRANES

Halophilic and thermophilic archaebacteria are growing under extreme conditions. Thus, *Halobacterium halobium* is living in saltworks, saturated with NaCl, and *Thermoplasma acidophilum* is growing at high temperatures (39–59°C) and low pH (1–2), even without any cell wall. That is the reason why the lipids of their cytoplasmic membranes are adapted to the harsh environment by linking the fatty acids of the isophytanyl-type to the glycerol backbone by an ether bond. The fatty acids of the lipids of the cytoplasmic membrane of *Streptomyces hygroscopicus* are also branch chained but ester linked. *S. hygroscopicus*, producing the macrolide antibiotic Turimycin, can be induced, like many other Gram-positive bacteria, to form protoplast type L-forms devoid of a cell wall structure. Here, three unique membrane structures, the purple membrane (PM) of *H. halobium,* the monolayer membrane of *T. acidophilum,* and the wafer structure of *S. hygroscopicus*, were simulated and interpreted by the help of liposomes, applied as a membrane model.

A. MODELING OF THE 2D CRYSTALLIZATION OF BACTERIORHODOPSIN FOUND IN THE PURPLE MEMBRANE OF *HALOBACTERIUM HALOBIUM* CELLS

Bacteriorhodopsin (BR), the light-driven proton pump of *Halobacterium halobium* cells (Figure 2A), is an extreme example of a protein that self-associates within the plane of the bilayer to form specialized paracrystalline patches of hexagonal symmetry.[4] BR is a protein of the α-helical barrel type and forms trimers, visible as protein particles in freeze-fracture electron micrographs of the bacterial cell (Figure 2A) as well as of proteoliposomes, where BR is reconstituted into the liposomal membrane (Figures 1 and 2B). To uncover which factors are responsible for the association properties of BR in membranes, two groups of reconstituted BR/phosphatidycholine (PC)–proteoliposome complexes, one without and one with added endogenous PM-lipids, have been investigated by FFEM under various conditions of temperature, BR content, salt type and concentration, and bilayer thickness. From these reconstitution experiments, minimal conditions were established for getting 2D-arrays of BR in lipid bilayers. These are (i) one of the essential lipid species of the PM bearing highly charged headgroups (as diphytanyl-phosphatidyl-glycerol-phosphate (DPhPGP) or diphytanyl-phosphatidyl-glycerol-sulfate (DPhPGS)), (ii) high content of BR (less than 100 lipid molecules per one BR molecule), and (iii) high salt (4*M* NaCl) (Figure 2B).[5,6] Under no circumstances it was possible to form 2D-arrays of BR in complexes not containing the PM-derived lipids DPhPGP or DPhPGS and composed of a synthetic lipid only, such as dimyristoyl-phosphatidylcholine (DMPC) (Figures 1, 6A and 6B).[7]

Additionally, it has been shown that the quality of the 2D BR arrays is dependent on the thickness of the lipid bilayer: In complexes where the bilayer thickness (d_L) is different from the hydrophobic thickness of BR (d_P) the quality of the 2D arrays is much better, showing even second- and third-order reflections in the calculated diffraction pattern, compared to complexes where d_L is similar to d_P (mismatching effect).[8] It turned out that the formation of well-ordered arrays is absolutely essential for structural elucidation of membrane proteins by electron microscopy in high resolution.[9] Thus, the factors essential for 2D array formation and responsible for improving the quality of the resulting crystals, which we found out for BR, may also help in getting well-ordered 2D arrays of other membrane proteins for high resolution structural work.[10]

B. MODELING OF THE CROSS-FRACTURE BEHAVIOR OF *THERMOPLASMA ACIDOPHILUM* CELL MEMBRANES

Thermoplasma acidophilum cells are protected from the acidic and hot environment by a cytoplasmic membrane only, without an additional cell wall (Figure 3A). Based on the extreme requirements, the lipids of the cytoplasmic membrane have to be specialized and very stable. Indeed, unique bipolar and membrane-spanning lipids of the "bola"-type were found containing the main glycophospholipid (MPL) up to 50% of the total membrane lipids. MPL has been characterized as a di-isopropanol-2,3-glycero-tetraether, modulated by the attachment of different head groups (phosphoryl- and monoglycosyl) at both 1-positions of the two glycerols.[11] Due to its bipolarity and membrane-spanning character of the saturated methyl-branched C_{40}-hydrocarbon (diphytanyl) chains, MPL forms a monolayer of about the dimensions of the common phospholipid bilayer when suspended into water. It is also able to form multilamellar (MLV, Figure 3B) as well as small unilamellar liposomes (SUV) down to diameters below 100 nm (Figure 3C).[12]

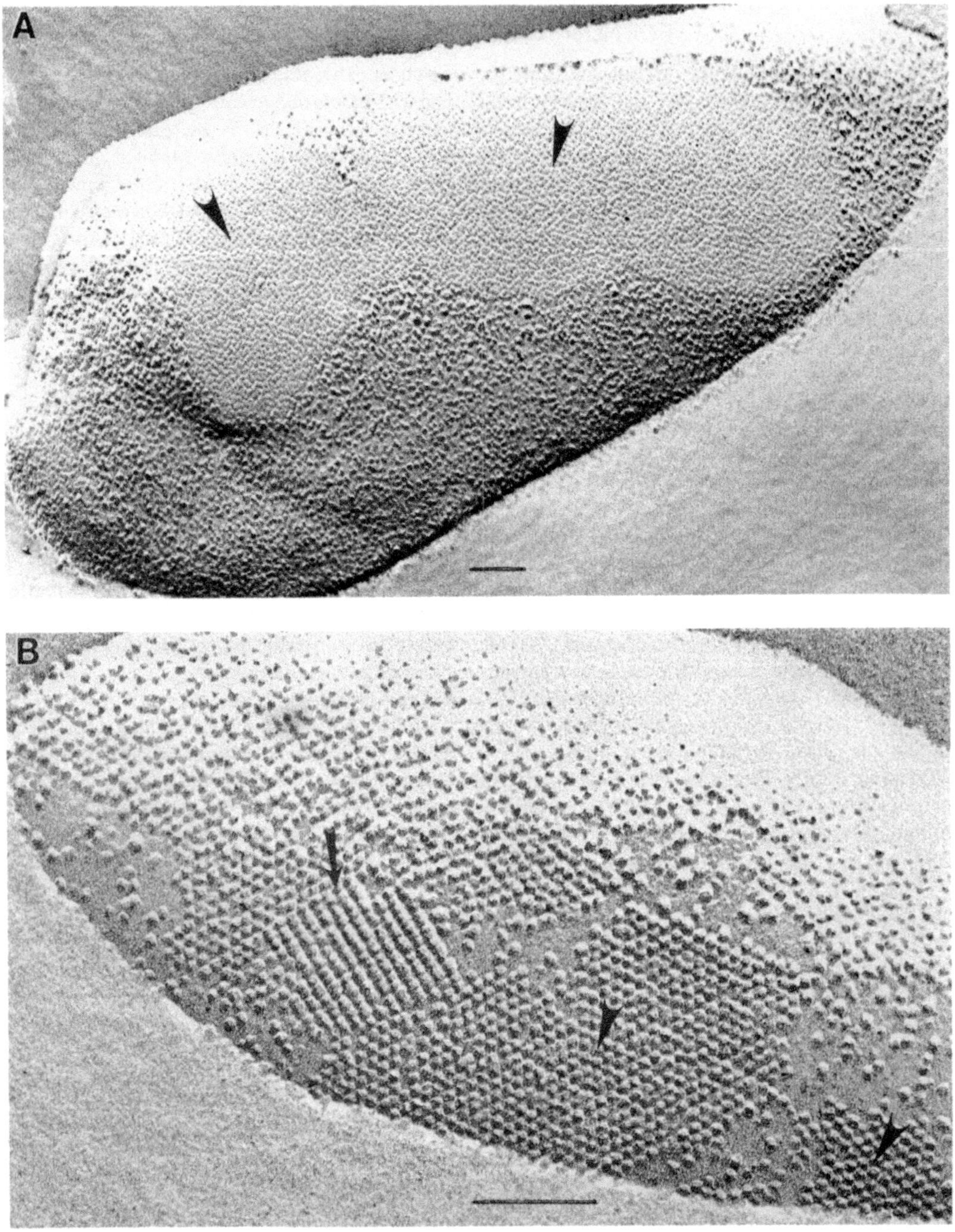

Figure 2 Protoplasmic fracture face (PF) of a *Halobacterium halobium* cell displaying purple membrane (PM)-areas, marked by an arrowhead, where bacteriorhodopsin (BR), represented by protein particles, is arranged in a hexagonal pattern (A). Proteoliposome, composed of BR/DMPC/PM-lipids (1:46:9, mole ratios) in 4 *M* NaCl, quenched at 55°C, displaying BR-arrays of hexagonal (marked by arrowheads) as well as orthogonal (marked by an arrow) symmetry (B). (B, from Sternberg, B., Watts, A., and Cejka, Z., *J. Struct. Biol.*, 110, 196, 1993. With permission.)

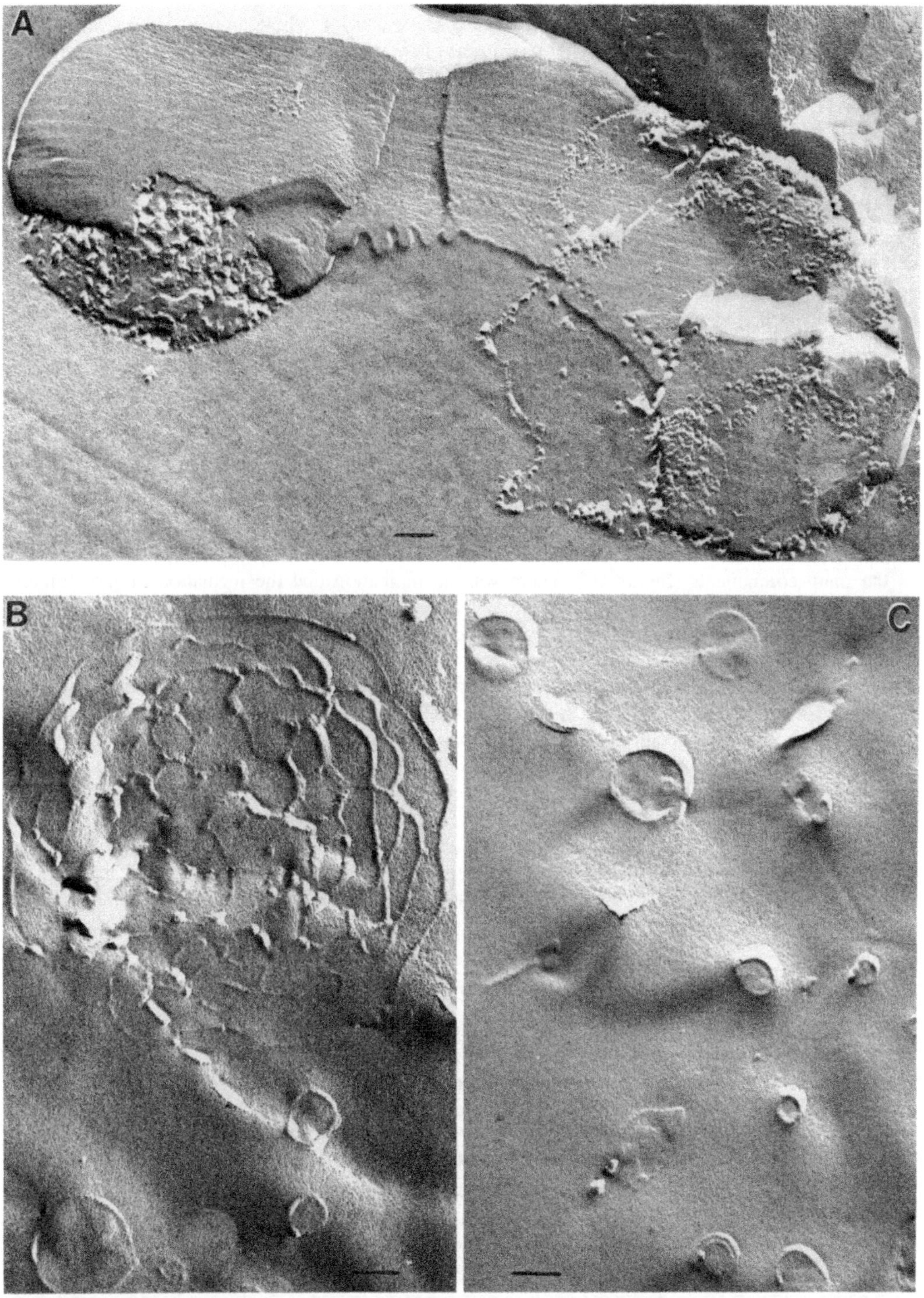

Figure 3 Freeze-fracture electron micrographs of a *Thermoplasma acidophilum* cell (A) and liposomes, composed of the main glycophospholipid (MPL), extracted from the cell membrane. Cell membrane (A) as well as a MLV (B) and some SUV (C) display cross-fracture behavior. (From Sternberg, B. and Rudolph, P., *Electron Microscopy,* 32 Vol. 3, Proceedings of the 10th European Congress on Electron Microscopy, A. Rios et al., Eds., Granada, Spain, 1992, 85. With permission.)

The bacterial cell membrane, containing the membrane-spanning bola-lipid MPL up to 50% by weight of the total membrane lipids, as well as liposomes made of MPL, show an unusual fracture behavior: they cannot be fractured within the apolar membrane plane and therefore show cross fractures exclusively (Figures 3A–C).[12] This is a late but additional proof that in membranes made of the normal monopolar and bilayer-forming lipids such as those isolated from egg, soybeans, or human skin,[13] the fracture plane is running along the hydrophobic interior of the lipid bilayers.[3,12,14]

C. MODELING OF A NONPLANAR LIPID STRUCTURE FOUND IN MEMBRANES OF *STREPTOMYCES HYGROSCOPICUS* CELLS

Many Gram-positive bacteria can be induced to form protoplasts and protoplast type L-forms devoid of cell wall structures. Protoplasts are usually unable to propagate and their cytoplasmic membrane corresponds to a great extent with the cytoplasmic membrane of normal vegetative cells. Stable protoplast type L-form cells, however, can grow and multiply, but the structural and physiological properties of their cytoplasmic membrane are altered. Cytoplasmic membranes of the bacterial strain *S. hygroscopicus* display a nonplanar, wafer-like membrane structure at their normal (Figure 4A) as well as their viable and stable L-form cells (Figure 4B). This wafer structure seems to exclude protein particles (Figure 4B) and shows a regular pattern of bulges with diameters of 30–40 nm (Figure 4A and B).[15,16]

To characterize this structure in more detail, liposomes were prepared from the extracted bacterial lipids and individual lipid fractions and the temperature dependency was investigated by FFEM, calorimetry, and ^{31}P-NMR spectroscopy. It was found that the phospholipid content of the membranes essential for the formation of that structure included cardiolipin (40–45%) and phosphatidylethanolamine (PE, 35–40%) bearing branched-chain (69–78%) and mainly anteiso $C_{15:0}$ and $C_{17:0}$ fatty acid residues as the main components (Figure 4C). There was no indication that the formation of the structure is related to temperature-induced lipid phase transition or nonbilayer structure formation processes. Together with the wafer structure, many small spherical vesicles were observed with diameters of about 20–40 nm (Figure 4B and C). Some of the vesicles (marked by an arrow) are situated directly under the bulges of the nonplanar structured liposome membrane (Figure 4C). It was suggested that both membrane features of the bacterial as well as liposomal membranes are induced by these vesicles, forming a hexagonal or cubic organization of vesicles on the cytoplasmic L-form membrane (Figure 4B) and between the cell wall and membrane of the normal cell and the lamellae of the multilamellar liposomes[15,16] (Figure 4A and C).

The formation of abundant microvesicles may be a consequence of the particular stereochemical properties of PE and cardiolipin with anteiso-fatty acid acyl chains showing small head groups and large chain area, especially at low pH. However, the function of the microvesicles and their relation to the composition of the lipids is not clear yet. They may play a role in encapsulation and transport of cell toxic metabolites such as antibiotics.

III. STRUCTURE TRANSFORMATIONS IN LIPOSOMAL BILAYER

Here, membrane structure transformations are described where the lamellar conformation is conserved throughout the whole process, such as lipid phase transition, domain formation, transformation of MLVs to SUVs, transformation between membrane-spanning and horseshoe structure, formation of vesicle assemblies, and formation of cochleated cylinders.

A. LIPID PHASE TRANSITION AND DOMAIN FORMATION

Lipids isolated from biological membranes have long been known to form liquid crystalline phases when mixed with excess water. The mixture of lipids taken from most membranes assemble under physiological conditions into the lamellar (Lα) phase, consisting of bilayer sheets of lipid stacked in a one-dimensional

Figure 4 Protoplasmic fracture faces of a normal *Streptomyces hygroscopicus* cell, cultivated for 4 days (A), and of L-form cells (B), display wafer structure in an orthogonal (A) and in a hexagonal (B) pattern. Multilamellar liposome, composed of the phospholipid fraction of the purified membranes of *S. hygroscopicus* L-form cells and prepared under acid conditions, displays a wafer-structure area and additionally many small vesicles, some of them (marked by an arrow) are directly situated under the bulges of the orthogonal pattern (C). (From Sternberg, B., Gumpert, J., Reinhardt, G., and Gawrisch, K., *Biochim. Biophys. Acta*, 898, 223, 1987. With permission.)

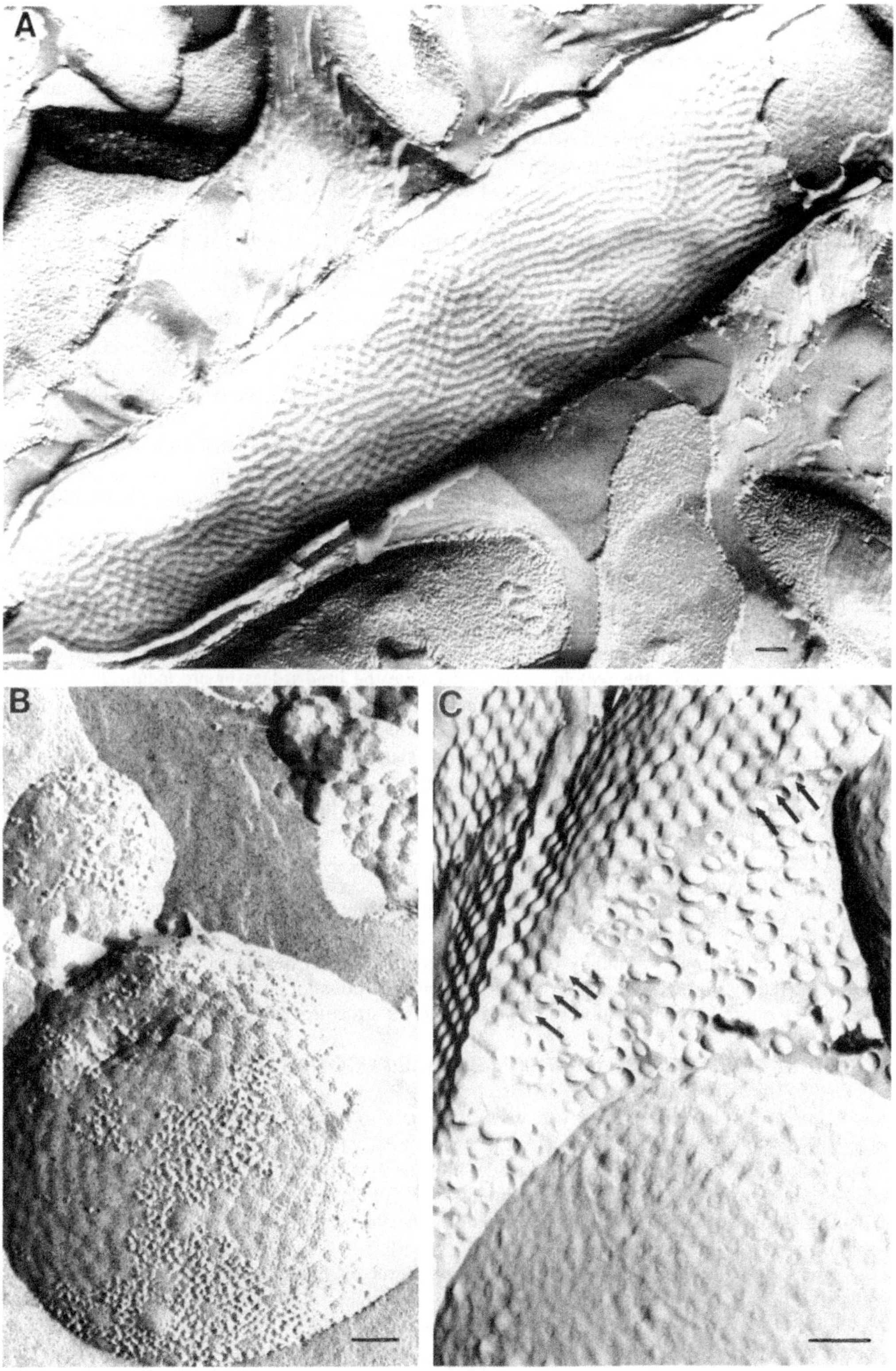
A
B
C

lattice separated by layers of water.[17] Larger liposomes with bigger diameters and less curved bilayers, made of lipid mixtures extracted from biological membranes as well as of a single lipid species are suitable models to study reversible thermally induced transitions of lipid bilayers from an ordered crystalline-like state at low temperature ($L\beta'$ and $P\beta'$ phases) to a disordered fluid state at high temperature ($L\alpha$ phase). This process consists of a change in the lipid hydrocarbon chains from largely all-trans conformation to a more disordered state with gauche-trans isomers accompanied by a lateral expansion and decrease in thickness of the bilayer.[18]

Order–disorder thermotropic transitions can be seen and studied by a wide variety of methods, among them FFEM. Smooth fracture faces are observed for liposomes made from the synthetic lipid 1,2-dipalmitoyl-*sn*-glycero-3-phosphocholine (DPPC) quenched at temperatures (T_q) above the phospholipid bilayer gel to liquid-crystalline transition temperature (T_c~41°C; Figure 5A)[3]. Quenched at temperatures below T_c, the fracture faces for similar liposomes made from 1,2-dimyristoyl-*sn*-glycero-3-phosphocholine (DMPC) display two kinds of ridges, termed $\lambda/2$ (zigzag) and λ (wave-like) ridges.[19] Figure 5B displays DMPC liposomes quenched at 4°C, which is below the pre-transition of DMPC (T_p~15°C) and indicates the formation of molecular ordering characteristic of the $L\beta'$ phase. Quite regular $\lambda/2$ ridges are visible, shown in a higher magnification at the inset of the left side, which appear modified to a golf ball pattern[20] at small liposomes with highly curved bilayers (Figure 5B and inset at the right side). Small vesicles produced by prolonged sonication or other methods are not suitable models for elucidating phase transitions in lipid bilayers. The main transition peak of synthetic lipids such as DPPC is split into two peaks, whereby the peak representing the cooperative phase transition of the lipids of the more curved inner shell of the bilayer is shifted to lower temperatures.[1,21]

When membrane proteins are reconstituted into liposomes to form proteoliposomes, the particles representing the intrinsic protein BR are randomly distributed when quenched above the T_c of the bilayer of saturated acyl chain phospholipids, when the cooling rate is rapid enough ($>10^4$ K·s^{-1}).[3,7] However, when quenched from $T_q>T_c$ but using a slower cooling rate ($<10^4$K·s^{-1}), domain formation is visible at the fracture faces displaying a pattern of particle-free areas (Figure 6A and Figure 1). When quenched at temperatures below T_c, the protein particles decorate the lipid ridges or are localized in structural defects of these lipid ridges, as shown in Figure 6B.[3,7] With increasing protein content, the lipid pool able to undergo the cooperative phase transition gets smaller, revealing at $T_q<T_c$ that the distances between the lipid ripples become wider until finally the ripples almost disappear, and at $T_q>T_c$ that the protein-free areas become very small or completely disappear.[7]

Domain formation can occur in bilayers containing a single lipid species at the transition temperature range and also in mixed lipid bilayers and biomembranes at temperatures where at least one of the lipid species adapts the gel state. This lateral phase separation can be initiated by divalent cations or surface proteins in bilayers containing negatively charged lipids. Since cell membranes usually contain lipids with quite unsaturated fatty acids and their transitions are completed well below body or growth temperature, bulk phase transitions seem not to play a unique physiological role.[22] However, microcrystalline regions (domains) are found to exist modulating the activity of a variety of membrane-associated enzymes and transport systems and providing optimal conditions for energy-gaining processes, and may also allow the formation of locally restricted non-bilayer structures within membranes.

B. TRANSFORMATION OF MULTILAMELLAR LIPOSOMES TO UNILAMELLAR ONES UNDER SHEAR PRESSURE

Small unilamellar liposomes (SUVs) are extensively used as model systems for permeability studies, in drug delivery, and for cosmetics. In the past they were mostly produced by prolonged sonication of lipid dispersions. Nowadays they are mainly made with the French press cell and extrusion, representing simple preparative and semiquantitative techniques.[23,24] For that, aqueous suspensions of lipid are placed within the chamber of a press cell at room temperature and rapidly extruded at 20,000 psi through a small orifice.

To get detailed information about the SUV-forming process, two types of MLVs, made from soybean lecithin (i) or MPL (ii) were treated by French press or nanojet for up to 40 passes and then investigated by FFEM. Under shear pressure, MLVs gradually disassemble to SUVs. After 10 passes, smaller liposomes are already preformed within the bigger MLVs (Figure 7A and C). After 20 passes, more and more of the SUVs become free from the assembly (Figure 7B and D), until after 40 passes almost all of the MLVs are disassembled to SUVs (not shown here). Since the fracture plane is running along the

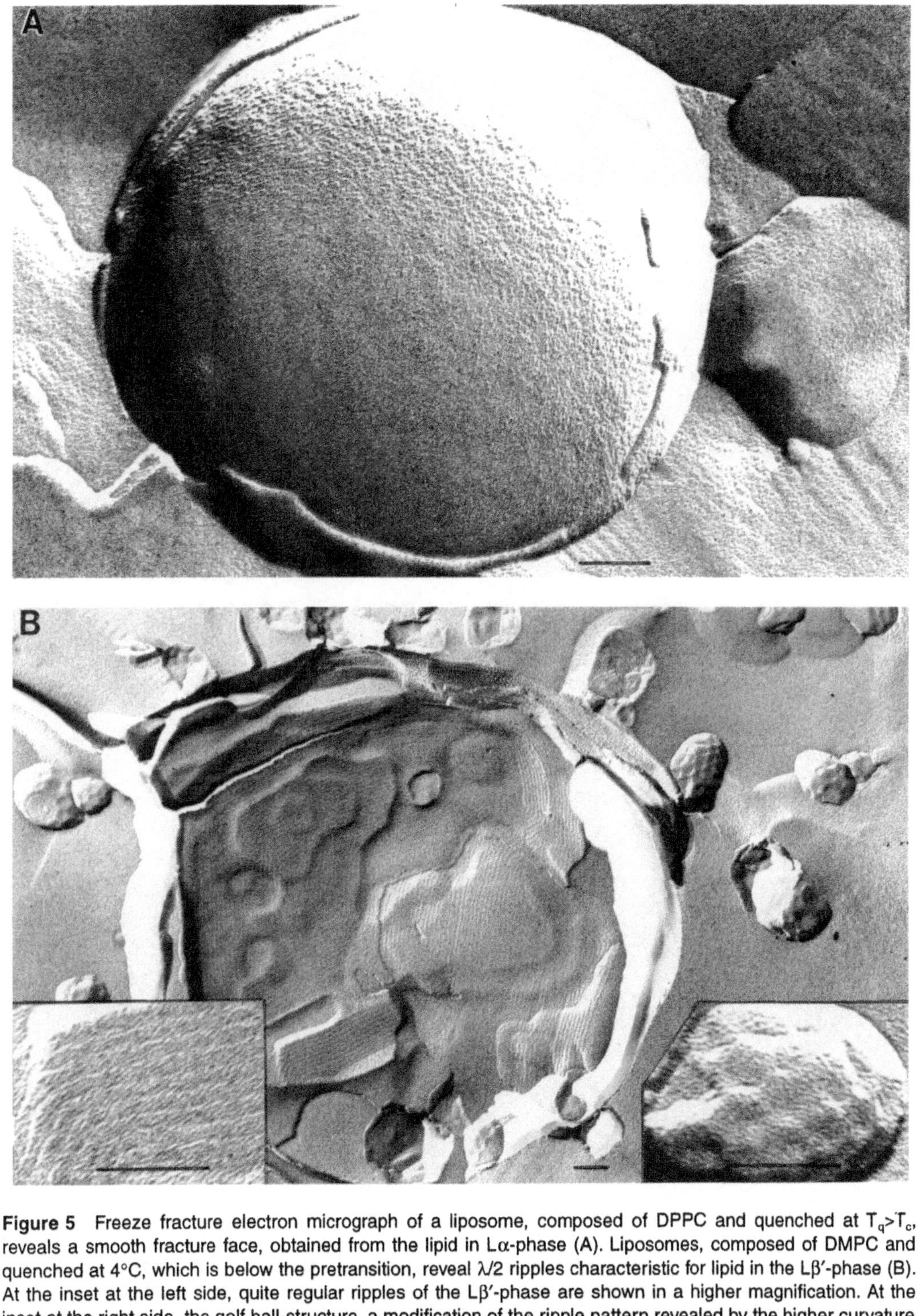

Figure 5 Freeze fracture electron micrograph of a liposome, composed of DPPC and quenched at $T_q > T_c$, reveals a smooth fracture face, obtained from the lipid in Lα-phase (A). Liposomes, composed of DMPC and quenched at 4°C, which is below the pretransition, reveal $\lambda/2$ ripples characteristic for lipid in the Lβ'-phase (B). At the inset at the left side, quite regular ripples of the Lβ'-phase are shown in a higher magnification. At the inset at the right side, the golf ball-structure, a modification of the ripple pattern revealed by the higher curvature of smaller liposomes, is shown in a higher magnification. (A, from Sternberg, B., in *Liposome Technology*, Vol. 1, Gregoriadis, G., Ed., CRC Press, Boca Raton, FL, 1993, 363; B, from Dempsey, C. E. and Sternberg, B., *Biochim. Biophys. Acta*, 1031, 143, 1990. With permission.)

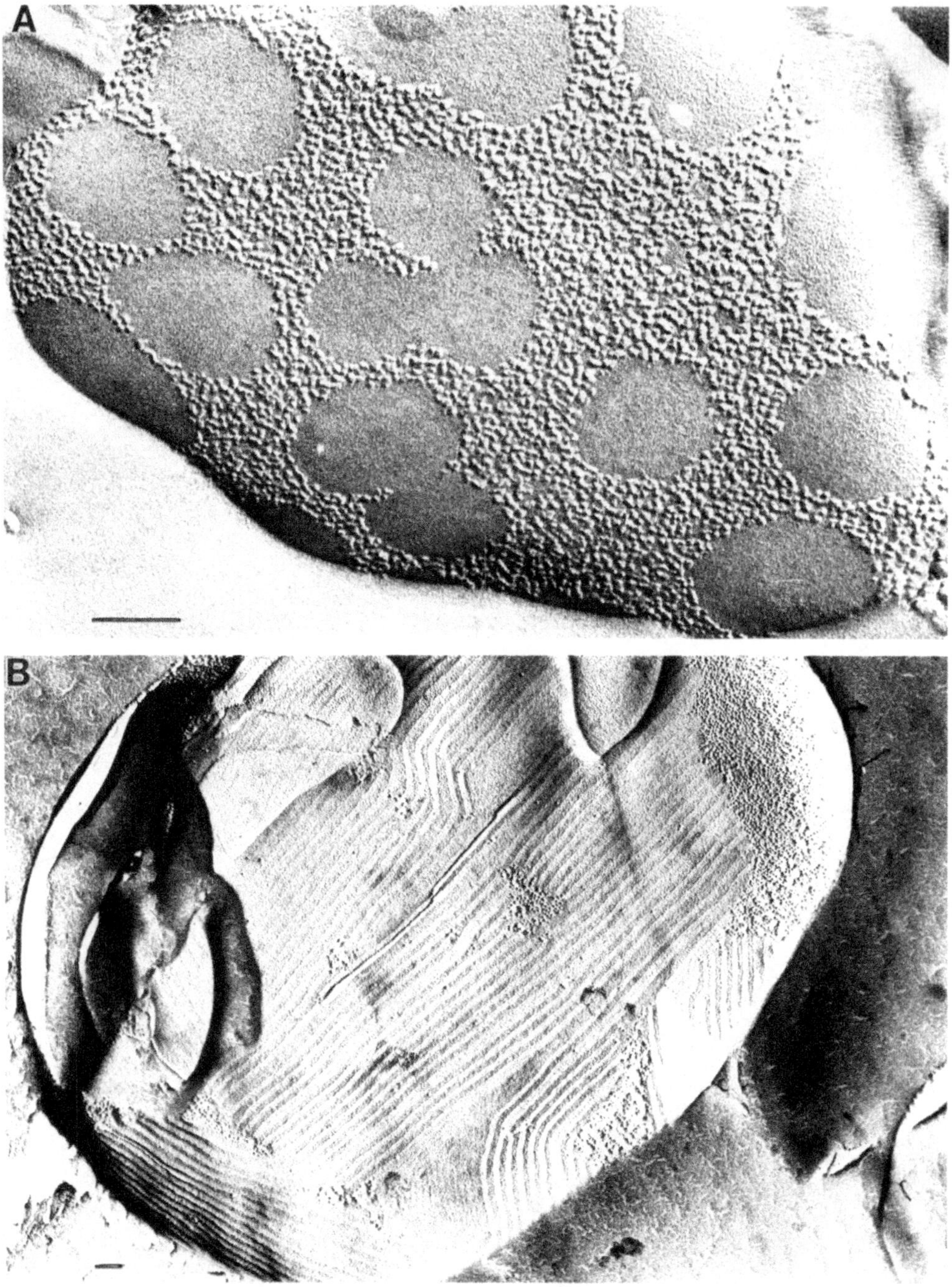

Figure 6 Proteoliposome, composed of BR and DMPC and quenched at $T_q > T_c$ with a slower cooling rate ($<10^4$ $K \cdot s^{-1}$), revealed a fracture face with particle-free areas obtained from the lipid in the fluid phase (A). BR/DMPC liposome quenched at $T_q < T_c$ displays a fracture face with protein particles decorating the lipid ridges or localized in structural defects of the lipid ridges (B).[7]

hydrophobic area of bilayers made of monopolar lipids such as soybean lecithin, we can get only some suggestions about the transformation process from those electron micrographs (Figure 7A and B). Liposomes made from the bipolar- and monolayer-forming MPL (ii), however, show cross-fracture behavior,[12] allowing us to have a look inside the liposomes. Under shear pressure the layers of the MLVs

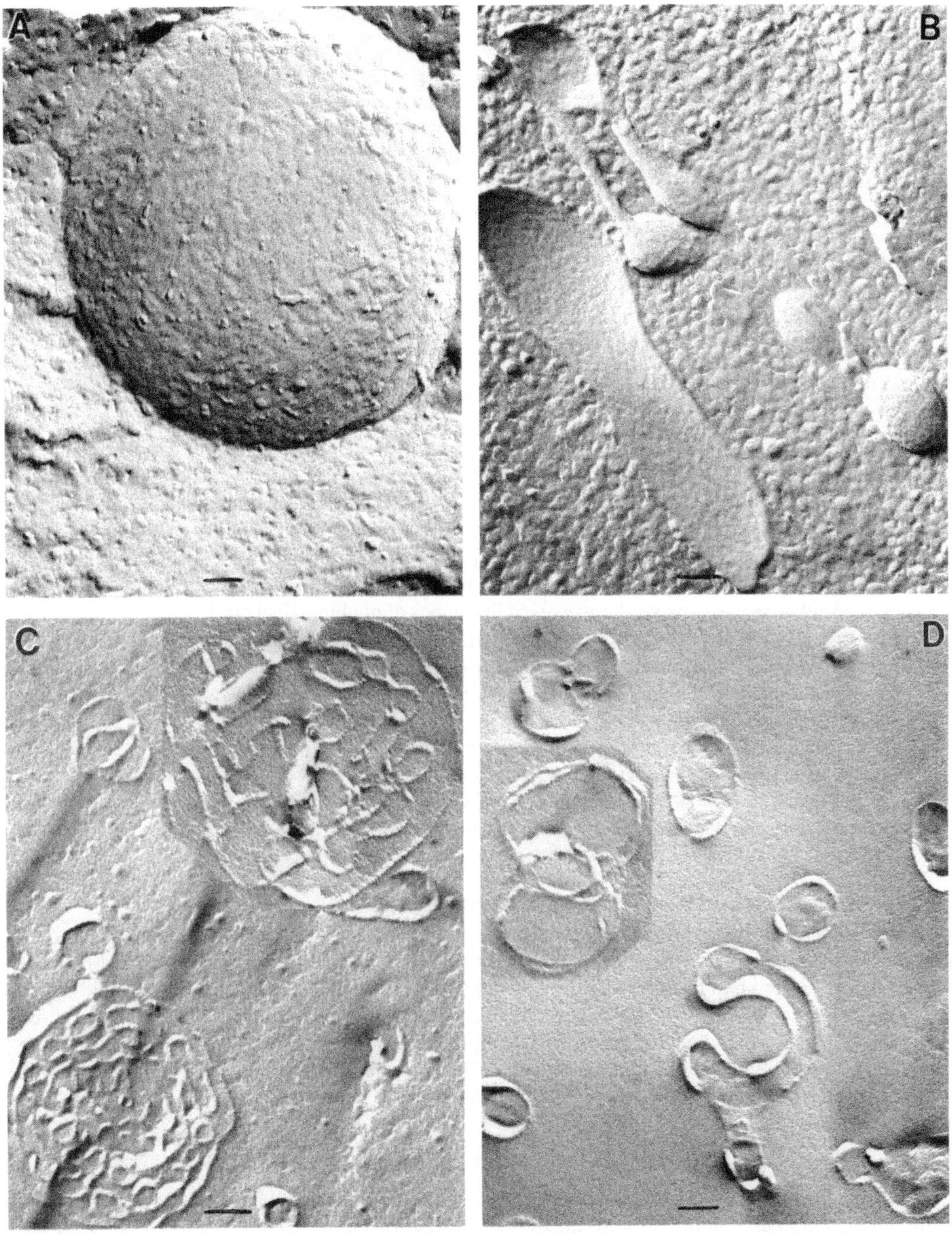

Figure 7 MLV, composed of soybean lecithin (A, B) and of MPL (C, D), were transformed to SUV under shear pressure. After 10 pressure cycles, smaller liposomes are already preformed within the bigger MLV (A, C). After 20 cycles, more SUV get free from the MLV assemblies (B, D). (C and D from Sternberg, B. and Rudolph, P. *Electron Microscopy,* Vol. 3, Proceedings of the 10th European Congress on Electron Microscopy, Granada, Spain, 1992, 85. With permission.)

have contact points and preformed SUVs are visible (Figure 7C) and become separated after more and more passes have been applied (Figure 7D). This procedure for forming SUVs seems to be a very mild procedure avoiding degradation of drugs or DNA encapsulated by the liposomes which might be degraded by ultrasonic treatment.

C. TRANSITION BETWEEN MEMBRANE-SPANNING- AND HORSESHOE STRUCTURE ADOPTED BY LIPIDS DERIVED FROM *THERMOPLASMA ACIDOPHILUM* CELLS

As described above in Sections IIB and IIIB, bipolar lipids such as MPL in excess water are adopting the membrane-spanning structure with one polar group at each side of the membrane, thus forming a monolayer membrane and showing cross-fracture behavior in FFEM (Figures 3A–C, 7C and D, and 8A, marked by an arrow).[12] In samples containing excess lipid, however, these "bola" lipids can also adopt the horseshoe structure, with both headgroups on one side of the monolayer now forming a bilayer and showing the common fracture behavior of bilayers (in Figure 8B, stacks of bilayer are marked by an arrowhead). At MPL concentrations of about 10 mg/ml combinations of cross (Figure 8A, marked by an arrow) and plane fractures (Figure 8A, marked by an arrowhead) are visible (sea shell-structure). Fracture planes derived from MPLs adopting horseshoe structure are marked by an arrowhead in Figure 8A and B. Excess lipid, addition of EDTA to these preparations, and higher temperatures seem to support the formation of horseshoe structure areas (Figure 8A and B).

D. FORMATION OF GEODESIC SPHERES MADE OF NONIONIC DETERGENTS

There is increasing interest in the structures formed by the self-association of amphiphiles, because of the potential uses of liquid crystalline arrays, vesicles, and micellar structures *per se* and as matrices from which new material can be fabricated. Nonionic surfactant vesicles (niosomes) are now widely studied as an alternative to liposomes. An increasing number of nonionic surfactants have been found to form vesicles, capable of entrapping hydrophilic and hydrophobic solutes.[25] These nonionic surfactant vesicles appear to be similar in terms of their physical properties to liposomes, being prepared in the same way and, under a variety of conditions, forming unilamellar or multilamellar structures. They may be regarded either as inexpensive alternatives to liposomes of nonbiological origin, or perhaps as a carrier system *in vivo*, physically similar to liposomes but undergoing less chemical degradation.

During formation of niosomes made from a series of sorbitan monoesters (Span 20, 40, 60, and 80) by a mechanical shaking technique without sonication,[26] we found in freeze fracture electron micrographs unusual structures, which we named geodesic spheres[27] (Figure 9). They seem to derive from a group of small niosomes (about 100 nm in diameter) close-packed, through some unknown attractive force, on the surface of one large vesicle (about 2 µm in diameter). If such fairly identical spheres are close-packed on a plane, a regular hexagonal lattice is formed. Wrapped around a large sphere, the resulting polyhedron cannot contain 6-fold vertices only; there must also exist other vertices. Indeed, on the resulting geodesic sphere pentagons and heptagons are also visible[27] (Figure 9).

E. FORMATION OF COCHLEATE LIPID CYLINDERS

Twenty years ago it was reported that SUVs made from phosphatidylserine (PS) are able to fuse in the presence of Ca^{2+}, creating large planar lamellae which roll up to form cylinders.[28] These cylinders seem to be formed from spirally folded lipid bilayers and therefore they got the name cochleate cylinders (Greek: κοχλιας = snail with spiral shell). We also found such cochleate cylinders, showing a length up to 1000 nm, formed by PS-liposomes, when prepared by freezing-thawing procedure, under acid conditions (pH 2.5, Figure 10A and B). By ^{31}P-NMR spectroscopy, it was confirmed that indeed bilayer sheets are rolled up to form the large spiral cylinders.

IV. BILAYER-TO-NONBILAYER TRANSFORMATIONS

The barrier property of bilayers has first priority among all membrane functions for regulating selectively the permeability between the internal milieu of a living cell and the external environment. Membranes also play a role in anchoring cytoskeletal elements, DNA replication and protein synthetic machinery, and in attaching cells to their extracellular matrix. Proteins localized in the lipid bilayer mediate these functions by serving as selective transport channels, energy and signal transducers, and attachment points. Although the composition of membranes is highly variable among organisms and between tissues of a single organism, there must be a strong control of the quantity and quality of the membranous components for the complex spectrum of cell membrane functions.[17,22,29]

Lipids isolated from biological membranes form a mesophase constructed from bilayer sheets highly impermeable even to small molecules such as glucose or inorganic ions. However, half of the naturally occurring lipids individually studied form nonlamellar phases, such as micelles, hexagonal (H_I or H_{II}),

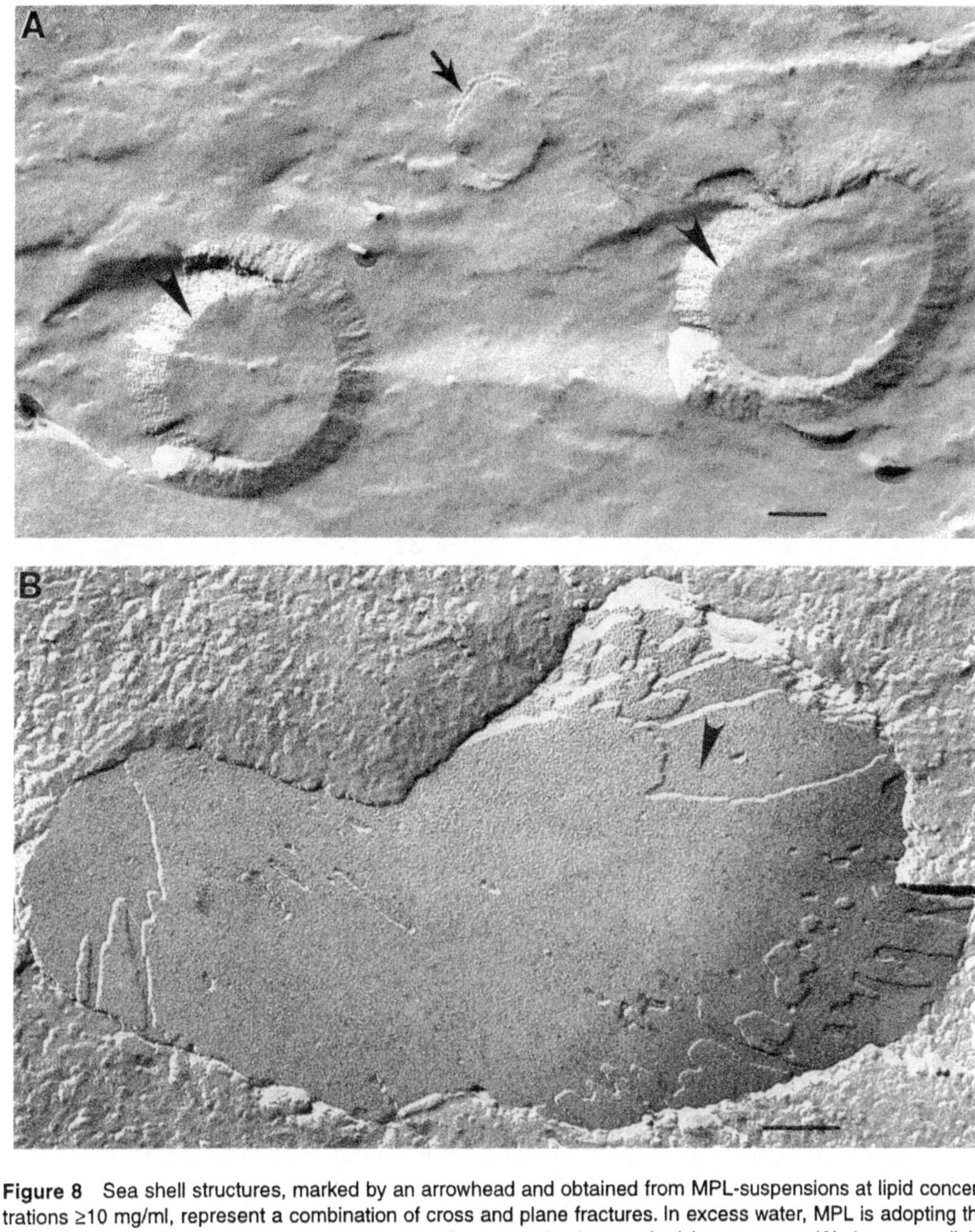

Figure 8 Sea shell structures, marked by an arrowhead and obtained from MPL-suspensions at lipid concentrations ≥10 mg/ml, represent a combination of cross and plane fractures. In excess water, MPL is adopting the membrane-spanning structure and shows cross-fracture behavior, marked by an arrow (A). In excess lipid, however, MPL is adopting horseshoe structure and reveals plane fracture behavior, marked by an arrowhead (B).

or cubic phases. Despite numerous systematic studies being undertaken on nonlamellar systems, their function in biology is not quite clear as yet. It has been hypothesized that hexagonal or other nonlamellar structures of lipids are essential in membrane fusion.[29]

A. LIPIDIC PARTICLE FORMATION AT LIPOSOME FUSION

Lipidic particles have been observed at the fusion sites of liposomes made either of negatively charged lipids such as PS and cardiolipin during interaction with Ca^{2+},[30] or by heating up liposomes containing H_{II} phase lipids.[31] Here, in Figure 10B, lipidic particles, marked by an arrow, are found during complex

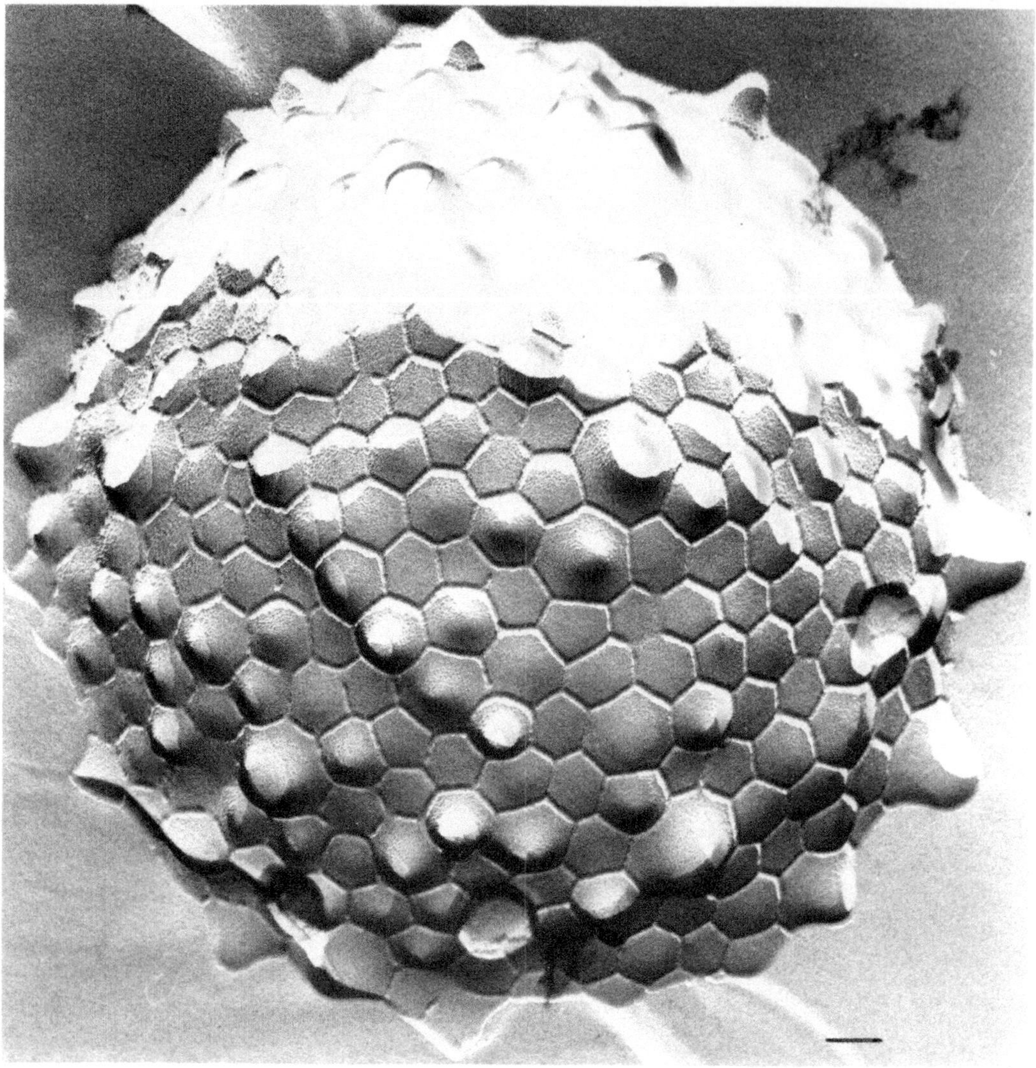

Figure 9 Geodesic sphere, derived from a group of small niosomes, close-packed on the surface of a large vesicle. (From Sternberg, B., Moody, M. F., Yoshioka, T., and Florence, A. T., *Nature*, forthcoming, with permission).

formation between DNA and cationic liposomes.[32] These structures, visualized as particles and pits in freeze-fracture micrographs have an approximate diameter of 110 Å and are accompanied by the appearance of a narrow, symmetric ^{31}P-NMR signal indicative of isotropic motional averaging. They have been suggested to reflect the presence of intrabilayer inverted micelles. Meanwhile, it turned out that neither the cochleate cylinders (Figure 10A and B) nor the lipidic particles (Figure 10C), both of which are observed at equilibrium, seems to play a role of an intermediate during fusion.[33]

B. H_I- AND H_{II}-LIPID TUBULES

In numerous cases the extracted lipid mixture of membranes has a strong tendency to form nonlamellar phases, such as H_I, H_{II}, or cubic phases, under small perturbations created by divalent ions, pH, and temperature. As a result, the lipid component of natural membranes is thermodynamically "close" to the lamellar to nonlamellar transition.[34] Hexagonal H_I phase consists of long cylindrical micelles arranged on a hexagonal lattice, separated by water (Figure 11A). Here, the hexagonal tubules are formed from DPhPGP, extracted from the PM and showing a large head group area compared to the hydrophobic

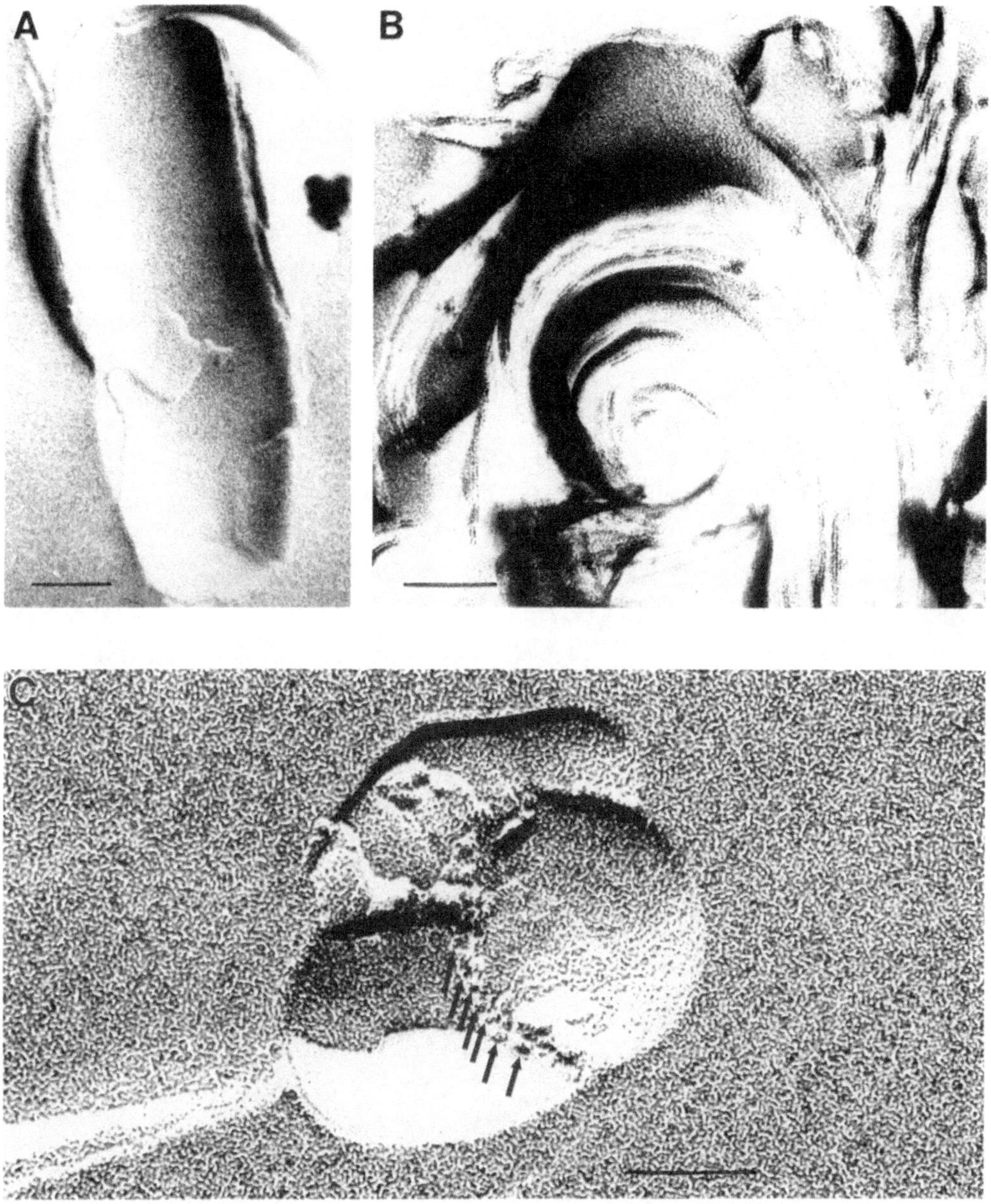

Figure 10 Cochleate lipid cylinders, formed during fusion of phosphatidylserine liposomes under acid conditions (pH 2.5), shown in top view (A) and cross-fractured (B). Lipidic particles, marked by arrows, are found during DNA-mediated fusion of cationic liposomes (C). (C, from Sternberg, B., Sorgi, F.L., and Huang, L., *FEBS Lett.*, 356, 361, 1994. With permission.)

part of the molecule, when suspended in excess water. In the inverse hexagonal H_{II} phase, long cylindrical cores of water are arranged on a hexagonal lattice. The polar head groups coat the exterior of these cores with the hydrocarbon chains filling the interstitial spaces between the cores.[34] Inverted hexagonal H_{II} tubules are made of a mixture of dioleoyl-phosphatidylethanolamine (DOPE), showing a small head group area compared to the hydrophobic part of the molecule, with PC (Figure 11B) or with the cationic lipid 3β[N-(N′,N′-dimethylaminoethane)-carbamoyl]cholesterol (DC-Chol) in a 4:1 molar ratio (Figure 11C). It has been hypothesized that hexagonal or other nonlamellar structures of lipids are essential to membrane fusion in biology.[29-31,33,34]

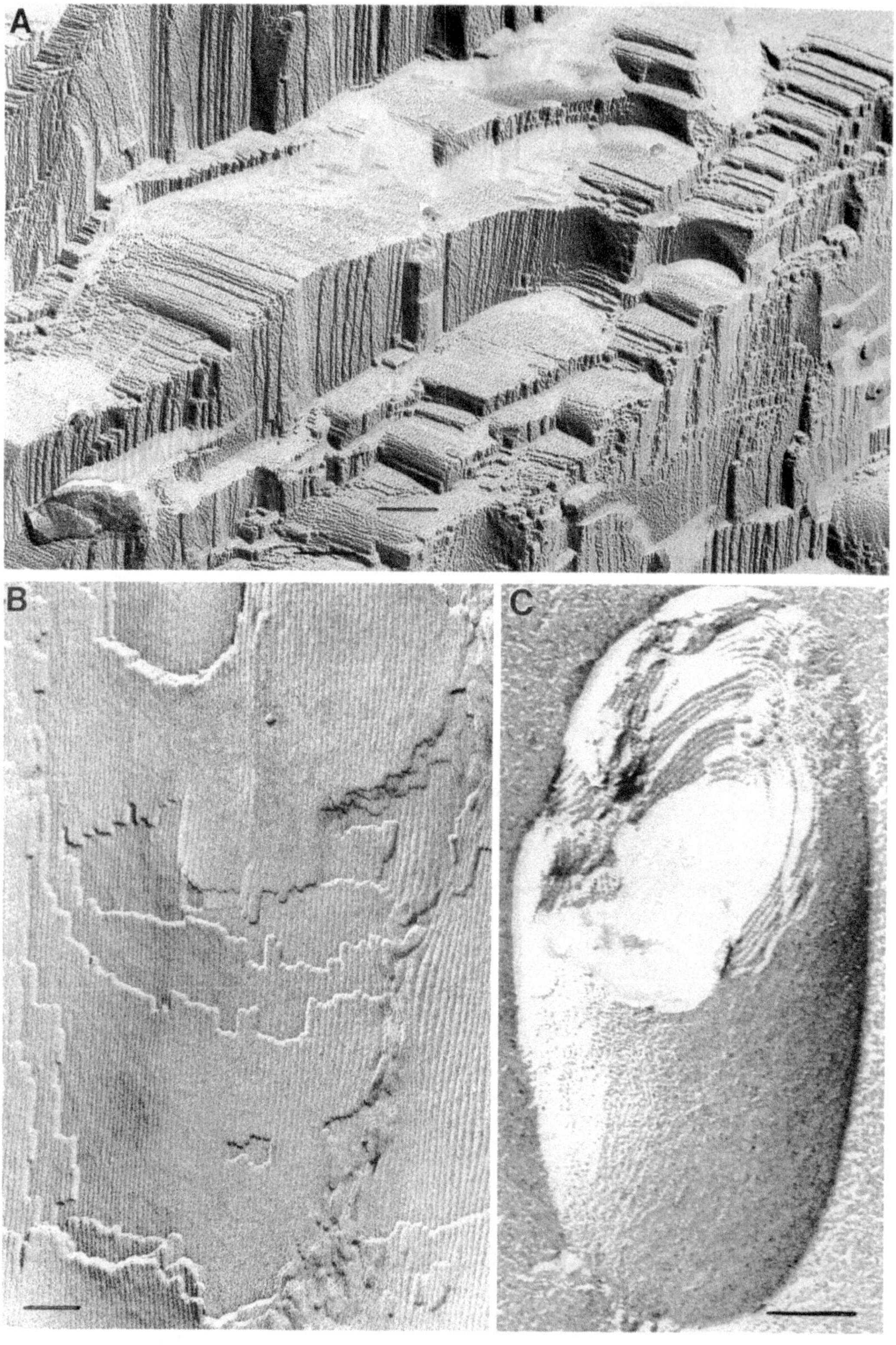

Figure 11 Freeze fracture electron micrographs of a hexagonal H_I-phase, adopted by DPhPGP, extracted from the PM (A), of an inverted hexagonal H_{II}-phase, adopted by a DOPE/DOPC mixture (B), and of a H_{II}-phase, adopted by DC-Chol/DOPE in a 4:1 molar ratio (C). (B, from Sternberg, B., in *Liposome Technology,* Vol. 1, Gregoriadis, G., Ed., CRC Press, Boca Raton, FL, 1993, 363, With permission.)

V. LIPOSOME STRUCTURE TRANSFORMATIONS DURING INTERACTION WITH MEMBRANE-ACTIVE AGENTS

Here, three examples are given for membrane structure transformations induced by incorporating specific molecules such as drugs, lytic peptides, or oligonucleotide that modify the behavior of liposomes. These induced properties may lead to liposomes suitable for a particular application.

A. RIPPLE-PATTERN DISAPPEARANCE BY INCORPORATION OF VITAMIN D_3 DERIVATIVES INTO DMPC-BILAYER

Vitamin D_3 derivatives are known to be effective in differentiation and proliferation of epidermal cells.[35] However, under certain circumstances, they also may show a hypercalcemic activity which can be a serious limitation in their use for dermatological application. For reducing the negative side effects of vitamin D_3 derivatives and for increasing the drug concentration in the skin, we investigated their incorporation in liposomes to optimize their use for psoriasis treatment.[36] Incorporation rates of more than 80% of the applied drug were found with significant variations related to the structural parameters of the steroid molecules. From our differential scanning calorimetry (DSC) measurements and freeze-fracture electron micrographs (Figure 12A–C) with increasing vitamin D_3 concentration, a depression in the main gel to liquid-crystalline phase transition of DMPC is observed: quenched at 15°C the samples with DMPC in the $P\beta'$ phase showed a very regular ripple structure on the fracture faces of empty liposomes (See insets in Figure 12A and 5B). At 3 mol% vitamin D_3, incorporated into DMPC liposomes, the ripple structures became wider and more irregular as shown clearly at the enlargement in the lower right corner of Figure 12B. At a concentration of 10 mol% vitamin D_3, a phase transition of DMPC molecules was not detectable any longer either by DSC or by FFEM. Here, the fracture faces were smooth, not showing any ripple structure (Figure 12C).[36] These findings support the idea that vitamin D_3 molecules, similarly to other amphipathic steroids such as cholesterol[19,37,38] and β-sitosterol,[21] are incorporated into the DMPC bilayer and intercalated between the hydrocarbon chains of phospholipid molecules, thereby disturbing the gel-to-liquid crystalline phase transition.[37,38]

The ability of liposomes to incorporate amphipathic or hydrophobic drugs into the bilayer, to encapsulate water-soluble drugs in the aqueous space inside the liposomes, and to anchor recognition groups to the surface of liposomes led to a great number of applications in medicine.[39-41] Initially the usefulness of liposomes as a drug-carrier system emerged in five discrete areas: drug solubilization, controlled release, site avoidance, uptake by the reticuloendothelial system, and site-specific targeting.[2] After a period of skepticism among some scientists in the field of drug delivery, interest in liposomes was rejuvenated by the introduction of new ideas from membrane biophysics, and this multidisciplinary approach has enhanced prospects for their use in medicine.[42] Liposomes can now be designed rationally, resulting in nonreactive (sterically stabilized) liposomes,[43] as well as polymorphic (cationic, fusogenic) liposomes.[32,44,45]

B. REVERSIBLE MICELLIZATION OF LIPID BILAYER CAUSED BY THE LYTIC PEPTIDE MELITTIN

Melittin (Mel), the 26-amino acid peptide from bee venom, binds to membranes in an amphipathic helical conformation, inducing voltage-gated anion-selective ion permeability and membrane lysis. It also causes the reversible micellization of bilayers composed of saturated acyl chain phosphatidylcholines when the temperature is below the bilayer T_c.[46] FFEM is a suitable technique to distinguish between disc micelles and liposomes and to investigate the bilayer micellization activity of Mel as well as of a chemically synthesized analog, Ala-14-melittin ($P^{14}A$), in some detail.[47,48]

Liposomes were prepared from the appropriate lipid to peptide mixtures, dispersed in puffer, by freezing-thawing cycles before experiments and showed quite different sizes, ranging from about 100 nm up to several microns. When quenched at temperatures above the bilayer T_c before fracturing, Mel:DMPC and $P^{14}A$:DMPC complexes yield electron micrographs showing liposomes displaying fracture faces decorated with intramembranous particles (IP; Figure 13A). The density of the IP is roughly proportional to the peptide:lipid ratio, but their size is independent of this ratio, indicating that under the conditions studied (50 m*M* Tris-HCL, pH 7.5, 5 m*M* EDTA)[47] the peptide is organized as discrete aggregates that penetrate deeply into the fluid bilayer. When quenched at temperatures below T_c and from a concentration of 3 mol% of the peptides on, each sample shows a massive collapse of liposomal structures into micelles. With increasing peptide concentration, fewer and fewer liposomes with bilayer in the gel

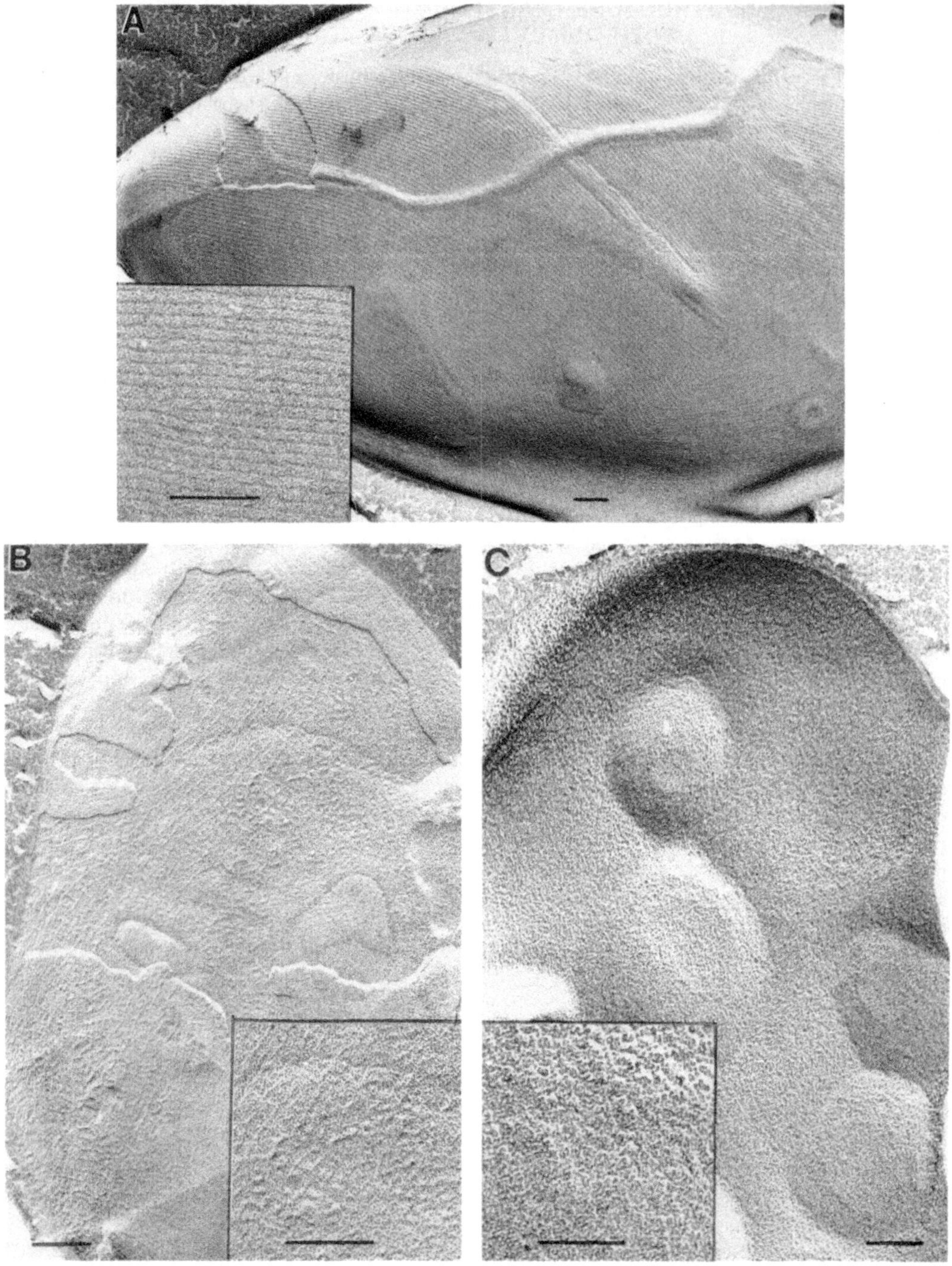

Figure 12 Freeze-fracture electron micrographs of empty DMPC liposomes (A), and vitamin D_3-loaded DMPC liposomes in a molar ratio of drug to lipid of 0.03:1 (B) or 0.1:1 (C). Insets, displaying selected membrane areas at a higher magnification, show, more clearly, the transformation of the ripple structures with increasing vitamin D_3 content. (From Merz, K. and Sternberg, B., *J. Drug Target.*, 2, 411, 1994. With permission.)

state are observed showing ripple structures ($P\beta'$ phase) identical to the peptide-free gel phase lipid. These residual bilayer areas, marked by an arrow in Figure 13B, do not exhibit IP on the fracture faces, indicating that the peptide is disassociated and/or expelled from a bilayer-inserted location in the gel phase lipid.[47,48] The synthetic analog, $P^{14}A$ where the proline at residue 14 near the center of the sequence

Figure 13 Freeze-fracture electron micrographs of Mel:DMPC complexes at peptide concentration of 5 mol% relative to lipid, quenched at 55°C (A) or 4°C (B). Whereas the fracture faces of the liposomes are decorated with IP in high concentration (A), residual bilayer areas, marked by an arrow, display ripple pattern but no IP (B). (B, from Dempsey, C.E. and Sternberg, B., *Biochim. Biophys. Acta,* 1061, 175, 199a; A, from Sternberg, B. and Dempsey, in *Structure and Conformation of Amphiphilic Membranes,* Lipowski, R. et al., Eds., Springer Verlag, Berlin, 1991, 80. With permission.)

is substituted by alanine, shows the same effect on the reversible micelle-to-bilayer transition as Mel, but the rates of both processes are quite different, ranging from a few minutes for Mel:DMPC complexes up to several hours for $P^{14}A$:DMPC complexes: it is decreased by 100-fold in $P^{14}A$:DMPC complexes compared to Mel:DMPC complexes. The synthetic analog has enhanced hemolytic activity, but is unable to form stable voltage-dependent ion channels, supporting the idea that discrete ion channel formation and reversible bilayer micellization are not required for Mel-induced hemolysis.[47]

The influence of EDTA, added to block residual phospholipase activity derived from the bee venom, on the bilayer stability is not quite clear yet. Without EDTA inclusion, even above T_c, a massive collapse of liposomal structures into membrane fragments is noticeable (not shown here). Some very few residual liposomes show any IP on the fracture faces, indicating that the peptide is localized at the bilayer surface probably as a monomer showing higher lytic activity.[48] Here, investigation of the effects of Mel on liposomes has given indications of the possible molecular mechanism of peptide membrane association and function.[3,47,48]

C. INTERACTION OF DNA WITH CATIONIC LIPOSOMES: FORMATION OF SPAGHETTI/MEATBALL-LIKE STRUCTURES

Among other nonviral vectors, cationic liposomes have been widely used as delivery agents for DNA and other polynucleotides.[44,49,50] In this context, liposomes offer several advantages over viral vectors, including the absence of viral components, the protection of the DNA/RNA from inactivation or degradation, and the possibility for cell-specific targeting. Since nucleic acids are highly negatively charged molecules, they can interact spontaneously with cationic liposomes. It is generally assumed that there is no true encapsulation of the DNA by the cationic vesicles, but rather complexing by electrostatic interactions accompanied by liposome aggregation and fusion.[51]

FFEM has been chosen to investigate the complex formation between plasmid DNA (Figure 14A) and preformed cationic liposomes, made of monovalent (Figure 14B) as well as polyvalent amphiphiles. FFEM provides the opportunity to look for structural modifications of the liposomes, to study the effect of DNA concentration and the dependency on exposure-time, and to study the correlation between morphology and transfection activity of the complexes. Negatively charged plasmid DNA seems to act as a fusogenic agent, drawing together the positively charged liposomes and forming partly fused liposome/DNA-complexes (formation of meatball-like structures, Figures 14C and 10C). Additionally, during the interacting process the DNA became clearly visible, possibly by amplifying its structure with a lipid bilayer coat (formation of spaghetti-like structures; some of them are marked by arrows in Figure 14C). These DNA-lipid strands are visible free in suspension but frequently also still attached to the liposomes (spaghetti/meatball complexes, Figure 14C). The observed convex and concave fracture faces of the DNA-lipid strands and their diameter of approx. 70 Å, measured by freeze-fracture technique, support the model of a DNA surrounded by a lipid bilayer.[32,52]

Spaghetti-like structures are formed in complexes with liposomes, made of DC-CHOL:DOPE at different molar ratios, ranging from 4:1 down to 2:3; but not at the composition of 1:4, where the liposomal bilayers are transformed to nonbilayer (H_{II}) structures as shown in Figure 11C. They are also found in similar complexes where DOPE was replaced by DOPC. The interaction process and formation of the spaghetti-like structures, described above, are found to be true for cationic liposomes containing amphiphiles bearing one positively charged group per molecule, such as DC-CHOL, LIPOFECTIN® and DMRIE®, but seems to be different for amphiphiles with more than one positive charge per mole, such as LIPOFECTAMINE®, and DOSPA®.[52] There seem to be a correlation between spaghetti-formation and transfection activity of the complexes made from monovalent cationic lipids. However, it is still an open question which of the morphological structures, found during the interaction process (weather the meatballs or the spaghetti), are active in transfection.

VI. LIPOSOME STRUCTURE TRANSFORMATION DURING INTERACTION WITH CELLS

Liposomes have raised considerable interest as carriers for the cytoplasmic delivery of biologically active molecules. Nucleic acids, cytotoxic agents, and a variety of macromolecules have been encapsulated in liposomes and delivered to cells in a functional form. Early studies implicated liposome–plasma membrane fusion, but more recent evidence suggested that liposomes do not fuse with the plasma membrane without perturbations such as polyethylene glycol treatment or the inclusion of viral proteins.[53] It has been demonstrated that negatively charged liposomes are endocytosed by the coated vesicle pathway. During this process liposome-encapsulated molecules, showing a tendency to become membrane permeant at low pH, can gain access to the cytoplasm. Large or highly charged molecules, unable to exploit the low-pH environment of the endocytic pathway, however, remain sequestered in intracellular vacuoles,

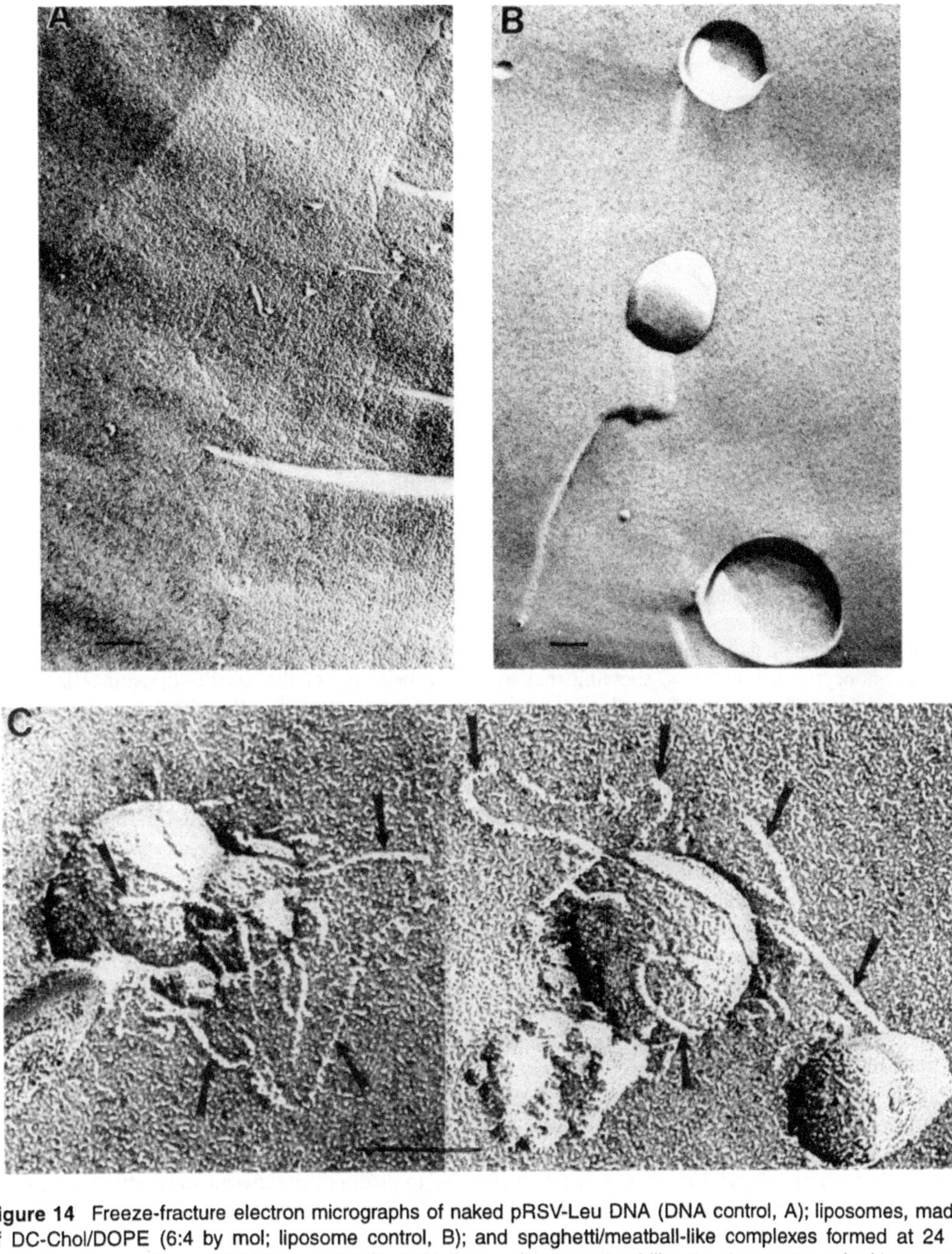

Figure 14 Freeze-fracture electron micrographs of naked pRSV-Leu DNA (DNA control, A); liposomes, made of DC-Chol/DOPE (6:4 by mol; liposome control, B); and spaghetti/meatball-like complexes formed at 24 h incubation time and 2 μg DNA/20 nmol DC-Chol (C). Some of the spaghetti-like structures are marked by arrows in (C). (From Sternberg, B., Sorgi, F. L., and Huang, L., *FEBS Lett.*, 356, 361, 1994. With permission.)

such as secondary lysosomes or dense bodies.[54] Endocytosis has not been regarded as a useful route by which liposomes could enter non-phagocytic cells or by which large or labile molecules could gain access to the cytoplasm.

A. ENDOCYTOSIS OF DMPC-LIPOSOMES BY SKIN CULTURE CELLS

Cultured human keratinocytes, such as HaCaT cells, were often used as a test system for drugs in dermatology and to characterize the interaction of liposomes (empty or drug loaded) with cells. We have investigated the influence of different properties of liposomes, loaded with a variety of vitamin D_3-analogs, on the proliferation of human keratinocytes and interleucine 1α-release. The effects of test liposomes are followed by fluorimetric and colorimetric measurements and compared to the effects of the free drugs as well as of the empty liposomes. The interaction between empty liposomes, with a diameter of less than 200 nm, and HaCaT cells was recorded by FFEM.[55]

Empty liposomes, composed of DMPC and prepared by extrusion, and quenched at room temperature (<20°C) display ripple structure of the Pβ′-phase on their fracture faces. One of them is marked by an arrow in Figure 15A. In smaller liposomes with diameter less than 100 nm and therefore higher curved bilayers, the ripple pattern can be modified, as shown in Figure 5B, revealing a golf ball structure.[20] One such liposome is marked by an arrow in Figure 15B.

Fresh HaCaT cells (1 day old) are loosely packed and display many vesicular structures not easily distinguishable from ordinary liposomes (Figure 16A and B). However, older HaCaT cells (4 days old), are growing like a lawn, are easy to manipulate, and display big, cylindrical, and tightly packed cells together with few vesicles only (Figure 15A and B). The fracture faces of the older cells (Figure 15A and 15B) as well as of the vesicles, derived from the younger cells (Figure 16A and 16B) are covered frequently but not always with protein-derived IPs.

Regular as well as modified ripple structures on the fracture faces of the DMPC liposomes (marked by arrows in Figure 15A and B) are used as morphological markers to keep track of the liposomes during interaction with the culture cells. During interaction with the HaCaT cells, the liposomes lost their ripple structures and appeared as endocytosed, small vesicles, displaying now smooth fracture faces (marked by arrowheads in Figure 15B), at the fracture faces of the culture cells.[55] Interaction between liposomes and keratinocytes has been described in terms of phagocytosis,[56] membrane fusion, and exchange of compounds. From the results of our structural investigations, endocytosis seems to be the most obvious mechanism of the interaction between liposomes and keratinocytes. Whereas after half an hour interaction time no endocytosed vesicles were visible (Figure 15A), the number of such vesicles is drastically increased after 3.5 h as seen in Figure 15B. Neither regular nor modified ripple structures were noticeable at the endocytosed liposomes, showing that the phase behavior of the DMPC liposomes is changed during endocytosis. As seen from fluorimetric studies, carried out in parallel, the liposomal uptake by HaCaT cells was saturated within 4 h. Macrophages are able to phagocyte liposomes within 6 h.[57] Our results confirm endocytotic abilities also for keratinocytes.[55]

B. INTERACTION OF DNA–LIPOSOME COMPLEXES WITH SKIN CULTURE CELLS

Liposomes have proven useful for gene therapy. The findings that cationic lipids can condense DNA and increase transfection yields *in vitro* by several orders of magnitude,[44,49] as well as reports on transfection *in vivo*[50] stimulated an intensive interest in using liposomes for gene therapy. For this new type of therapy, cationic liposome/DNA complexes are added to cells *in vitro*, and injected or aerosolized for *in vivo* applications. Although neither the morphology and the related physicochemical properties of the complexes nor their interactions with cells are understood in detail, the extent of transfection for a variety of cells and tissues *in vivo* is reasonably good. Endocytosis and/or fusion of the complexes with cell membranes may facilitate the nuclear localization of DNA.[42]

To study the interaction of cationic liposome/DNA complexes with cells in more detail, we treated HaCaT cells with complexes made from plasmid DNA and DC-Chol/DOPE and displaying spaghetti/meatball-like structures. In a preliminary study we looked at the time-dependency for structural transformations of all components involved in the interaction process. By using the spaghetti-like structure as morphological marker, it is relatively easy to distinguish between spaghetti/meatball complexes (two of them are marked by an arrow in Figure 16B) and 1-day-old HaCaT cells. Although free spaghetti, not attached to liposomes, are very small structures with a diameter of about 7 nm, measured by freeze-fracture technique, we could frequently observe them, after short incubation times of 10 or 30 min, intact and inside the cells. Here, some of them are marked by arrowheads at the left side of Figure 16A, inside a cross-fractured HaCaT cell. They are also attached intact to the fracture faces of the cell membrane as shown at the right side of Figure 16A. After longer incubation time (2 h), endocytosis of the cationic liposome/DNA complexes seems to take place (not shown clearly in Figure 16B). The high content of the fusogenic lipid DOPE in the cationic mixture might promote fusion between spaghetti-like structures and cell membrane. Additionally, spaghetti-like structures as well as the whole complexes, observed during our investigation, may still bear residual positive charges on their surfaces. This may also be important for the interaction and fusion with cell- and probably nuclear membranes, thereby promoting the transfer of the DNA into the cytoplasm and eventually into the nucleus of the cells.[32]

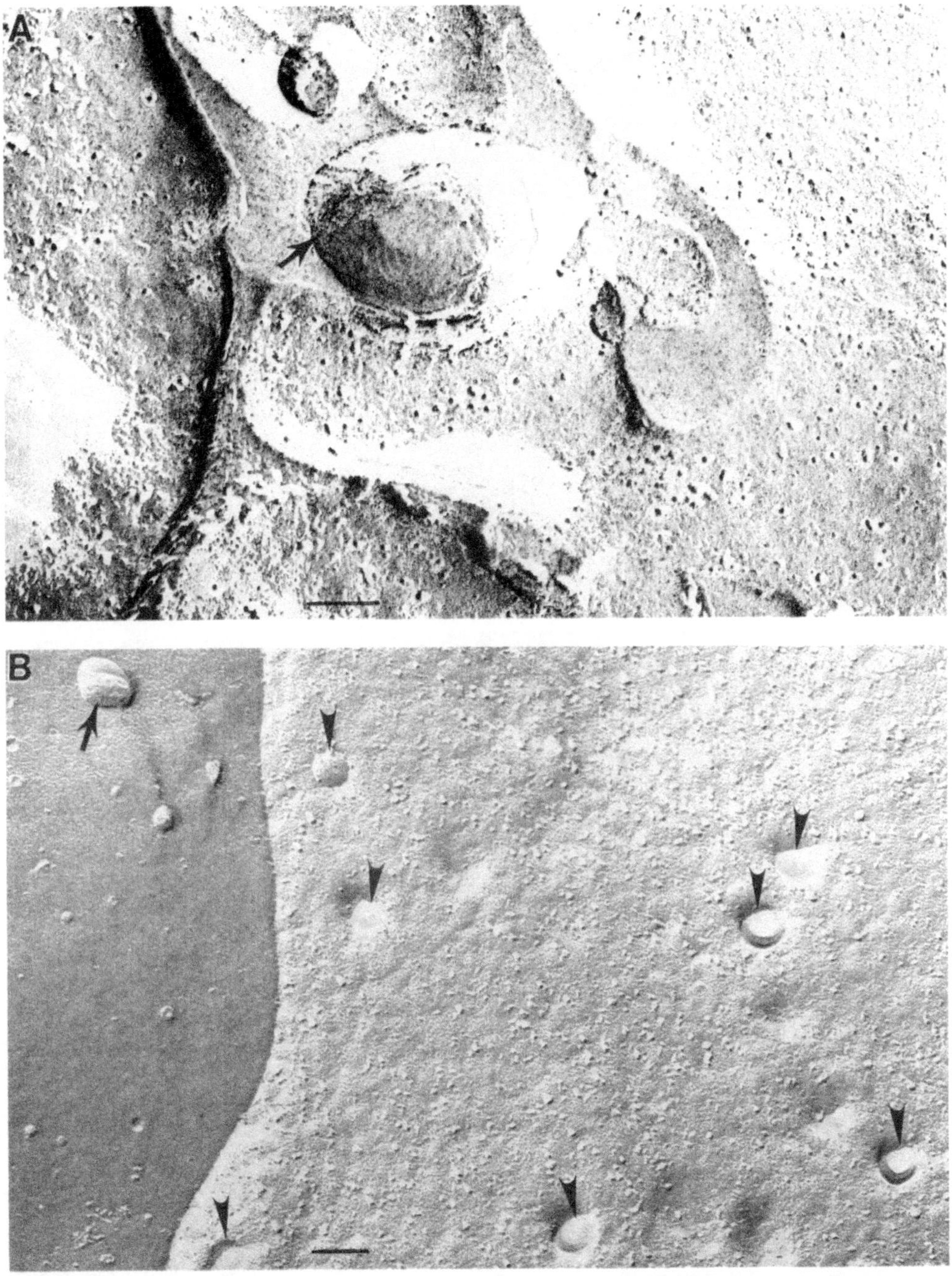

Figure 15 Freeze-fracture electron micrographs of 4-day-old HaCaT cells, treated with DMPC liposomes for 1 (A) and 3.5 hs (B). Arrows mark liposomes with ripple pattern (A) or modified ripple structure, resembling golf ball structure (B). Arrowheads in (B) mark liposomes endocytosed by the HaCaT cells. (From Prüfer, K., Merz, K., Barth, A., Wollina, U., and Sternberg, B., *J. Drug Target.*, 2, 419, 1994. With permission.)

VII. CONCLUSION

Considering the examples given here for reconstituted, intrinsic, as well as induced membrane structures, it is reasonable to say that liposomes are excellent models for membrane structures and for studying structural transformations to elucidate the underlying mechanism of a variety of cellular processes in more detail.

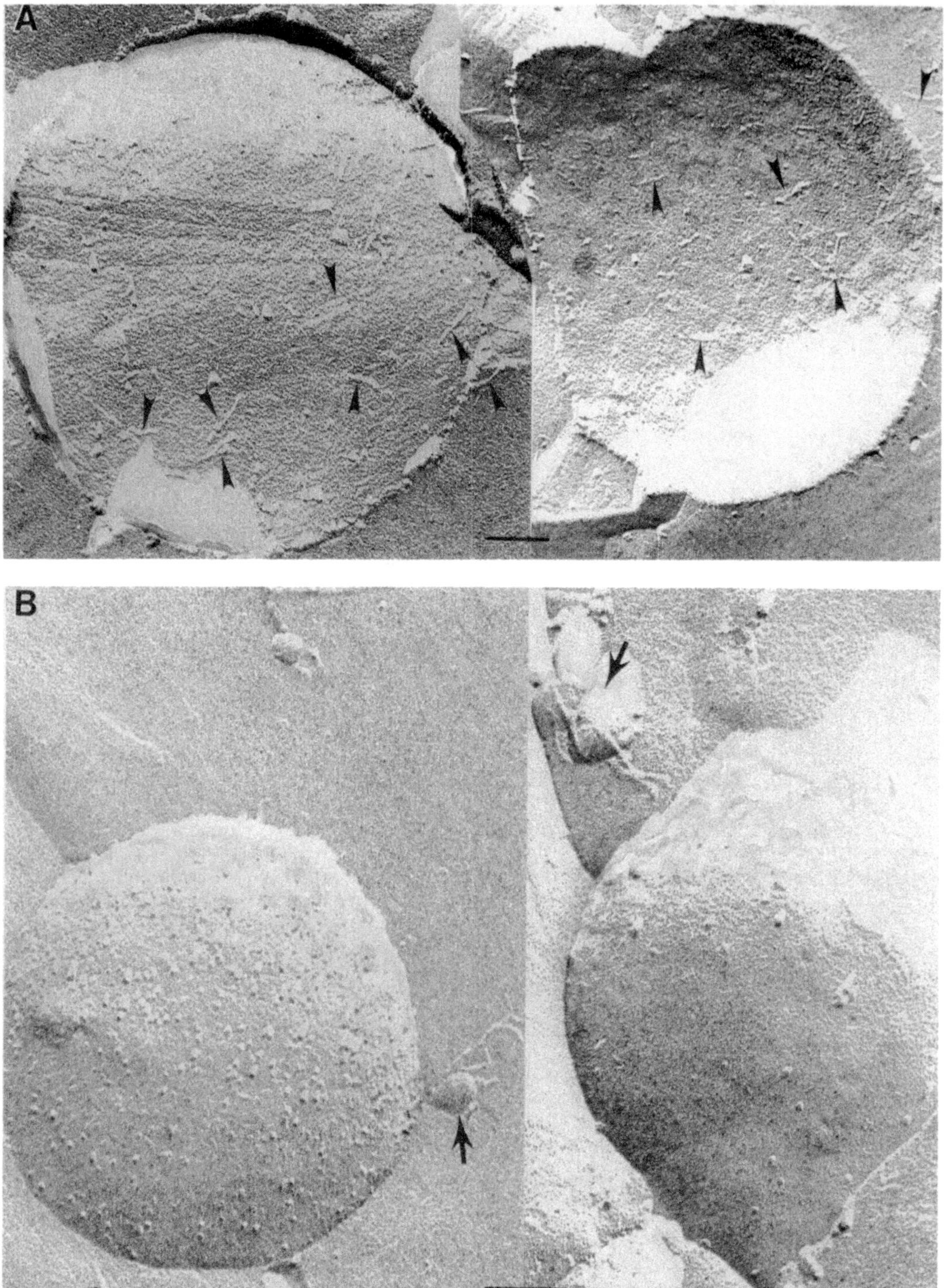

Figure 16 Freeze-fracture electron micrographs of 1-day-old HaCaT cells, treated with DNA–liposome complexes, showing spaghetti/meatball-structures, marked by arrows in (B) and similar to those of Figure 14C, for 30 min (A) or 2 h (B). Free spaghetti-like structures, marked by arrowheads, are visible inside a cross-fractured HaCaT cell (left side of A) as well as at the extracellular fracture face of a HaCaT cell (right side of A).

ACKNOWLEDGMENTS

I wish to thank Mrs. G.Engelhardt and Mrs. G. Vöckler for their phototechnical work. I am grateful to Professor D. Papahadjopoulos for reading the manuscript and for many helpful discussions.

REFERENCES

1. **Gruner, S. M.**, Material properties of liposomal bilayers, in *Liposomes, From Biophysics to Therapeutics*, Ostro, M. J., Ed., Marcel Dekker, New York, 1987, 1.
2. **Papahadjopoulos, D.**, Liposome formation and properties: an evolutionary profile, *Biochem. Soc. Trans.*, 6, 910, 1988.
3. **Sternberg, B.**, Freeze-fracture electron microscopy of liposomes, in *Liposome Technology* Vol. 1, Gregoriadis, G., Ed., CRC Press, Boca Raton, FL, 1993, 363.
4. **Blaurock, A. E. and Stoeckenius, W.**, Structure of the purple membrane, *Nature New Biol.*, 233, 152, 1971.
5. **Sternberg, B., Hostis, C. L., Whiteway, C. A., and Watts, A.**, The essential role of specific *Halobacterium halobium* polar lipids in 2D-array formation of bacteriorhodopsin. *Biochim. Biophys. Acta*, 1108, 21, 1992.
6. **Sternberg, B., Watts, A., and Cejka, Z.**, Lipid induced modulation of the protein packing in two-dimensional crystals of bacteriorhodopsin, *J. Struct. Biol.*, 110, 196, 1993.
7. **Sternberg, B., Gale, P., and Watts, A.**, The effect of temperature and protein content on the dispersive properties of bR from *H. halobium* in reconstituted DMPC complexes free of endogenous purple membrane lipids: a freeze-fracture electron microscopy study, *Biochim. Biophys. Acta*, 980, 117, 1989.
8. **Sternberg, B., Whiteway, C., Seifert, G., Strohbach, U., Engelhardt, R., and Watts, A.**, Effect of bilayer thickness on 2D-arrays of bacteriorhodopsin, *ICEM 13-Paris*, 3A, 615, 1994.
9. **Jap, B. K., Zulauf, M., Scheybani, T., Hefti, A., Baumeister, W., Aebi, U., and Engel, A.**, 2D crystallization: from art to science, *Ultramicroscopy*, 46, 45, 1992.
10. **Watts, A., Venien-Bryan, C., Sami, M., Whiteway, C., Boulter, J., and Sternberg, B.**, Lipid-protein interactions in controlled membrane protein array and crystal formation, in *New Comprehensive Biochemistry*, Vol. 25, Watts, A., Ed., Elsevier, Amsterdam, 1993, 351.
11. **Langworthy, T. A.**, Lipids of archaebacteria, in *The Bacteria*, Vol. 8, Woese, C. R. and Wolfe, R. S., Eds., Academic Press, New York, 1985, 459.
12. **Sternberg, B. and Rudolph, P.**, Unusual fracture behavior of membranes made of bipolar lipids of *Thermoplasma acidophilum*, *Electron Microscopy*, Vol. 3, Proceedings of the 10th European Congress on Electron Microscopy 92, Granada, Spain, 1992, 85.
13. **Lasch, J., Schmitt, U., Sternberg, B., and Schubert, R.**, Human stratum corneum lipid-based liposomes (hSCLLs), *J. Liposome Res.*, 4 (1), 93, 1994.
14. **Branton, D.**, Fracture faces of frozen membranes, *Proc. Natl. Acad. Sci. U.S.A.*, 55, 1048, 1966.
15. **Sternberg, B., Gumpert, J., Meyer, H. W., and Reinhardt, G.**, Structures of liposome membranes as models for similar features of cytoplasmic membranes of bacteria, *Acta Histochem.*, Suppl.-Vol. 33, 139, 1986.
16. **Sternberg, B., Gumpert, J., Reinhardt, G., and Gawrisch, K.**, Electron microscopic and biophysical studies of liposome membrane structures to characterize similar features of the membranes of *Streptomyces hygroscopicus*, *Biochim. Biophys. Acta*, 898, 223, 1987.
17. **McElhaney, R. N.**, The structure and function of the *Acholeplasma laidlawii* plasma membrane, *Biochim. Biophys. Acta*, 779, 1, 1984.
18. **Papahadopoulos, D., Jacobson, K., Nir, S., and Isaac, T.**, Phase transitions in phospholipid vesicles: fluorescence polarization and permeability properties concerning the effect of temperature and cholesterol, *Biochim. Biophys. Acta*, 311, 330, 1973.
19. **Copeland, B. R. and McConnell, H. M.**, The rippled structure in bilayer membranes of phosphatidylcholine and binary mixtures of phosphatidylcholine and cholesterol, *Biochim. Biophys. Acta*, 599, 95, 1980.
20. **Düzgünes, N., Wilschut, J., Hong, K., Praley, R., Perry, C., Friend, D. S., James, T. L., and Papahadopoulos, D.**, Physicochemical characterization of large unilamellar phospholipid vesicles prepared by reverse-phase evaporation, *Biochim. Biophys. Acta*, 732, 289, 1983.
21. **Sternberg, B.**, Das Phasen- und Barriereverhalten von Liposomen-Modellmembranen und deren Beeinflussung durch β-Sitosterin. *Dissertationsschrift*, Humboldt University, Berlin, 1976.
22. **Melchior, D. L. and Steim, J. M.**, Thermotropic transitions in biomembranes, *Annu. Rev. Biophys. Bioeng.*, Vol. 5, 205, 1976.
23. **Hamilton, R. L., Goerke, J., Guo, L. S. S., Williams, M. C., and Havel, R. J.**, Unilamellar liposomes made with the French pressure cell: a simple preparative and semiquantitative technique, *J. Lipid Res.*, Vol. 21, 981, 1980.
24. **Olson, F., Hunt, C. A., Szoka, F. C., Vail, W. J., and Papahadjopoulos, D.**, Preparation of liposomes of defined size distribution by extrusion through polycarbonate membranes, *Biochim. Biophys. Acta*, 557, 9, 1979.
25. **Bailie, A. J., Florence, A. T., Hume, L., Muirhead, G. T., and Rogerson, A.**, The preparation and properties of niosomes-nonionic surfactant vesicles, *J. Pharm. Pharmacol.*, 37, 863, 1985.

26. **Yoshioka, T., Sternberg, B., and Florence, A.,** Preparation and properties of vesicles (niosomes) of sorbitan monoester (Span 20, 40, 60, 80) and a sorbitan triester (Span 85), *Int. J. Pharm.*, 105, 1, 1994.
28. **Papahadjopoulos, D., Vail, W. J., Jacobson, K., and Poste, G.,** Cochleate lipid cylinders: formation by fusion of unilamellar lipid vesicles, *Biochim. Biophys. Acta,* 394, 483, 1975.
29. **Cullis, P. R. and de Kruijff, B.,** Lipid polymorphism and functional roles of lipids in biological membranes, *Biochim. Biophys. Acta,* 559, 399, 1979.
30. **Papahadjopoulos, D., Poste, G., Schaeffer, B. E., and Vail, W. J.,** Membrane fusion and molecular segregation in phospholipid vesicles, *Biochim. Biophys. Acta*, 352, 10, 1974.
31. **Verkleij, A. J., van Echtfeld, C. J. A., Gerritsen, W. J., Cullis, P. R., and de Kruijff, B.**, The lipidic particle as an intermediate structure in membrane fusion processes and bilayer to hexagonal H_{II} transitions, *Biochim. Biophys. Acta*, 600, 620, 1980.
32. **Sternberg, B., Sorgi, F. L., and Huang, L.**, New structures in complex formation between DNA and cationic liposomes visualized by freeze-fracture electron microscopy, *FEBS Lett.*, 356, 361, 1994.
33. **Bearer, E. L., Düzgünes, N., Friend, D. S., and Papahadjopoulos, D.,** Fusion of phospholipid vesicles arrested by quick-freezing: the question of lipidic particles as intermediates in membrane fusion, *Biochim. Biophys. Acta,* 693, 93, 1982.
34. **Tate, M. W., Eikenberry, E. F., Turner, D. C., Shyamsunder, E., and Gruner, S. M.**, Nonbilayer phases of membrane lipids, *Chem. Phys. Lipids,* 57, 147, 1991.
35. **Kragballe, K. and Widfang, I. L.**, Calcipotriol (MC 903), a novel vitamin D_3 analogue stimulates terminal differentiation and inhibits proliferation of cultured human keratinocytes, *Arch. Dermatol. Res.*, 282, 164, 1990.
36. **Merz, K. and Sternberg, B.**, Incorporation of vitamin D_3-derivatives in liposomes of different lipid types, *J. Drug Target.*, 2, 411, 1994.
37. **Papahadjopoulos, D. and Kimelberg, H. K.,** Phospholipid vesicles (liposomes) as models for biological membranes: their properties and interactions with cholesterol and proteins, in *Progress in Surface Science,* Vol. 4, Davison, S. G., Ed., Pergamon Press, Oxford, 1973, 141.
38. **Yeagle, P. L.**, Cholesterol and cell membranes, in *Membranes of Cells*, Vol. 5, Academic Press, San Diego, 1993, 139.
39. **Mayhew, E. and Papahadjopoulos, D.,** Therapeutic applications of liposomes, in Liposomes, Ostro, M. J., Ed., Marcel Dekker, New York, 1983, 289.
40. **Prüfer, K. and Sternberg, B.,** Liposomen in der Medizin–Eine aktuelle Bestandsaufnahme; Z. *Ärztl. Fortbild.*, 88, 257, 1994.
41. **Gregoriadis, G.**, Liposomes: a tale of drug targeting, *J. Drug Target.*, 1, 3, 1993.
42. **Lasic, D. D. and Papahadjopoulos, D.**, Liposomes revisited, *Science,* 267, 1275, 1995.
43. **Papahadjopoulos, D., Allen, T., Gabizon, A., Mayhew, E., Matthay, K., Huang, S. K., Lee, K. D., Woodle, M. C., Lasic, D. D., Redemann, C., and Martin, F. J.**, Sterically stabilized liposomes: improvements in pharmacokinetics, and anti-tumor therapeutic efficacy, *Proc. Natl. Acad. Sci. U.S.A.*, 88, 11460, 1991.
44. **Behr, J. P.**, DNA strongly binds to micelles and vesicles containing lipopolyamines or lipointercalants, *Tetrahedron Lett.*, 27, 5861, 1986.
45. **Schoen, P., Bron, R., and Wilschut, J.,** Delivery of foreign substances to cells mediated by fusion-active reconstituted influenza virus envelopes (virosomes), *J. Liposome Res.*, 3, 767, 1993.
46. **Dempsey, C. E.**, The action of melittin on membranes, *Biochim. Biophys. Acta*, 1031, 143, 1990.
47. **Dempsey, C. E. and Sternberg, B.,** Reversible disc-micellization of DMPC bilayers induced by melittin and (Ala-14)-melittin, *Biochim. Biophys. Acta,* 1061, 175, 1991.
48. **Sternberg, B. and Dempsey, C. E.,** Melittin-induced reversible micelle↔bilayer transition, in *Structure and Conformation of Amphiphilic Membranes*, Lipowsky, R., Richter, D., and Kremer, K., Eds., Springer Verlag, Berlin, 1991, 80.
49. **Felgner, P. L., Gadek, T. K., Holm, M., Roman, R., Hardy, W. C., Wenz, M., Northrop, J. P., Ringold, G. P., and Danielsen, M.**, Lipofection: a highly efficient, lipid-mediated DNA-transfection procedure, *Proc. Natl. Acad. Sci. U.S.A.* 84, 7413, 1987.
50. **Debs, R. J.**, Cationic liposome-mediated gene transfer in vitro and in vivo, in *Liposomes as Tools in Basic Research and Industry,* Philippot, J. R., and Schuber, F., Eds., CRC Press, Boca Raton, FL, 1995, 171.
51. **Gershon, H., Ghirlando, R., Guttman, S. B., and Minsky, A.**, Mode of formation and structural features of DNA-cationic liposome complexes used for transfection, *Biochemistry,* 32, 7143, 1993.
52. **Sternberg, B., Sorgi, F. L., and Huang, L.,** Complex formation between DNA and cationic liposomes visualized by freeze-fracture electron microscopy, *J. Mol. Med.*, 73, 4, B12, 1995.
53. **Szoka, F., Magnusson, K.-E., Wojcieszyn, J., Hou, Y., Derzko, Z., and Jacobson, K.**, Use of lectins and polyethylene glycol for fusion of glycolipid-containing liposomes with eukaryotic cells, *Proc. Natl. Acad. Sci. U.S.A.*, 78, 1685, 1981.
54. **Straubinger, R. M., Hong, K., Friend, D. S., and Papahadjopoulos, D.,** Endocytosis of liposomes and intracellular fate of encapsulated molecules: encounter with low pH compartment after internalization in coated vesicles, *Cell,* 32, 1069, 1983.
55. **Prüfer, K., Merz, K., Barth, A., Wollina, U., and Sternberg, B.**, Interaction of liposomal incorporated vitamin D_3-analogues and human keratinocytes, *J. Drug Target.*, 2, 419, 1994.

56. **Korting, H. C., Schmidt, M. H., Hartinger, A., Maierhofer, G., Stolz, W., and Braun-Falco, O.,** Evidence for phagocytosis of intact oligolamellar liposomes by human keratinocytes in vitro and consecutive intracellular disintegration, *J. Microencapsul.,* 10, 223, 1993.
57. **Verma, J. N., Wassef, N. M., Wirtz, R. A., Atkinson, C. T., Aikawa, M., Loomis, L. D., and Alving, C. R.,** Phagocytosis of liposomes by macrophages: intracellular fate of liposomal malaria antigen, *Biochim. Biophys. Acta,* 1066, 229, 1991.

Chapter 21

Liposomes: Past, Present, and Future

Danilo D. Lasic and Yechezkel Barenholz

CONTENTS

I. INTRODUCTION

In this final chapter we shall briefly review the history of liposomes in the last 150 years and especially in the last 30 years, describe their present status and speculate on the future of this field.

Liposomes were described a long time before their "official" recognition, which was reported in 1965.[1] In 1995, we celebrated the 30th Liposome Birthday with a Conference in Cambridge, England which was organized by their discoverer, Dr. A. D. Bangham. Earlier researchers, which were studying lipids at least since the mid 1800s and used names such as lecithin emulsions, sols, micelles, suspensions, etc. to describe aqueous lecithin phases, however, did not recognize that the lipid colloidal structures ("detached myelin figures") are self-closed and that their membrane(s), which separate inner compartment(s) from the external bulk solution, represent(s) a significant permeability barrier to various molecules, especially ions. After the realization of the self-closed nature of liposomes, many of important results were obtained on these "Bangasomes", now called large multilamellar vesicles. But it was only after the introduction of better defined small and large unilamellar liposomes that very well-defined experiments were performed.

After many thorough studies and an optimistic beginning in the 1970s, the field, especially in the liposome applications in drug delivery fell into a period of skepticism and pessimism due to (unexpected) difficulties in *in vivo* behavior of liposomes and encapsulated agents in the late 1980s and early 1990s. After a sobering period, more optimistic expectations are starting to re-emerge. This time, however, this optimism is based not on naive scenarios and dreams of some scientists and businessmen, but on solid experimental data and vastly improved theoretical and technological understanding of the field.

0-8493-4013-6/96/$0.00+$.50

II. PAST

A. A BRIEF HISTORY

We have already mentioned that scientists have been interested in lipid systems for quite some time.[2] Briefly, in the last 150 years numerous researchers were interested in lipid systems and many earlier reports on the swelling behavior and other properties of lecithin films and suspensions and myelin figures have been published.

Lecithin was discovered in the middle of the last century by Gobley in 1846/47 in Paris, who isolated it from brain and egg yolks, and later named these fractions "lécithine" from the Greek for egg yolk, *lekitos* (λεκιθοσ). Its chemical formula was determined in 1868 by Strecker.

The swelling of lipids was first described by Virchow (1854), who observed the swelling when he transferred nerve core into an aqueous phase.[3] He later described the swelling of dried nerves in water. Since it was very likely that some liposomes detached from the hydrating mass, either due to external agitation or some crystal defect, Virchow failed to recognize not only liposomes, but also liquid crystals of myelin figures during his microscopic observation. More work followed and as early as 1888 Thudicum in London recognized the importance of colloidal associates of phosphatides.

Lehmann's book on liquid crystals (1911) shows a figure of the growth of myelin figures schematically,[4] as described by Virchow. Lehmann (Leipzig) and Reinitzer (Vienna), who were both studying anisotropic liquids, discovered liquid crystals around 1888. In 1911 Lehmann showed a micrograph of liposomes and called them "Künstlische Zellen", artificial cells.

In the period between the two World Wars, several scientists were studying various chemical and physical properties of lecithin molecules and their aqueous suspensions. These studies included measurements of isoelectric points of lecithin, filtering of lecithin suspensions, and the like, as reviewed in Reference 2. Thickness of the bilayer was estimated to be 5 nm (Trilliat, 1920) and a model of the cell membrane based on a bilayer was proposed (Gorter and Wendel, 1926).

In the 1950s several scientists, mostly in England, studied aqueous lecithin dispersions by light scattering, centrifugation, and some other techniques. Their papers speak about lecithin micelles formed upon sonication of lecithin sols[5] and estimates of molecular mass of "micelles" being in the order of 5 million Da.[6] Despite many careful experiments, these scientists failed to realize (or to appreciate the importance of the fact) that liposomes enclose an aqueous phase. Bangham and his co-workers challenged the concept of planar and/or "rolled up" (swiss roll-type) micelles what eventually led to the recognition of special properties of these phases and liposome science.

B. DISCOVERY OF LIPOSOMES

In the late 1950s, several scientists living in or near Cambridge, England studied the physical properties of blood, blood cells, and phospholipases. Bangham was primarily interested in the physical properties of different cellular constituents in blood. For these purposes, he used electrophoretic measurements, since he believed that surface properties alone determined their behavior. Rex Dawson was trying to understand the function of phospholipases. For these reasons, he needed pure phospholipids as substrates. As a consequence, the best characterized lipid samples at that time were found in Dawson's laboratory. To his surprise, the purer the lipids, the less enzymatic activity could be measured. In collaboration with Bangham, Dawson realized that the substrate had to be negatively charged to show enzymatic activity.[7] Together, they used electrophoresis of various lipid dispersions, and many measurements of mobility of these particles in various conditions were reported. Today we would call these multilamellar vesicles having a variety of lipid compositions.

Bangham wanted to characterize the lipid suspensions he had used to model the surfaces of cellular membranes. His main goal was to quantify dispersal through size and number of particles to estimate the surface area of various preparations. In collaboration with Horne, he studied negatively stained preparations of briefly sonicated lecithin suspensions using negative stain electron microscopy, and they were the first to show the spherical, self-enclosed, multilamellar nature of dispersed particles. Bangham started to measure the permeability properties of the various membranes for various ions. This work showed that liposomes are (quite) impermeable to ions and large molecules, but permeable to water and small nonelectrolytes. Papers written by Bangham and his collaborators in 1964 and 1965 clearly established the existence of liposomes as multilamellar vesicles, characterized their properties and resulted in a quick development of liposomes into a major system in biophysics, colloidal science, chemistry, and cell biology.[8]

The subsequent development was very fast. One reason was that the state of the bilayers had been well characterized by Luzzati and Dervichian (Paris), Chapman (London), and several others who studied black lipid membranes. Luzzati studied phase diagrams of phospholipid water systems in the early 1960s,[9] while Stockenius investigated these liquid crystalline phases by electron microscopy.[10] In addition, phase diagrams of soap-water systems had been known since the 1920s because of the work of McBain and Lawrence (England).

The 1970s witnessed a fast development of various liposome preparation and characterization methods. After developing liposome preparation by sonication in a largely forgotten paper in 1967[11] Papahadjopoulos with collaborators also introduced the formation of liposomes from reverse phases[12] and by extrusion,[13] which is the best method for the preparation of homogeneous liposomes with the tightest size distribution. C. Huang did a series of important experiments in which unilamellar vesicles produced by sonication were fractionated by gel exclusion chromatography to obtain homogeneous fractions of minimal size unilamellar vesicles.[14] This fractionation enabled treating these SUV as macromolecules and the detailed study of topology and structural organization of vesicles. This work introduced the concepts of asymmetrical lipid distribution between the two bilayer leaflets in vesicles and was important for subsequent studies by Thompson, Barenholz and Lichtenberg on the importance of bilayer curvature (reviewed in Reference 15). Barenholz and colleagues realized that one can down-size liposomes also by other high energy treatments, and they introduced the French press extrusion.[16] The next step was the use of homogenizers.[17] In-parallel detergent depletion techniques, which were performed by Racker in 1974, were further developed by Hauser and others. In addition to many biophysical and biochemical studies in the laboratories of Thompson, Tanford, and Nozaki, and many others many novel lipids were introduced into liposome preparations by Barenholz and others, including nonionic lipids by Vanlerberghe of what eventually resulted in the first cosmetic formulations. In-parallel ethanol and ether injection techniques for liposome preparation were introduced.[18,19]

The introduction of many physical methods was very important to the characterization of liposomes and their use as membrane model systems. Techniques such as various spectroscopies (NMR, EPR, IR, R, etc.,) fluorescence methods, thermodynamic (DSC, calorimetry), hydrodynamic (chromatographies, viscosity, etc.), electrodynamic, diffraction, and scattering techniques (QELS,X-rays) as well as ever improving microscopies were introduced and further developed in the 1970s and 1980s by Chapman, Thompson, Tanford, Deamer, Barenholz, McConnell, deKruijff, Cullis, Parsegian, and later by E. Sackmann, Bloom, Hauser, Mantsch, Frederik, and others.

In the 1980s, new concepts, such as bilayer interdigitation, tilted bilayered phases, ripple phases and the like were introduced into liposomes and their correlation with rich polymorphic phase behavior in liquid crystalline mesophases resulted in improved theoretical understanding of the bilayers.[20] Also, theoretical understanding of vesicle formation, their stability and thermodynamics started to emerge.[21]

Many fundamental concepts of liposome preparation and characterization were developed, including control over vesicle leakage (membrane permeability rate), stability of encapsulation of agents and electrostatic stabilization.[22-24]

In 1980s and 1990s liposomes became a very important tool in biophysics, biochemistry, and cell biology. Properties of biological membranes were studied by using liposomes, and many membrane bound systems, such as proteins, ion pumps, transport, and redox proteins, which were studied by reconstitution into liposomes which became also a useful system to study fusion of cells.

In parallel liposomes were becoming a very important drug delivery system due to their natural origin, biocompatibility, biodegradability, low toxicity, and nonimmunogenicity and is the properties such as solubilization power, a microencapsulation system with programmable leakage rate release kinetics, and protection properties. Work in 1970s generated so much optimism that in 1981 three liposome enterprises were founded in the U.S.

III. PRESENT

A. GENERAL

This "hype", however, was premature, since many scientific and methodological issues which are cardinal to medical application were unresolved. These include: full understanding the biofate of the liposome-drug assembly,[25] means to control pharmacokinetics and biodistribution, especially to minimize liposome uptake in the reticuloendothelial system (RES), detailed information on tolerability of liposomal formulations,[26] methodologies to have large-scale pharmaceutically acceptable production of liposomal formulations with consistent raw materials; and, finally, the aspect of quality control which is crucial for production follow-up was not sufficient to give the needed support.[27-29] As a matter of fact, issues such as long-term stability — which is a must for having liposome-based products — were not of interest to the scientists in basic liposome research. Only very recently these issues, including sterility and pyrogenicity of products, have been resolved, and the way to having liposomal products is now open.[8,30]

Nowadays, many thousands of scientists and technicians are involved in liposome research in academia and industry. This work has resulted in more than 20,000 publications, close to 20 books including one monograph (see addendum), and about 1100 patents. Dozens of small research and development companies are devoted solely to the application of liposomes, mainly in medicine. Three of them (The Liposome Company, Inc., Princeton, NJ; Liposome Technology, Inc. (previously SEQUUS Pharmaceuticals, Inc.), Menlo Park, CA; and Vestar, Inc. (previously Nextar, Inc.), San Dimas, CA) have been in existence since the early 1980s, while twelve major pharmaceutical companies are involved in liposome research, mainly in medicine, veterinary medicine, or medical diagnostics. In addition, there are several very small companies which are exploiting particular applications. Many universities and research institutions are engaging in extensive liposome research and a number even have dedicated research units. All these research efforts have resulted in only a very few parenteral products (see "Liposomes in Drug Delivery", below). The most significant, perhaps in the present situation, is that there are several hundred cosmetic enterprises which do produce liposomal products. It is here where liposomes have made the greatest impact on our every day lives. The term has become almost an everyday term; as an executive from C. Dior (G. Redziniak) stated, the name "liposome" has become a widely used phrase like transistor or laser, and means something to ordinary shoppers and buyers.

Recent developments include the development of long-circulating liposomes (Allen, Gabizon, Papahadjopoulos) and sterically stabilized liposomes (Lasic, Woodle, Huang, Cevc, Gregoriadis), and improved methods for encapsulation of some drugs into liposomes (Deamer, Forte, Cullis, Barenholz, Lasic). An extremely rapidly developing field are cationic liposomes (see this volume), which may become the major delivery system in gene therapy. DNA encapsulation in classical liposomes started in the late 1970s (Papahadjopoulos) but due to cumbersome procedures these studies largely died out. In the mid 1980s Behr was condensing DNA with cationic surfactants and liposomes, while Felgner with collaborators synthesized a novel cationic lipid and used it for transfection in 1987.

A definitive understanding of the mechanism of liposome formation is lacking. Various theories are available (reviewed in References 8,15,31,32). We hope that in the near future some missing crucial experiments will be performed and the "mystery" of liposome formation using the various approaches and methodologies will be revealed. What is clear today is that both key steps in liposome formation — the hydration and downsizing (Barenholz and Lasic, in Volume III of this series) — have to be performed at temperatures above the gel-to-liquid-crystalline phase transition. Also, it seems that the use of ultrasound and of high-pressure extrusion have a common denominator of causing very high local temperatures (of a few thousand degrees Celsius) which, due to cavitation, lead to the formation of unilamellar liposomes of small size.

A better understanding of the forces leading to liposome formation may eventually trigger the development of new methodologies which are cheaper, as well as resulting in a more homogeneous product.

Mechanisms of Liposome Formation. Currently, we believe that liposomes can be formed by fragmentation of preformed bilayers (MLV formed by hydration of lipids by thin lipid film, freeze-dried lipid cake, spray-dried lipid powder, or injection of an organic lipid solution method) or by budding off of daughter vesicles from mother liposomes (Figure 1). The driving force in the first process is the exposure of the nonpolar edges of the fragments (produced upon high energy treatment of MLV, such as sonication, homogenization, etc.) and the system closes into a spherulite to eliminate this unfavorable energy. Its balance with unfavorable bending energy determines range of liposome sizes. In the fission mechanism asymmetry across the bilayer (due to different pH, ionization, presence of interacting hydrophilic or hydrophobic molecules) causes area expansion, induction of curvature, and budding off. Reverse phase methods are similar: lipid molecules are present at the polar–nonpolar interface and the mean dielectric constant as well as volume ratio dictate the curvature. Upon removal of the organic phase, the curvature changes sign and vesicles are formed. Spontaneous vesicles, which due to their characteristics (either very soft membranes or quick exchange rates of lipid molecules) can be formed with systems with asymmetric membranes or systems with very soft membranes where excess bending energy is compensated by entropy. Figure 1 shows the two major mechanisms and text to Figure 2 (which shows energy considerations of a growing disklike lipid fragment) describes a simple scaling argument to show vesicle formation from fusing fragments and explain broad size distribution. If we postulate longitudinal (within the bilayer) and transversal (normal to the bilayer) perturbation, we can see that fragmentation and fission mechanisms can be unified. The first correlates with lysis tension and the second with bending elasticity. In some methods, such as extrusion or injection, both mechanisms are coupled.[8,32]

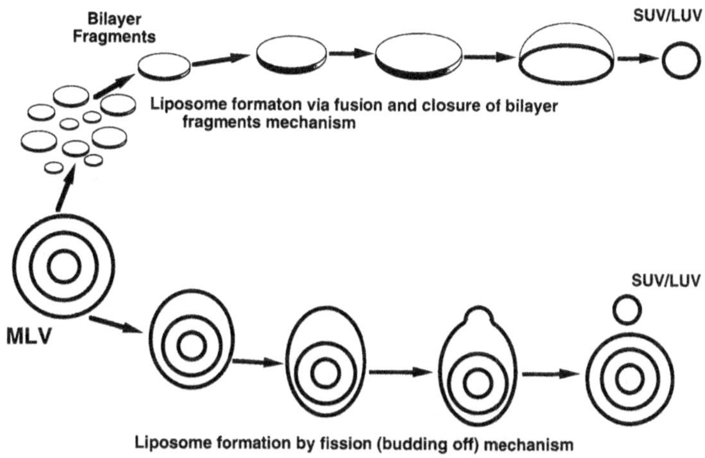

Figure 1 Schematic presentation of vesicle formation process: Top: fragmentation of pre-formed bilayers; bottom: fission (budding off) of "daughter vesicles" from "mother vesicle". (Sizes of structures are not to scale.) The first process occurs in all high energy treatments of pre-formed MLV dispersions, such as sonication, homogenization, microfluidization and French press extrusion, as well as in detergent dialysis. Budding off occurs in pH cycling, addition/exchange of counterions, and asymmetric addition of hydrophilic or hydrophobic (amphipihlic) molecules. Some processes, such as extrusion and injection of organic lipid solution into an aqueous phase probably operate via both mechanisms. In reverse phase methods the curvature is dictated by the average dielectric constant of the medium (volume ratios of polar and nonpolar phase) and therefore the vesiculation process resembles the fission pathway in which topological changes occur due to changes in curvature. Spontaneous vesiculation can follow the induced curvature pathway (driven by nonzero spontaneous curvature) or resemble micellization process (entropically stabilized).

Theoretical studies of vesicle shapes and vesicle fusion and fission became very popular in the early 1990s. The first catalogue of vesicle shapes by minimizing bending elasticity was published by Helfrich in 1970s. In the 1980s Svetina and Žekš calculated various vesicle shapes by minimizing bending elasticity at various ratios of lipid molecules in the inner and outer monolayer (fixed surface area, variable volume). Nowadays several groups study vesicle shapes, both experimentally and theoretically, as reviewed in the beginning of Volume I. One of the reasons is also the fact that these theories can be relatively easily checked experimentally.

B. NONMEDICAL APPLICATIONS OF LIPOSOMES

The major application is in cosmetics. With the development of inexpensive lipids, we believe, liposomal systems will become more important also in other areas, from diagnostics, ecology, and the coating industry to bioreactors and preparation of gels and other colloidal systems.

Many other nonmedical applications of liposomes, such as their use in agriculture, veterinary medicine, food and wine processing, and wool dyeing, are included in Volumes III and IV. In all these cases the added value of liposomes is related to their ability either to encapsulate an agent in the intraliposomal aqueous phase or to associate with the liposomal membrane(s), thereby achieving a controlled release of the agent into the milieu in which it is desired to act.

Similar advantages apply to another powerful application of liposomes — the field of *in vitro* diagnostics. Liposomes enable increasing the sensitivity of diagnostic assays due to their ability to obtain a large amplification of the signal, as described in Volume IV. However, before the commercial application of liposomes in diagnostics can be introduced, one has to validate liposome long-term stability over a

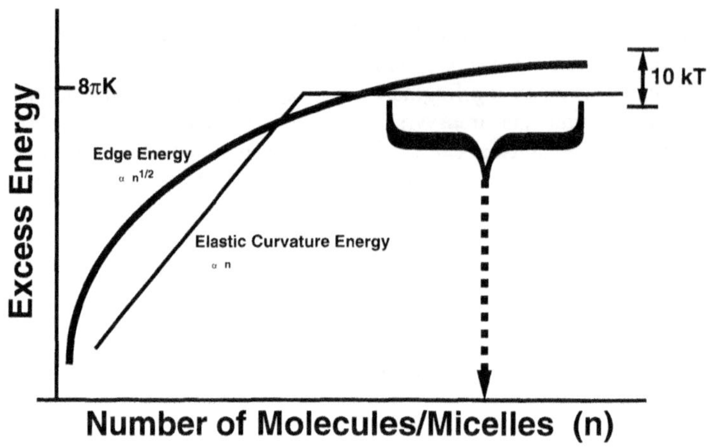

Figure 2 Energy considerations of the fusion of fragments/monomers and bending into the sphere for the transformations in the vesicle fission (budding off) see papers by Svetina and Žekš and Seifert and Lipowsky (Volume I). In this case, the driving force is the mismatch of the areas of polar heads in each of the opposed leaflets due to asymmetric conditions. In the case of fragmentation, the lipid molecules, or smaller fragments produced upon high energy treatment, fuse. The whole system is reducing the excess edge energy by fusion. Assuming equal size of fusing circular fragments the energy of the system scales is $n^{-1/2}$ (two equal disklike micelles reduce their total circumference by 1/2 by fusion), where n is the number of micelles. Each individual growing micelle, however, is increasing its unfavorable edge energy proportional to $n^{1/2}$ and a parabolic dependence is predicted. The elastic curvature, however, scales proportionally to n and linear dependence is predicted. When it becomes larger than unfavorable energy associated with bending of a symmetric membrane, the fragment starts to bend and finally closes upon itself. When the system closes, edge energy vanishes and excess energy becomes constant, independent of the radius.[8] The model can explain the driving force for vesiculation and, because both energies may be similar for quite a range in n, the heterogeneous population of vesicles formed may be qualitatively explained.

broad range of temperatures and demonstrate superior signal-to-noise ratio over conventional radio- or enzyme-immunoassays (Emanuel et al., Volume IV).

Other advantages of having an intraliposome aqueous phase are related to the small volume of this compartment relative to the large volume of the bulk extraliposomal aqueous phase. Therefore, by moving rather small numbers of ions across the liposome membrane, large changes in the intraliposomal concentrations of the moving species are obtained, leading to large gradients which can be used as a driving force for accumulation of amphipathic weak bases (reviewed in Ceh, Volume III; Parr and Cullis, Volume II) or weak acids.[50]

For some molecules, their large accumulation inside the liposome aqueous phase leads to their precipitation, gelation, or even crystallization.[34,35] The small defined intraliposomal aqueous phase may also be advantageous for homogeneous and heterogeneous catalysis, as described in Volumes III and IV. The ability to control surface properties of liposomes and to include amphiphilic molecules in the liposomal membrane makes liposome a tool to improve transformation of these amphiphiles to products of interest using chemical and enzymic procedures, as demonstrated in this series by Bar (Volume III).

C. LIPOSOMES IN COSMETICS

In addition to hundreds of cosmetic liposome products which were introduced in 1986 and 1987 by L'Oreal (Niosomes) and C. Dior (Capture), respectively, we have several products which are on the borderline between cosmetic and pharmaceutical/medical. These are various topical liposome formulations which contain antifungals and antibiotics, as well as various proteins, growth stimulating factors and, possibly, including genes and antisense oligonucleotides.

One of the most important issues in cosmetics is not only the physical, chemical, and biological properties of active substances and their action but also the way to formulate, apply, and deliver them. Very often these substances are water incompatible. However, Nature as well as mankind have found several ways to mix immiscible oily and watery substances.

In cosmetic formulations, one of the important problems is formulation of water-incompatible substances which has been in most cases successfully solved by using amphiphilic colloidal systems.

Such systems in cosmetic applications include mostly gels, ointments, creams, tinctures, and lotions. With development of surfactant science micelles and microemulsions were introduced, and in the late 1980s, following an explosion in liposome research on medical applications, liposomes were also introduced in cosmetic formulations.

Despite the fact that theoretical understanding of these phenomena is beginning to emerge only now, practice has already produced many successful formulations. Trial and error methods were implemented with large empirical knowledge, while in the last few years rational design of active ingredient carrier systems has begun to emerge. In addition to conventional phospholipid liposomes, sphingosomes, liposomes made from "skin-lipids" and nonionic surfactant vesicles (Niosomes, and many others) are used.

Rationale: The rationale of a delivery system for the active ingredients in cosmetics is similar to the use of liposomes in drug delivery. In addition, as we shall show later, in cosmetic applications vehicles themselves may have beneficial effects and can therefore exhibit double action.

Advantageous effects of liposomes and encapsulated ingredients may be summarized into several different categories. Mainly they are

1. Improved solubilization (dispersion) of difficult-to-solubilize compounds,
2. Microencapsulation in a vehicle which can enhance penetration into the skin,
3. Improved adhesion on the surface and sustained release system,
4. Beneficial properties of the carrier itself, and
5. Reduced skin toxicity/irritation of the carrier/solubilizer system.

As stated above, hydrophobic substances are often difficult to dissolve. Nonliposomal systems use alcohols, detergents, and/or oils to dissolve them and this frequently can lead to skin irritation and allergy. On the other side, liposomes can dissolve these substances in a system based solely on aqueous dispersion of natural, biocompatable, and biodegradable molecules, for instance, lecithin and water. Microencapsulation can protect the encapsulated molecules, provide a sustained release system, and enhance penetration into the skin. In some cases a formation of the protective film on the surface of the skin is needed and the liposome surface can be tailored to improve adhesion. The same is true in ophthalmic applications of liposomes.

In contrast to many carrier systems, which may exhibit irritation or toxicity, liposomes per se can provide beneficial effects. If properly selected, lipids themselves can bring water into the skin and their composition may contain some essential fatty acids, predominantly linolenic acid, which is a very important constituent of a healthy skin. Therefore some skin treatments can include empty liposomes. A very simple formulation with potentially beneficial effects would be liposomes containing antioxidant α-tocopherol (vitamin E) and using well hydrated lipids with large fraction of linolenic chains. Care has to be taken in the selection of lipid composition because it was shown that some lipids, such as phosphatidylethanolamine can actually dehydrate skin.

Skin forms a very tight barrier and, in general, compounds cannot permeate through skin or penetrate into it. Pharmaceutical applications include mechanical, electrical, and chemical means to improve drug delivery through the skin. Obviously, the usage of these techniques is not desired in cosmetic applications and a carrier which could improve skin permeation is still being sought. Many researchers believe that these may be liposomes.

Classical chemical penetration enhancers use either the principle of occlusion or skin softening to increase permeability. Upon occlusion with a nonpermeant coating such as silicone, skin softens because water cannot exit, humidity increases, and this results in the swelling of the lamellar bodies. However, all these decrease the function of vital epidermis, resulting in the skin irritation. Another concept is temporary dissolution of barriers by alcohols, fatty acids, dimethylsulfoxide, and the like. Needless to say, these treatments result in skin damage and contact allergy.

First results of liposome diffusion into skin and the concomitant hypothesis of the penetration of intact liposomes into the skin was not supported by subsequent experiments. A geometrical equivalent would be to push a basketball through a chain-link fence, actually through a stack of several of them. Research that followed showed some improvement in the penetration of both, liposome lipids and encapsulated hydrophilic and hydrophobic molecules into the skin. For liposomes it was postulated that they can increase hydration of the skin (lecithin can contain up to 23 water molecules bound to its polar head) as well as, upon fusion with lamellar bodies, fluidize these bilayers, which all together results in the increased transport through the skin. In this case active ingredients do not have to be encapsulated into liposomes which promote natural moisturizing action of the skin. Better miscibility with the stratum corneum lipids was claimed in the case of liposomes prepared from the very same, the so-called skin lipids.

Despite almost two decades of work, it is still not clear if liposomes do improve penetration substantially enough to warrant commercial therapeutic applications. For cosmetics, however, this is enough or in some cases even unnecessary, and a variety of applications were undertaken. Currently, the majority of researchers believe that some improvement in skin penetration and/or its hydration due to still unknown transport mechanism can be achieved. As in the field of medical applications of liposomes, the researchers include everyone from skeptics to optimists.

Apart from the very successful launch of liposomes in cosmetics (nowadays more than 500 cosmetic liposomal products world-wide account for more than a billion dollars in sales), and some pharmaceutical-cosmetic formulations (the first example of this rapidly growing area was the antifungal topical cream Prevaryl,® launched in Switzerland in 1987), a quick analysis of the present situation shows a stagnation and saturation in the development of novel liposome preparation methods and their characterization (with few exceptions such as cryoelectron microscopy), a few successful and promising medical applications, a handful of medical formulations at various stages of clinical trials, and their promising start in gene therapy.

Animal vaccines, made by Microvesicular Systems, have also been commercially available.

D. LIPOSOMES IN TERTIARY OIL RECOVERY

Amphiphiles composing small unilamellar liposomes, especially above gel-to-liquid-crystalline phase transition temperature (Tm), when spread over an interface between media of high and low dielectric constants such as air/water or oil/water, will almost instantaneously cover the interface with a monomolecular layer composed of the amphiphile. This monolayer at the interface of petroleum will modify the physical properties of the interface. This results[36] in very low interfacial tensions, reaching values in the millidyne/centimeter range. For example, small unilamellar liposomes (<100nm) at a phospholipid concentration of 0.5% will reduce the interfacial tension between hexadecane and 15% NaCl at 60°C to 3 mdyn/cm a value which is most suitable to tertiary oil recovery.[36-38]

Increasing liposome size or inclusion of phospholipids such as phosphatidylinositol (negatively charged) or phosphatidylethanolamine (which at high temperature and/or high salt concentration induces transformation from lamellar- to hexagonal-typeII phase) prevents this very large reduction in the interfacial tension and reduces efficiency of tertiary oil recovery (using a model coreflood system).[36] Our studies using these corefloods were done using hexadecane as oil and 15% NaCl as brine at a temperature of 60°C. Barea sandstone cores (10 × 1.5 in.) were preheated at 800°F for 24 h and saturated with brine, and the "pore volume" was calculated. The brine-saturated core was then oil-flooded and subsequently water-flooded, reaching a stage at which the core simulated an oil reservoir that had been exhaustively water-flooded. Liposomes were then injected at a rate of 1–1.5 ft/day and the displaced oil quantified. Using a liposomal preparation which provided an interfacial tension of 4.8 mdyn/cm resulted in a recovery of 33% of the oil remaining in the core after the water flood. We plan to extend these studies to an actual field test. Experiments could also be done to test tertiary oil recovery using liposomes for a combined effect of releasing the entrapped oil as well as facilitating microbial growth in the reservoir.

Another related application is dislodging of deposits from bores or tubing strings in oil wells. Solid materials are frequently deposited in oil-well bores and on tubing strings; this occurs when the oil is cooled below the cloud point of the crude oil. Treatment of deposit build-up can be mechanical, thermal, or chemical. To simulate dispersion of paraffin compounds by phospholipid-liposomes, several waxes were selected, e.g., polywaxes (Polywax500, approximately C_{16}-C_{58} or Polywax655, approximately C_{24}-C_{100}); polymers of ethylene that are similar to petroleum waxes; commercial canning paraffin, and scrapings of tubing strings near the perforation in a producing well in San Andreas, Texas. Experiments done with liposomes suggested that the best dispersion occurs above the melting points of the waxes. When a light hydrocarbon was removed by the liposomes, probably by the same mechanism of modifying the surface properties of the oil and wetting it by a monolayer of phospholipid molecules, these liposomes were also effective in preventing the re-deposition of the suspended paraffins.

The main drawback in using liposomes for tertiary oil recovery is the current low price of crude oil, which makes most procedures of tertiary oil recovery uneconomical. The need for a phospholipid fraction which is free of phosphatidylethanolamine and charged lipids makes this raw material too expensive for such use and development of cheap sources for suitable raw materials is crucial to achieve economic feasibility (see on raw materials in Section III.A and E). Another problem which has to be kept in mind for such use is the almost complete biodegradability of the liposomes composed of phospholipids, which is advantageous from the environmental point of view, though very problematic for the application and requires addition of preservatives (especially bacteriostatic compounds).

The potential use of liposomes for tertiary oil recovery led to development of other applications which are described below (see Section IV, "Future").

E. RAW MATERIALS

As mentioned before, one of the most important parameters in large-scale applications of liposomes is the price of raw materials and liposome manufacturing. With the advent of spray drying, injections of lipids from ethanol and isopropyl alcohol solutions, hydration can be simplified and the presence of undesired solvents (chloroform, ether, etc.) can be avoided. Using large-scale homogenizers (as developed in dairy and emulsion industries), the hydration process as described above can be made into a continuous and scalable process.

In other nonmedical applications the cost for the beneficial use of liposomes was simply too great to justify economy of many applications, from ecological to coating industries. With cheap sources of liposome forming lipids, from natural extracts to synthetic amphiphilic compounds and the potential of large scale continuous production, as already excersized in the cosmetic and pharmaceutical industries, most of these obstacle are removed.

While for medical applications the price of liposome raw materials (lipids) is not a major obstacle, for most nonmedical applications the price of the raw material will determine feasibility.

The larger the volumes of liposomes needed, the cheaper the raw materials have to be. The best example is tertiary oil recovery for which liposomes are very effective. However, their current high price relative to price of the recovered crude oil makes their use uneconomical. Even cosmetic products cannot afford to include a large amount of very expensive lipids. This was one of the reasons for introducing alternative amphiphiles to phospholipids, such as neutral single-chain polyoxyethylene alkyl or acyl ethers and the saccharose diesters having two hydrocarbon chains (reviewed in Reference 30). For industrial and ecological use the raw materials have to be very cheap (<$10/kg). The cost can be lowered (a) by finding cheaper sources than soy for the raw materials, for example, the phospholipids obtained during the production of edible canola oil; or (b) by simplifying the processes of phospholipid purification. A good example of the latter is developing cheap chemical processes (such as methylation) by which the major contaminant of plant phosphatides, phosphatidylethanolamine, can be transformed into phosphatidylcholine. It will be advantageous if at the same time phosphatidylinositol will be removed by precipitation as an insoluble salt. This combination will result in a fraction enriched with PC.

For medical applications the major efforts will be to improve steric stabilization. A few molecules which serve as alternatives to pegilated phospholipids were recently developed. These include polyoxazolidines and polyacrylamide and some other lipids (see Reference 39, and Torchilin in Vol. I). We believe that this trend will continue.

Synthetic amphiphiles such various polyglycerols, polyethyleneoxide, and similar polar head containing lipids may also become the major source of novel lipids. (See Chapters 7 and 9 by Vanlerberghe and Wallach and Mathar in this volume).

Major efforts will also be invested in the development and characterization of new cationic lipids. These will focus mainly on agents which enable the formation of stable complexes with DNA or oligonucleotides. Another important property of these lipids should be minimal toxicity *in vivo*. Current studies indicate that it should be advantageous for this agent to induce DNA condensation (see Chapter 1 by Lasic in this volume).

Some of the commercially available lipid transfection kits include Lipofectin (DOTMA:DOPE 1:1 w/w dioleyloxy propyl trimethylammonium chloride: dioleoylphosphatidylethanolamine, SUV suspension at 1 mg/ml), LipofectACE (DODAB:DOPE 1:2.5, dioctadecyl dimethyl ammonium bromide), LipofectAMINE (DOSPA:DOPE 3:1 w/w, dioleyloxy sperminecarboxamido ethyl dimethyl propanaminium trifluoroacetate), DOTAP (dioleoyloxy propyl trimethyl ammonium sulfate), CellFectin (tetramethyl tetrapalmitylspermine: DOPE, 1:1.5 w/w) and Transfectam (synthetic cationic lipid with spermine headgroup). Most of the solutions are presumably SUV at 1–1.5 mg/ml. Microscopy reveals, however, the presence of MLV (Lipofectamine and Lipofectace) and giant uni- and oligolamellar liposomes (DOTAP) (Lasic, unpublished).

In laboratory experiments on cell function and in recombinant DNA technology, cationic liposomes serve as an effective gene transfer system. So, liposomes are indirectly involved in the studies of functionality of DNA. Already now, but especially in future, liposomes will become increasingly important as a model system, tool, and reagent in the studies in cell biology. Now, they are used in reconstitution studies of structure and function of various membrane proteins (see Volume II) while their role in the study of synaptic vesicles, inter- and intramolecular trafficking, and secretion are becoming more and more important, as well as by the introduction of sterically stabilized liposomes (nonspecific repulsion) and incorporation of various ligands for specific interactions.[8]

F. LIPOSOMES IN DRUG DELIVERY

Overly optimistic expectations of medical applications of liposomes did not materialize, as discussed in numerous articles. While primary problems, such as availability and quality of raw materials, scalability of reproducible preparation techniques, and liposome stability and sterility were finally successfully solved, biological reasons prevented more widespread use of liposomes in drug delivery.[8] The most critical issues (after large-scale preparation, stability, and sterility were solved) are still inadequate control of liposome biodistribution and sufficient and stable entrapment of the encapsulated agent. Nevertheless three parenteral liposomal products are now available, one for the treatment of parasitic infections, the second a vaccine and the third, an anticancer formulation.

Although the present situation, also due to economical and political trends in biotechnology, is not very encouraging, we believe that some of these companies and/or products will survive and that liposomal drugs will become standard treatment in some diseases.

In medical applications we shall distinguish parenteral from topical products because the federal requirements for the former ones are obviously much more demanding.

Currently there are two true parenteral liposomal formulations on the market in Europe: Amphotericin B in small unilamellar liposomes (Ambisome[40]) and antihepatitis A vaccine.[41,42] Surprisingly, formulations containing viral proteins (virosomes) can contain several antigenes. Both are based on unilamellar vesicles. Two products which are not liposomal, but which evolved from liposomal research, namely, antifungal parenteral dosage forms in which AmphotericinB is the active ingredient are also on the market in Europe. One, Amphocil®, is a stable disk-like mixed micelle assembly composed of amphotericin B and cholesteryl sulfate.[43] The other, ABLC®, is a less well-defined complex between Amphotericin B and phosphatidylcholine/phosphatidylglycerol.[44]

Recently, the first approval of a liposomal pharmaceutical occurred also in the U.S.A. After ODAC meeting in February, another liposome product was unanimously approved for accelerated approval[45] in the much more strict U.S. market: the anticancer agent doxorubicin encapsulated in stealth liposomes by the ammonium sulfate gradient method.[46] There are some others, such as Daunorubicin in lecithin-cholesterol SUV, which may be soon commercially available. In Europe there is a topical anticancer formulation, developed by Hansjörg Eibl (produced by AstaMedica, Degussa), which is based on specially designed phospholipids, which also make liposomes, as active ingredient. It contains also propyl, hexyl, and nonyl glycerol as penetration enhancers.

The long-time dream of a magic bullet is still far from the realization. The first problem, the nonspecific uptake of liposomes was solved by the development of sterically stabilized liposomes[46,47] but the remaining problems of accessibility of a particular tissue and cells as well as overlooked severity of triggering of an immune response of the host organism by antibody or lectin-coated liposomes make this goal rather remote at present, with the possible exception of a few particular applications and *in vitro* diagnostic and basic research tools.

IV. FUTURE

A. GENERAL

As it seems now, we can expect in the near future several more pharmaceutical liposomal products, an increase in their cosmetic applications from mostly skin care creams to lotions, aftershaves, dentifrices, skin tanning/whitening creams, perfumes, and other beautifying cosmetic/medical topical formulations.

Work in the future will undoubtedly use rational designs based on structure–activity–performance relations, as was already used in the doxorubicin encapsulated in sterically stabilized liposomes. In this system, stability of vesicles in liposomicidal environment was increased by optimizing polymer coating and the stability of encapsulated drug was enhanced by precipitation by encapsulated salt. Both mechanisms are well understood and can be theoretically calculated and predicted by scaling and classical solution thermodynamic theories, respectively.[48,49] Within current ranges we can expect the following new developments in drug delivery.

In pharmaceutical applications we anticipate few more anticancer formulations, several products against parasitic infections, and vaccines. Also, several diagnostic kits may make use of liposomes as carriers of signal molecules and signal amplifiers.

One of the most difficult problems in liposome applications is stable encapsulation of the active ingredient into liposomes. Contrary to the opinion of many researchers, even the hydrophobic, membrane-associated drugs are very difficult to retain in liposomes and they may simply pop out upon dilution. The same is true for the membrane-bound drugs, either electrostatically or hydrophobically (see Volume III, chapters by Barenholz, Barenholz and Lasic).

Most water-soluble drugs are entrapped passively during liposome formation. If they are not too big (hydrodynamic diameter <1/4 of the core diameter) the interior concentration may match the bulk concentration. Obviously, unilamellar liposomes are preferred to maximize the use of lipid envelopes. Also, it is clear that working at higher lipid concentrations this may drastically improve the encapsulation. A dense liposome solution can in theory encapsulate about 72% of the aqueous volume (even more if liposomes of different sizes are used). Contrary to expectations, many of these molecules leak out after dilution. Exceptions are larger molecules, which on the other hand are more difficult to entrap. In general, if one prepares liposomes at concentrations above 200 m*M*, one can encapsulate about 50% of the molecules dissolved in the aqueous phase.

This effect can be in some cases reproduced by freezing/thawing or dehydration/rehydration cycles. In the first case, it is hoped that secluded lipid (by ice crystals) and solute will be closer together when reswelling will occur and more material will get encapsulated, while in the later case rehydration with a smaller amount of water further improves the loading. Unfortunately, these liposomes are normally rather large and heterogeneous in size.

Reverse phase liposome preparation methods are characterized by high encapsulation efficiencies of water soluble molecules. This is simply due to the nature of preparation which starts from the water-in-oil phase, and if the gel is broken properly the majority of the primary aqueous droplets become water cores of liposomes formed. Of course, the leakage problems, as discussed above, apply to these systems as well.

Active loading methods are based on the distribution of weak acids and bases according to the pH gradient. They can, in addition use pre-encapsulated ions to increase the encapsulation and retention. The same mechanism of pre-encapsulating chelators, ion traps, hydrophobic traps and the like can be used to drive in metal ions and other molecules which can interact with traps.

Liposome loading by itself is not sufficient unless it is stable, and at the same time the agent loaded can become available and released at the desired time or site. Therefore, pH gradients by themselves may not be sufficient, and other mechanisms such as reversible binding to intraliposomal polymers or precipitation inside the liposomes are needed. For agents in which the energy of activation of their release of the liposome is high, performing the loading at temperatures higher than the storage temperature will also improve loading stability.[33-35,50]

As we see it now, this work will cement the (quantitative) structure–activity relations for particular lipids, lipid composition,s and rational design of novel lipids, including polymer–bearing and polymer–sensitive lipids and liposomes. We anticipate a far greater role of synthetic organic lipid chemistry in the rational design of novel liposome systems.

The control of encapsulation and its leakage rate will also be extremely important. This work will concentrate on current remote loading technologies supplemented with special agents which will be able to reduce/collapse transmembrane gradients or interacting molecules which will specifically interact with appropriate membranes, such as particular liposomes. Present approaches, such as temperature, pH stability, controllable chemical stability, and triggered response will be further improved. Perhaps, liposome-specific lytic agents, from polypeptides to polysoaps, will be synthetized as specific antiliposomal agents. The novel lipid design may be the most effective in gene delivery and transfer.

In addition to many commercially available liposome kits for *in vitro* transfection of cells with various genes, we anticipate several liposome-based formulations in human gene therapy. After preliminary work, mostly by L. Huang and G. Nabel, clinical studies began last year[51] and more are on the way. Although the first results are not too encouraging (good tolerability but low efficacy of gene expression), we believe that improved liposomal formulations and gene designs will ultimately justify large-scale production of such drugs. First trials in humans, some of which have already started, anticipate treatment of cystic fibrosis, muscular distrophy, cancer, AIDS, and cardiovascular diseases. Formulations will eventually contain more than one gene and more complex diseases may be treated.

All these studies, however, will require further basic research because not much is understood about DNA condensation, protection, transfer to nucleus, and decondensation, not to mention interactions in plasma, biological fluids, mechanisms of cellular uptake, and the like, which are crucial for improvements of formulations.

With the development of cheap sources of large quantities of industrial liposomes, we anticipate an increase in their use in the coating industry, bioreclamation, and ecological uses. In nonmedical applications we expect several alleles of development. Very cheap liposomes will probably be used (studied) in ecological applications, such a bioreclamation, an alternative to detergent cleaning of contaminated (porous) surfaces, etc. Nonoxidizable lipid will be used in liposomes in coating industries from water-soluble paints, thin surface coatings with adjustable hydrophobicity/hydrophilicity etc.

B. FACILITATION OF RECLAMATION AND BIOREMEDIATION BY LIPOSOMES

Recently it was demonstrated that small unilamellar liposomes (SUV composed of PC- enriched soy phosphatides) are very effective in combatting oil (petroleum) spills on water and on soils. The mechanism is either the physical removal of oil or a combination of this and modification of the oil surface, making it available to microorganisms, and thereby facilitating oil biodegradation and soil bioremediation. Liposomes composed of phospholipids may also serve as a source for nutrients for the microorganisms (see below).

Physical Effects of Liposomes in Combatting Oil Spills on Water and on Soil (Based on Refs. 52 and 37)

a. General properties
   1. Effective in various degrees of purity and compositions of phosphatides.
   2. Effective at extremely low concentrations: 0.001% for water/oil and 0.05% for soil/oil.
   3. Effective with very heavy (weathered), crude or light oil.
   4. Liposome effect differs from many detergents and surfactants.
   5. High efficiency is pH independent (range pH2 to 11.0).
   6. Effective on oil spread on fresh water or sea water.
   7. More effective as SUV than as multilamellar liposomes (MLV).
   8. More efficient if phosphatides are unsaturated rather than saturated.
   9. More effective if SUV have low contents of phosphatidylethanolamine (PE) and phosphatidylinositol (PI).
   10. Exhibit properties of herding, dispersing, and gelling at considerably lower doses than the conventional agents.
b. Properties as herding agent
   1. Addition of traces of the liposomal dispersion to the surface of water covered with oil immediately pushes away thin layers of oil, forming viscous, condensed oil drops.
   2. Addition of oil onto water which contains liposomes results in coalescing of the oil into viscous, condensed droplets.
   3. The liposome-treated droplets stick to many objects, such as wetted paper, cloth, glass, and ceramics, and can be collected from them.
c. Properties as dispersing agent
   1. Gentle shaking after addition of liposomes to water covered with a thin oil layer results in smaller droplets which rise to the surface but do not coalesce.
   2. Vigorous shaking results in smaller droplets which rise slowly to surface and do not coalesce.
   3. Extensive ultrasonic irradiation results in formation of much smaller (though still visible to the naked eye) droplets which stay dispersed for at least several weeks.

d. Properties as gelling agent

Treating oil on water with liposomes as described in b. above, followed by gentle shaking or stirring, results in tight adherence of the condensed, viscous droplets to the surface of the container (e.g., glass, plastic, or ceramic). Up to nearly 100% of the oil can thus be adsorbed onto the surface of the container or any device used to remove the oil (i.e., paper, plastic, ceramic, etc.). This property may improve the skimming of water contaminated with oil.

The oil which adheres to the surface of the container can be dislodged by a light hydrocarbon, such as hexane, and thereby be recovered almost free of phospholipids.

C. MECHANISM OF PHYSICAL ACTION OF THE LIPOSOMES IN COMBATING OIL SPILLS ON WATER AND SOIL

The properties summarized in the above table led Barenholz and co-workers[37,38,52] to propose that SUV acts as a reservoir for a phospholipid monolayer in the air/water, oil/water, and sand/oil interfaces. In an early field trial it was found that 10 g of liposomes (and probably less) effectively acts against 10 kg of crude oil spread over an area of 10 m^2 of sea water. This effectiveness is unaffected by the water volume. This suggests that the oil/water partition coefficient is very high, which explains the long-term wetting activity (months to years). The presence of phospholipids at the surface of the oil is responsible for the gelling which occurs at this low agent-to-oil ratio. Theoretically, a monolayer composed of 1 g of soy PC can fully cover an area of 1000 m^2 of water (at 1 nm^2/PC molecule).

The above physical effects can be explained by the two physical properties described below:

1. The liposomes affect the spreading pressure of the oil spill on water.

$$F = Sw - So - So/w$$

where F is spreading pressure of oil on water (dynes/cm), Sw and So are the surface tension (dynes/cm) of the desired water phase, and the desired oil phase, respectively, and So/w is the interfacial tension (dynes/cm) of the oil/water interface.[54]

For example, for sea water and Kuwait crude oil without liposomes $F = 61 - 28 - 22 > 0$, and, therefore, the oil will be spread over the sea. The PC monolayer derived from the PC SUV lowers the Sw to ~40 dyne/cm so $F < 0$ and, therefore, no oil spreading occurs.

Liposomes may have a dual effect: they reduce Sw to low values of ~40 dynes/cm; some liposomal compositions drastically decrease the interfacial tension between oil and water (So/w) to the millidyne range (Harnoy et al., 1989). The latter property provides a liposome-mediated interaction of oil with the water surface, which may explain how the treated oil can easily be removed from the surface as well as being accessible to microorganisms.

2. Lecithin has hydrophilic-to-lipophilic balance (HLB) values in the range of 7–9, which is in the range of wetting agents.[53] Therefore, it lacks detergent activity. As demonstrated in Gatt et al.,[52] as expected from compounds of this HLB range, it does not induce, and actually prevents, oil spreading on the water surface (see (1) above).[53,54]

1. Demonstration of Efficiency of Cleaning Soil Contaminated by Petroleum Fractions

Cleaning of tinnic sand contaminated with jet fuel

Jet fuel is contaminating various fresh water aquifers near many airports. This contamination results from the contaminated soil around the aquifer. The cleaning of this soil is a prerequisite for ridding the aquifer of the contaminating jet fuel. To evaluate the potency of liposomes composed of PC-enriched soy phosphatides (S-45),[59] Barenholz et al. (in preparation) used tinnic sand saturated with jet fuel. This sand was treated with: (a) water only; (b) a phospholipid fraction enriched with PC (S-45) dispersed in water; (c) MLV prepared from S-45 by thin-layer lipid hydration and prolonged vortexing; (d) dry granules of S-45 added to water without prior dispersion; and (e) SUV (<50 nm) of PC-enriched S-45 fraction. Each system contained 30 g tinnic sand contaminated with 33 mg jet fuel. 150 ml of bidistilled water, or bidistilled water containing the S-45, in forms described in b–e (Table 1), was added and shaken for 20 min at room temperature. The dispersion was then centrifuged at 3,000 g for 15 min. The water was removed. Dry $MgSO_4$ (3 g) was added to adsorb the residual water, followed by 150 ml of Freon, 1,1,2-trichloro,1,2,2-trifluroethane (boiling point 47°C). The dispersion was mixed using a magnetic stirrer for 30 min. After the precipitation of the sand, 75 ml of the Freon was transferred to a sealed Erlenmeyer flask containing 5 g silicic acid, particle size 0.063–0.2 mm, pH 6.5–7.5, and mixed for 5 min (to adsorb polar lipids), and then the Freon was filtered, using filter paper (Whatman No. 3) covered with 2.5 g dry $MgSO_4$, into a round-bottom flask. The Freon was then evaporated under reduced pressure using a flash evaporator.

The percentage of residual oil in the sand was determined by comparing the weight of oil extracted from untreated soil (control) to the various treatments (a–e), as described in Table 1. Each value in Table 1 is an average of three determinations. Standard deviation was not larger than 10%.

Table 1 Effect of Liposomes on Remediation of Sand Contaminated with Jet Fuel

	Treatment	Residual jet fuel (mg per 30g tinnic sand)	%
—	Control	33.0	—
a	Water	30.0	91
b	Dispersed PC	15.2	46
c	PC MLV	14.6	44
d	Dry PC	13.1	39.5
e	PC SUV	5.1	15.4

Table 1 demonstrates that PC SUV are the most effective in removing jet fuel from the sand. Similar results were obtained for other soil types.[59]

2. Facilitation of Bioremediation

Liposomes enhance bioremediation by exogenous bacteria

Each sample included 50g soil (contaminated by 117,684 ppm crude oil), to which 10 ml water (20%) was added together with the specified additive. The exogenous bacterial mixture was added to all samples. The presence of fertilizer improved bacterial growth. The bacterial count in the presence of the liposomes was higher by 50 to 100% than in the same sample lacking the liposomes. The effect of treatment for 21 days on the level of residual oil and its aromatic compounds was determined.[59]

Table 2 demonstrates that the most efficacious bioremediation was achieved in a combination of fertilizer + liposomes. It is worth noting that the specific fertilizer used in this experiment is designed to induce enhancement of only the added oil-degrading bacteria.

Table 2 Effect of Liposomes on Bioremediation of Soil Contaminated with Crude Oil[a]

System	Residual Oil after Bioremediation (%)	Residual Aromatic Components (%)
Soil at time 0	(100)	100.0
Soil alone	52.5	90.0
Soil + fertilizer	18.0	60.0
Soil + fertilizer + liposomes	10.0	30.5

[a] The amount of oil in 10g soil at time 0 was 200 mg.

A very important aspect of the bioremediation was to test to what extent the bioremediation includes the aromatic components of the contaminating oil because they are the most problematic. This was determined using a fluorescence assay. Table2 demonstrates that the bioremediation of the soil from the aromatic oil components is minimal in soil incubated without the additives. The best bioremediation regarding the aromatic compounds was obtained in the presence of fertilizer and liposomes.

To sum up, under conditions in which the liposomes were added to soil as minor components (0.6% lipid [w/w]), their presence enhanced the degradation of both paraffinic and aromatic oil components.

D. CONCLUSION AND PREDICTIONS

The food industry has relied on the use of lecithin for mixing oily and watery substances since the origin of homo sapiens. Present studies, despite showing some improvement in particular applications due to microencapsulation effects, better dispersivity of various enzymes or (natural) preservatives (vitamins, antioxidants) still do not support larger uses.[55,56] With further development this may change.

In basic science, we expect an increase in the use of liposomes to understand cell function and the origin of its mulfunction. Such studies will include reconstitution studies and interaction studies as well as molecular biology studies of the function of various genes and their modifications on the cell function. In the latter studies, liposomes can serve only as a vehicle to transfect cell *in vitro* or *in vivo*. In the latter, in addition to therapeutic, veterinarian, and agricultural application, gene transfer via liposomes (genosomes) may serve as a unique powerful research tool to get a better understanding of major biological processes such as immune responses, signal transduction, differentiation, tumorgenicity, aging, and processes alike.

Interactions between colloidal particles and cells, fusion, and fission are still not well understood and liposomes will continue to serve as a valuable tool in this research. Efforts to couple various ligands, receptors, antibodies, functional polymers and other (macro)molecules and the like on the liposome surface will continue to grow. Recently, for instance, clusters of liposomes, each having a specific function (one, for instance carrying toxin or drug, the other a gene, the third targeting moiety, etc.) and coupled together via biotynylated lipids and streptavidin, were introduced.[57] Polymer coating of liposomes will result in intelligent liposomes which will be able to trigger fusion, fission, release of the encapsulated cargo, etc. Currently, a big emphasis is in the studies of specific interactions and lipids containing ligands such as streptavidin and applications of these. Scientists, like Ringsdorf, are synthesizing such molecules with variable spacers and several groups, including Israelachvili, are measuring forces between ligands in the bilayers. With addition of PEGylated lipids they may mimic natural conditions where one has nonspecific repulsion and specific attraction. Improved understanding of liposome stability and interactability will undoubtedly result in quick growth of this area.

With respect to medical formulations, one has to be aware of a large difference with cosmetic ones, where the standards with respect to reproducibility, sterility, stability, as well as efficacy/activity are not so strict. Medical applications of conventional liposomes have culminated with one liposomal formulation of an antifungal agent, Amphotericin; show promise in vaccines, with some promise in some diseases of liver, spleen; and yet there are unexploited parasitic diseases of immune system which are perhaps the easiest to treat, but in which Third World economies cannot support research or attract interest of developed nations. Sterically stabilized liposomes will reach their culmination in anticancer chemotherapy with a possibility of the treatment of infectious disease and inflammations, while targeting of these liposomes is, in our view, rather restricted due to geometrical limitations, inaccessibility of sites, generation of immune response, and not-always controllable cellular uptake of liposomes as well as demanding preparation. Cationic lipids and liposomes are currently the most rapidly developing field with promising new concepts in gene therapy, as shown in the beginning of this book. In the future we foresee a far greater role of synthetic organic chemistry which can via controllably unstable bonds in lipids or polymers induce various changes, from liposome disintegration, controlled release of its cargo, and so forth. Polymer decorated liposomes will undoubtedly become very important, not only because of steric stabilization but also because they can induce drug release, fusion with cells etc. In addition to these, many other nonmedical applications, as these volumes may testify, may become very important, from diagnostics, biomaterials, ecology, and food industry. Theoretical studies will perhaps concentrate more on real systems where sample heterogeneity and defects are present than on simple model systems. All the interactions described above, from steric stabilization, to specific interactions and interactions of macromolecules, polyelectrolytes (DNA) with liposomes will undoubtedly be able to feed theoreticians with complex but highly interesting and relevant systems for decades to come.

In conclusion, liposomes have reached our everyday lives in less than 30 years, and we hope that with improved understanding of their properties (stability, interaction characteristics, etc.) on the fundamental level new important applications will be achieved[58] and they will become an even greater part of our everyday life.

REFERENCES

1. **Bangham, A.D., Standish, M.M., and Watkins, J.C.,** Diffusion of univalent ions across the lamellae of swollen phospholipids, *J. Mol. Biol.* 13, 238, 1965.
2. **Lasic, D.D.,** On the History of Liposomes, This series, Bareholz, Y. and Lasic, D. D., Eds.,Vol. 1, Chap. 1.
3. **Virchow, R.,** Über das ausgebreitete Vorkommen einer dem Nervenmark analogen Substanz in den Thierische Geweben, *Virchows Arch. Pathol. Anat. Physiol.*, 6, 563, 1854.
4. **Lehmann, O.,** *Die flussigen Kristalle,* Leipzig, 1911.
5. **Saunders, L., Perrin, J., and Gammack, D.,** Ultrasonic irradiation of some phospholipid sols, *J. Chem. Soc.* 14, 567, 1962; and **Attwood, D. and Saunders, L.,** A light scattering study of ultrasonically irradiated lecithin sols, *Biochim. Biophys. Acta* 98, 344, 1965.
6. **Robinson, N.,** A light scattering study of lecithin, *J. Chem. Soc.* 12, 1260, 1960.
7. **Bangham, A.D.,** Liposomes in Nuce, *Biol. Cell* 17, 1, 1983.
8. **Lasic, D.D.,** *Liposomes: from Physics to Applications,* Elsevier, Amsterdam, 1993.
9. **Luzzati, V. and Husson, F.,** The structure of liquid crystalline phases of lipid water systems, *J. Cell. Biol.* 12, 207, 1962.
10. **Stockenius, W.,** Some EM observations on liquid crystalline phases in lipid water systems, *J. Cell. Biol.* 12, 221, 1962.
11. **Papahadjopoulos, D. and Miller, N.,** Phospholipid model membranes: 1. Structural characterization of hydrated liquid crystals, *Biochim. Biophys. Acta* 135, 624, 1967; and 2. Permeability properties of hydrated liquid crystals, *Biochim. Biophys. Acta* 135, 639, 1967.
12. **Szoka, F.C. and Papahadjopoulos D.,** Procedure for preparation of liposomes with large internal aqueous space and high capture by reverse phase evaporation, *Proc. Natl. Acad. Sci U.S.A.*, 75, 4194, 1978.
13. **Olson, F., Hunt, C.A., Szoka, F., Vail, W.J., and Papahadjopoulos, D.,** Preparation of liposomes of defined size by extrusion through polycarbonate membranes, *Biochim. Biophys. Acta* 557, 9, 1979.
14. **Huang, C.,** Studies on phosphatidylcholine vesicle formation and physical characteristics, *Biochemistry* 8, 344, 1969.
15. **Lichtenberg, D. and Barenholz, Y.,** Liposomes: preparation, characterization and preservation, in *Methods in Biochemical Analysis,* Vol. 33, Glick, D., Ed., Wiley, New York, 1988, 337.
16. **Barenholz, Y., Amselem, S., and Lichtenberg, D.,** A new method for preparation of phospholipid vesicles, *FEBS Lett.* 99, 210, 1979.
17. **Mayhew, E., Lazo, R., Vail, W., King, J., and Green, A.M.,** Characterization of liposomes prepared using a microemulsifier, *Biochim. Biophys. Acta* 557, 9, 1984.
18. **Batzri, R. and Korn, E.,** Single bilayer liposomes prepared without sonication, *Biochim. Biophys. Acta* 298, 1015, 1973.
19. **Deamer, D. and Bangham, A.D.,** Large volume liposomes by an ether vaporization method, *Biochim. Biophys. Acta* 443, 629, 1976.
20. **Knight, C.G., Ed.,** *Liposomes: from Biophysics to Therapeutics,* Elsevier, Amsterdam, 1981.
21. **Lasic, D.D.,** A molecular model of vesicle formation, *Biochim Biophys. Acta* 692, 501, 1982
22. **Papahadjopoulos, D., Ed.,** Liposomes, *Ann. N.Y. Acad. Sci.* 308, 1978.
23. **Bangham, A.D., Ed.,** *Liposome Letters,* Academic Press, London, 1983.
24. **Carmona-Ribeiro, A.M.,** Does DLVO account for interactions between charged spherical vesicles, *J. Phys. Chem.* 96, 9555, 1992.
25. **Barenholz, Y., Eds.,** Quality control of liposomes, *Chem. Phys. Lipids,* 64, 1, 1993.
26. **Storm, G., Oussoren, C., Peters, P.A., and Barenholz, Y.,** Tolerability of liposomes *in vivo*, in *Liposome Technology,* 2nd ed., Vol. III, Gregoriadis, G., Ed., CRC Press, Boca Raton, FL, 1993, 527.
27. **Barenholz, Y. and Amselem, S.,** Quality control assays in the develpment and clinical use of liposome based pharmaceuticals, in *Liposome Technology,* Gregoriadis, G., Ed., CRC Press, Boca Raton, FL, 1993, 527.
28. **Barenholz, Y.,** Liposome production: historic aspects, in *Liposome Dermatics,* Braun-Falco, O., Korting, H.C., and Maibach, H., Eds., Springer Verlag, Berlin, 1992, 69.
29. **Barenholz, Y., Ed.,** *Chem. Phys. Lipids,* Special issue: Quality Control of Liposomes 64, 1993.
30. **Barenholz, Y. and Crommelin, D.J.A.,** Liposomes as pharmaceutical dosage forms, in *Encyclopedia of Pharmaceutical Technology,* Swarbick, J. and Boylan, C., Eds., Marcel Dekker, New York, Vol. 9, 1994, 1.
31. **Lasic, D.D.,** Review: The mechanism of liposome formation, *Biochem. J.* 256, 1, 1988.
32. **Lasic, D.D.,** Kinetic and thermodynamic effects in the formation of amphiphilic colloidal particles, *J. Liposome Res.* 3, 257, 1993; Mechanisms of vesicle formation, *J. Liposome Res.*, 5, 431, 1995.
33. **Haran, G., Cohen, G., Bar, R., L., and Barenholz, Y.,** Transmembrane ammonium sulfate gradients in liposomes produce efficient and stable entrapment of amphiphatic weak bases, *Biochim. Biophys. Acta* 1151, 201, 1993.
34. **Lasic, D.D., Frederik, P., Stuart, M., Barenholz, Y., and MacIntosh, T.,** Gelation of liposome interior: a novel method for drug encapsulation, *FEBS Lett.* 312, 255, 1992.
35. **Lasic, D.D, Ceh, B., Stuart, M.M., Guo, L., Frederik, P., and Barenholz, Y.,** Transmembrane driven phase transitions within vesicles: lessons for drug delivery, *Biochim. Biophys. Acta,* in press.
36. **Harnoy, G., Gatt, S., and Barenholz, Y.,** Enhanced oil recovery, U.S. Patent 4 811 791, 1989
37. **Gatt, S., Barenholz, Y., and Bercovier, H.,** Method for enhancing the biodegradation of biodegradable organic waste, U.S. Patent 5 401 413, 1995.

38. **Gatt, S., Bercovier, H., and Barenholz, Y.,** Method for treating oil spills on water, U.S. Patent 5 244 574, 1993.
39. **Woodle, M.C., Engbers, C.M., and Zalipsky, S.,** New amphiphatic polymer-lipid conjugates forming long-circulating liposomes, *Bioconjugate Chem.* 5, 493, 1994.
40. **Adler-Moore, J.P., and Proffitt, R.T.,** Development, characterization, efficacy and mode of action of AmBisome, a unilamellar liposomal formulation of amphotericin B, *J. Liposome Res.* 3, 429, 1993.
41. **Gluck, R.,** Combined vaccine — the European contribution, *Biologicals* 22, 347, 1994.
42. **Louton, L., Bovier, P., Althaus, B., and Gluck, R.,** Inactivated virosome hepatits A vaccine, *Lancet* 343, 322, 1994.
43. **Lasic, D.D.,** Mixed micelles in drug delivery, *Nature* 355, 279, 1992
44. **Janoff, A.S., Perkins, W.R., Saletan, S., and Swenson, C.E.,** Amphotericin B lipid complex: a molecular rationale for attenuation of amphotericin B related toxicities, *J. Liposome Res.* 3, 451, 1993.
45. **Piercey, L.,** Liposome Technology's DOX-SL gets nod for accelerated approval, *BioWorld Today,* Feb. 15, 1995, p.1.
46. **Lasic, D.D. and Martin, F., Eds.,** *Stealth Liposomes,* CRC Press, Boca Raton, FL, 1995.
47. **Lasic, D.D.,** Sterically stabilized vesicles, *Angew. Chem. Int. Ed. Engl.* 33, 1675, 1994.
48. **Kuhl, T., Leckband, D., Lasic, D.D., and Israelachvili, J.,** Modification of interaction forces between bilayers exposing short-chained ethylene oxide headgroups, *Biophys. J.* 66, 1479, 1994.
49. **Ceh, B., and Lasic, D.D.,** A rigorous theory of drug loading into liposomes, Langmuir, in press
50. **Clerc, S.M. and Barenholz, Y.,** submitted.
51. **Caplen, J.C., Alton, E., Huang, L., Williamson, R., Geddes, D.M. et al.,** Liposome mediated CFTR gene transfer to the nasal epithelium of patients with cystic fibrosis, *Nature Medicine* 1, 39, 1995.
52. **Gatt, S., Bercovier, H., and Barenholz, Y.,** Use of liposomes for combating oil spills and their potential application to bioreclamation, in *On Site Bioreclamation,* San Diego, CA.
53. **Becher, P.,** *Emulsions, Theory and Practice,* Rheinhold, New York, 1965, 231.
54. API Dispersant Task Force, The role of oil spill dispersant in oil spill control, American Petroleum Institute, 1986, 39.
55. **Kirby, C.,** Delivery systems for enzymes, *Chem. Britain* 26, 847, 1990.
56. **Abra, R.M. and Lasic, D.D.,** Liposomes in the agro-food community, *Agro-Industry,* hi-tech, Nov/Dec 1992, p. 12.
57. **Chirivolu, S., Walker, S., Israelachvili, J.N., Schmitt, F., Leckband, D., and Zaszadinski, J.A.,** Higher order self-assembly of vesicles by site specific binding, *Science* 264, 1753, 1995.
58. **Lasic, D.D. and Papahadjopolos, D.,** Liposomes revisited, *Science,* 267, 1275, 1995.

LIST OF MAJOR BOOKS AND REVIEWS IN LIPOSOMOLOGY *(APRIL 1995)*

1. **Papahadjopoulos, D., Ed.,** *Liposomes and Their Uses in Biology and Medicine,* Vol. 308: Annl. N.Y. Acad. Sci., Vol. 38, 1978.
2. **Baldwin, T. and Howard, S. R.,** *Liposomes and Immunobiology,* Elsevier, North Holland, Amsterdam, 1980.
3. **Knight, C. G., Ed.,** *Liposomes: From Physical Structure to Therapeutic Application,* Elsevier, North Holland, Biomedical Press, Amsterdam, 1981.
4. **Gregoriadis, G., Ed.,** *Liposome Technology,* Vols. 1, 2, 3, CRC Press, Boca Raton, FL, 1984.
5. **Yagi, K., Ed.,** *Medical Application of Liposomes,* Special issue: Scientific Societies Press, Tokyo, 1986.
6. **Cullis, P. R. and Hope, M. J., Eds.,** *Liposomes, Chem. Phys. Lipids,* 40(2–4), **87**, 1986.
7. **Schmidt, K. H., Ed.,** *Liposomes as Drug Carriers,* George Thiem Verlag, Stuttgart, 1986.
8. Ostro, M., *Liposomes: From Biophysics to Therapeutics,* Marcel Dekker, New York, 1987.
9. **Gregoriadis, G., Ed.,** *Liposomes as Drug Carriers,* Wiley, New York, 1988.
10. **Lichtenberg, D. and Barenholz, Y.,** Liposomes: preparation, characterization and preservation, in *Methods in Biochemical Analysis,* Vol. 33, Glick, D., Ed., Wiley, New York, 1988, 337.
11. **Lopez-Berestein, G. and Fidler, I. J., Eds.,** *Liposomes in the Therapy of Infectious Diseases and Cancer,* Alan R. Liss, New York, 1989.
12. **New, R. R. C., Ed.,** *Liposomes — A Practical Approach,* IRL Press, Oxford, 1990.
13. **Marsh, M. A. and Phill, D.,** *Handbook of Lipid Bilayers,* CRC Press, Boca Raton, FL, 1990.
14. **Szoka, F. C.,** Liposomal drug delivery: Current status and future prospects, in *Membrane Fusion,* Wilschut, J. and Hoekstra, D., Eds., Plenum, New York, 1991, 845.
15. **Braun-Falco, O., Korting, H. C., and Maibach, H. I., Eds.,** *Liposome Dermatics,* Springer Verlag, Berlin, 1992.
16. **Gregoriadis, G., Ed.,** *Liposome Technology,* 2nd ed., Vols. 1, 2, 3, CRC Press, Boca Raton, FL, 1993.
17. **Lasic, D. D.,** *Liposomes: from Physics to Applications,* Elsevier, Amsterdam, 1993.
18. **Barenholz, Y., Ed.,** Special issue: *Quality Control of Liposomes,* Chem. Phys. Lipids, Vol. 64, 1993.
19. **Cevc, G., Ed.,** *Phospholipid Handbook,* Marcel Dekker, New York, 1993.
20. **Barenholz, Y. and Crommelin, D.,** Liposomes as pharmaceutical dosage forms, in *Encyclopedia of Pharmaceutical Technology,* Vol. 9, Swarbrick, J. and Boylan, J.C.C., Eds., Marcel Dekker, New York, 1994.
21. **Philippot, J. R. and Schuber, F., Eds.,** *Liposomes as Tools in Basic Research and Industry,* CRC Press, Boca Raton, FL, 1995.
22. **Lasic, D. D., and Martin, F. J.,** *Stealth Liposomes,* CRC Press, Boca Raton, FL, 1995.

INDEX

D

E

H

I

K

L

M

N

O

P

Q

R

S

T

For Product Safety Concerns and Information please contact our EU representative GPSR@taylorandfrancis.com Taylor & Francis Verlag GmbH, Kaufingerstraße 24, 80331 München, Germany

Batch number: 10731069

Printed by Printforce, the Netherlands